Biofuels: Advanced Technologies and Applications

Biofuels: Advanced Technologies and Applications

Edited by **Robbie Larkin**

SYRAWOOD
PUBLISHING HOUSE

New York

Published by Syrawood Publishing House,
750 Third Avenue, 9th Floor,
New York, NY 10017, USA
www.syrawoodpublishinghouse.com

Biofuels: Advanced Technologies and Applications
Edited by Robbie Larkin

© 2016 Syrawood Publishing House

International Standard Book Number: 978-1-68286-076-2 (Hardback)

Contents

Preface

Biofuels have become prominent source of fuel in the last few decades. They are primarily produced from plants, agricultural and domestic waste. This book contains some interesting topics that offer an exhaustive insight into the field of biofuels such as enzyme formation, degradation of biomass, cellulose and alcohol production, genomics, etc. Researches and case studies by eminent scientists and experts on biofuels have been included in this book. This book is an essential guide for both professionals and those who wish to pursue this discipline further.

Various studies have approached the subject by analyzing it with a single perspective, but the present book provides diverse methodologies and techniques to address this field. This book contains theories and applications needed for understanding the subject from different perspectives. The aim is to keep the readers informed about the progress in the field; therefore, the contributions were carefully examined to compile novel researches by specialists from across the globe.

Indeed, the job of the editor is the most crucial and challenging in compiling all chapters into a single book. In the end, I would extend my sincere thanks to the chapter authors for their profound work. I am also thankful for the support provided by my family and colleagues during the compilation of this book.

Editor

Comparative analysis of sugarcane bagasse metagenome reveals unique and conserved biomass-degrading enzymes among lignocellulolytic microbial communities

Wuttichai Mhuantong[1], Varodom Charoensawan[2,3*], Pattanop Kanokratana[1], Sithichoke Tangphatsornruang[4] and Verawat Champreda[1]

Abstract

Background: As one of the most abundant agricultural wastes, sugarcane bagasse is largely under-exploited, but it possesses a great potential for the biofuel, fermentation, and cellulosic biorefinery industries. It also provides a unique ecological niche, as the microbes in this lignocellulose-rich environment thrive in relatively high temperatures (50°C) with varying microenvironments of aerobic surface to anoxic interior. The microbial community in bagasse thus presents a good resource for the discovery and characterization of new biomass-degrading enzymes; however, it remains largely unexplored.

Results: We have constructed a fosmid library of sugarcane bagasse and obtained the largest bagasse metagenome to date. A taxonomic classification of the bagasse metagenome reviews the predominance of Proteobacteria, which are also found in high abundance in other aerobic environments. Based on the functional characterization of biomass-degrading enzymes, we have demonstrated that the bagasse microbial community benefits from a large repertoire of lignocellulolytic enzymes, which allows them to digest different components of lignocelluoses into single molecule sugars. Comparative genomic analyses with other lignocellulolytic and non-lignocellulolytic metagenomes show that microbial communities are taxonomically separable by their aerobic "open" or anoxic "closed" environments. Importantly, a functional analysis of lignocellulose-active genes (based on the CAZy classifications) reveals core enzymes highly conserved within the lignocellulolytic group, regardless of their taxonomic compositions. Cellulases, in particular, are markedly more pronounced compared to the non-lignocellulolytic group. In addition to the core enzymes, the bagasse fosmid library also contains some uniquely enriched glycoside hydrolases, as well as a large repertoire of the newly defined auxiliary activity proteins.

Conclusions: Our study demonstrates a conservation and diversification of carbohydrate-active genes among diverse microbial species in different biomass-degrading niches, and signifies the importance of taking a global approach to functionally investigate a microbial community as a whole, as compared to focusing on individual organisms.

Keywords: Lignocellulose degradation, Metagenomics, Comparative genomics, Biorefinery, Biofuels

* Correspondence: varodom.cha@mahidol.ac.th
[2]Department of Biochemistry, Faculty of Science, Mahidol University, Bangkok 10400, Thailand
[3]Integrative Computational BioScience (ICBS) Center, Mahidol University, Nakhon Pathom 73170, Thailand
Full list of author information is available at the end of the article

Background

Lignocellulose is a basic constituent of plant biomass and represents one of the most abundant sources of renewable carbon in the biosphere. Its complex structure consists mainly of carbohydrate polymers: cellulose, hemicellulose, and lignin. In nature, the degradation of lignocellulose requires multiple enzymes produced by diverse microorganisms, which act corporately and attack the complex structure of lignocellulosic biomass [1,2]. The growing number of studies on the complex pathways of lignocellulose degradation not only allows us to comprehensively understand the mechanisms and interplay between microbes in maintaining carbon balance in geobiochemical cycles, but may also lead to potential discovery of uncharacterized microbes and novel enzymes, which in turn, could improve the conversion of underused plant biomass to biofuels, chemicals, and other materials for biorefinery industries [3-7].

An industrial bagasse collection site at sugar mills represents a unique ecological niche for lignocellulose decomposition due to its high enrichment of lignocellulosic materials under high temperature and low nitrogen, with varying microenvironmental conditions, from the aerobic pile surface to the anoxic interior region. Recent studies have showed complexity in the bagasse microbial community and inherent metabolic potential in plant biomass decomposition, which provides novel genetic resources for biotechnological exploration [5,8,9]. However, it remains to be seen how the phylogenetic diversity and biomass-degrading enzyme repertoire of this microbial community compare to those of previously characterized lignocellulose-degrading environments.

Comparative metagenomic studies have been used to investigate the microbial communities in different environments in terms of taxonomy, gene contents, and also biochemical and metabolic potentials [10-14]. Culture-independent high-throughput sequencing has previously been used to explore the complexity of metagenomes obtained from several lignocellulose-degrading environments, including peat swamp forest [15], cow rumen [16,17] wallaby gut [18], and termite gut [19]. A comparison of soil metagenomes from distinct geographical locations, including cold and hot deserts, forests, grasslands, and tundra, has demonstrated the uniqueness of microbial communities in terms of taxonomic diversity and also the high relative abundance of functional genes that can be linked to the metabolic capability required to cope with specific environmental conditions [13]. Other comparative metagenomic analyses performed in different biomass-degrading environments also showed variation in metabolic potentials and enzymatic profiles related to decomposition of plant biomass in various ecological niches with different temperature, pH, and oxygen availability, for example, composts from a tropical zoo park [20], animal guts [21], and structurally stable symbiotic biomass-degrading consortia [22]. These findings thus shed light onto the highly complex mechanism of plant biomass decomposition through cooperative interactions between multiple microbial species and their enzymes in different environments. Metagenomic studies from biomass-degrading environments also serve as a useful starting point to discover new uncharacterized enzymes. As demonstrated in a study using ultra-deep sequencing of switchgrass degraded in cow rumens [17], only 12% of carbohydrate-active genes sequenced have 75% or more identity to known genes, suggesting a great potential of new enzyme discovery.

In this study, a fosmid library of a microbial community extracted from industrial sugarcane bagasse was constructed and analyzed by shotgun pyrosequencing to characterize and catalog the biodiversity of the microbe community, as well as its lignocellulolytic enzyme potential in biomass decomposition. The metagenome sequenced from this bagasse fosmid library, called the bagasse metagenome herein, was analyzed and compared with several reported metagenomic datasets from both lignocellulolytic and non-lignocellulolytic ecological niches. As well as elucidating the taxonomic compositions of microorganisms in different metagenomes, we have identified the biomass-degrading genes conserved among different microbial communities and the unique genes that could be related to specific metagenomes, thus providing a basis for understanding the roles and interplays of different microbes and their enzymes in biomass degradation in different environments.

Results and discussion

Constructing the fosmid library and pyrosequencing of sugarcane bagasse metagenome

We first constructed a fosmid library from the microbial DNA sequences obtained from the soil-contacting region of sugarcane bagasse collected from an industrial collection site (see Methods for more details). The fosmid clones were pooled and pyrosequenced on one full lane of a 454 Genome Sequencer FLX (Roche, Branford, CA, USA). Approximately one million raw reads were obtained, with an average read length of 570 bp (Table 1). Low quality sequences including short reads (<100 bp) and repetitive sequences were filtered out. The sequences contaminated by the vector and host genome used in the fosmid library construction were also removed at this step. After this data filtering, 726,980 reads remained with an average read length of 580 bp, and were subsequently assembled for longer overlapping sequences. This resulted in a total of 17,829 assembled contigs and 185,543 non-redundant singletons, which were then used for functional and comparative genomic analyses (see Additional file 1: Figure S1 for summary of

Table 1 Summary of bagasse fosmid pyrosequencing data

Raw reads

Dataset	Number of sequences	Number of nucleotides	Sequence length			
			Average	SD	Minimum	Maximum
1. Raw reads	1,038,205	591,656,071	569.9	173.3	40	1,595
2. Read screen repeats	982,383	569,556,388	579.8	164.7	40	1,595
3. Read screen repeats and trim vector	726,980	421,491,438	579.8	166.0	40	1,595

Assembled sequences

Dataset	Number of sequences	Number of nucleotides	Sequence length			
			Average	SD	Minimum	Maximum
1. Contigs	17,829	32,867,905	1,843.5	2,394.6	100	46,577
2. Singletons (non-redundant)	185,543	109,290,202	589.0	163.5	40	1,595

The bagasse fosmid library was sequenced on one full lane of the 454 GS-FLX Titanium, resulting in approximately one million raw reads. The reads with contaminating sequences of vector or host genome were removed before contig assembling and redundant sequence cleaning.

data analyses). The entire bagasse metagenomic library has been deposited to the National Center for Biotechnology Information (NCBI) Sequence Read Archive (SRA) (SRX493840).

Taxonomic classification and microbial diversity of the bagasse fosmid library

To explore the phylogenetic diversity and complexity of the microbial community in the bagasse metagenome, we first identified taxonomic classification of singletons and contigs before removing duplicates, using BLASTN against the NCBI non-redundant nucleotide database (NT) [23] (Figure 1 and Additional file 2: Table S1). Based on the NCBI attributes, most of the mapped sequences from our bagasse fosmid library are of bacterial origin (94.4%, with a total of 1,164 assigned unique bacterial species), together with a small amount of eukaryotic DNA (4.3%), which are mainly from plants and fungi, and only trace of archaeal DNA (0.6%). We note; however, that functional genes of eukaryotes were predicted with a smaller degree of confidence than those of prokaryotes, as the genes are normally longer and frequently contain isoforms. The rest of the sequenced DNA comprises traces of DNA from viruses and other unidentified sequences.

Focusing on bacteria, the majority of reads were mapped to the Proteobacteria phylum (approximately two thirds of all mapped reads), which is metabolically versatile and spanned a wide range of bacterial taxa capable of aerobic as well as fermentative anaerobic metabolisms. The predominance of Proteobacteria in the sample collected from the exterior of a bagasse pile is in line with our previous observation in tagged 16S rRNA of the bagasse samples [5]. In our bagasse metagenome, most of the Proteobacteria have been assigned to one of three major classes: Alpha-, Beta-, and Gammaproteobacteria. Alphaproteobacteria is the largest class (22.5% of mapped reads) of microbes found in the bagasse

metagenome, comprising both the aerobic and anaerobic bacterial orders Rhizobiales, Rhodospirillales, Sphingomonadales, and Caulobacterales. The majority of Gammaproteobacteria (19.1%) found belong to the Pseudomonadales, Chromatiales, Xanthomonadales, Methylococcales, and Enterobacteriales orders, whereas almost all Betaproteobacteria are from the genus *Burkholderia* (13.6%).

The next largest phyla are Bacteroidetes (10.2%) and Actinobacteria (7.9%), followed by relatively smaller amounts of DNA from Acidobacteria, Chloroflexi, and Firmicutes. Bacteroidetes are mostly anaerobic and are widely distributed in soil, sediment, aquatic habitats, and animal guts [6,24-27]. Actinobacteria are active biomass degraders under aerobic conditions and either mesophilic or thermophilic temperature ranges, and they have a significant role in lignocellulose decomposition in soil and aquatic environments [28,29].

Biomass-degrading metabolic potential in bagasse fosmid library

We then explored the repertoire of lignocellulose-degrading enzymes in the bagasse microbial community by assigning the predicted open reading frames (ORFs) with three carbohydrate-active enzyme families from the CAZy database [30]: glycoside hydrolases (GHs), carbohydrate-binding modules (CBMs), and the recently introduced auxiliary activities (AAs), to the non-redundant reads (see Methods). Of all the predicted ORFs, 1,774 (approximately 1%) have hits to 72 GH, 18 CBM, and 7 AA families (as summarized in Figure 2).

The microbial community found in bagasse is capable of producing various types of enzymes required to convert cellulose, hemicellulose, and lignin into different types of monosaccharides that are essential energy sources for aerobic (via the tricarboxylic acid, or TCA, cycle) as well as anaerobic bacteria (through fermentation processes). Of all the ORFs mapped to the GH

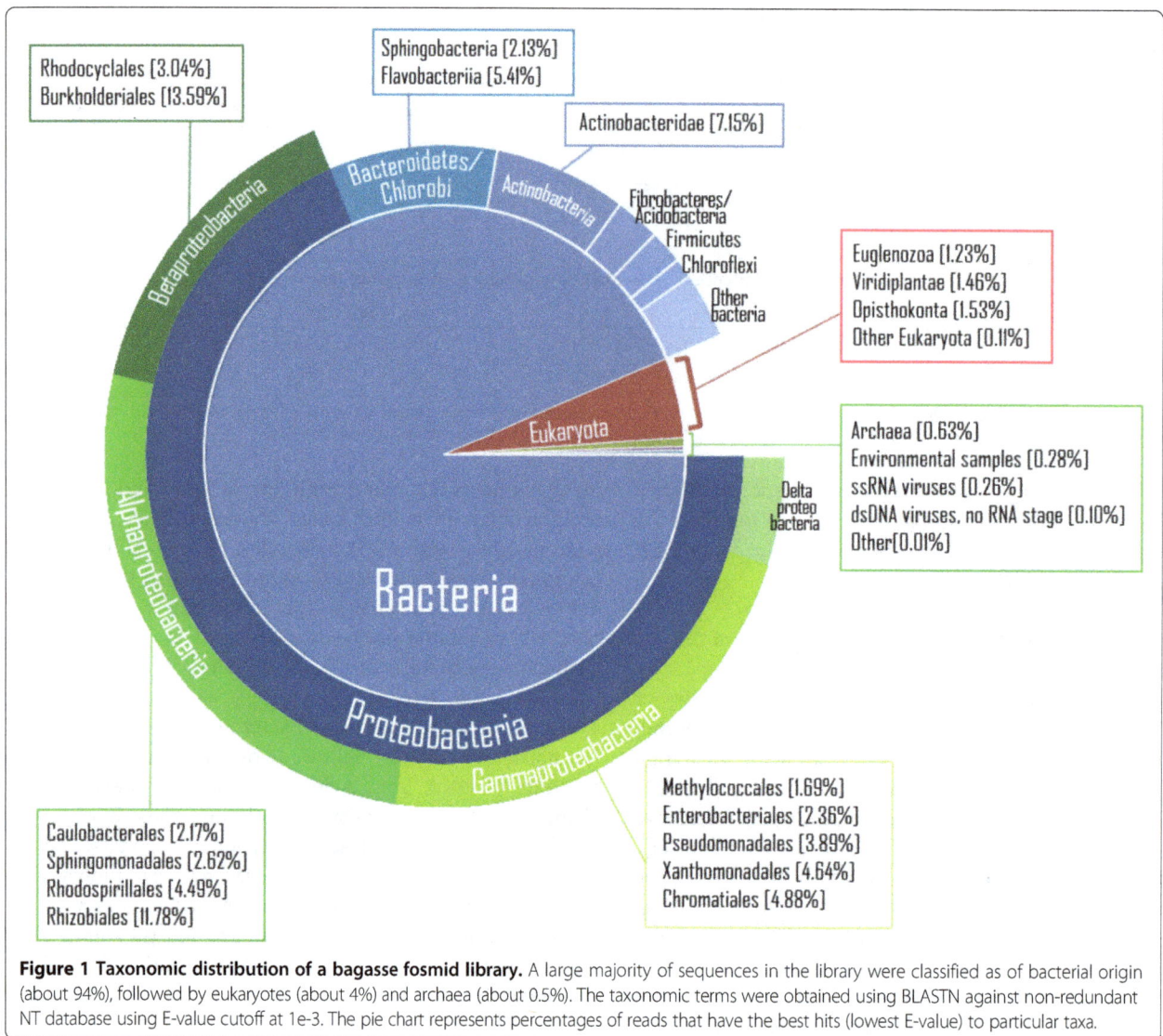

Figure 1 Taxonomic distribution of a bagasse fosmid library. A large majority of sequences in the library were classified as of bacterial origin (about 94%), followed by eukaryotes (about 4%) and archaea (about 0.5%). The taxonomic terms were obtained using BLASTN against non-redundant NT database using E-value cutoff at 1e-3. The pie chart represents percentages of reads that have the best hits (lowest E-value) to particular taxa.

families, 679 ORFs (about 42%) are related to 27 GH families that have lignocellulose-degrading enzymatic activities (Table 2). The majority of enzymes that degrade cellulose belong to two main families: GH5 and GH9, which contain cellulases including endoglucanases, exoglucanases, and beta-glucosidases. The exo-acting cellobiohydrolases are involved in initiating the attack on the highly ordered cellulose fraction comprising crystalline and amorphous regions. The cello-oligosaccharides and cellobiose are further processed by the enzymes involving the hydrolysis of beta-linked dimers of oligosaccharides such as beta-glucosidases from the GH1, 2, and 3 families.

Hemicellulose contains a greater variety of carbohydrate compositions and thus requires a broader range of endo-acting enzymes to degrade, including endo-1,4-beta-xylanase (GH10) for hydrolysis of xylan, the most abundant hemicellulose in bagasse; endo-1,4-beta-mannosidase

(GH26) for mannan; and endo-1,4-beta-galactosidase (GH16) for galactan. The genes encoding these enzymes are all present in the bagasse metagenome. The higher percentage of reads mapped to GH10 (2.6%) might reflect the requirement to digest the greater abundance of xylan, as compared to mannan (GH26, 0.4%) and galactan (1.5%). Further downstream, xylan is degraded by xylo-oligosaccharide hydrolyzing enzymes and side chain cleaving hydrolases such as beta-glucosidases and beta-xylosidases from multiple families, including GH3, 39, 43, and 52. The downstream decomposition of manno-oligosaccharides and their dimeric sugars is catalyzed by alpha-mannosidase (GH38), whereas the dimeric sugars from galactan are degraded by beta-galactosidases from GH2 and 35. We also observed a wide range of debranching enzymes such as alpha-L-arabinofuranosidase (GH62), alpha-glucuronidase (GH67), and alpha-L-rhamnosidase (GH78, relating to pectin degradation),

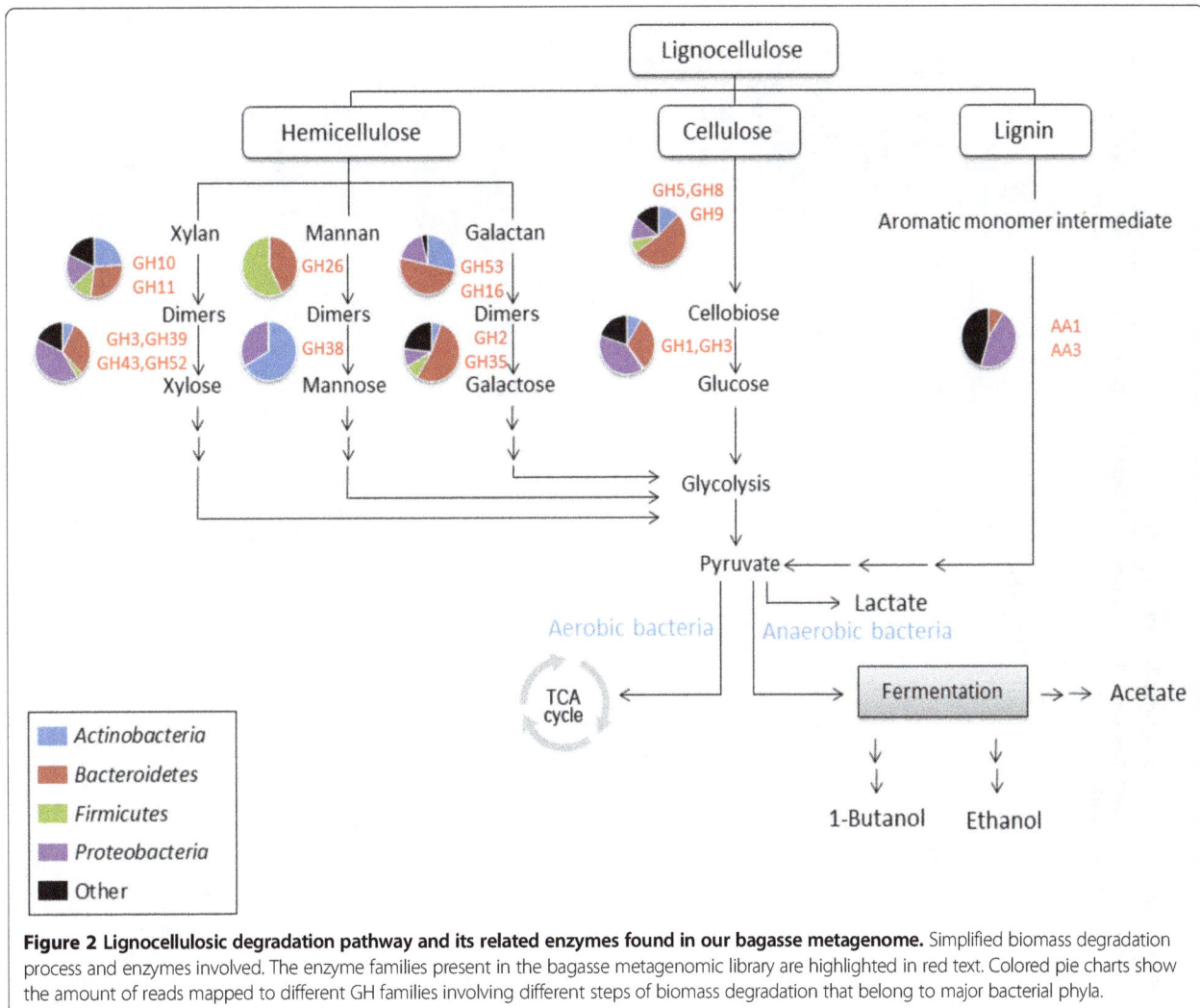

Figure 2 Lignocellulosic degradation pathway and its related enzymes found in our bagasse metagenome. Simplified biomass degradation process and enzymes involved. The enzyme families present in the bagasse metagenomic library are highlighted in red text. Colored pie charts show the amount of reads mapped to different GH families involving different steps of biomass degradation that belong to major bacterial phyla.

reflecting corporate enzymatic functions in the degradation of hemicellulose.

In addition to a variety of glycoside hydrolases that degrade cellulose and hemicellulose, we also observed a moderate fraction of ORFs mapped to carbohydrate-binding modules (CBMs), a group of non-catalytic proteins that promote the association of the enzymes and substrates. The majority of the CBM enzymes found in the metagenome were identified as CBM50 (24.2%) and CBM48 (17.4%), peptidoglycan-binding and glycogen-binding proteins, respectively, although they are not known to be directly related to lignocellulosic hydrolysis. We also found a number of a recently defined CAZy class of lignin-degrading enzymes known as auxiliary activities (AAs), which contains eight families of ligninolytic enzymes and three families of lytic polysaccharide mono-oxygenases. The majority of lignin-breakdown enzymes found are multicopper oxidase (AA1, 25.0% of all AAs) and choline dehydrogenase (AA3, 30.0%), followed by smaller amounts of AA4, 5, 6, 7, and 9.

In terms of the microorganisms producing carbohydrate-degrading enzymes, our results show that heterogeneous hemicellulose and cellulose are degraded by specific endo-acting enzymes produced by all major bacterial phyla: Actinobacteria, Bacteroidetes, Firmicutes, and Proteobacteria (Figure 2), except for mannan, which specifically requires the GH26 (beta-mannanase) family from Bacteroidetes and Firmicutes. Mannobiose is then broken down into a single-molecule sugar by GH38 (alpha-mannosidase) from Actinobacteria and Proteobacteria. Other oligodimers from hydrolysis of lignocelluloses are subsequently degraded by specific exo-acting oligosaccharide-degrading and debranching enzymes produced from Actinobacteria, Bacteroidetes, and Proteobacteria, mainly from the bacterial orders of Sphingobacteriales, Actinomycetales, Caulobacterales, Cytophagales, and Ignavibacteriales. Lignin-degrading enzymes, on the other hand, are mostly generated by Bacteroidetes and Proteobacteria.

Proteobacteria is the most abundant phylum in the sugarcane bagasse community; however, smaller phyla

Table 2 Summary of the number of reads from the bagasse metagenome mapped to lignocellulose-degrading genes

Enzyme_group	Family	Actinobacteria	Bacteroidetes/ Chlorobi	Chlamydiae/ Verrucomicrobia	Chloroflexi	Fibrobacteres/ Acidobacteria	Firmicutes	Planctomycetes	Proteobacteria	Otherbacteria	Other organisms	Activity
Cellulases	GH5	6	18	1	0	3	4	0	4	0	0	Cellulase; endoglucanase; beta-glucosidase
Cellulases	GH6	3	0	0	0	0	0	0	0	0	0	Endoglucanase; cellobiohydrolase
Cellulases	GH9	1	9	0	0	0	0	0	2	3	1	Endoglucanase; beta-glucosidase
Cell wall elongation	GH16	8	12	0	1	0	0	0	5	0	0	Xyloglucan; licheninase
Cell wall elongation	GH17	0	0	0	0	0	0	0	28	0	0	Exo-beta-1,3-glucanase; licheninase
Cell wall elongation	GH74	0	0	0	0	0	2	0	0	0	0	Endoglucanase; xyloglucanase
Oligosaccharide-degrading enzymes	GH1	9	3	0	2	0	0	0	9	2	1	Beta-glucosidase; beta-galactosidase
Oligosaccharide-degrading enzymes	GH2	6	64	5	9	5	10	0	9	4	0	Beta-mannosidase; beta-galactosidase
Oligosaccharide-degrading enzymes	GH3	8	57	1	4	8	2	0	68	19	2	Beta-glucosidase; beta-glucosylceramidase
Oligosaccharide-degrading enzymes	GH29	0	14	1	0	7	4	1	0	0	0	Alpha-L-fucosidase
Oligosaccharide-degrading enzymes	GH35	1	0	3	1	0	0	0	3	0	1	Beta-galactosidase
Oligosaccharide-degrading enzymes	GH38	4	0	0	0	0	0	0	2	0	0	Alpha-mannosidase
Oligosaccharide-degrading enzymes	GH39	7	0	0	2	0	1	0	4	0	0	Beta-xylosidase
Oligosaccharide-degrading enzymes	GH42	0	0	0	0	0	0	0	0	2	0	Beta-galactosidase
Oligosaccharide-degrading enzymes	GH43	1	12	1	2	1	0	0	20	0	0	Beta-xylosidase
Oligosaccharide-degrading enzymes	GH52	0	0	0	0	0	5	0	1	0	0	Beta-xylosidase

Table 2 Summary of the number of reads from the bagasse metagenome mapped to lignocellulose-degrading genes *(Continued)*

Category	GH Family										Function
Endohemicellulases	GH8	0	2	0	0	0	0	1	0	0	Endo-1,4-D-glucanase; chitosanase
Endohemicellulases	GH10	11	14	2	5	6	0	8	0	0	Xylanase; beta-1, 4-xylanase; endo-1, 4-beta-xylanase
Endohemicellulases	GH11	1	0	0	0	0	0	1	1	1	Endo-1,4-beta-xylanase; xylanase
Endohemicellulases	GH12	0	0	0	0	0	0	2	0	1	Endoglucanase; xyloglucan hydrolase
Endohemicellulases	GH26	0	3	0	0	4	0	0	0	0	Beta-mannanase; endo-1,4-beta-mannosidase
Endohemicellulases	GH28	0	15	0	4	9	0	2	4	0	Polygalacturonase; pectate lyase; endopolygalacturonase
Endohemicellulases	GH53	0	2	0	0	0	0	0	0	0	Endo-1,4-beta-galactosidase
Debranching enzymes	GH51	4	7	2	13	1	0	3	0	0	Alpha-L-arabinofuranosidase; endoglucanase
Debranching enzymes	GH62	2	0	2	0	0	0	1	0	0	Alpha-L-arabinofuranosidase
Debranching enzymes	GH67	0	6	0	0	0	0	3	0	0	Alpha-glucuronidase
Debranching enzymes	GH78	0	11	0	1	0	0	0	0	0	Alpha-L-rhamnosidase

Summary of the number of reads in the bagasse metagenome mapped to genes encoding lignocellulose-degrading enzyme homologs annotated by the CAZy database.

such as Actinobacteria, Bacteroidetes, and Firmicutes contribute as much to the production of lignocellulolytic enzymes for the microbial community. Bacteroidetes, for instance, is the second largest bacterial phylum in our bagasse metagenome based on the number of mapped reads (10.2%), but is still far behind the most abundant phylum, Proteobacteria (66.1%). Intriguingly, the phylum Bacteroidetes produces the largest repertoire of many carbohydrate-degrading enzymes, especially cellulases (GH5, 9), oligosaccharide-degrading enzymes (GH2), and endo-hemicellulases (GH10, 28) (Table 2). This illustrates a complex interactivity and synergism of the microbial community in the decomposition of lignocellulosic biomass in the environment.

Comparative genomic analysis of lignocellulolytic and non-lignocellulolytic metagenomes

Having explored the newly assembled metagenome from the sugarcane bagasse fosmid library, we now seek to investigate the shared and unique characteristics of the bagasse microbial community with other publicly available metagenomes. We have selected representative metagenomes from five other lignocellulolytic and six non-lignocellulolytic environments available from the NCBI Whole Genome Shotgun (WGS) and Sequence Read Archive (SRA) projects [31], based on comparable numbers of sequences and average lengths (Additional file 3: Table S2). The average number of reads is 98,000; the largest is approximately 200,000 reads (sugarcane bagasse from this study, and compost [32]), and the smallest is about 25,000 reads (human distal gut [33] and sludge [34]). The average read length of the combined dataset is approximately 1,000 bp. To minimize a potential bias from different analytic strategies previously used by different groups, we obtained assembled reads for each dataset and reanalyzed them using the same pipeline, as used in our sugarcane bagasse dataset (Additional file 1: Figure S1). We summarize the bacterial taxonomic distributions, which represent the largest superkingdom by far in these 12 metagenomes, in Figure 3 and Additional file 4: Table S3.

In brief, bagasse pile, farm soil, and compost are classified as active lignocellulose-degrading environments, comprising both aerobic and anoxic regions [5,32]. Decomposition of lignocellulose in bagasse in an open field is a slow process characterized by its low nitrogen, low moisture, and relatively high temperature conditions [5,8]. Farm soil is a complex and nutrient-rich environment active in decomposition of plant biomass and a wide range of organic materials [14]. The compost system is considered a relatively aerobic environment with dominant microorganisms in multi-phase composting processes, in which the highest biomass decomposition activity is found at the high-temperature thermophilic

phase [32,35,36]. The gut ecosystems are closed anaerobic systems. Termite and wallaby guts represent examples of animal guts that are highly capable of digesting plant cell walls, and are thus considered as anaerobic lignocellulolytic environments [18,19]. However, we classified human and mouse guts as non-lignocellulolytic metagenomes in this study because the environments are relatively less effective in digesting a large amount of lignocellulosic biomass, although some symbiotic bacteria that produce polysaccharide-degrading enzymes can be found [25,37,38]. The rest of the metagenomes are considered inactive lignocellulose-degrading ecological niches under various physical and environmental conditions.

We first assessed the diversity of the above-mentioned microbial communities using the Shannon diversity index, based on 16S rRNA extracted from the metagenomes. The bagasse metagenome has a Shannon index of 2.08, comparable to the average of 2.76 ± 0.73 (SD) (Additional file 3: Table S2). There are 1,164 different bacterial species detected in the bagasse metagenome, whereas the average is $1,035.25 \pm 201.45$ (SD). Using the combined dataset from all 12 metagenomes as a reference, "all-combined" dataset herein, we observed that the microbial community in bagasse is more enriched in Proteobacteria than the average (all P-values $< 2.2 \times 10^{-16}$, Fisher's exact test, unless indicated otherwise). This is still true even when compared with other lignocellulolytic datasets combined (Additional file 5: Table S4). By contrast, the bagasse metagenome has smaller proportions of reads identified as Actinobacteria, Cyanobacteria, and Firmicutes than the all-combined and lignocellulolytic datasets.

Proteobacteria dominates "open environment" lignocellulolytic and non-lignocellulolytic metagenomes

Overall, the 12 microbial communities analyzed contain different combinations and abundances of the bacterial phyla, with five or more phyla in all the datasets analyzed (Figure 3A). Note that we present and discuss the abundances of reads belonging to taxonomic groups or functional categories as percentages of mapped reads in that dataset, as well as numbers of unique species or genes that any read in the dataset mapped to. The most distinguishable characteristic in the taxonomic profile is the prevalence of reads assigned to Proteobacteria between oxygenated "open" environments (for example, sugarcane bagasse, compost, and sludge) and anoxic "close" environments (such as animal guts). We observed the domination of reads from aerobic Proteobacteria in all the open-environment metagenomes, which account for more than half of all the mapped reads, whereas they are almost entirely absent from the metagenomes of animal guts, in agreement with previous studies [5,22]. Of all the four gut

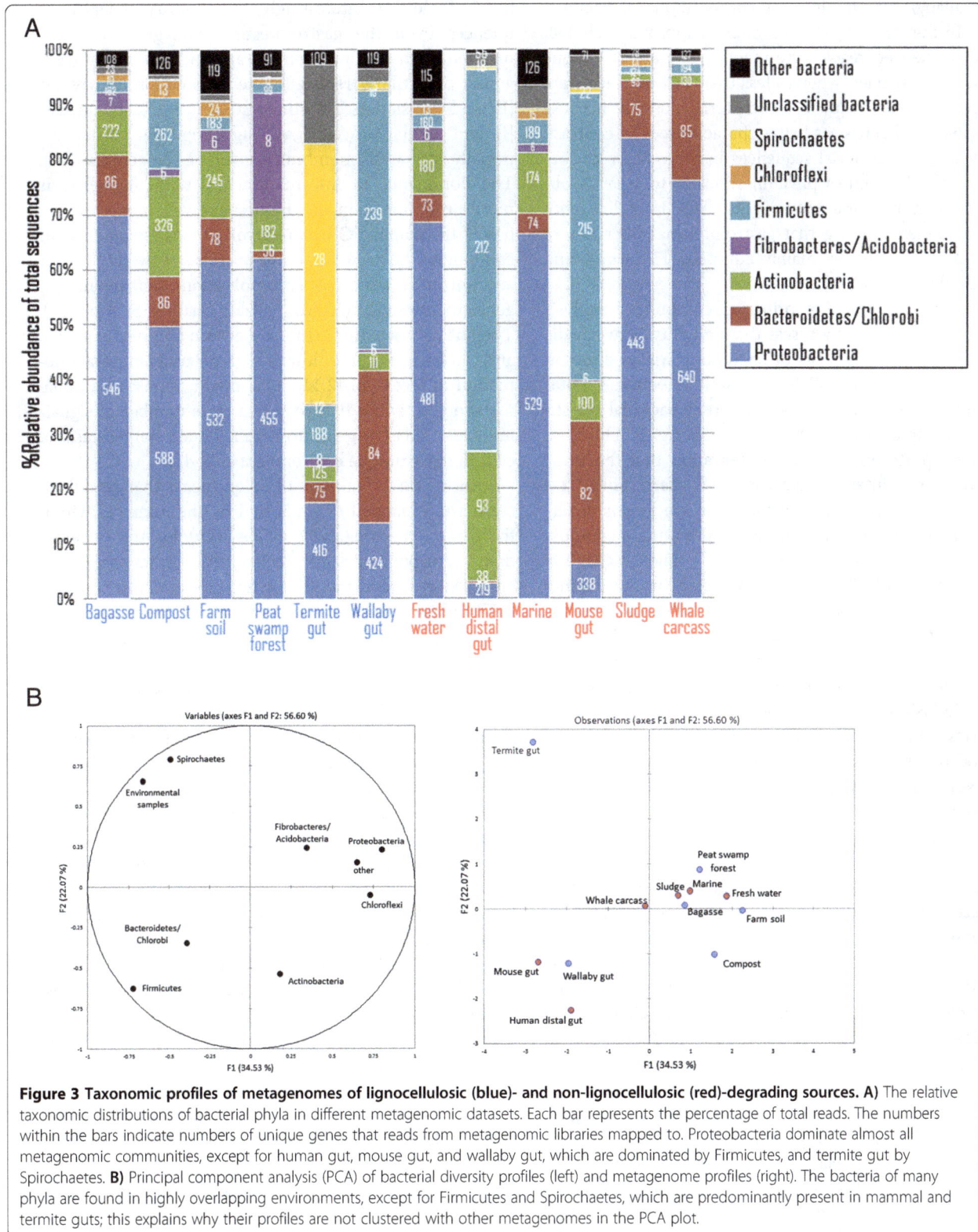

Figure 3 Taxonomic profiles of metagenomes of lignocellulosic (blue)- and non-lignocellulosic (red)-degrading sources. A) The relative taxonomic distributions of bacterial phyla in different metagenomic datasets. Each bar represents the percentage of total reads. The numbers within the bars indicate numbers of unique genes that reads from metagenomic libraries mapped to. Proteobacteria dominate almost all metagenomic communities, except for human gut, mouse gut, and wallaby gut, which are dominated by Firmicutes, and termite gut by Spirochaetes. **B)** Principal component analysis (PCA) of bacterial diversity profiles (left) and metagenome profiles (right). The bacteria of many phyla are found in highly overlapping environments, except for Firmicutes and Spirochaetes, which are predominantly present in mammal and termite guts; this explains why their profiles are not clustered with other metagenomes in the PCA plot.

metagenomes included in this study, the animal guts that have relatively less effective cellulose-degrading function (human and mouse) predominantly contain anaerobic Firmicutes such as Clostridia, which have been shown to possess a number of lignocellulose-degrading genes [25,37-39]. Interestingly, the two herbivorous guts (termite and wallaby) contain different compositions of bacterial phyla. In the termite gut metagenome, approximately half of the sequenced reads were mapped to only 28 unique Spirochaetes species, whereas less than 20% of reads were mapped to 416 unique Proteobacteria, possibly because Proteobacterial genes are better characterized. By contrast, the majority of bacterial sequences in the wallaby gut belong to the Firmicutes phylum, similarly to what is observed in human and mouse guts. This is in line with earlier studies showing that gastrointestinal microbes of warm-blooded animals mainly comprise Firmicutes and Bacteroidetes [24,40,41].

We then performed a principal component analysis (PCA) to quantitatively assess the similarity between the presence/absence patterns of bacterial phyla in different environments (Figure 3B left), as well as between metagenomic datasets that contain different bacterial constitutions using a loading plot (Figure 3B right). The PCA result supports our previous observation that the three mammal gastrointestinal environments share similar sets of bacterial phyla compositions, which are mostly anaerobic. The termite gut, by contrast, has a unique composition of microbes, most likely due to a much higher pH environment [42]. This is also reflected by the distinct prevalence of anaerobic Spirochaetes, which are mostly found in the termite gut, but almost entirely disappear from other guts and in the open environments. Our comparative genomic analysis thus demonstrates that lignocellulosic and non-lignocellulosic biomass-degrading lifestyles are not necessarily linked to the taxonomic diversity of the microbial communities. For comprehensive analyses of genomes and their functions across multiple gastrointestinal metagenomes, we refer the reader to an earlier comparative genomic study [21].

Meta-analysis of functional gene contents reviews high abundances of metabolic genes in biomass-degrading metagenomes

To gain an overall picture of the functions of genes possessed by the 12 metagenomes, we first classified the assembled reads to different Clusters of Orthologous Groups (COGs) [43] (Figure 4 and Additional file 4: Table S3). Globally, we observed a large proportion (approximately 40% or more) of reads mapped to genes involving metabolic processes (green bars) in all the open aerobic environments, especially energy production and conservation [COG class C] and amino acid transport and metabolism [E]. However, metabolic COGs are present in only 20 to 30% of the reads from the animal gut metagenomes, which are hierarchically clustered together. As expected, bagasse and other lignocellulolytic metagenomes are more enriched in carbohydrate transport and metabolism [G] genes than the all-combined dataset, with the exception of wallaby guts (Additional

file 5: Table S4). Interestingly, the majority of DNA sequences from the gastrointestinal metagenomes were mapped to the information storage and processing (red) genes, particularly replication recombination and repair [L] and translation, ribosomal structure, and biogenesis [J], and cellular processes and signaling (blue) genes, especially cell wall/membrane/envelope biogenesis [M]. The dominance of information and signaling genes is most pronounced in the mouse gut, where more than half of the mapped ORFs are involved in these two classes combined. The mouse gut also possesses a twofold higher amount of replication recombination and repair [L] genes than average ($P < 10^{-40}$, Additional file 5: Table S4). The gut microenvironments are anoxic and nutrient-rich, and can have extremely low or high pH and temporal fluctuation of feces [40,42,44]. This might impose additional metabolic activities that require a large number of signaling and regulatory genes to help maintain homeostasis of cells in these unique environments [45-47].

Focusing on the number of unique genes that assembled reads were mapped to (indicated by the numbers within the bars), many metabolic COGs including the carbohydrate transport and metabolism [G] genes are most enriched in three lignocellulolytic environments: compost (577 unique genes), bagasse (541), and farm soil (319), suggesting a greater diversity of carbohydrate-active genes in these three metagenomes. However, one might consider that the numbers of total reads in these datasets are slightly larger than in other datasets (Additional file 3: Table S2), and thus contribute to the larger numbers of unique genes observed. We believe this is only partly true, as the number of reads from the peat swamp forest dataset is as large, but the numbers of unique genes are similar to those from the datasets with lower numbers of total reads.

We then focused on metabolic potential of the metagenomes using the Kyoto Encyclopedia of Genes and Genomes (KEGG) pathways [48] (Additional file 4: Table S3 and Additional file 6: Figure S2). All the KEGG classes involved in carbohydrate metabolism can be found in all the 12 metagenomes, and this is also true for most of the enzymes related to the metabolism of amino acids, and cofactors and vitamins. The gastrointestinal tract environments are, again, more similar to one another. We observed a number of KEGG classes specifically absent or present at much lower percentages in animal guts, including lipid metabolic classes such as alpha-linolenic acid metabolism and fatty acid elongation, whereas sphingolipid metabolic genes are more pronounced (Additional file 4: Table S3). Looking at the number of unique genes mapped to the KEGG carbohydrate metabolism pathways, a wide range of carbohydrate metabolism enzymes are detected in all the datasets impartially, except for the termite gut, which seems to have a lower number of enzymes than other datasets.

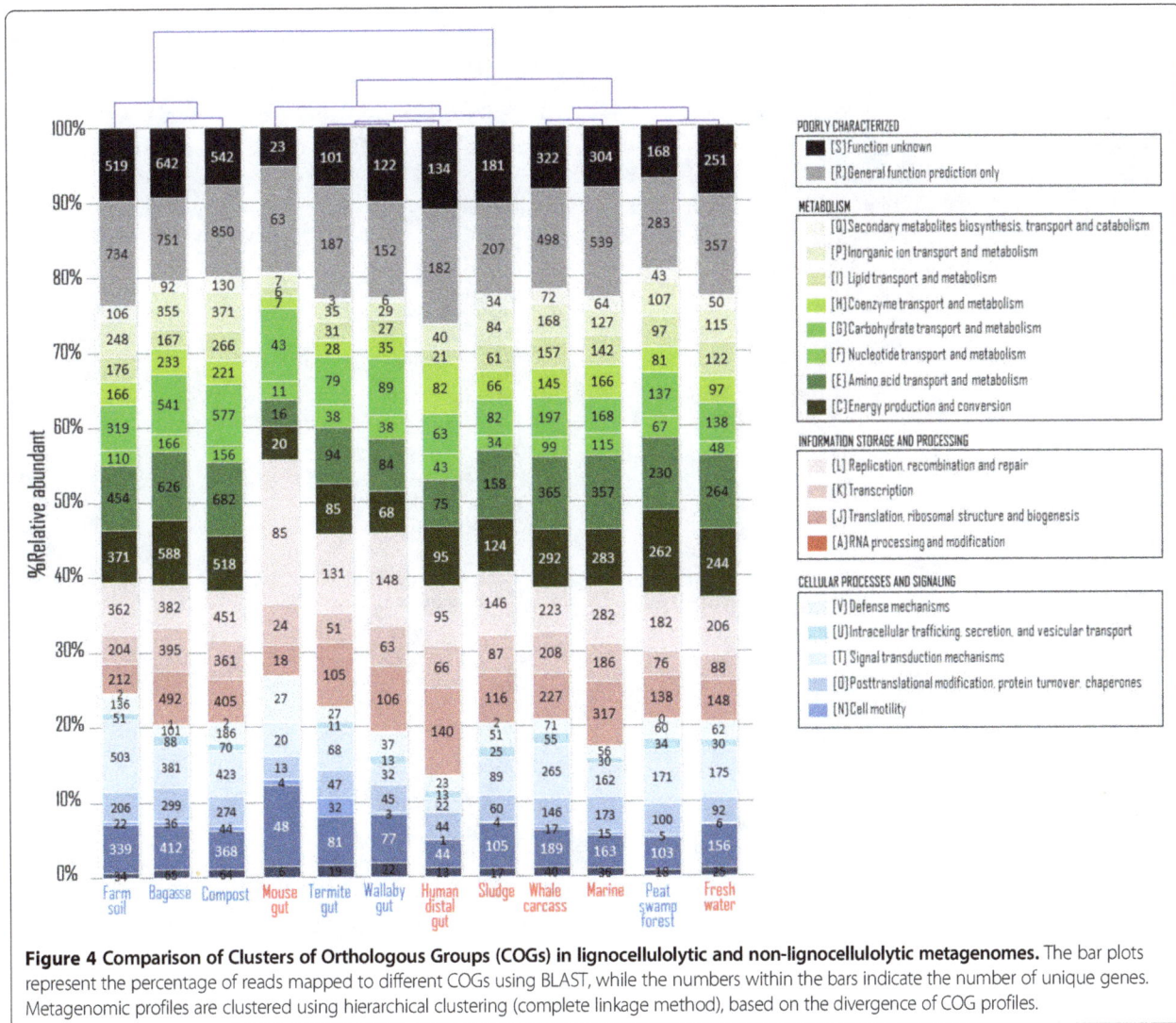

Figure 4 Comparison of Clusters of Orthologous Groups (COGs) in lignocellulolytic and non-lignocellulolytic metagenomes. The bar plots represent the percentage of reads mapped to different COGs using BLAST, while the numbers within the bars indicate the number of unique genes. Metagenomic profiles are clustered using hierarchical clustering (complete linkage method), based on the divergence of COG profiles.

Lignocellulolytic environments are enriched in enzymes required for degrading large carbohydrate molecules

So far we have addressed the taxonomic and functional genomic similarities between the sugarcane bagasse and other selected lignocellulolytic and non-lignocellulolytic metagenomes. In this section, we focus on the carbohydrate enzyme classifications from the CAZy database [30], which provides manually curated information for all characterized carbohydrate-active enzymes, covering cellulases, hemicellulases, and pectinases.

As expected, lignocellulolytic metagenomes contain a greater number of unique GH genes with lignocellulose-degrading enzymatic activities (Figure 5, red), as well as the numbers of reads mapped to these GH families (Additional file 4: Table S3), especially those encoding enzymes acting on large carbohydrate molecules, further up the lignocellulose degradation pathway (see Figure 2 for a simplified pathway). To illustrate the point, the lignocellulolytic metagenomes have a larger repertoire of

the so-called "true cellulases" (GH5, 9), as the numbers of unique genes are markedly higher than those of the non-lignocellulolytic metagenomes. Interestingly, although the lignocellulolytic metagenomes contain similar numbers of unique GH5 and 9 genes, they are most enriched in the termite gut in terms of read abundance (ninefold and fivefold of the all-combined dataset, respectively, $P < 10^{-19}$ Additional file 5: Table S4). Similarly, several endo-acting hemicelluloses including GH10, 16, 26, 51, and 53 are all more abundant in the lignocellulolytic metagenomes based on unique genes as well as mapped reads, whereas GH11, a xylanase family, is almost entirely absent from the non-lignocellulolytic environments. However, major oligosaccharide-degrading families such as GH2 and 3, which are required at the later stage to break down disaccharides into monosaccharides, are present in nearly all the metagenomes analyzed at comparable gene numbers and percentages, with the exception of sludge, marine, and carcass, where GH2 is present at relatively

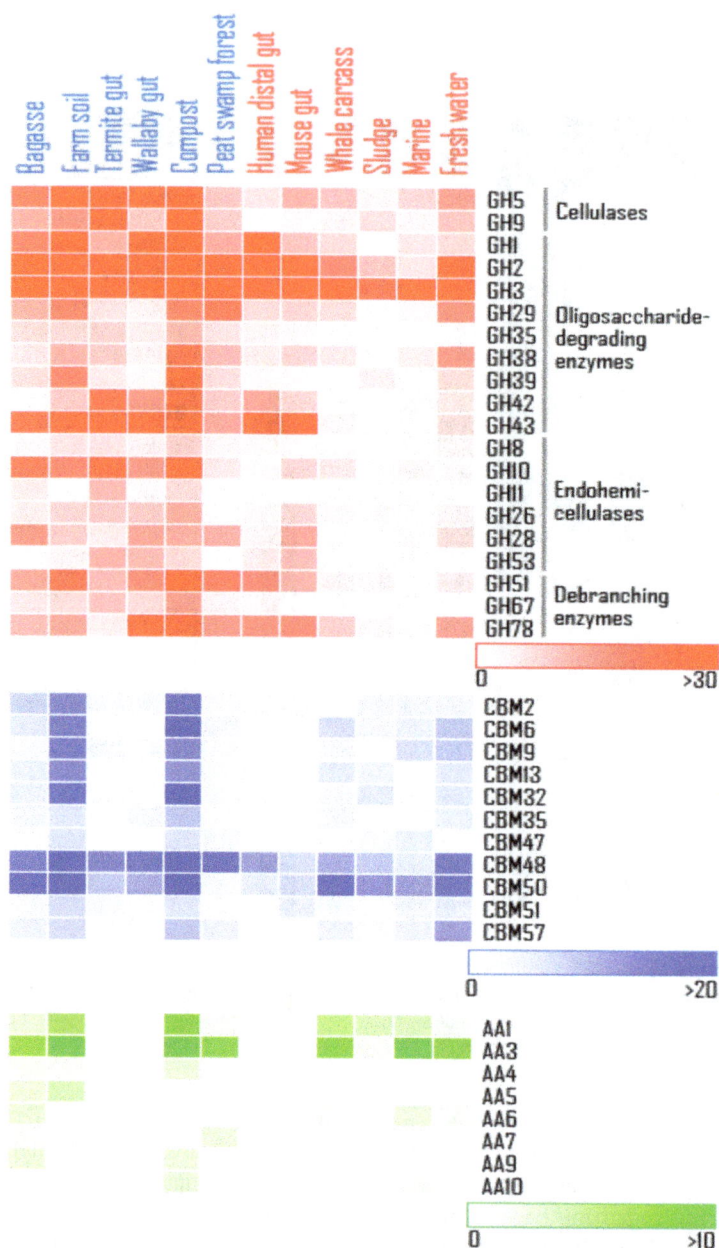

Figure 5 Comparison of reads mapped to different KEGG metabolic pathways and CAZy enzyme families. Comparative genomic analysis of selected glycoside hydrolase (GH), carbohydrate-binding module (CBM), and auxiliary activities (AA) families. Color shades indicate the numbers of unique genes in the families to which metagenomic reads were mapped.

lower abundances ($P < 10^{-19}$ Additional file 5: Table S4). This suggests a remarkable ability of the microorganism communities in lignocellulolytic metagenomes to break down large carbohydrate molecules. Accessory enzymes involved in cleavages of hemicellulose side chains, for example, beta-galactosidases and alpha-arabinofuranosidases (GH43), are found in all open lignocellulolytic environments and in the guts of herbivores and omnivores. In addition to these lignocellulolytic "core" enzymes, we also observed a number of GH families most pronounced in

bagasse, namely GH15 (glucoamylase), GH17 (glucan endo-1,3-beta-glucosidase) GH65 (maltose phosphorylase, rehalose phosphorylase), and GH115 (xylan alpha-1,2-glucuronidase) ($P < 10^{-3}$ when compared with the all-combined dataset).

In a similar manner to the majority of the GH families, CBMs (for example, CBM2, 6, 9, and 32) are evidently most enriched in the farm soil and compost metagenomes, with the exception of CBM48 and 50, which are highly present in all the environments analyzed (Figure 5, blue). Note that although the numbers of raw reads and mapped ORFs obtained from the farm soil and compost environments are higher than the average of the 12 metagenomes, these numbers are still comparable to those of the bagasse and peat swamp metagenomes. The CBM2 family has been shown to bind to cellulose and xylan, while CBM9 is found only in xylanase. CBM6 has been demonstrated to function in binding of amorphous cellulose and xylan, as well as beta-1,3 and beta-1,4 glucan. A high abundance of CBM32, involved in binding to galactose and lactose, is also found in these two environments. The CBM48 family, on the other hand, binds to GH13, a large GH family that includes alpha-amylase, which is also highly enriched in both groups of environments. Similarly, CBM50 binds to a number of GH families including GH23, 24, 25, and 73, which are all found in both lignocellulolytic and non-lignocellulolytic metagenomes.

The AA class represents families of ligninolytic enzymes and lytic polysaccharide mono-oxygenases [30]. As lignin is intimately associated with the carbohydrates in the plant cell wall, these ligninolytic enzymes cooperate with the classical GHs in decomposition of lignocelluloses. Intriguingly, the AA families are absent from animal guts altogether (Figure 5, green). The two major families AA1 and 3, for instance, are present in all the microbial communities, except for the closed anaerobic environments, possibly because most AAs identified to date are related to aerobic fungi and bacteria. Importantly, the bagasse microbial community had the most complete set of AA families (seven out of eight families analyzed: AA1, 3, 4, 5, 6, 7, and 9). AA9 (formerly GH61) in particular, has received growing attention recently, as the remarkable synergism between AA9 and GHs in boosting enzymatic cleavages of lignocellulosic biomass has been reported and patented [49-52]. The AA9 proteins are copper-dependent lytic polysaccharide monooxygenases (LPMOs), which function in cleaving cellulose chains with oxidation of various carbons (C-1, C-4, and C-6) [53]. The AA9 family found in the bagasse metagenome originates from fungi, as in compost, the only other metagenome in this study where AA9 is found.

To quantify the similarity between different metagenomic profiles, we have computed Pearson and Spearman correlations among all the metagenomes based on the three patterns of the four characteristics: taxonomic, COG, KEGG, and CAZy profiles (Figure 6 and Additional file 7: Table S5). As described in the previous sections, Figure 6 recapitulates our observation that the two groups of selected metagenomes: lignocellulolytic and non-lignocellulolytic, are hardly distinguishable based on the taxonomic, COG, or KEGG profiles. However, here we show that the lignocellulolytic metagenomes possess more similar sets of CAZy families, and also a significantly greater similarity of proportions of reads

mapped to different families, than those within the non-lignocellulolytic group and also those between the metagenomes from different groups. The lignocellulolytic group also possesses a larger number of unique CAZy genes (1,118.5 ± 538.0, SD, 931 in bagasse) compared to the non-lignocellulolytic group (501.8 ± 140.8, SD). This signifies the common carbohydrate-degrading gene repertoire and composition in the lignocellulolytic metagenomes, which enable the microbial communities as a whole to harvest energy and nutrients from lignocellulosic biomass, regardless of the taxonomy and enrichment of individual organisms in each microbial community.

Conclusions

Sugarcane bagasse is one of the most abundant agricultural biomasses, with a global production of over 250 million tons per year [5,54]. Microbial communities in industrial bagasse piles provide a useful starting point for the exploration and characterization of new biomass-degrading enzymes, which are stable and active at relatively high temperatures, in low amounts of nitrogen, and under the varying microenvironmental conditions commonly found in different regions of the piles. To the best of our knowledge, the phylogenetic distribution of microorganisms in the bagasse metagenome has previously been characterized using 16S rRNA and a restricted number of shotgun sequencing reads (70,000 reads) [5,8], and thus the metagenome constructed from the fosmid library in this study provides the largest collection of metabolic genes found in this ecological niche to date (approximately one million raw reads and over 200,000 assembled contigs plus singletons). This, for the first time, allows us to investigate the functions, prevalence, diversity, and abundance of biomass-degrading enzymes, all of which was not possible with the previous 16S rRNA and smaller shotgun sequencing libraries. The bagasse fosmid library also serves as a useful resource for subsequent enzymatic assays of prospective biomass-degrading enzymes, which could be developed further for industrial use. The fosmid library also allows easy recovery of genes of interest for further cloning and detailed expression study [5,55].

We have characterized our newly constructed bagasse metagenome and demonstrated that the microbe community is taxonomically diverse, and at the same time, functionally capable of converting large polysaccharides to monomeric sugars, as all the major cellulolytic and hemicellulolytic enzymes were found. Based on our comparative genomic analyses of the bagasse metagenome with other publicly available lignocellulolytic and non-lignocellulolytic metagenomes, we have shown that the phylogenetic distributions of the microbes are separated mainly by their aerobic/anoxic lifestyles. Intriguingly,

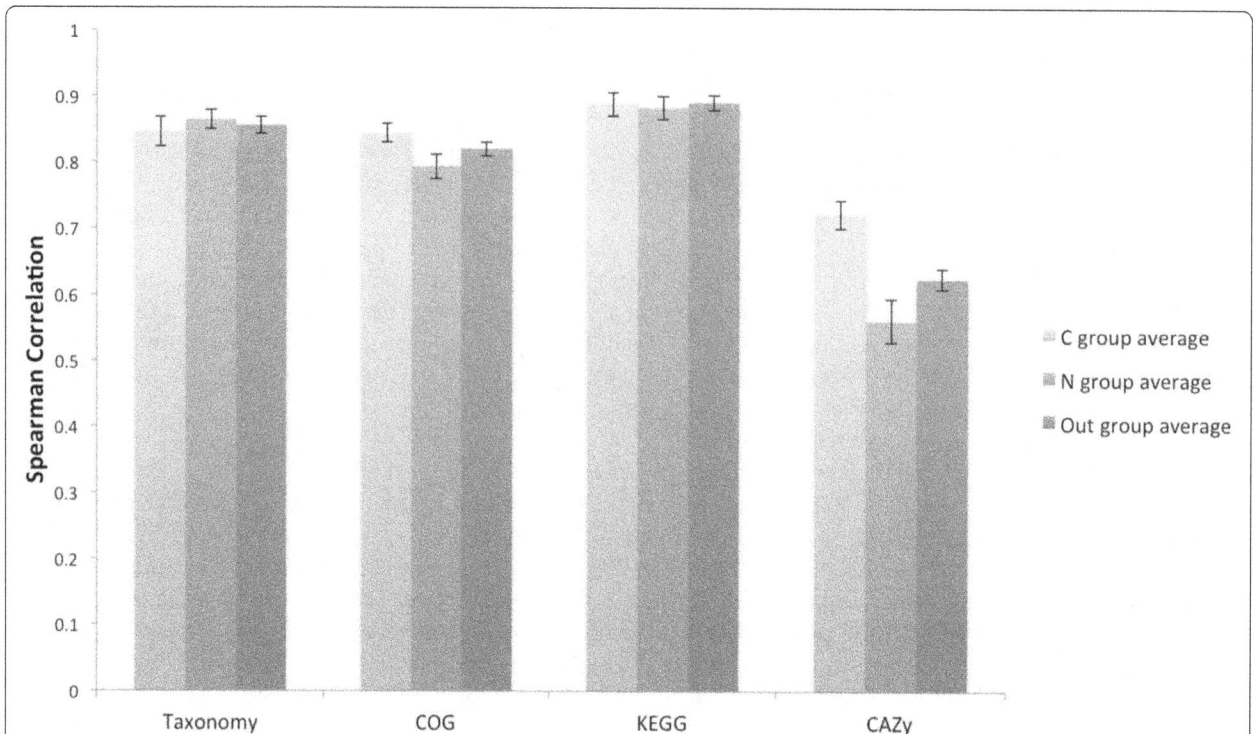

Figure 6 Lignocellulolytic metagenomes are taxonomically diverse, but their carbohydrate-active enzymes are conserved.
Spearman correlations were computed for metagenomic libraries within the lignocellulose-degrading environment group (C), non-lignocellulose-degrading group (N), or metagenomes from different groups (Out group), based on the taxonomy, COG, KEGG, and CAZy profiles. Error bars represent standard errors of means.

although the lignocellulolytic and non-lignocellulolytic groups are not distinguishable by their taxonomic contents or by their high-level functional classifications (COGs) and metabolic genes (KEGG), the lignocellulolytic group possesses highly similar lignocellulose-degrading core genes, which are produced by different types and abundances of microbes within different lignocellulose-degrading communities. That is, even though the species compositions of lignocellulolytic metagenomes are no more similar than when they are compared across the two groups, their carbohydrate-active enzyme compositions are significantly more conserved than those in non-lignocellulolytic groups. This exemplifies an important interplay between diverse microorganisms in the communities that contribute to the enzyme repertoires required to degrade lignocelluloses under mixed microenvironmental conditions, in different ecological systems.

Methods

Sample collection and DNA extraction

The sugarcane bagasse sample was collected from soil-contacting regions of a bagasse pile (one meter from the edge of the pile) at Phu Khieo Bio-Energy Chaiyaphum province, Thailand (N 16°28'54", W 102°07'05"). The bagasse piles are normally approximately 10 m in height covering an area of several acres. The sample was rapidly

frozen in liquid nitrogen and kept at -80°C for subsequent experiments. Metagenomic DNA was directly extracted from a sample by the SDS-based DNA extraction procedure [56], with slight modifications [57]. Briefly, five grams of sample was subjected to direct cell lysis with DNA extraction buffer, proteinase K, and sodium dodecyl sulfate (SDS). Protein contamination was removed by chloroform extraction, and then the DNA was precipitated with isopropanol. High molecular weight DNA of size ranging from 30 to 50 kb was selected and purified using pulse field gel electrophoresis and electroelution techniques. The extracted DNA was separated by electrophoresis in a CHEF DRIII system (Bio-Rad, Hercules, CA, USA) in 0.5X Tris/borate/EDTA (TBE) at 14°C, using a 0.1 to 14 sec switch time at 6 V/cm for 12 h. The high molecular weight DNA (30 to 50 kb) was excised from the gel and recovered by electroelution techniques. The gel slice was electroeluted in a dialysis bag (Spectra/Por 4, Spectrum Laboratories, Rancho Dominguez, CA, USA) using a field strength of 70 V at 4°C for 2 h. The purified DNA solution was collected and subsequently concentrated using an Amicon Ultra filter unit (Millipore, Billerica, MA, USA).

Fosmid library construction

A metagenomic fosmid library from the bagasse sample was constructed using a CopyControl™ Fosmid Library

Production Kit (Epicentre Biotechnologies, Madison, WI, USA) according to the manufacturer's instructions, with slight modifications. The purified DNA was end-repaired to generate blunt 5'-phospholyrated ends and then ligated to the pCC1FOS vector at 25°C for 3 h. The ligated DNA was packaged using the lambda packaging extract supplied and subsequently transformed into *Escherichia coli* EPI300-T1R. The transformants were selected on LB agar plates supplemented with 12.5 µg/ml of chloramphenicol. The library was stored at -80°C in 15% glycerol in the form of individual clones as well as pool libraries.

Shotgun pyrosequencing and data pre-processing

A total of 3,300 randomly selected fosmid clones were sequenced on one full lane of the 454 GS-FLX Genome Sequencer System using the Titanium platform (Roche, Brandford, CT, USA) following the manufacturer's protocol. Repeats in raw sequenced reads obtained were removed using RepeatMasker (http://www.repeatmasker.org). The vector and host sequences were filtered by BLASTN, with an E-value cutoff of 1e-3. The filtered reads were assembled using the Newbler assembly software, developed by 454 Life Sciences (version 2.6, Roche). Non-overlapping fragment singletons were clustered using the CD-HIT software [58] to minimize redundant sequences. The overall process of metagenomic data preparation and analysis is summarized in Additional file 1: Figure S1. The entire sequences of the bagasse fosmid library have been deposited to the NCBI Sequence Read Archive (SRA), which can be accessed using the accession number: SRX493840.

Functional gene annotation and metabolic pathway analysis

The taxonomic classifications were performed on assembled contigs and singletons using BLASTN against the NT database. The E-value cutoff was set to 1e-3, and the best BLAST hit was used to refer the taxonomic rank of each sequence. The non-redundant singletons and contigs were predicted for open reading frames (ORFs) by MetaGeneMark [59]. The Shannon diversity index was computed using mothur [60] on 16S rRNA sequences extracted by BLASTN against the NCBI 16S microbial database using E-value cutoff 1e-5 and a minimal alignment length of 50 bp. The functional annotation was initially performed by stand-alone BLAST on predicted ORFs against the Non-Redundant protein database (NR) [23] using an E-value cutoff of 1e-6. The BLAST results containing the best hits were subsequently processed using the Blast2GO program [61] to assign their functional gene contents and enzymes based on the Kyoto Encyclopedia of Genes and Genomes (KEGG) [48].

Orthologous genes were identified using Clusters of Orthologous Groups (COGs) [43]. The carbohydrate-active enzymes were predicted using BLAST against the CAZy database using an E-value cutoff of 1e-10.

Comparative metagenomic analysis of lignocellulose- and non-lignocellulose-degrading sources

We compared our bagasse metagenome to publicly available metagenomic projects obtained from NCBI Whole Genome Shotgun (WGS) and Sequence Read Archive (SRA) projects [31]. All additional datasets obtained were reanalyzed using the same procedures as described previously (as seen in Additional file 1: Figure S1) for an unbiased comparison. The publicly available datasets for lignocellulose-degrading and non-lignocellulose-degrading environments used in this study include metagenomic profiles from carcass [14], compost [32], farm soil [14], fresh water [62], human distal gut [33], marine water, mouse gut [39], peat swamp forest [15], sludge [34], termite gut [19], and wallaby gut [18] (summarized in Additional file 3: Table S2). Taxonomic distributions and functional genomic profiles of each metagenomic dataset are presented as unique hits (that is species or functional categories) or relative abundances, which are read counts normalized by the total number of mapped reads in each particular metagenome. Enrichment of read abundances was assessed using Fisher's exact test against the all-combined dataset of the 12 metagenomes, or lignocellulolytic and non-lignocellulolytic groups. The *P*-values from Fisher's exact test and odds ratios were derived using the module SciPy in Python (http://www.scipy.org/).

Additional files

Additional file 1: Figure S1. Summary of data analyses and comparisons. The bagasse metagenomic fosmid library was pyrosequenced, the vector and host sequences removed, and it was assembled and deposited to the NCBI Sequence Read Archive (SRA). Additional metagenomic libraries of both lignocellulosic and non-lignocellulosic sources and their SRAs were obtained, and the subsequent functional analyses were performed using the same procedures.

Additional file 2: Table S1. Summary of the numbers of hits that reads from the bagasse metagenomes were matched to for different taxonomic classifications.

Additional file 3: Table S2. Numbers of total and mapped reads in lignocellulolytic and non-lignocellulolytic metagenomes retrieved and analyzed in this study.

Additional file 4: Table S3. Numbers and percentages of reads, mapped to different NCBI taxonomy, COG, KEGG, and CAZy terms.

Additional file 5: Table S4. Enrichment of taxa, COGs, and CAZy families in different metagenomes, as compared to reference set (lignocellulolytic and non-lignocellulolytic groups and all-combined group). *P*-values were computed using Fisher's exact test, together with odds ratios, as described in the last tab in the table.

Additional file 6: Figure S2. Percentage of metagenomic sequences mapped to the KEGG pathways, relative to all the reads in each metagenomic dataset.

Additional file 7: Table S5. Correlations based on similarity of taxonomy, COGs, KEGG, and CAZy between the lignocellulose-degrading environment group (C), non-lignocellulose-degrading group (N), or libraries from the two different groups.

Abbreviations
AA: auxiliary activity; bp: base pairs; CAZy: Carbohydrate-Active enZYmes; CBM: carbohydrate-binding module; COGs: Clusters of Orthologous Groups; GH: glycoside hydrolase; KEGG: Kyoto Encyclopedia of Genes and Genomes; ORF: open reading frame; PCA: principal component analysis; SD: standard deviation; SDS: sodium dodecyl sulfate; TCA cycle: tricarboxylic acid cycle.

Competing interests
The authors declare that they have no competing interests.

Authors' contributions
WM, V Charoensawan, and V Champreda conceived and designed the study. PK collected the samples and constructed the fosmid library. ST performed shotgun sequencing. WM and V Charoensawan performed the analyses. WM, V Charoensawan, and V Champreda wrote the manuscript. All authors read and approved the final manuscript.

Acknowledgements
This project was supported by a research grant from the National Center for Genetic Engineering and Biotechnology, National Science and Technology Development Agency. PK was supported by a Royal Golden Jubilee scholarship (PHD/0260/2549). V Charoensawan was supported by a Mahidol University's New Researchers grant and a Research Fellowship from Clare Hall, Cambridge, UK. We also thank the anonymous reviewers for their excellent suggestions and support.

Author details
[1]Enzyme Technology Laboratory, Bioresources Technology Unit, National Center for Genetic Engineering and Biotechnology (BIOTEC), Thailand Science Park, Pathumthani 12120, Thailand. [2]Department of Biochemistry, Faculty of Science, Mahidol University, Bangkok 10400, Thailand. [3]Integrative Computational BioScience (ICBS) Center, Mahidol University, Nakhon Pathom 73170, Thailand. [4]Genome Institute, National Center for Genetic Engineering and Biotechnology (BIOTEC), Thailand Science Park, Pathumthani 12120, Thailand.

References
1. Van Dyk JS, Pletschke BI. A review of lignocellulose bioconversion using enzymatic hydrolysis and synergistic cooperation between enzymes - factors affecting enzymes, conversion and synergy. Biotechnol Adv. 2012;30(6):1458–80.
2. Lynd LR, Weimer PJ, van Zyl WH, Pretorius IS. Microbial cellulose utilization: fundamentals and biotechnology. Microbiol Mol Biol Rev. 2002;66(3):506–77.
3. Rubin EM. Genomics of cellulosic biofuels. Nature. 2008;454(7206):841–5.
4. Lee J. Biological conversion of lignocellulosic biomass to ethanol. J Biotechnol. 1997;56(1):1–24.
5. Kanokratana P, Mhuantong W, Laothanachareon T, Tangphatsornruang S, Eurwilaichitr L, Pootanakit K, et al. Phylogenetic analysis and metabolic potential of microbial communities in an industrial bagasse collection site. Microb Ecol. 2013;66(2):322–34.
6. Weimann A, Trukhina Y, Pope PB, Konietzny SG, McHardy AC. De novo prediction of the genomic components and capabilities for microbial plant biomass degradation from (meta-)genomes. Biotechnol Biofuels. 2013;6(1):24.
7. Li LL, McCorkle SR, Monchy S, Taghavi S, van der Lelie D. Bioprospecting metagenomes: glycosyl hydrolases for converting biomass. Biotechnol Biofuels. 2009;2:10.
8. Rattanachomsri U, Kanokratana P, Eurwilaichitr L, Igarashi Y, Champreda V. Culture-independent phylogenetic analysis of the microbial community in industrial sugarcane bagasse feedstock piles. Biosci Biotechnol Biochem. 2011;75(2):232–9.
9. Alvarez TM, Goldbeck R, dos Santos CR, Paixao DA, Goncalves TA, Franco Cairo JP, et al. Development and biotechnological application of a novel endoxylanase family GH10 identified from sugarcane soil metagenome. PLoS One. 2013;8(7):e70014.
10. Sangwan N, Lata P, Dwivedi V, Singh A, Niharika N, Kaur J, et al. Comparative metagenomic analysis of soil microbial communities across three hexachlorocyclohexane contamination levels. PLoS One. 2012;7(9):e46219.
11. Shi W, Xie S, Chen X, Sun S, Zhou X, Liu L, et al. Comparative genomic analysis of the microbiome [corrected] of herbivorous insects reveals eco-environmental adaptations: biotechnology applications. PLoS Genet. 2013;9(1):e1003131.
12. Gianoulis TA, Raes J, Patel PV, Bjornson R, Korbel JO, Letunic I, et al. Quantifying environmental adaptation of metabolic pathways in metagenomics. Proc Natl Acad Sci U S A. 2009;106(5):1374–9.
13. Fierer N, Leff JW, Adams BJ, Nielsen UN, Bates ST, Lauber CL, et al. Cross-biome metagenomic analyses of soil microbial communities and their functional attributes. Proc Natl Acad Sci U S A. 2012;109(52):21390–5.
14. Tringe SG, von Mering C, Kobayashi A, Salamov AA, Chen K, Chang HW, et al. Comparative metagenomics of microbial communities. Science. 2005;308(5721):554–7.
15. Kanokratana P, Uengwetwanit T, Rattanachomsri U, Bunterngsook B, Nimchua T, Tangphatsornruang S, et al. Insights into the phylogeny and metabolic potential of a primary tropical peat swamp forest microbial community by metagenomic analysis. Microb Ecol. 2011;61(3):518–28.
16. Brulc JM, Antonopoulos DA, Miller ME, Wilson MK, Yannarell AC, Dinsdale EA, et al. Gene-centric metagenomics of the fiber-adherent bovine rumen microbiome reveals forage specific glycoside hydrolases. Proc Natl Acad Sci U S A. 2009;106(6):1948–53.
17. Hess M, Sczyrba A, Egan R, Kim TW, Chokhawala H, Schroth G, et al. Metagenomic discovery of biomass-degrading genes and genomes from cow rumen. Science. 2011;331(6016):463–7.
18. Pope PB, Denman SE, Jones M, Tringe SG, Barry K, Malfatti SA, et al. Adaptation to herbivory by the Tammar wallaby includes bacterial and glycoside hydrolase profiles different from other herbivores. Proc Natl Acad Sci U S A. 2010;107(33):14793–8.
19. Warnecke F, Luginbuhl P, Ivanova N, Ghassemian M, Richardson TH, Stege JT, et al. Metagenomic and functional analysis of hindgut microbiota of a wood-feeding higher termite. Nature. 2007;450(7169):560–5.
20. Martins LF, Antunes LP, Pascon RC, de Oliveira JC, Digiampietri LA, Barbosa D, et al. Metagenomic analysis of a tropical composting operation at the Sao Paulo zoo park reveals diversity of biomass degradation functions and organisms. PLoS One. 2013;8(4):e61928.
21. Lamendella R, Domingo JW, Ghosh S, Martinson J, Oerther DB. Comparative fecal metagenomics unveils unique functional capacity of the swine gut. BMC Microbiol. 2011;11:103.
22. Wongwilaiwalin S, Laothanacharoen T, Mhuantong W, Tangphatsornruang S, Eurwilaichitr L, Igarashi Y, et al. Comparative metagenomic analysis of microcosm structures and lignocellulolytic enzyme systems of symbiotic biomass-degrading consortia. Appl Microbiol Biotechnol. 2013;97(20):8941–54.
23. Pruitt KD, Tatusova T, Brown GR, Maglott DR. NCBI Reference Sequences (RefSeq): current status, new features and genome annotation policy. Nucleic Acids Res. 2012;40(Database issue):D130–135.
24. Morrison M, Pope PB, Denman SE, McSweeney CS. Plant biomass degradation by gut microbiomes: more of the same or something new? Curr Opin Biotechnol. 2009;20(3):358–63.
25. Larsbrink J, Rogers TE, Hemsworth GR, McKee LS, Tauzin AS, Spadiut O, et al. A discrete genetic locus confers xyloglucan metabolism in select human gut Bacteroidetes. Nature. 2014;506(7489):498–502.
26. Fernandez AB, Vera-Gargallo B, Sanchez-Porro C, Ghai R, Papke RT, Rodriguez-Valera F, et al. Comparison of prokaryotic community structure from Mediterranean and Atlantic saltern concentrator ponds by a metagenomic approach. Front Microbiol. 2014;5:196.
27. Patel DD, Patel AK, Parmar NR, Shah TM, Patel JB, Pandya PR, et al. Microbial and Carbohydrate Active Enzyme profile of buffalo rumen metagenome and their alteration in response to variation in the diet. Gene. 2014;545(1):88–94.
28. Park SK, Jang HM, Ha JH, Park JM. Sequential sludge digestion after diverse pre-treatment conditions: sludge removal, methane production and microbial community changes. Bioresource Technol. 2014;162:331–40.
29. Hollister EB, Forrest AK, Wilkinson HH, Ebbole DJ, Tringe SG, Malfatti SA, et al. Mesophilic and thermophilic conditions select for unique but highly parallel microbial communities to perform carboxylate platform biomass conversion. PLoS One. 2012;7(6):e39689.

30. Levasseur A, Drula E, Lombard V, Coutinho PM, Henrissat B. Expansion of the enzymatic repertoire of the CAZy database to integrate auxiliary redox enzymes. Biotechnol Biofuels. 2013;6(1):41.

31. Kodama Y, Shumway M, Leinonen R. The Sequence Read Archive: explosive growth of sequencing data. Nucleic Acids Res. 2012;40(Database issue):D54–56.

32. Allgaier M, Reddy A, Park JI, Ivanova N, D'Haeseleer P, Lowry S, et al. Targeted discovery of glycoside hydrolases from a switchgrass-adapted compost community. PLoS One. 2010;5(1):e8812.

33. Gill SR, Pop M, Deboy RT, Eckburg PB, Turnbaugh PJ, Samuel BS, et al. Metagenomic analysis of the human distal gut microbiome. Science. 2006;312(5778):1355–9.

34. Garcia Martin H, Ivanova N, Kunin V, Warnecke F, Barry KW, McHardy AC, et al. Metagenomic analysis of two enhanced biological phosphorus removal (EBPR) sludge communities. Nature Biotechnol. 2006;24(10):1263–9.

35. Dougherty MJ, D'Haeseleer P, Hazen TC, Simmons BA, Adams PD, Hadi MZ. Glycoside hydrolases from a targeted compost metagenome, activity-screening and functional characterization. BMC Biotechnol. 2012;12:38.

36. Gladden JM, Park JI, Bergmann J, Reyes-Ortiz V, D'Haeseleer P, Quirino BF, et al. Discovery and characterization of ionic liquid-tolerant thermophilic cellulases from a switchgrass-adapted microbial community. Biotechnol Biofuels. 2014;7(1):15.

37. Dassa B, Borovok I, Ruimy-Israeli V, Lamed R, Flint HJ, Duncan SH, et al. Rumen cellulosomics: divergent fiber-degrading strategies revealed by comparative genome-wide analysis of six ruminococcal strains. PLoS One. 2014;9(7):e99221.

38. Martens EC, Kelly AG, Tauzin AS, Brumer H: The devil lies in the details: how variations in polysaccharide fine-structure impact the physiology and evolution of gut microbes. J Molecular Biol. 2014.

39. Turnbaugh PJ, Ley RE, Mahowald MA, Magrini V, Mardis ER, Gordon JI. An obesity-associated gut microbiome with increased capacity for energy harvest. Nature. 2006;444(7122):1027–31.

40. Ley RE, Peterson DA, Gordon JI. Ecological and evolutionary forces shaping microbial diversity in the human intestine. Cell. 2006;124(4):837–48.

41. Thomas F, Hehemann JH, Rebuffet E, Czjzek M, Michel G. Environmental and gut Bacteroidetes: the food connection. Frontiers Microbiol. 2011;2:93.

42. Brune A, Emerson D, Breznak JA. The termite gut microflora as an oxygen sink: microelectrode determination of oxygen and pH gradients in guts of lower and higher termites. Appl Environmental Microbiol. 1995;61(7):2681–7.

43. Tatusov RL, Fedorova ND, Jackson JD, Jacobs AR, Kiryutin B, Koonin EV, et al. The COG database: an updated version includes eukaryotes. BMC Bioinformatics. 2003;4:41.

44. Sekelja M, Rud I, Knutsen SH, Denstadli V, Westereng B, Naes T, et al. Abrupt temporal fluctuations in the chicken fecal microbiota are explained by its gastrointestinal origin. Appl Environmental Microbiol. 2012;78(8):2941–8.

45. Charoensawan V, Wilson D, Teichmann SA. Genomic repertoires of DNA-binding transcription factors across the tree of life. Nucleic Acids Res. 2010;38(21):7364–77.

46. Charoensawan V, Wilson D, Teichmann SA. Lineage-specific expansion of DNA-binding transcription factor families. Trends Genetics : TIG. 2010;26(9):388–93.

47. Maslov S, Krishna S, Pang TY, Sneppen K. Toolbox model of evolution of prokaryotic metabolic networks and their regulation. Proc Natl Acad Sci U S A. 2009;106(24):9743–8.

48. Nakaya A, Katayama T, Itoh M, Hiranuka K, Kawashima S, Moriya Y, et al. KEGG OC: a large-scale automatic construction of taxonomy-based ortholog clusters. Nucleic Acids Res. 2013;41(Database issue):D353–357.

49. Hu J, Arantes V, Pribowo A, Saddler JN. The synergistic action of accessory enzymes enhances the hydrolytic potential of a "cellulase mixture" but is highly substrate specific. Biotechnol Biofuels. 2013;6(1):112.

50. Langston JA, Shaghasi T, Abbate E, Xu F, Vlasenko E, Sweeney MD. Oxidoreductive cellulose depolymerization by the enzymes cellobiose dehydrogenase and glycoside hydrolase 61. Appl Environmental Microbiol. 2011;77(19):7007–15.

51. Berka RM, Grigoriev IV, Otillar R, Salamov A, Grimwood J, Reid I, et al. Comparative genomic analysis of the thermophilic biomass-degrading fungi Myceliophthora thermophila and Thielavia terrestris. Nature Biotechnol. 2011;29(10):922–7.

52. Eastwood DC, Floudas D, Binder M, Majcherczyk A, Schneider P, Aerts A, et al. The plant cell wall-decomposing machinery underlies the functional diversity of forest fungi. Science. 2011;333(6043):762–5.

53. Harris PV, Welner D, McFarland KC, Re E, Navarro Poulsen JC, Brown K, et al. Stimulation of lignocellulosic biomass hydrolysis by proteins of glycoside hydrolase family 61: structure and function of a large, enigmatic family. Biochemistry. 2010;49(15):3305–16.

54. Kiatkittipong W, Wongsuchoto P, Pavasant P. Life cycle assessment of bagasse waste management options. Waste Manag. 2009;29(5):1628–33.

55. Kanokratana P, Eurwilaichitr L, Pootanakit K, Champreda V: Identification of glycosyl hydrolases from a metagenomic library of microflora in sugarcane bagasse collection site and their cooperative action on cellulose degradation. J Bioscience Bioengineering. 2014.

56. Zhou J, Bruns MA, Tiedje JM. DNA recovery from soils of diverse composition. Appl Environ Microbiol. 1996;62(2):316–22.

57. Kanokratana P, Chanapan S, Pootanakit K, Eurwilaichitr L. Diversity and abundance of Bacteria and Archaea in the Bor Khlueng Hot Spring in Thailand. J Basic Microbiol. 2004;44(6):430–44.

58. Fu L, Niu B, Zhu Z, Wu S, Li W. CD-HIT: accelerated for clustering the next-generation sequencing data. Bioinformatics. 2012;28(23):3150–2.

59. Zhu W, Lomsadze A, Borodovsky M. Ab initio gene identification in metagenomic sequences. Nucleic Acids Res. 2010;38(12):e132.

60. Schloss PD, Westcott SL, Ryabin T, Hall JR, Hartmann M, Hollister EB, et al. Introducing mothur: open-source, platform-independent, community-supported software for describing and comparing microbial communities. Appl Environmental Microbiol. 2009;75(23):7537–41.

61. Conesa A, Gotz S. Blast2GO: a comprehensive suite for functional analysis in plant genomics. Int J Plant Genomics. 2008;2008:619832.

62. Kalyuzhnaya MG, Lidstrom ME, Chistoserdova L. Real-time detection of actively metabolizing microbes by redox sensing as applied to methylotroph populations in Lake Washington. ISME J. 2008;2(7):696–706.

2

Genome-wide transcriptional analysis suggests hydrogenase- and nitrogenase-mediated hydrogen production in *Clostridium butyricum* CWBI 1009

Magdalena Calusinska[1,5†], Christopher Hamilton[2†], Pieter Monsieurs[3†], Gregory Mathy[4], Natalie Leys[3], Fabrice Franck[4], Bernard Joris[1], Philippe Thonart[2], Serge Hiligsmann[2] and Annick Wilmotte[1*]

Abstract

Background: Molecular hydrogen, given its pollution-free combustion, has great potential to replace fossil fuels in future transportation and energy production. However, current industrial hydrogen production processes, such as steam reforming of methane, contribute significantly to the greenhouse effect. Therefore alternative methods, in particular the use of fermentative microorganisms, have attracted scientific interest in recent years. However the low overall yield obtained is a major challenge in biological H_2 production. Thus, a thorough and detailed understanding of the relationships between genome content, gene expression patterns, pathway utilisation and metabolite synthesis is required to optimise the yield of biohydrogen production pathways.

Results: In this study transcriptomic and proteomic analyses of the hydrogen-producing bacterium *Clostridium butyricum* CWBI 1009 were carried out to provide a biomolecular overview of the changes that occur when the metabolism shifts to H_2 production. The growth, H_2-production, and glucose-fermentation profiles were monitored in 20 L batch bioreactors under unregulated-pH and fixed-pH conditions (pH 7.3 and 5.2). Conspicuous differences were observed in the bioreactor performances and cellular metabolisms for all the tested metabolites, and they were pH dependent. During unregulated-pH glucose fermentation increased H_2 production was associated with concurrent strong up-regulation of the nitrogenase coding genes. However, no such concurrent up-regulation of the [FeFe] hydrogenase genes was observed. During the fixed pH 5.2 fermentation, by contrast, the expression levels for the [FeFe] hydrogenase coding genes were higher than during the unregulated-pH fermentation, while the nitrogenase transcripts were less abundant. The overall results suggest, for the first time, that environmental factors may determine whether H_2 production in *C. butyricum* CWBI 1009 is mediated by the hydrogenases and/or the nitrogenase.

Conclusions: This work, contributing to the field of dark fermentative hydrogen production, provides a multidisciplinary approach for the investigation of the processes involved in the molecular H_2 metabolism of clostridia. In addition, it lays the groundwork for further optimisation of biohydrogen production pathways based on genetic engineering techniques.

Keywords: Dark fermentation, *Clostridium butyricum*, [FeFe] hydrogenase, Nitrogenase, RNA-seq, 2D-DIGE

* Correspondence: awilmotte@ulg.ac.be
†Equal contributors
[1]Centre for Protein Engineering, Bacterial Physiology and Genetics, University of Liège, Allée de la Chimie 3, B-4000 Liège, Belgium
Full list of author information is available at the end of the article

Background

Molecular hydrogen has great potential as a clean energy vector given its pollution-free combustion and the ease with which it can be converted into electricity via fuel cells. However, current industrial hydrogen production processes, such as steam reforming of natural gas, release large quantities of CO_2 and thereby contribute substantially to the greenhouse effect [1]. Consequently, scientific interest in recent years has focused on alternative methods of hydrogen production, in particular on the use of photosynthetic and fermentative microorganisms for CO_2-neutral H_2 production from renewable energy sources, such as solar energy and biomass [2]. Nevertheless, a major challenge when using microorganisms for H_2 production is the low yield generally obtained, with typical mesophilic fermentation of carbohydrates supplying only 10 to 20% of the H_2 potentially available in the substrate. Consequently, research in the field of metabolic engineering has investigated different approaches with a view to optimising the yield of well-characterised biohydrogen production pathways [3]. However, such strategies require a thorough and detailed understanding of the relationships between genome content, gene expression patterns, pathway utilisation and metabolite synthesis.

In dark anaerobic fermentation microorganisms break down carbohydrate-rich substrates into organic acids and alcohols while releasing H_2. Strict anaerobes such as clostridia have been the most widely studied among the various anaerobic and facultative anaerobic bacteria capable of fermentative hydrogen production [4]. H_2 production in living organisms is always dependent on the presence of H_2-producing enzymes such as hydrogenases and nitrogenases. [FeFe] hydrogenases, which are especially abundant in clostridia, are well recognised as the main H_2-producing enzymes in this genus [5,6]. By contrast, nitrogenase-mediated hydrogen production has never been proposed for clostridia, even though this is known to be an intrinsic metabolic property of many cyanobacteria and photosynthetic bacteria [2]. Interestingly, in 1960 Carnahan showed that the free-living soil microorganism *Clostridium pasteurianum* is an efficient N_2 fixator [7]. Later *nif* operons were described in *C. acetobutylicum* and *C. beijerinckii* [8], but the contribution of nitrogenase to overall H_2 production in clostridia has not yet been reported in the literature.

Clostridium butyricum CWBI 1009, a strain recently isolated from an anaerobic sludge, was previously shown to ferment different carbon substrates at acidic pH to H_2 and CO_2 with formate, butyrate and acetate as the main end-products [9,10]. It possesses four different [FeFe] hydrogenases, three of which are monomeric belonging to clusters A2, B2 and B3, and one which is a trimeric enzyme representing cluster A8 [6,11]. Despite the fact

that optimum growth for this bacterium occurs at pH 7.3, it only starts to produce H_2 when the pH declines due to the natural acidification of the medium as fermentation proceeds. The optimal pH value to produce H_2 under fixed-pH culture conditions was found to be 5.2 and has been discussed in previous work [9]. Although many studies have described the fermentative activity of *C. butyricum* associated with H_2 production [12,13], a comprehensive analysis of the fermentative pathways at the genomic and proteomic levels has not yet been reported. Relative gene expression profiles, together with the associated proteomic and metabolite data, can now be used to provide the visibility needed for well-targeted metabolic engineering. Moreover, a better understanding of the shifts in gene and protein expression, which occur in response to pH changes and during the growth phase, should facilitate the optimisation of bioreactor performance.

Therefore, in this study three parallel approaches were used to investigate the changes at the molecular level associated with pH-dependent hydrogen production in *C. butyricum* CWBI 1009, namely metabolite analysis, transcriptomics and proteomics. The effect of the naturally decreasing pH was studied with glucose in 20-L batch bioreactors. Additionally, the effect of fixed-pH fermentations was evaluated to provide comparative data under optimal pH conditions for cellular growth (pH 7.3) and H_2 production (pH 5.2), respectively. The genome of *C. butyricum* CWBI 1009, which was unknown until now, was also sequenced to provide better mapping of the RNA-sequencing (RNA-seq) reads during natural acidification of the medium. In addition the expression levels of the H_2-producing enzymes, namely the [FeFe] hydrogenases and the nitrogenase, were determined under different pH conditions.

Results and discussion
Experimental design

Three 20-L glucose fermentations were carried out under unregulated-pH conditions (Figure 1, A, D and G), allowing characterisation of the impact of naturally decreasing pH on fermentative H_2 production by *C. butyricum* CWBI 1009. Additional glucose fermentations (three replicates for each condition) performed at two fixed pH values, namely 7.3 (Figure 1, B, E and H) and 5.2 (Figure 1, C, F and I), were carried out to provide comparative data for the interpretation of the results obtained during the unregulated-pH fermentations. These fixed pH values of 7.3 and 5.2 were previously determined to be optimal for *C. butyricum* CWBI 1009 growth and H_2 production, respectively [9]. The fermentation results presented below for the triplicated experiments were characterised by standard deviations ranging from 5 to 10% of the absolute value. Therefore, the reproducibility of the results may be

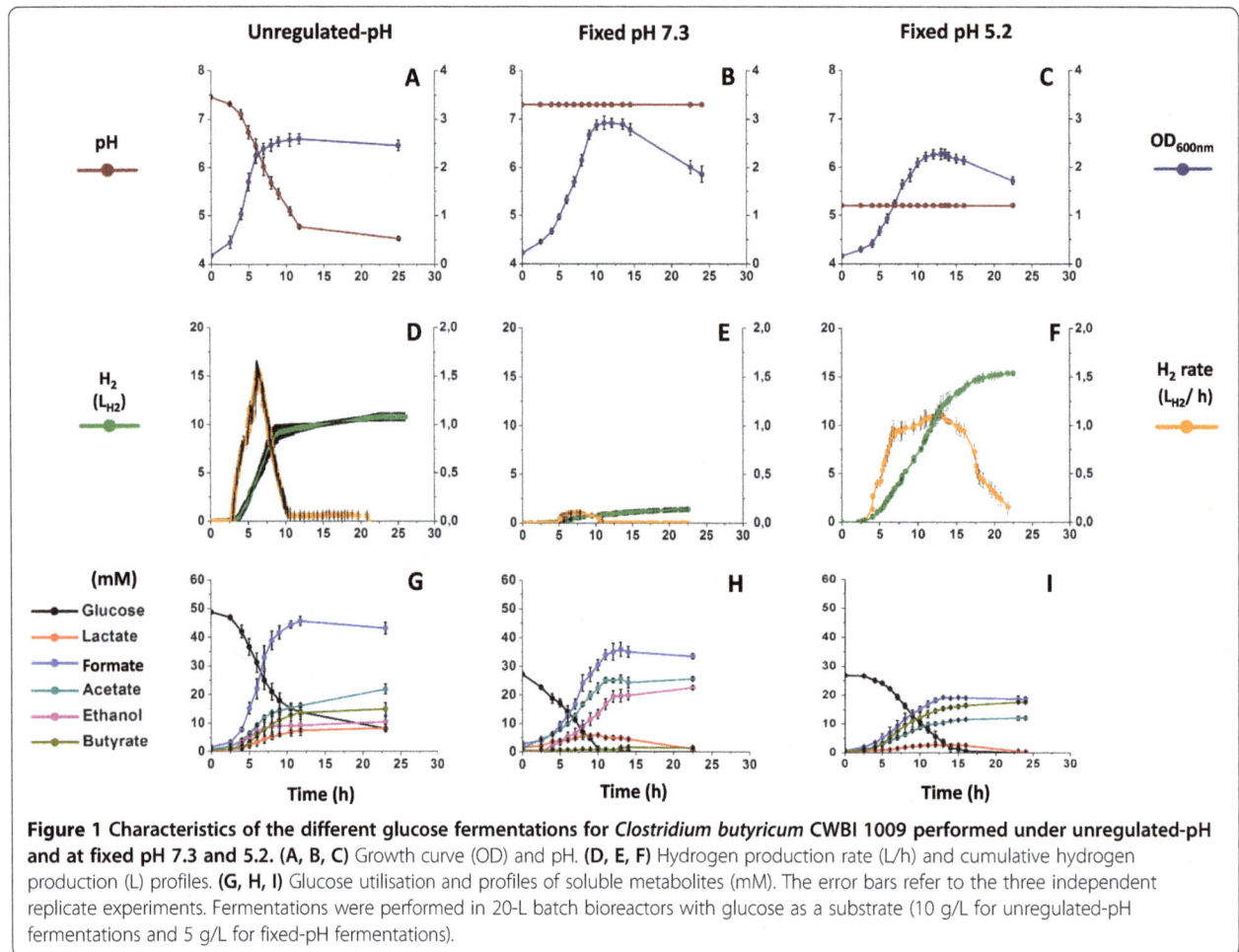

Figure 1 Characteristics of the different glucose fermentations for *Clostridium butyricum* CWBI 1009 performed under unregulated-pH and at fixed pH 7.3 and 5.2. (A, B, C) Growth curve (OD) and pH. **(D, E, F)** Hydrogen production rate (L/h) and cumulative hydrogen production (L) profiles. **(G, H, I)** Glucose utilisation and profiles of soluble metabolites (mM). The error bars refer to the three independent replicate experiments. Fermentations were performed in 20-L batch bioreactors with glucose as a substrate (10 g/L for unregulated-pH fermentations and 5 g/L for fixed-pH fermentations).

considered as sufficient and typical for these kinds of experiments in 20-L bioreactors.

For all the above fermentations, analyses were carried out to determine whether the H_2 production and the cellular metabolism for all the tested metabolites varied and, if so, to what extent these variations were pH dependent. The genome-wide transcriptional response of *C. butyricum* CWBI 1009 to naturally decreasing pH (associated with increasing H_2 production) was characterised during glucose fermentation without pH regulation. The gene expression levels of the H_2-producing enzymes, namely the [FeFe] hydrogenases and the nitrogenase, were then examined, and the data were related to the H_2-production profiles under the different conditions studied. Finally the proteomic response of *C. butyricum* CWBI 1009 to declining pH was analysed.

H_2 and metabolite production under unregulated-pH and fixed pH conditions

The effect of naturally decreasing pH on the growth, H_2-production and fermentation profiles of *C. butyricum* CWBI 1009 cultivated with glucose (10 g/L) was evaluated

in a 20-L batch anaerobic bioreactor maintained at 30°C and N_2 atmosphere. The culture was monitored during the 25 h of fermentation by which time the pH had dropped naturally to about 4.5 (Figure 1A). Directly after the lag phase there was a rapid consumption of glucose and an increase in growth. In total 17.16 ± 0.84 L of biogas were produced. Given that the average H_2 content of the biogas was $63 \pm 4\%$, it was calculated that 10.80 ± 0.44 L of H_2 were produced (Additional file 1: Table S1). The highest H_2 flow rate, 1.56 ± 0.15 L H_2/h, was recorded after 6 to 7 h of fermentation (Figure 1D) by which time the pH of the medium had dropped to around 6.3. The highest calculated H_2 yield was 1.78 ± 0.11 mol H_2/mol glucose. The primary soluble metabolites at the end of fermentation were formate, acetate and butyrate, followed by ethanol and lactate (Figure 1G).

To highlight the impact of pH on the gene expression for the H_2-producing enzymes, that is, the hydrogenases and the nitrogenase, the comparative results for H_2 and metabolite production with glucose (5 g/L), using the same bioreactor, are briefly presented for two different fixed-pH conditions: pH 7.3 and 5.2 ± 0.15. At fixed pH 7.3 a lag phase of 3 h was observed (Figure 1B), compared to a lag phase of 4.5 h at fixed pH 5.2 (Figure 1C). Exponential growth followed the lag phase under both pH conditions. At pH 7.3 rapid consumption of glucose was observed (Figure 1H), and the glucose uptake rate peaked at 0.96 ± 0.08 g glucose/h. Biogas production started after 5.5 h of fermentation, yielding a total of 2.90 ± 0.23 L of biogas after 15 h of fermentation (Additional file 1: Table S2). At pH 5.2, the glucose uptake rate was much lower (0.44 ± 0.03 g glucose/h) and biogas production had already started after 3.5 h of fermentation. After 20 h of fermentation a total of 23.32 ± 1.01 L of biogas was produced. At pH 7.3 the H_2 production rate and yield peaked at 0.21 L ± 0.03 H_2/h (Figure 1E) and 0.23 ± 0.02 mol H_2/mol glucose, respectively. By contrast, at pH 5.2 the corresponding H_2 production rate and yield were much higher; 1.11 L ± 0.06 H_2/h (Figure 1F) and 1.95 ± 0.09 mol H_2/mol glucose. The cellular metabolisms for all the tested metabolites varied between the unregulated-pH fermentations and those when the pH was fixed at 5.2 and 7.3. Moreover, the lactate produced during the early stages of fermentation was later consumed, which was not the case with the fermentations carried out under unregulated-pH conditions (Figure 1H and I). The ability of clostridia to reconsume lactate produced during the early stages of fermentation indicates the existence of novel metabolic pathways, an observation that has already been discussed in the literature [10].

Transcriptional response of *C. butyricum* CWBI 1009 to decreasing pH

As an initial step towards understanding how the bacterium responds to naturally decreasing pH, which is associated with increasing H_2 production, the transcriptional response of *C. butyricum* CWBI 1009 was analysed during unregulated-pH glucose fermentation. The genome of *C. butyricum* CWBI 1009 was also sequenced to provide better mapping of the RNA-seq reads. The RNA-seq data were obtained from rRNA-depleted mRNA samples isolated from two independent reactor cultures (biological replicates). RNA-seq data were acquired from samples taken at pH 7.3 (early exponential growth phase: control sample) and at pH 6.3 (late exponential growth phase: test samples), and are shown in Additional file 2: Table S4 and Additional file 3: Table S5. The selection of these values was based on the fact that pH 7.3 and 6.3 corresponded respectively to the minimum and maximum H_2 production phases of the three fermentations without pH regulation. Additionally, the gene expression profiles during the stationary phase, corresponding in this experiment to pH 5.2, were analysed and are shown in Additional file 2: Table S4. The reproducibility of the transcriptomic data between the two

biological replicates was high with an R^2 ranging from 0.873 for the control sample (pH 7.3) to 0.78 for the samples taken at pH 6.3 (Additional file 1: Figure S1). For each sample both the reads mapping to rRNA sequences and those not mapping uniquely to the genome of *C. butyricum* CWBI 1009 were omitted from further analysis (Additional file 1: Table S3). The RNA-seq expression data have been presented in two ways. Firstly, the total number of reads for each coding DNA sequence (CDS) was calculated and converted to reads per kilobase per million mapped reads (RPKM numbers), and secondly the genes that were differentially regulated between the stages corresponding to pH 6.3 and 7.3 were identified (Figures 2 and 3).

Calculation of the RPKM enabled a comparison between the relative mRNA abundances of different genes for a given condition and also between the relative abundances of a specific gene under the different pH conditions. Based on the RPKM numbers, the genes coding for the glycolytic enzymes and the auxiliary proteins involved in the core metabolic reactions (for example, ferredoxin, NAD-dependent glyceraldehyde-3-phosphate dehydrogenase, pyruvate formate-lyase, acetaldehyde dehydrogenase, pyruvate kinase or flavodoxin) were among the most highly expressed, and their expression did not vary significantly throughout the fermentation (Additional file 2: Table S4). Unexpectedly, multiple genes encoding different subunits of the nitrogenase (for example, *nifN*, *nifH*, *nifD*, *nifS*) and the urease (α, β and γ subunits) were also very highly expressed at pH 6.3 (RPKM $\geq$ 500). To our surprise the mRNA moieties of the [FeFe] hydrogenases were among the least abundant transcriptional units, which contrasts strongly with the increased H_2 production associated with this fermentation stage (Figure 2).

RNA-seq data were also used to identify the genes that were significantly up- or down-regulated under the

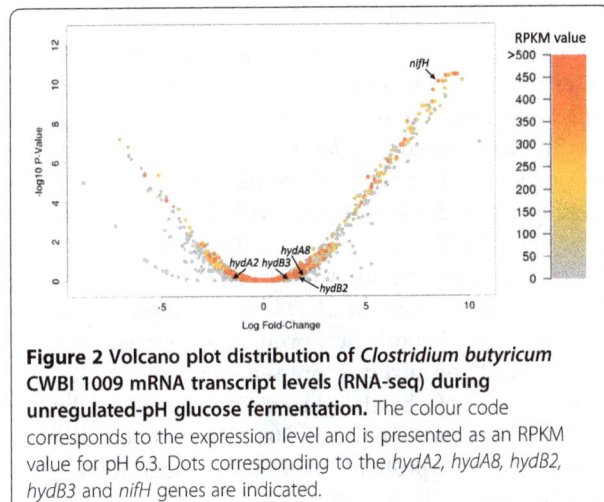

Figure 2 Volcano plot distribution of *Clostridium butyricum* CWBI 1009 mRNA transcript levels (RNA-seq) during unregulated-pH glucose fermentation. The colour code corresponds to the expression level and is presented as an RPKM value for pH 6.3. Dots corresponding to the *hydA2*, *hydA8*, *hydB2*, *hydB3* and *nifH* genes are indicated.

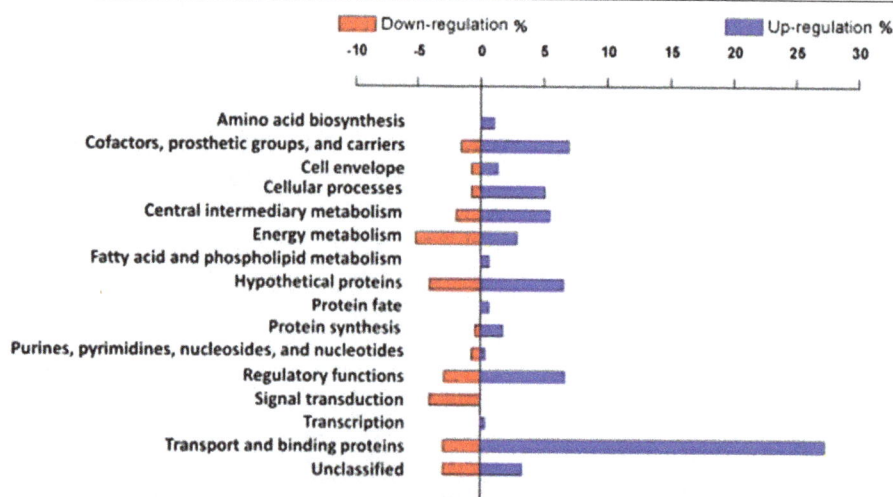

Figure 3 Differentially regulated pathways of *Clostridium butyricum* CWBI 1009 during unregulated-pH glucose fermentation based on the RNA-seq data (pH 6.3 versus pH 7.3).

different pH conditions studied. The relative expression levels were presented as a fold change ($\log_2$) between the control sample (pH 7.3) and the test sample (pH 6.3). In total more than 290 genes were found to be differentially expressed, with 72% being up-regulated and only 28% down-regulated at pH 6.3 (Figure 3, Additional file 3: Table S5). Many of these genes were located in close proximity on the chromosome, and are therefore likely to represent polycistronic operons encoding proteins with similar functions. The differentially regulated genes were automatically assigned functional annotations using the Clusters of Orthologous Groups (COG) database; these annotations were then corrected based on the Pathema-*Clostridium* assigned categories (http://pathema.jcvi.org).

Over 30% of the annotated and differentially regulated transcripts were associated with transport proteins (28% of which were up-regulated), suggesting that transport plays a crucial role in maintaining cell homeostasis at acidic pH. Sequences linked to the biosynthesis of cofactors, prosthetic groups and carriers, as well as the central intermediary metabolism and regulatory function proteins were also found to be differentially regulated (Figure 3). Genes encoding for conserved hypothetical proteins constituted around 10% of all the differentially regulated genes, and a few of them had very high RPKM values at pH 6.3 (RPKM ≥ 1,000, Additional file 2: Table S4), suggesting their importance for cell metabolism. In line with this result, Wang *et al.* [14] reported that many genes encoding for hypothetical proteins accounted for a large fraction of the highly expressed genes during the different stages of batch glucose fermentation by *Clostridium beijerinckii* NCIMB 8052. In addition, the genes coding for the proteins involved in protein folding and stabilisation, such as heat shock proteins, were not

differentially expressed, but were nevertheless very highly expressed throughout the fermentation (RPKM ≥ 1,000 for *dnaK*, *groEL* and *groES*). For comparison, genes such as *groESL*, *hsp90* or *dnaK* were previously reported to be induced by acetate and butyrate shock in *C. acetobutylicum* [15] and also by butanol in this bacterium [16] and were confirmed as being important in the general stress response.

Is H_2 production under unregulated-pH conditions nitrogenase-mediated?

The genome of *C. butyricum* CWBI 1009 encodes four [FeFe] hydrogenases. However, the RNA-seq analysis unexpectedly indicated that none of them showed any signs of being up-regulated at pH 6.3, when the H_2 production rate was maximum (Figure 2, Additional file 3: Table S5). Furthermore, this result was confirmed by RT-qPCR with hydrogenase-specific primers for each gene (Figure 4A and B, Additional file 1: Figure S2).

Surprisingly, the genes that were among the most strongly induced at pH 6.3 included the genes encoding nitrogenase and other proteins related to N_2 fixation, ammonium transport and molybdenum transport (Additional file 3: Table S5). The physiological electron donors for nitrogenases, ferredoxins (RPKM > 8,800) and flavodoxins (RPKM > 4,700) were also very highly abundant at pH 6.3 (Additional file 2: Table S4). In addition various nitrogen regulatory P-II proteins were significantly up-regulated at pH 6.3 and accounted for 30% of the regulatory factors that were differentially regulated between the two pH conditions studied (Additional file 4: Table S8).

Nitrogenase-mediated H_2 production is a property of several microorganisms including cyanobacteria and photosynthetic bacteria [2], and has recently been revealed in marine

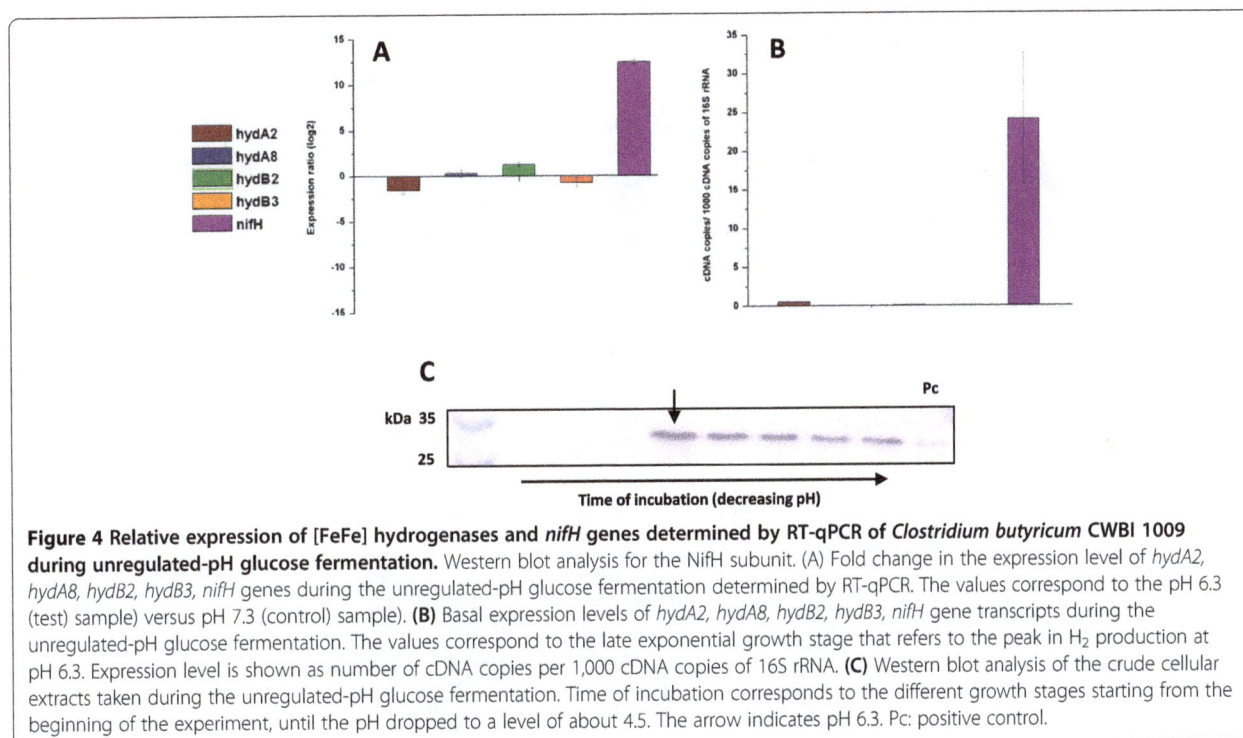

Figure 4 Relative expression of [FeFe] hydrogenases and *nifH* genes determined by RT-qPCR of *Clostridium butyricum* CWBI 1009 during unregulated-pH glucose fermentation. Western blot analysis for the NifH subunit. (A) Fold change in the expression level of *hydA2, hydA8, hydB2, hydB3, nifH* genes during the unregulated-pH glucose fermentation determined by RT-qPCR. The values correspond to the pH 6.3 (test) sample versus pH 7.3 (control) sample. **(B)** Basal expression levels of *hydA2, hydA8, hydB2, hydB3, nifH* gene transcripts during the unregulated-pH glucose fermentation. The values correspond to the late exponential growth stage that refers to the peak in H_2 production at pH 6.3. Expression level is shown as number of cDNA copies per 1,000 cDNA copies of 16S rRNA. **(C)** Western blot analysis of the crude cellular extracts taken during the unregulated-pH glucose fermentation. Time of incubation corresponds to the different growth stages starting from the beginning of the experiment, until the pH dropped to a level of about 4.5. The arrow indicates pH 6.3. Pc: positive control.

Enterobacteriaceae, for example, *Pantoea agglomerans* [17]. However, it has never before been proposed for clostridia. Therefore, to validate the up-regulation of the nitrogenase genes, an RT-qPCR analysis was carried out for the *nifH* gene, which encodes a nitrogenase reductase subunit. The results showed a significant up-regulation of this gene at pH 6.3 and were in agreement with the RNA-seq results (Figure 4A, Additional file 1: Figure S2). Additionally, by using anti-NifH antibodies, the presence of the nitrogenase H subunit in crude cellular extracts was confirmed by Western blot analysis (Figure 4C). Since the H_2 production rate peaked at 1.56 ± 0.15 L/h at pH 6.3 and was below 0.21 ± 0.03 L/h at pH 7.3 (Figure 1D), the very strong up-regulation of the nitrogenase coding genes at the lower pH, combined with the concurrent absence of up-regulation of the [FeFe] hydrogenase genes, led us to conclude that the H_2 production in *C. butyricum* CWBI 1009 during glucose fermentation with unregulated- pH may be nitrogenase-mediated.

Interestingly, though nitrogenase is known to produce H_2 as a by-product of N_2 fixation, already in the early 1980s it was reported that the enzyme may act as an ATP-powered hydrogenase and produce only H_2 in the absence of N_2 [18]. Therefore, to check if this was the case with clostridia, *C. butyricum* CWBI 1009 was cultured in an N_2-free atmosphere with unregulated- pH, using argon instead of nitrogen to initiate the anaerobic conditions in the bioreactor. The preliminary results demonstrated that, in contrast to the fermentation with unregulated- pH under N_2, under an Ar atmosphere there

was an induction of three [FeFe] hydrogenase genes (*hydA8, hydB2* and *hydB3)* at low pH values (results discussed in Additional file 6). A better understanding of the differences between the mechanisms leading to H_2 production under these two conditions would nevertheless require a more detailed analysis of the Ar sample at the transcriptomic and proteomic levels. Additionally, as a follow-up to the present work, a study of the physiological activity of the [FeFe] hydrogenases and the nitrogenase could be carried out to evaluate the exact contribution of each enzyme to the overall H_2 production under different experimental conditions.

The physiological function of hydrogen and ammonia-generating enzymes under unregulated-pH conditions may be to maintain the pH homeostasis of the cell

N_2 fixation (reduction of N_2 to two ammonia molecules) is a highly energy-intensive process, consuming at least 16 ATP molecules per molecule of nitrogen fixed. It has been shown that while ammonium nitrogen can repress the formation of nitrogenase in different species [19], amino acid nitrogen can actually stimulate it (the main N source in this study was the amino acids contained in casein peptone and yeast extract) [20]. The release of molecular H_2, a by-product of N_2 fixation, enables the disposal of excess protons, thereby preventing acidification of the cytoplasm. Clearly proton disposal is the main physiological function of [FeFe] hydrogenases, but in this case nitrogenase activity may provide an additional buffering molecule, namely ammonia (Figure 5).

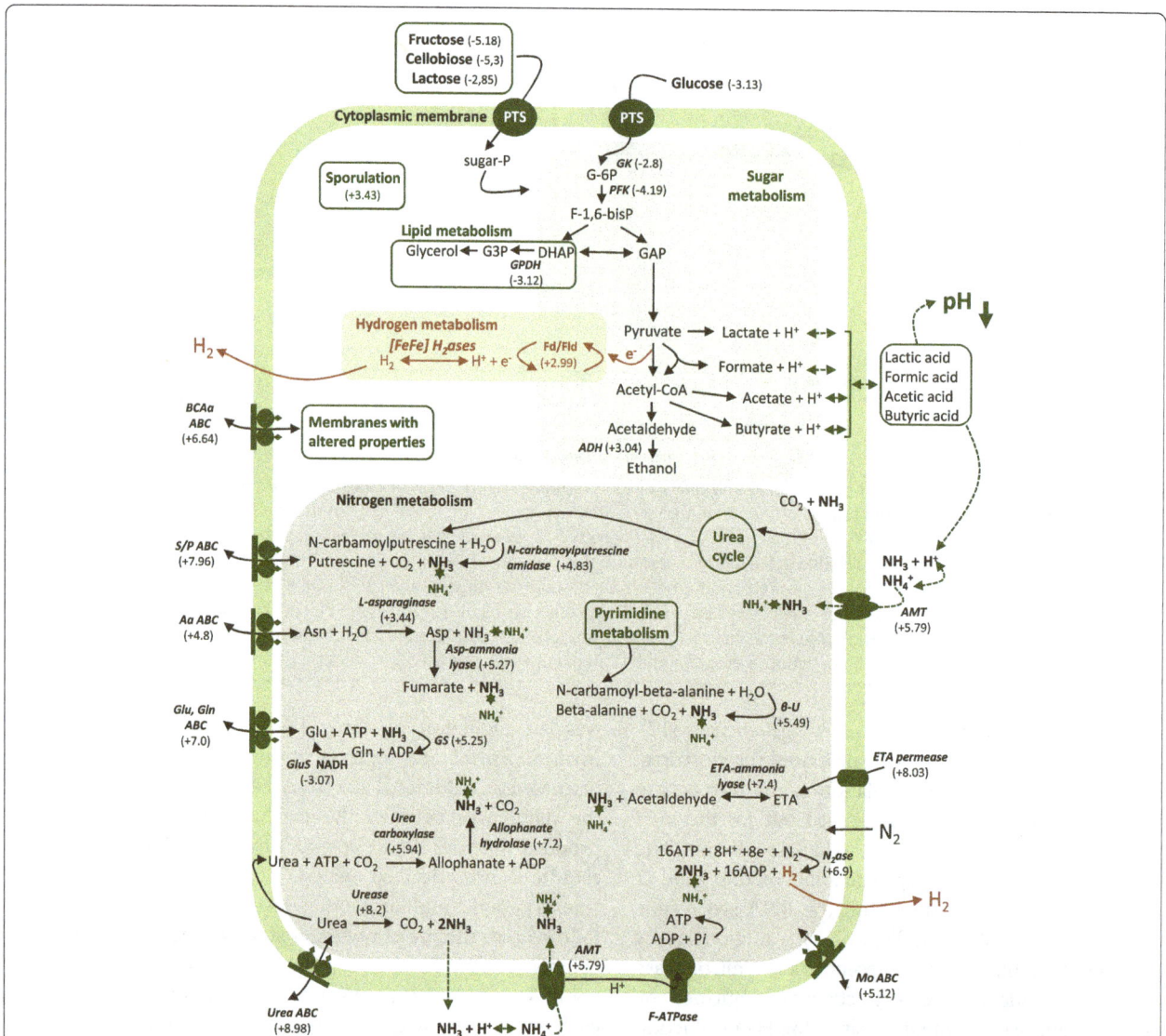

Figure 5 Metabolic pathways for *Clostridium butyricum* CWBI 1009 affected by decreasing pH during the unregulated-pH glucose fermentation. The numbers are the calculated averages of the individual calculated expression ratios when several subunits encoding the same enzyme were detected to be differentially expressed by RNA-seq. Pathways directly related to H_2 production are indicated in red. The dashed lines refer to pathways involved in pH regulation. *AaABC*: amino acid ABC transporter; *ADH*: alcohol dehydrogenase; *AMT*: ammonium transporter; *β-U*: Beta-ureidopropionase; *BCAa ABC*: branched-chain amino acid ABC transporter; *ETA*: ethanolamine permease; *Fd/Fld*: ferredoxin, flavodoxin; *GK*: glucokinase; *Glu/Gln ABC*: glutamate/glutamine ABC transporter; *GluS*: NADH-glutamate synthase; *GS*: glutamine synthetase; *[FeFe] H_2ase*: [FeFe] hydrogenase; *MoABC*: molybdenum ABC transporter; *N_2ase*: nitrogenase; *PFK*: 1-phosphofructokinase; *S/P ABC*: spermidine/putrescine ABC transporter; *Urea ABC*: urea ABC transporter.

Cytoplasmic pH buffering is one of the strategies employed by many microorganisms to maintain pH homeostasis, involving the production of various different buffer molecules such as amino acids, ammonia and polyamines, [21]. With declining external pH (the pK for ammonia is 9.25), the ammonia produced may buffer the cytoplasm via the $H^+ + NH_3 \leftrightarrow NH_4^+$ reaction. Another enzyme for which various subunits were up-regulated sixfold to ninefold (log$_2$ scale) was urease. It is a nickel-containing enzyme which catalyses the hydrolysis of urea to ammonia and carbamic acid, with

the latter spontaneously hydrolysing to carbonic acid and an additional ammonia molecule [22]. Most bacteria use the products of this reaction for anabolic processes, but at low pH these moieties may be used for cytoplasmic buffering. Additionally, the up-regulation of urea carboxylase and allophanate hydrolase at pH 6.3 may indicate the existence in *C. butyricum* CWBI 1009 of another pathway for urea breakdown that also leads to the formation of ammonia and carbon dioxide. Moreover, at pH 6.3 several additional pathways that could contribute to cytoplasmic pH buffering via ammonia synthesis (for example, L-asparaginase, ethanolamine ammonia-lyase, beta-ureidopropionase, aspartate ammonia-lyase and N-carbamoylputrescine amidase) were also significantly up-regulated (Figure 5, Additional file 4: Table S6).

Cellular response to decreasing pH involves differential regulation of ABC transporters and other transport proteins

Molecular pumps, which are unidirectional efflux systems that actively expel various chemical substances or ions from the cytoplasm to the extracellular space, have been shown to play a role in acid tolerance in yeast [23]. ABC transporters are membrane-bound molecular pumps that utilise ATP hydrolysis energy to translocate a large variety of solutes across cellular membranes. In our study numerous genes related to the various ABC transporters were up-regulated following the progressive drop in pH, that is, ABC transporters for spermidine, putrescine, arginine, glutamine, branched-chain amino acids, molybdate and other metal ions (Figure 5, Additional file 4: Table S7). Due to their positive charges, polyamines, including spermidine and putrescine, bind to macromolecules such as DNA, RNA and proteins, exerting a protective effect. In bacteria they have been shown to be involved in stress responses, particularly acid tolerance [24]. They are also involved in various different processes, including regulation of gene expression, cell proliferation, cell signaling and membrane stabilisation [25]. Other more frequently transported compounds were the branched-chain amino acids which are involved in the biosynthesis of membranes with altered properties (fluidity). Their increased transport and biosynthesis in *C. acetobutylicum* in response to butyrate and butanol stresses have been previously described [15].

Expression of sporulation genes is strongly induced by decreasing pH

During the late exponential growth phase, when the pH of the medium decreased to 6.3, several genes associated with sporulation were significantly induced, constituting around 5% of the differentially regulated genes (Additional file 4: Table S9). Furthermore, the RNA polymerase sporulation-specific sigma factor, SigE, that was reported to be responsible for the expression of stage II sporulation-specific genes in *Bacillus* [26], was also up-regulated by a factor of 3.34 ($\log_2$ scale) at pH 6.3. The initiation of endospore formation is usually accompanied by reduced chemotaxis and motility, and such a down-regulation was observed here for two genes, the methyl-accepting chemotaxis protein (MCP) signalling domain and a putative methyl-accepting chemotaxis protein. In contrast to our observations, the expression of sporulation genes in some solventogenic clostridia, for example,

C. acetobutylicum, has been reported as being largely unaffected by a low pH [15]. Instead, to prevent the collapse of the transmembrane pH gradient, some solventogenic bacteria react by solvent production which allows the external pH to increase. Solventogenic *Clostridium beijerinckii* NCIMB 8052 was reported to initiate sporulation concurrently with the onset of solventogenesis [14]. Although *C. butyricum* CWBI 1009 is also capable of producing solvents, mainly ethanol, it does so concomitantly with the production of acids during the exponential growth phase and at a lower concentration (Figure 1G). Therefore, to keep the internal pH close to the optimum during fermentative growth, *C. butyricum* CWBI 1009 employs various different mechanisms for proton disposal as described above. Additionally, the bacterium appears to initiate sporulation even during the early stages of its fermentative growth to prevent cell lysis and death due to the re-uptake, via the membrane, of the fatty acids produced. This typically occurs when the pH decreases below the pKa value (for example, formate pKa 3.77, acetate pKa 4.76, butyrate pKa 4.83) [27].

Proteomic response to decreasing pH

While mRNAs are mediators of certain biological functions and provide information on transcriptional patterns, most functions are carried out by proteins. Therefore, to gain a better understanding of clostridial hydrogen metabolism, the changes in the relative protein abundance profiles for *C. butyricum* CWBI 1009 were analysed in response to naturally decreasing pH. For this purpose liquid samples were harvested at pH 7.3 (control sample, Figure 1A) and pH 5.2 (test sample) during the unregulated-pH fermentation. The experiment was successfully carried out on two separate biological replicates (two independent cultures). The protein abundances were analysed by two-dimensional difference in gel electrophoresis (2D-DIGE) (Additional file 1: Figure S3). Overall 2,750 spots exhibiting differences (>20%) in the normalised spot-volume ratios were detected; 1,423 of these spots increased and 1,327 decreased in size at pH 5.2 versus pH 7.3. More than 500 protein spots varied significantly and were subjected to in-gel tryptic digestion, followed by mass spectrometry fingerprinting of the resulting peptides. The experiment was successfully carried out on two separate biological replicates (two independent cultures). A total of 166 proteins (97 unique and 69 redundant) with significant Mascot probability-based scores were identified and categorised according to their metabolic functions (Additional file 5: Table S10 and S11). Overall 16.5% of the identified proteins were associated with energy metabolism, 10.3% with protein synthesis, 8.2% with signal transduction and 7.2% with amino acid biosynthesis. Though none of the differentially abundant protein spots were identified as [FeFe] hydrogenases or nitrogenase, a differential abundance of these two enzymes between the two pH conditions cannot be excluded since not all of the statistically differentially abundant proteins were identified after matrix-assisted laser desorption/ionisation time of flight (MALDI-TOF)/TOF analysis.

Central metabolic enzymes and the proteins involved in fermentative pathways appear more abundant at lower pH

C. butyricum CWBI 1009 utilises the Embden-Meyerhof-Parnas pathway for the conversion of glucose to phosphoenolopyruvate (PEP, Figure 5). The core metabolic proteins, such as glucose kinase (+2.52-fold more abundant at pH 5.2), phosphofructokinase 1 (+2.21), phosphopyruvate hydratase (+2.02) and pyruvate kinase (+1.88), that predominantly determine the carbon and electron flow from the carbohydrate substrate to the end-products were more abundant at pH 5.2 compared to pH 7.3 (Additional file 5: Table S11). Electrons derived from the main glycolytic nodes, namely pyruvate-ferredoxin oxidoreductase (PFOR) and to a lesser extent NADH-ferredoxin oxidoreductase (NFOR), are passed on by electron acceptors to the hydrogenases, thereby enabling the reversible reduction of the protons accumulated during the fermentation process to molecular hydrogen [28]. Also, the reducing equivalents necessary for N_2 fixation are mainly obtained by the nitrogenase via reduced ferredoxin, which can be generated by the action of PFOR [29]. In line with these reports, seven redundant proteins identified as PFOR were all more abundant at pH 5.2 (on average + 2.37). By contrast, the relative abundance of type I glyceraldehyde-3-phosphate dehydrogenase, a tetrameric NAD-binding protein, was decreased at lower pH (on average -1.74), suggesting a limited flow of electrons via the NFOR node and towards the putative bifurcating [FeFe] hydrogenase (Hyd A8) [12,30]. Electron transferring flavoprotein (+1.73) and flavodoxin (+1.69) were also more abundant at lower pH, which was consistent with the RNA-seq data for the low pH condition.

Among the other proteins that were more abundant at acidic pH, a cysteine desulphurase NifS (+4.89) and an iron-sulphur cluster-binding protein (+1.92) were identified (Additional file 5: Table S11). The function of the former is to mobilise sulphur atoms for the biosynthesis of iron-sulphur (FeS) clusters. Both proteins are involved in the maturation process of various FeS proteins, such as hydrogenases and nitrogenases [31].

As regards the nitrogen metabolism, the 2D-DIGE analysis did not confirm the differential abundance of the nitrogenase, but did however show that glutamine synthetase was more abundant (by an average factor of +1.72) at lower pH. The mRNA level for the corresponding gene was also higher at the low-pH stage of the fermentation (Additional file 2: Table S4 and

Additional file 3: Table S5). In free-living diazotrophs fixed N_2 is assimilated by the organism in the form of ammonium to produce glutamine and glutamate. This occurs via the glutamine synthetase/glutamate synthase pathway in accordance with the equation: glutamate + $ATP + NH_3 \leftrightarrow$ glutamine + ADP + Pi [32]. Therefore, the ability to synthesise glutamate and glutamine is essential to the cellular metabolism, since they are involved in the incorporation of inorganic nitrogen into cell material; that is, the synthesis of new proteins [32,33]. Further ammonia incorporation occurs through the action of NAD(P)-specific glutamate dehydrogenase, and this protein also was found to be 1.81-fold more abundant at lower pH. Moreover, amino acids such as glutamate have been described as known osmoprotectants and have been shown to play a role in acid tolerance in, for example, *C. acetobutylicum* [15].

The HPLC analyses for metabolites indicated that, in addition to formate, acetate and butyrate were the main fermentation end-products obtained when *C. butyricum* CWBI 1009 was cultured under unregulated-pH conditions (Figure 1G). Therefore, it was not surprising that acetate and butyrate kinases were respectively 1.49- and 1.68-fold more abundant at the lower pH compared to pH 7.3. Additionally, acetyl-CoA acetyltransferase (thiolase), which catalyses the first steps of the butyrate synthesis pathway [32], was also more abundant at the lower pH (+1.51). Of the other core fermentative enzymes, type II acetaldehyde/alcohol dehydrogenase was also more abundant, suggesting that increased alcohol production was associated with decreasing pH.

H_2 production from glucose at fixed pH 5.2 is likely to be hydrogenase-mediated

Maintaining the *C. butyricum* CWBI 1009 culture at fixed pH 5.2 (Figure 1F) led to a 50% increase in cumulative H_2 production (in comparison to the unregulated-pH culture; Figure 1D). Furthermore, bacterial fermentations at fixed acidic pH are commonly used for efficient H_2 production. This is why glucose fermentations were also performed at two different fixed pH values, namely pH 7.3 (control fermentation with best growth) and pH 5.2 (test fermentation with best H_2 production). In addition to the description of the fermentative profiles characteristic for the two fixed pH values, RT-qPCR was used to study the gene expression levels for the [FeFe] hydrogenases and the nitrogenase.

At pH 7.3 no differential gene expression occurred throughout the fermentation (Figure 6A), as was expected given the barely detectable H_2 production (Figure 1E). Surprisingly, at fixed pH 5.2, although the H_2 production rate constantly increased during the fermentation, no change in the gene expression pattern was detected between the early and the late exponential stages of

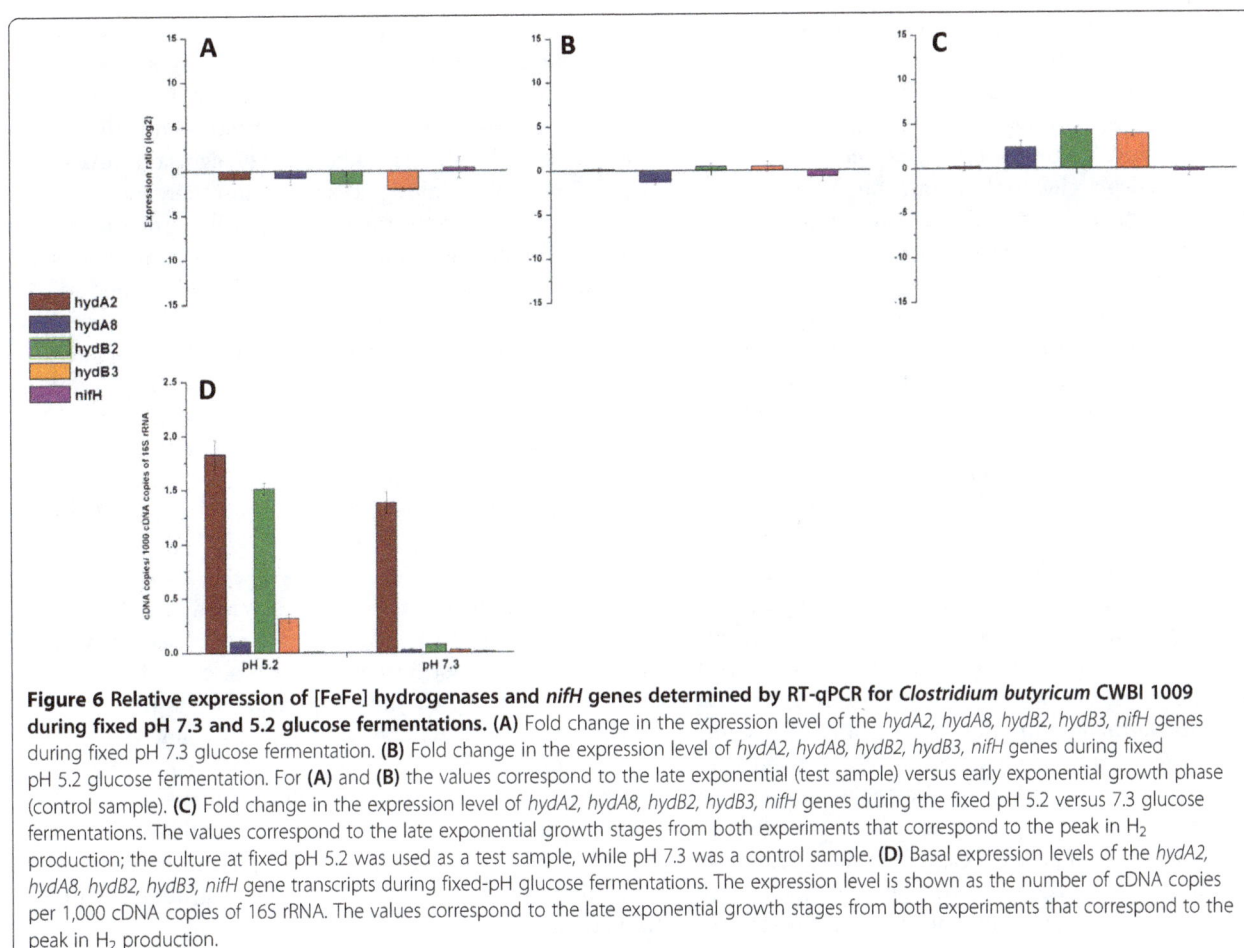

Figure 6 Relative expression of [FeFe] hydrogenases and *nifH* genes determined by RT-qPCR for *Clostridium butyricum* CWBI 1009 during fixed pH 7.3 and 5.2 glucose fermentations. (A) Fold change in the expression level of the *hydA2, hydA8, hydB2, hydB3, nifH* genes during fixed pH 7.3 glucose fermentation. **(B)** Fold change in the expression level of *hydA2, hydA8, hydB2, hydB3, nifH* genes during fixed pH 5.2 glucose fermentation. For **(A)** and **(B)** the values correspond to the late exponential (test sample) versus early exponential growth phase (control sample). **(C)** Fold change in the expression level of *hydA2, hydA8, hydB2, hydB3, nifH* genes during the fixed pH 5.2 versus 7.3 glucose fermentations. The values correspond to the late exponential growth stages from both experiments that correspond to the peak in H_2 production; the culture at fixed pH 5.2 was used as a test sample, while pH 7.3 was a control sample. **(D)** Basal expression levels of the *hydA2, hydA8, hydB2, hydB3, nifH* gene transcripts during fixed-pH glucose fermentations. The expression level is shown as the number of cDNA copies per 1,000 cDNA copies of 16S rRNA. The values correspond to the late exponential growth stages from both experiments that correspond to the peak in H_2 production.

fermentation, neither for the nitrogenase nor for the [FeFe] hydrogenases (Figure 6B). This apparent lack of change in the temporal gene expression profile could be due to the fact that the bacterial pre-culture, used to inoculate the reactor, was pre-incubated at pH 5.2 as well. Therefore, the lower H_2 production observed at the beginning of the fixed pH 5.2 fermentation (Figure 1F) was more likely attributable to low cell density rather than any differential [FeFe] hydrogenase gene expression. Since the pH was kept constant during the whole fermentation, the [FeFe] hydrogenases could not have been activated by a change in pH [34].

A direct comparison of the gene expression profiles for the same growth stage at the two different fixed pHs showed that the [FeFe] hydrogenase gene transcripts were indeed differentially abundant. During the late exponential growth phases (10 h incubation, Figure 1B and C), when the H_2 production rate reached the maximum for both cultures (Figure 1E and F), the monomeric hydrogenase gene *hydA2* showed similar expression levels at fixed pH 5.2 and at fixed pH 7.3 (Figure 6C). Surprisingly, its basal expression at pH 5.2 was between 1.2- and 6 -fold higher than that for *hydB2* and for *hydB3* respectively, and

as much as 18 times higher than that for the *hydA8* gene (Figure 6D). Hydrogenase A8 is a putative trimeric hydrogenase that requires both reduced ferredoxin and NADH to efficiently catalyse H_2 production [30]. Its relatively low expression level compared to the other [FeFe] hydrogenases may suggest that there was less electron flow through the NFOR node than through the PFOR node (in line with the 2D-DIGE data). During the same late exponential growth phase the remaining two monomeric hydrogenase gene transcripts (*hydB2* and *hydB3*) were respectively 4.28- and 3.9-fold more abundant ($P < 0.05$, $\log_2$ scale) at fixed pH 5.2 compared to fixed pH 7.3.

The data obtained suggest two main conclusions. Firstly, based on the cDNA copy numbers, *hydA2* is the most abundant hydrogenase and seems to be insignificantly regulated under the two different fixed pH conditions. Secondly, both *hydB2* and *hydB3* are very strongly up-regulated at fixed pH 5.2 versus fixed pH 7.3; however, the basal expression level of *hydB2* at fixed pH 5.2 was around five times higher than that for *hydB3* (Figure 6D). Of particular interest also is the observation that during the unregulated-pH fermentation the basal expression levels of the four [FeFe] hydrogenase coding genes were lower than their expression levels at fixed pH 5.2 (10 h incubation, late exponential growth phase, Figure 4B and Figure 6D). For the two fixed-pH fermentations the expression level of the *nifH* gene was similar (Figure 6C), and was significantly lower than during the unregulated-pH glucose fermentation (Figure 4B). A Western blot analysis carried out on samples corresponding to the different fermentation stages for both fixed-pH cultures gave no positive signal for the NifH subunit (data not shown). This observation suggests that in contrast to the unregulated-pH glucose fermentation, where a strong up-regulation of nitrogenase genes was observed, H_2 production at fixed pH 5.2 may be essentially hydrogenase-mediated.

To better understand the phenomenon of possibly differential nitrogenase- and hydrogenase-mediated H_2 production in clostridia under different pH conditions, further investigation is required. Moreover, the genetic elements for the transcriptional control of the *nif* operon have not yet been well defined in clostridia [27]. Nevertheless, it is tempting to speculate about the significance of the differential carbon metabolisms (Figure 1G and I) observed for fermentation under unregulated- pH compared with fermentation under fixed pH 5.2 condition. This may be an indication of metabolic adaptations affecting the H_2-producing enzymes, thereby leading to distinct H_2 production profiles (Figure 1D and F). Interestingly, it has recently been shown that the nitrogenase-mediated H_2 evolution in a photoheterotrophic bacterium *Rhodobacter sphaeroides* was enhanced by ethanol [35,36]. Oh *et al.* [36] reported a 60% increase in H_2 production in the presence of ethanol in the ammonium-containing medium. Moreover, the authors showed that the increased nitrogenase activity was regulated at the level of *nifHDK* transcription and that ethanol was not used as a carbon source by the bacterium. Though the study discusses another type of bacterium, it can be compared with the unregulated-pH fermentation in our study, where both an increase in the production of ethanol and transcription of *nif* genes were detected and were concurrent with a higher H_2 production. In contrast, during fixed pH 5.2 glucose fermentation, where no ethanol production was observed, the expression level of *nifH* was significantly lower as well.

Nevertheless, there are still unresolved questions concerning the metabolic fluxes and regulations in clostridia that need to be studied before a more grounded hypothesis can be made, especially as our study is the first to suggest, based on transcriptional analysis, a putative involvement of nitrogenase to the overall H_2 production in this bacterial genus.

Conclusions

This paper presents a biomolecular overview of the changes occurring when the metabolism of *C. butyricum* CWBI

1009 shifts to H_2 production. The results show that the cellular metabolism for all the tested metabolites varied under all conditions and was pH dependent. The primary soluble metabolites were formate, acetate and butyrate, followed by ethanol and lactate. The highest H_2 yield (1.78 ± 0.11 mol H_2/mol glucose) was in accordance with previous studies with the same strain [9,10]. Further investigations into the genomic/proteomic changes associated with naturally decreasing pH and with H_2 production indicated the differential regulation of numerous genes/ proteins; some were known to be directly associated with H_2 production (for example, hydrogenases and pyruvate-ferredoxin oxidoreductase). However the up-regulation of others, such as those involved in N_2 assimilation, were more surprising, since H_2 production in clostridia has never before been described as nitrogenase-mediated.

As the transcription levels of the [FeFe] hydrogenases and the nitrogenase coding genes varied significantly between different fermentations, transcriptional analyses indicated the need for further biochemical characterisation of the H_2-producing enzymes, especially for the [FeFe] hydrogenases HydA8, B2 and B3, which have not yet been studied [6]. Additionally, the identification of multiple hypothetical genes/proteins (Additional file 2: Table S4) that were differentially regulated suggests that other unknown mechanisms may be governing H_2 production in clostridia. This could therefore be fertile ground for future studies.

This work contributes to the field of dark fermentative hydrogen production by providing a multidisciplinary approach for the investigation of the processes involved in the molecular H_2 metabolism of clostridia, which could in time lay the groundwork for further optimisation based on genetic engineering techniques [3].

Methods

Microorganism and growth conditions

The strain *Clostridium butyricum* CWBI 1009 was isolated from an anaerobic sludge and cultivated in a modified MDT medium as previously described [9]. The modified MDT medium contained, per litre of deionised water: glucose monohydrate (10 g), casein pepton (5 g), yeast extract (0.5 g), Na_2HPO_4 (5.1 g), KH_2PO_4 (1.2 g), $MgSO_4.7H_2O$ (0.5 g), and cystein hydrochloride (0.5 g). The PCA (Plate Count Agar) medium, used to verify the absence of aerobic and facultative aerobic contaminants, contained per litre of deionised water: glucose monohydrate (1 g), casein peptone (5 g), yeast extract (2.5 g), and agar (15 g). All the chemicals used were of analytical or extra pure quality and were supplied by Merck, UCB and Sigma. Casein peptone and yeast extract were supplied by Organotechnie (La Courneuve, France).

Reactor setup and experimental procedure

Fermentations were carried out in a 20-L laboratory-scale bioreactor (Biolafitte Niort, F) consisting of a double envelope and a stainless steel lid equipped with a butyl septum, 0.20-μm gas filters (Midisart, Sartorius) and tubing (for gas inlet, gas outlet and medium removal). Prior to inoculation the bioreactor and the medium were sterilised at 120°C for 20 min. Glucose and cystein were autoclaved separately in 2.5-L flasks to prevent Maillard reactions, and were then added sterilely to the tank before being cooled and purged with N_2 or Ar. After inoculation with 1.7 L of pre-culture (obtained in 2-L hermetic bottles incubated for 24 hours at 30°C), the bioreactor had a final working volume of 17 L. Finally the pH was adjusted to either 7.3 or 5.2 ± 0.1 via automatic addition of sterile 1.5 N KOH combined with a Mettler Toledo probe (465-35-SC-P-K9/320) and needles inserted through the septum. For the unregulated-pH fermentation, after setting the initial pH value at 7.6 ± 0.1, the controller for base addition was turned off. Three biological replicates corresponding to each pH condition (unregulated pH, fixed pH 7.3 and fixed pH 5.2) were prepared and monitored for 25 hours. Throughout the fermentation, the bioreactor was maintained at 30°C and stirred at 100 rpm.

Biogas and metabolite monitoring

The biogas flow rate was measured by a MilliGasCounter-1 PMMA Ritter flow meter connected to a computer (Rigamo V1.30-K1 software, acquisition every 30 seconds). The biogas production was also regularly checked with a second drum-type flow meter (Ritter TG01) connected in series. The proportion of hydrogen gas was determined using a gas chromatograph (GC) (Hewlett-Packard 5890 Series II) fitted with a thermal conductivity detector (TCD) and a 30 m × 0.32 mm GasPro GSC capillary column (Altech) in series with a 20 m × 0.25 mm Carbo-PLOT P7 column (Chrompak). The temperatures of the injections, the TCD chambers and the oven were maintained at 90°C, 110°C and 55°C, respectively. Nitrogen was used as the carrier gas in the column at a flow rate of 20 ml min^{-1}. The harvested liquid samples were centrifuged at 13,000 g for 1 min, and the supernatant was filtered through a 0.2-μm cellulose acetate membrane (Minisart, Sartorius) before HPLC analysis for glucose, ethanol, lactate, acetate, formate and butyrate. HPLC was carried out using an Agilent 1110 series chromatograph (HP Chemstation software) with a Supelcogel C-610H column preceded by a Supelguard H pre-column (oven temperature 40°C). 0.1% H_3PO_4 (in Milli-Q water) was used for the isocratic mobile phase (flow rate of 0.5 ml min^{-1}) using a differential refraction index detector (RID, heated at 35°C). The method lasted for 35 min at a maximum pressure of 60 bar. The data for the glucose and metabolite concentrations were used to calculate the mass balance (MB) of the glucose conversion as described previously [37].

RNA extraction

For RNA extraction, the RiboPure™ -Bacteria (Ambion) extraction kit was used. The cells in a 2-ml suspension were harvested by centrifugation (16,000 g for 1 min), frozen in liquid nitrogen and stored at -80°C. The RNA extraction was performed according to the instruction manual. The total RNA was eluted by the addition of 50 μl of the preheated elution solution. Before reverse transcription, any contaminating DNA was removed by a double treatment with TURBO DNase (TURBO DNA-free™, Ambion), according to the instruction manual. In each step, the reaction mixture was incubated at 37°C for 30 min. After the second incubation step, the DNase was inactivated by the addition of the DNase inactivation reagent at a concentration of 20% of the volume of the treated RNA. The mix was then incubated for 2 min at room temperature and subsequently centrifuged for 1 min at 10,000 g to pellet the inactivation reagent. The absence of the genomic DNA contamination was confirmed by qPCR directly using 5 ng of total RNA. Total RNA concentrations were determined with a NanoVue spectrophotometer (GE Healthcare), and the RNA integrity was checked on a formaldehyde-agarose gel. Before RNA-seq, the RNA quality was checked on an automated electrophoresis system (Experion, BioRad). The RNA was stored at -80°C before analysis. First-strand cDNA was synthesised with the Reverse Transcription System (Promega, Madison, WI, USA) according to the manufacturer instructions. The total volume of the reaction mixture was 20 μl and contained 500 ng of the total RNA and 500 ng of the control mouse RNA (Quantum RNA™ ß-actin Internal Standards, Ambion). Control mouse RNA was used as an external standard. The reaction was primed with the random primers supplied with the kit. Obtained cDNAs were diluted in diethylpyrocarbonate (DEPC)-treated water to a final concentration of 5 ng initial total RNA/μl.

mRNA sample preparation and RNA-seq analysis

10 μg of extracted total RNA was treated with the MICROBExpress™ Kit (Ambion) to enrich for mRNA, by removing the 16S and 23S rRNA. Paired-end libraries were prepared according to the TruSeq™ RNA Sample Preparation Guide (Illumina). The library preparation and Illumina RNA sequencing was performed by the GIGA transcriptomics platform (Liège, Belgium). Obtained reads were aligned using the BWA software using default parameters [38]. Raw counts per gene were calculated based on the genome annotation of *Clostridium butyricum* CWBI 1009. Reads were allowed to map 50 bp upstream of the start codon or 50 bp downstream of

the stop codon. Reads mapping to ribosomal or transporter RNA were removed from the raw counts data to prevent bias in detecting differential expression. Differential expression was calculated using the edgeR package (version 3.2.4) [39] in BioConductor (release 2.12, R version 2.15.0), resulting for each gene in a fold change and a corresponding P-value corrected for multiple testing. Genes with a 2-fold ($\log_2$) up or down-regulation and a corrected P-value lower than 0.05 were assigned as being differentially expressed.

Clostridium butyricum CWBI 1009 genome sequencing

The *Clostridium butyricum* CWBI 1009 genome was sequenced by BaseClear (Leiden, The Netherlands) using the pair-end sequencing on the Illumina Hiseq 2000. Genome assembly was performed using Velvet version 1.2.10 [40], using a hash length of 29, a minimum contig length of 500 and a minimal coverage of 20. The Whole Genome Shotgun project was deposited at [DDBJ/EMBL/GenBank:ASPQ00000000]. The accession version described in this paper is version ASPQ01000000 (Additional file 1: Table S12).

RT-qPCR

Species-specific primers and probes are listed in Table 1. ß-actin-specific primers were taken from the kit (QuantumRNA™ ß-actin Internal Standards, Ambion). The quantitative PCR amplifications were carried out with a Mini Opticon (BioRad). The DNA template used for a standard curve was prepared as previously described [41]. For the gene expression analysis, 1 µl of cDNA was used. The total volume of the PCR mix was 25 µl. Each reaction consisted of 1 × PCR mix (ABsolute™ Blue QPCR SYBR® Green Fluorescein Mix or ABsolute™Blue QPCR mix, Thermo Scientific), each primer and/or hydrolysis probe (HPLC cleaned, Biomers, Germany) at a final concentration of 150 nM. Each sample was analysed in triplicate. A 'no template' control was included in each run. The specificities of the primers were verified at the end of each qPCR reaction by performing the melting curve analysis (for a SYBR Green-based quantification). The standard curve preparation and the cycling conditions for SYBR Green chemistry were as previously described [41]. For the probe-based chemistry, the initial denaturation of 15 min was followed by 40 cycles of denaturation at 95°C for 15 s and primer annealing/amplification step at 60°C for 30 s. The reaction efficiency was calculated as factor specific [42] according to the equation: $E = 10^{-1/\text{slope}}$. For gene expression analysis, the relative expression levels were calculated with a Relative Expression Software Tool 2009, REST© [43]. To estimate the up- and down-regulation of analysed genes, the obtained Cqs were compared to those of the reference genes and an external standard control gene. Two internal reference genes, 16S rRNA and *recA*, and one external RNA control gene (ß-actin) were used. An external mouse RNA was added to the sample RNA before the cDNA synthesis to correct for intrinsic and technical variations introduced throughout the experimental process [44]. The stability of the chosen reference genes and the integrity of each RNA sample were evaluated using the BestKeeper Excel-based tool [45] (data not shown).

Table 1 List of *C. butyricum* CWBI 1009-specific primers and probes used in this study

Target gene	Primer/probe	Sequence 5′ → 3′	Length (bp)	Ref.
recA	RecA-ButF	AAGCATTAGTGCGTTCTGGAG	97	[41]
	RecA-ButR	GAATCTCCCATTTCCCCTTC		
hydA2	ButA2F	ATAGTTGCAATGGCTCCTGC	250	This study
	ButA2R	TTTCTGCTTGCCTAACCCAT		
hydA8	ButA8F	TCTTTGGAGTTACAGGGGGA	188	This study
	ButA8R	TTCAGCATTTGCAAGACCAC		
hydB2	ButB2F	TGGTGGTGTATCAACTGCTG	168	This study
	ButB2R	TTGCATCCCATTCCTTCAAT		
hydB3	ButB3F	CAATGGTTGCTACAGGCAGA	168	This study
	ButB3R	CAAAAGCATCGAATAACGCA		
16S rRNA	16SButF	CCTGCCTCATAGAGGGGAAT	143	This study
	16SButR	GAGCCGTTACCTCACCAACT		
	HP16SBut[a]	CCGCATAAGATTGTAGTACCGCATGGTACA		
nifH	NifHButF	CATCAGCATTGGCTGAGATG	206	This study
	NifHButR	TGGTTCTGGTCCTCCTGATT		

[a]HP: hydrolysis probe.

Western blot analysis

Whole cell protein extracts were prepared by sonicating the bacteria in TpW buffer (100 mM TRIS, 150 mM NaCl, 1mM EDTA). The cell lysates were centrifuged (16,000 g for 20 min) and protein concentrations in lysates were estimated using the Bradford assay, performed according to the manufacturer protocol (Fermentas). Western blot analysis was performed using a hen polyclonal affinity purified IgY raised against the NifH subunit of nitrogenase (Agrisera, Sweden). As a positive control, cell lysate of *Anabaena variabilis* was used. 5 μg of the whole protein extracts were directly subjected to 10% SDS-PAGE and then blotted onto Immobilon-P membrane (Millipore). Membranes were blocked for 1 h with TBST (20 mM Tris, 140 mM NaCl, 0.1% Tween, pH 7.6) containing 3% BSA, and then incubated with 1:2,000 diluted antibody in TBST, overnight at 4°C. As secondary antibody, a 1:10,000 dilution of rabbit anti-chicken IgG coupled to alkaline phosphatase (Sigma) was used. Reactive protein bands were detected using NBF/BdP reagents (Sigma).

Sample preparation for 2D-DIGE

Freshly harvested cells were resuspended in a denaturation buffer (7 M urea, 2 M thiourea, 2% ASB-14, 20 mM DTT, Complete EDTA free [Roche], 1 mM EDTA pH 8.5, 50 mM Tris-HCl pH 7.5) and intensively vortexed for 30 min at room temperature. Subsequently, the suspensions were briefly sonicated and centrifuged at 10,000 g for 10 min to remove any insoluble material. In order to discard the remaining salts, fatty acids and nucleic acids, the protein extracts were precipitated three times and cleaned twice with a 2D-clean up Kit (GE Healthcare). Protein pellets were then resuspended in a DIGE labelling buffer (7 M urea, 2 M thiourea, 2% ASB-14, EDTA free anti-protease cocktail [Roche], 0.5 mM EDTA 50 mM TRIS adjusted at pH 8.5). The protein concentration was estimated with an RC/DC protein assay kit (BioRad Laboratories), and was adjusted to a value between 5 and 10 mg/ml for optimal CyDye labelling. Each sample (containing 25 μg of protein) was labelled with 0.2 nmol of either Cy3 or Cy5 (minimal labelling). At the same time, an internal standard consisting of equimolar amounts of the two samples was labelled with Cy2 (GE Healthcare). The labelling reaction was stopped after 30 min by adding 5 nM of lysine. The Cy2, Cy3- and Cy5-labelled proteins were pooled together prior to isoelectrofocussing with the IPGphor3 IEF System (GE Healthcare). Pooled samples were reduced by adding 20 mM DTT, resuspended in a Drystrip rehydration buffer (7 M urea, 2 M thiourea, 2% ASB-14 w/v, 0.6% IPG Buffer [GE Healthcare] v/v), and then supplemented with Destreak solution (GE Healthcare), to provide a final volume of 450 μl which was spread on a 24-cm regular strip holder. The 3-11 NL IPG Drystrips (GE Healthcare) were passively rehydrated

in the strip holder for 10 h at 20°C prior to running the IEF under the following conditions: 50 V for 2 h (step), 200 V for 200 Vh (step), 500 V for 150 Vh (gradient), 500 V for 500 Vh (step), 1,000 V for 500 Vh (gradient), 8,000 V for 13,500 Vh (gradient), 8,000V for 8,0000 Vh (step) and 500 V for 10 h (step); with a maximum current setting fixed at 50 μA [46]. After the first electrophoretic migration (first dimension), the strips were reduced in an equilibration buffer (50% glycerol v/v, 4% SDS w/v, 6 M urea, 50 mM Tris-HCl adjusted to pH 8.8) and incubated with 300 mM DTT for 15 min. The strips were then alkylated for 15 min in the same equilibration solution with DTT replaced by 350 mM of iodoacetamide. After equilibration, the strips were put on top of a 12.5% acrylamide gel in a Laemmli SDS electrophoresis buffer (25 mM Tris, 192 mM glycine, 1% SDS w/v). Electrophoresis was carried out overnight at 1 W/gel. Following migration, the gels were scanned (Typhoon 9400, GE Healthcare) at a resolution of 100 μm for the excitation wavelengths corresponding to each CyDye. The scanner generated 18 gel images for each biological replicate (9 images for the comparison between the two pH values and 6 images for the internal standard) which were analysed with the DeCyder V7.0 software (GE Healthcare). The co-detection of the three CyDye-labeled forms for each spot was done using the Differential In-gel Analysis (DIA) module. Statistical analysis was carried out in the Biological Variation Analysis (BVA) module after inter-gel matching. Protein spots that resulted in a statistically significant Student's *t*-test value ($P < 0.05$, n = 6) were considered as being differentially abundant at pH 5.2 compared to pH 7.3.

In-gel digestion and mass spectrometry

Protein identification was performed using preparative gels (150 μg of loaded material) prepared under the same experimental conditions, but with only one CyDye to label the protein sample (Cy5). In addition, the gel plates were treated with a Bind-silane solution for spot picking. The resulting scanned gels were matched with the BVA module. Matched spots presenting a statistical difference were detected using the Ettan Dalt Spot Picker (GE Healthcare). Subsequently, the proteins in the gel pieces were washed (three times with Milli-Q water and 100% ACN and three times with 25 mM NH_4HCO_3 and 100% ACN) to remove excess detergent and buffer. After a final dehydration step in ACN, the gel pieces were rehydrated and treated with 2 μl of a 5 ng/μl trypsin proteomic grade solution (Roche) for 2 h at 4°C, to ensure sufficient diffusion in the gel. The temperature was then raised to 37°C for an overnight digestion. After tryptic digestion, the resulting peptides were extracted from the gel pieces by adding 5 μl of a 1% trifluoroacetic acid (TFA) v/v, 30% ACN v/v solution, and vortexing for 30 min. 1 μl of the resulting extract was dropped on a 384-600 MTP Anchorship MALDI target plate (Bruker Daltonic), previously spotted with a 3% w/v HCCA matrix (Sigma) dissolved in acetone. Each drop was washed 3 times with a 10 mM $(NH_4)_2(HPO_4)$ solution. Protein identification was carried out with MALDI-TOF/TOF instrumentation (Ultraflex II, Bruker Daltonic) in MS and MS/MS modes and the Mascot search engine was configured with a maximal mass error rate at 100 ppm [47]. Protein identification was performed with the Biotools software (Bruker) using the Mascot search engine on the *Clostridium butyricum* 5521 protein database (Additional file 5: Table S10).

Additional files

Additional file 1: Figure S1. Double-log scatter of sequence reads and the coefficient of determination (R^2) for the biological replicates of the RNA-seq reads mapped to the genome of *Clostridium butyricum* CWBI 1009. **Figure S2.** Correlation of RNA-seq data with RT-qPCR for *Clostridium butyricum* CWBI 1009 cultivated in a 20L batch bioreactor with glucose (10 g/L) under unregulated-pH conditions. **Figure S3.** 2D-gel pattern of the *Clostridium butyricum* CWBI 1009 proteome. **Table S1.** Bioreactor performance of *Clostridium butyricum* CWBI 1009 cultivated in a 20L batch bioreactor with glucose (10 g/L) under unregulated-pH conditions. **Table S2.** Bioreactor performance of *Clostridium butyricum* CWBI 1009 cultivated in a 20L batch bioreactor with glucose (5 g/L) at fixed pH 7.3 and 5.2. **Table S3.** Summary of *Clostridium butyricum* CWBI 1009 RNA-seq data results. **Table S12.** Summary of *Clostridium butyricum* CWBI 1009 genome information.

Additional file 2: RNA-seq data. Table S4. List of the most highly expressed genes at pH 6.3 (column indicated in bold type) during glucose fermentation under unregulated-pH conditions.

Additional file 3: RNA-seq data. Table S5. Differentially regulated genes in *C. butyricum* CWBI 1009 in response to the decline in pH (unregulated-pH glucose fermentation) as analysed by RNA-seq.

Additional file 4: RNA-seq data. Table S6. Acid stress response: alkalisation of the internal/external environment. Genes and pathways regulated at acidic pH 6.3 versus 7.3 during glucose fermentation under unregulated-pH conditions (RNA-seq data). **Table S7.** Acid stress response: cellular transport. Genes differentially regulated at acidic pH 6.3 versus 7.3 during glucose fermentation under unregulated-pH conditions (RNA-seq data). **Table S8.** Acid stress response: transcriptional regulators and other regulatory proteins. Genes differentially regulated at acidic pH 6.3 versus 7.3 during glucose fermentation under unregulated-pH conditions (RNA-seq data). **Table S9.** Acid stress response: activation of sporulation. Genes differentially regulated at acidic pH 6.3 versus 7.3 during glucose fermentation under unregulated-pH conditions (RNA-seq data).

Additional file 5: 2D-DIGE data. Table S10. Protein identification table generated with the Biotools software (Bruker) using the Mascot search engine on the *Clostridium butyricum* 5521 protein database. **Table S11.** Proteins identified as differentially abundant in *C. butyricum* CWBI 1009 in response to the decline in pH during unregulated-pH glucose fermentation (2D-DIGE analysis).

Additional file 6: Argon discussion. Figure S4. Characteristics of the glucose fermentation for *Clostridium butyricum* CWBI 1009 performed under unregulated-pH conditions and Ar atmosphere. **Figure S5.** Relative expression of [FeFe] hydrogenases and *nifH* genes determined by RT-qPCR. Western blot analysis for NifH subunit of *Clostridium butyricum* CWBI 1009 during unregulated-pH glucose fermentation under N_2 and Ar atmospheres. **Table S13.** Bioreactor performance of *Clostridium butyricum* CWBI 1009 cultivated in a 20L batch bioreactor with glucose (10 g/L) under unregulated-pH conditions and N_2 or Ar atmospheres.

Abbreviations
2D-DIGE: Two-dimensional difference in gel electrophoresis; CDS: Coding DNA sequence; NFOR: NADH-ferredoxin oxidoreductase; PEP: Phosphoenolopyruvate; PFOR: Pyruvate-ferredoxin oxidoreductase; RPKM: Reads per kilobase per million mapped reads; RT-qPCR: Reverse transcription quantitative real-time PCR.

Competing interests
The authors declare that they have no competing interests.

Authors' contributions
MC designed the experiments, carried out the molecular genetic studies, analysed the data and drafted the manuscript. CH designed the experiments, carried out the fermentations and the 2D-DIGE study, analysed the data and drafted the manuscript. PM analysed the RNA-seq data and helped to draft the manuscript. GM participated in the 2D-DIGE study and helped to analyse the data. NL, FF, BJ and SH helped to design the experiments and coordinated the study. PT and AW conceived the study and participated in its design and coordination, and helped to draft the manuscript. All authors read and approved the final manuscript.

Acknowledgements
The research was funded by an ARC project (Action de Recherche Concertée, ARC -07/12-04), granted by the French Community of Belgium. A. Wilmotte is Research Associate of the FRS-FNRS of Belgium. We thank Pierre Leprince from the University of Liège for his assistance during the protein spots picking procedure. We would also like to acknowledge Julien Masset and Laurent Beckers from the University of Liège for their advice concerning the anaerobic fermentations and Gregory Hex for his advice in proteomics. We would like to thank Ian Hamilton for his English correction of the manuscript.

Author details
[1]Centre for Protein Engineering, Bacterial Physiology and Genetics, University of Liège, Allée de la Chimie 3, B-4000 Liège, Belgium. [2]Walloon Centre of Industrial Biology, University of Liège, Boulevard du Rectorat 29, B-4000 Liège, Belgium. [3]Microbiology Unit, Expertise Group for Molecular and Cellular Biology, Institute for Environment, Health and Safety, Belgian Nuclear Research Centre (SCK-CEN), Boeretang 200, B-2400 Mol, Belgium. [4]Bioenergetics Laboratory, University of Liège, Boulevard du Rectorat 27, B-4000 Liège, Belgium. [5]Environmental Research and Innovation Department, Luxembourg Institute of Science and Technology, Rue du Brill 41, L-4422 Belvaux, Luxembourg.

References
1.   Lens P, Westermann P, Haberbauer M, Moreno A. Biofuels for fuel cells: renewable energy from biomass fermentation. IWA Publishing; 2005
2.   Hallenbeck PC. Microbial paths to renewable hydrogen production. Biofuels. 2011;2:285–302.
3.   Hallenbeck PC, Ghosh D. Improvements in fermentative biological hydrogen production through metabolic engineering. J Environ Manag. 2012;95, Supplement:S360–4.
4.   Levin DB, Pitt L, Love M. Biohydrogen production: prospects and limitations to practical application. Int J Hydrogen Energ. 2004;29:173–85.
5.   Vignais PM, Colbeau A. Molecular biology of microbial hydrogenases. Curr Issues Mol Biol. 2004;6:159–88.
6.   Calusinska M, Happe T, Joris B, Wilmotte A. The surprising diversity of clostridial hydrogenases: a comparative genomic perspective. Microbiology. 2010;156:1575–88.
7.   Carnahan JE, Mortenson LE, Mower HF, Castle JE. Nitrogen fixation in cell-free extracts of *Clostridium pasteurianum*. Biochim Biophys Acta. 1960;44:520–35.
8.   Chen JS, Toth J, Kasap M. Nitrogen-fixation genes and nitrogenase activity in *Clostridium acetobutylicum* and *Clostridium beijerinckii*. J Ind Microbiol Biotechnol. 2001;27:281–6.
9.   Masset J, Hiligsmann S, Hamilton C, Beckers L, Franck F, Thonart P. Effect of pH on glucose and starch fermentation in batch and sequenced-batch

mode with a recently isolated strain of hydrogen-producing *Clostridium butyricum* CWBI1009. Int J Hydrogen Energ. 2010;35:3371–8.

10. Masset J, Calusinska M, Hamilton C, Hiligsmann S, Joris B, Wilmotte A, et al. Fermentative hydrogen production from glucose and starch using pure strains and artificial co-cultures of *Clostridium spp.* Biotechnol Biofuels. 2012;5:35.

11. Calusinska M, Joris B, Wilmotte A. Genetic diversity and amplification of different clostridial [FeFe] hydrogenases by group-specific degenerate primers. Lett Appl Microbiol. 2011;53:473–80.

12. Liu IC, Whang LM, Ren WJ, Lin PY. The effect of pH on the production of biohydrogen by clostridia: thermodynamic and metabolic considerations. Int J Hydrogen Energ. 2011;36:439–49.

13. Cai G, Jin B, Saint CP, Monis PT. Metabolic flux analysis of hydrogen production network by *Clostridium butyricum* W5: Effect of pH and glucose concentrations. Int J Hydrogen Energ. 2010;35:6681–90.

14. Wang Y, Li X, Mao Y, Blaschek H. Genome-wide dynamic transcriptional profiling in *Clostridium beijerinckii* NCIMB 8052 using single-nucleotide resolution RNA-Seq. BMC Genomics. 2012;13:102.

15. Alsaker KV, Paredes C, Papoutsakis ET. Metabolite stress and tolerance in the production of biofuels and chemicals: gene-expression-based systems analysis of butanol, butyrate, and acetate stresses in the anaerobe *Clostridium acetobutylicum.* Biotechnol Bioeng. 2010;105:1131–47.

16. Schwarz KM, Kuit W, Grimmler C, Ehrenreich A, Kengen SWM. A transcriptional study of acidogenic chemostat cells of *Clostridium acetobutylicum* - cellular behavior in adaptation to *n*-butanol. J Biotechnol. 2012;161:366–77.

17. Ma Y, Huang A, Zhu D, Pan G, Wang G. Biohydrogen production via the interaction of nitrogenase and anaerobic mixed-acid fermentation in marine bacteria. Int J Hydrogen Energ. 2015;40:176–83.

18. Burgess BK, Wherland S, Newton WE, Stiefel EI. Nitrogenase reactivity: insight into the nitrogen-fixing process through hydrogen-inhibition and HD-forming reactions. Biochemistry. 1981;20:5140–6.

19. Igarashi RY, Seefeldt LC. Nitrogen fixation: the mechanism of the Mo-dependent nitrogenase. Crit Rev Biochem Mol Biol. 2003;38:351–84.

20. Yoch DC, Pengra RM. Effect of amino acids on the nitrogenase system of *Klebsiella pneumoniae.* J Bacteriol. 1966;92:618–22.

21. Slonczewski JL, Fujisawa M, Dopson M, Krulwich TA. Cytoplasmic pH measurement and homeostasis in bacteria and archaea. In: Robert KP, editor. Advances in microbial physiology. Volume 55. Academic Press; 2009:1–317.

22. Sachs G, Kraut JA, Wen Y, Feng J, Scott DR. Urea transport in bacteria: acid acclimation by gastric *Helicobacter spp.* J Membrane Biol. 2006;212:71–82.

23. Piper P, Mahe Y, Thompson S, Pandjaitan R, Holyoak C, Egner R, et al. The Pdr12 ABC transporter is required for the development of weak organic acid resistance in yeast. EMBO J. 1998;17:4257–65.

24. Wortham B, Oliveira M, Patel C. Polyamines in bacteria: pleiotropic effects yet specific mechanisms. In: Perry R, Fetherston J, editors. The genus *Yersinia.* 603rd ed. New York: Springer; 2007. p. 106–15.

25. Kusano T, Berberich T, Tateda C, Takahashi Y. Polyamines: essential factors for growth and survival. Planta. 2008;228:367–81.

26. Higgins D, Dworkin J. Recent progress in *Bacillus subtilis* sporulation. FEMS Microbiol Rev. 2012;36:131–48.

27. Dürre P. Handbook on Clostridia. 2005. Taylor & Francis Group.

28. Vignais PM, Billoud B. Occurrence, classification, and biological function of hydrogenases: an overview. Chem Rev. 2007;107:4206–72.

29. Bothe H, Schmitz O, Yates MG, Newton W: Nitrogenases and hydrogenases in cyanobacteria. In: Peschek GA, Obinger C, Renger G, editors. Bioenergetic processes of Cyanobacteria. Springer Netherlands; 2011. p. 137–157.

30. Schut GJ, Adams MWW. The iron-hydrogenase of *Thermotoga maritima* utilizes ferredoxin and NADH synergistically: a new perspective on anaerobic hydrogen production. J Bacteriol. 2009;191:4451–7.

31. Meyer J. Clostridial iron-sulphur proteins. J Mol Microbiol Biotechnol. 2000;2:9–14.

32. White D. The physiology and biochemistry of prokaryotes. 3rd ed. New York: Oxford University Press; 2006.

33. Amon J, Titgemeyer F, Burkovski A. Common patterns - unique features: nitrogen metabolism and regulation in Gram-positive bacteria. FEMS Microbiol Rev. 2010;34:588–605.

34. Adams MW, Mortenson LE. The physical and catalytic properties of hydrogenase II of *Clostridium pasteurianum.* A comparison with hydrogenase I. J Biol Chem. 1984;259:7045–55.

35. Kim DH, Lee JH, Kang S, Hallenbeck P, Kim EJ, Lee J, et al. Enhanced photo-fermentative H$_2$ production using *Rhodobacter sphaeroides* by ethanol addition and analysis of soluble microbial products. Biotechnol Biofuels. 2014;7:79.

36. Oh EK, Kim EJ, Hwang HJ, Tong X, Nam JM, Kim MS, et al. The photoheterotrophic H$_2$ evolution of *Rhodobacter sphaeroides* is enhanced in the presence of ethanol. Int J Hydrogen Energ. 2012;37:15886–92.

37. Hamilton C, Hiligsmann S, Beckers L, Masset J, Wilmotte A, Thonart P. Optimization of culture conditions for biological hydrogen production by *Citrobacter freundii* CWBI952 in batch, sequenced-batch and semicontinuous operating mode. Int J Hydrogen Energ. 2010;35:1089–98.

38. Li H, Durbin R. Fast and accurate short read alignment with Burrows-Wheeler transform. Bioinformatics. 2009;25:1754–60.

39. Robinson MD, McCarthy DJ, Smyth GK. edgeR: a Bioconductor package for differential expression analysis of digital gene expression data. Bioinformatics. 2010;26:139–40.

40. Zerbino DR, Birney E. Velvet: algorithms for de novo short read assembly using de Bruijn graphs. Genome Res. 2008;18:821–9.

41. Savichtcheva O, Joris B, Wilmotte A, Calusinska M. Novel FISH and quantitative PCR protocols to monitor artificial consortia composed of different hydrogen-producing *Clostridium spp.* Int J Hydrogen Energ. 2010;36:7530–42.

42. Rasmussen R. Quantification on the LightCycler. In: Meuer S, Wittwer C, Nakagawara K, editors. Rapid cycle real-time PCR, methods and applications. Heidelberg: Springer Press; 2001. p. 21–34.

43. Pfaffl MW, Horgan GW, Dempfle L. Relative expression software tool (REST-®) for group-wise comparison and statistical analysis of relative expression results in real-time PCR. Nucleic Acids Res. 2002;30:e36.

44. Ellefsen S, Stenslekken KO, Sandvik GK, Kristensen TA, Nilsson GE. Improved normalization of real-time reverse transcriptase polymerase chain reaction data using an external RNA control. Anal Biochem. 2008;376:83–93.

45. Pfaffl M, Tichopad A, Prgomet C, Neuvians T. Determination of stable housekeeping genes, differentially regulated target genes and sample integrity: BestKeeper-Excel-based tool using pair-wise correlations. Biotechnol Lette. 2004;26:509–15.

46. Gorg A, Weiss W, Dunn MJ. Current two-dimensional electrophoresis technology for proteomics. Proteom. 2004;4:3665–85.

47. Shevchenko A, Jensen ON, Podtelejnikov A, Sagliocco F, Wilm M, Vorm O, et al. Linking genome and proteome by mass spectrometry: Large-scale identification of yeast proteins from two dimensional gels. Proc Natl Acad Sci U S A. 1996;93:14440–5.

3

Downregulation of *GAUT12* in *Populus deltoides* by RNA silencing results in reduced recalcitrance, increased growth and reduced xylan and pectin in a woody biofuel feedstock

Ajaya K Biswal[1,3,4], Zhangying Hao[2,3,4], Sivakumar Pattathil[3,4], Xiaohan Yang[4,5], Kim Winkeler[4,6], Cassandra Collins[4,6], Sushree S Mohanty[3,4], Elizabeth A Richardson[2], Ivana Gelineo-Albersheim[3,4], Kimberly Hunt[3,4], David Ryno[3,4], Robert W Sykes[4,7], Geoffrey B Turner[4,7], Angela Ziebell[4,7], Erica Gjersing[4,7], Wolfgang Lukowitz[2], Mark F Davis[4,7], Stephen R Decker[4,7], Michael G Hahn[2,3,4] and Debra Mohnen[1,3,4*]

Abstract

Background: The inherent recalcitrance of woody bioenergy feedstocks is a major challenge for their use as a source of second-generation biofuel. Secondary cell walls that constitute the majority of hardwood biomass are rich in cellulose, xylan, and lignin. The interactions among these polymers prevent facile accessibility and deconstruction by enzymes and chemicals. Plant biomass that can with minimal pretreatment be degraded into sugars is required to produce renewable biofuels in a cost-effective manner.

Results: GAUT12/IRX8 is a putative glycosyltransferase proposed to be involved in secondary cell wall glucuronoxylan and/or pectin biosynthesis based on concomitant reductions of both xylan and the pectin homogalacturonan (HG) in *Arabidopsis irx8* mutants. Two GAUT12 homologs exist in *Populus trichocarpa*, *PtGAUT12.1* and *PtGAUT12.2*. Knockdown expression of both genes simultaneously has been shown to reduce xylan content in *Populus* wood. We tested the proposition that RNA interference (RNAi) downregulation of *GAUT12.1* alone would lead to increased sugar release in *Populus* wood, that is, reduced recalcitrance, based on the hypothesis that GAUT12 synthesizes a wall structure required for deposition of xylan and that cell walls with less xylan and/or modified cell wall architecture would have reduced recalcitrance. Using an RNAi approach, we generated 11 *Populus deltoides* transgenic lines with 50 to 67% reduced *PdGAUT12.1* transcript expression compared to wild type (WT) and vector controls. Ten of the eleven RNAi lines yielded 4 to 8% greater glucose release upon enzymatic saccharification than the controls. The *PdGAUT12.1* knockdown (*PdGAUT12.1*-KD) lines also displayed 12 to 52% and 12 to 44% increased plant height and radial stem diameter, respectively, compared to the controls. Knockdown of *PdGAUT12.1* resulted in a 25 to 47% reduction in galacturonic acid and 17 to 30% reduction in xylose without affecting total lignin content, revealing that in *Populus* wood as in *Arabidopsis*, GAUT12 affects both pectin and xylan formation. Analyses of the sugars present in sequential cell wall extracts revealed a reduction of glucuronoxylan and pectic HG and rhamnogalacturonan in extracts from *PdGAUT12.1*-KD lines.

(Continued on next page)

* Correspondence: dmohnen@ccrc.uga.edu
[1]Department of Biochemistry and Molecular Biology, University of Georgia, B122 Life Sciences Bldg., Athens, GA 30602, USA
[3]Complex Carbohydrate Research Center, University of Georgia, 315 Riverbend Road, Athens, GA 30602, USA
Full list of author information is available at the end of the article

(Continued from previous page)

Conclusions: The results show that downregulation of GAUT12.1 leads to a reduction in a population of xylan and pectin during wood formation and to reduced recalcitrance, more easily extractable cell walls, and increased growth in *Populus*.

Keywords: Biofuel, Growth, Pectin, *Populus*, Saccharification, Secondary cell wall, Xylan, Wood development

Background

Populus is a woody feedstock for biofuel and bioproduct formation. The major challenge of using woody feedstock as a source for biofuels is the rigid cell wall, which is recalcitrant to degradation by bacterial and fungal enzymes [1-3]. The identification of genes and proteins involved in the formation of secondary cells wall is necessary to understand and overcome the recalcitrance of woody feedstocks. Towards this aim, we have manipulated the expression of putative 'recalcitrance' genes in *Populus* for use in studying the genetic basis of recalcitrance in this biomass feedstock.

Wood formation in *Populus* starts with the differentiation of secondary cell walls. Cellulose, hemicellulose, and lignin are the three major components of *Populus* secondary walls, with pectin being a minor component. In *Populus* wood, the hemicelluloses are largely xylans which provide 18 to 28% of the total dry weight [4]. Xylans are polysaccharides with linear backbones of β-$(1 \rightarrow 4)$-linked D-xylosyl residues. The major xylan in dicot wood, glucuronoxylan (GX), is decorated with side chains of O-2-linked α-D-glucuronic acid (GlcA) and/or 4-O-methyl-α-D-glucuronic acid (MeGlcA). The xylosyl backbone in secondary wall xylan is also highly acetylated at C-2 and C-3.

Multiple types of *Arabidopsis* xylan synthesis mutants [5], including xylan backbone mutants *irx9 irx9-L* [6-9], *irx14 irx14-L* [9-11], *irx10 irx10-L* [12,13], and *irx15 irx15-L* [14,15], have been extensively studied in an effort to understand xylan biosynthesis. The recovery of xylan xylosyltransferase activity from heterologously expressed Arabidopsis IRX10-L [16] and *Plantago ovate* IRX10 [17], and the demonstration of xylan acetyltransferase activity from heterologously expressed Arabidopsis ESK1/TBL29 [16], confirmed a role for these enzymes in xylan backbone synthesis and acetylation, respectively. *Arabidopsis* xylan substitution mutants *gux1* and *gux2* have reduced α-glucuronidation of the xylan backbone [18,19] while the level of methylation of the GlcA residues is reduced in *gxmt1* mutants [20]. The respective genes have been shown to encode functional xylan glucuronosyltransferases [18] and xylan 4-O-methyltransferases [20]. Additional *Arabidopsis* mutants have also been identified that have defects in both xylan and other cell wall polymers. For example, xylan and cellulose deposition are affected in *irx7*

(fra8) and *irx7-L* (F8H) [10,21,22] while *qua1* [23-25], *parvus-3* [10,26-28], and *irx8/GAUT12* [8,29,30] have defects in both pectin and xylan.

The *Arabidopsis irx8* mutant has been extensively characterized in Arabidopsis [5]. The *IRX8/GAUT12* gene belongs to the *GAUT1* (GAlactUronosylTransferase1)-related gene family. The *GAUTs* constitute one clade of the glycosyltransferase 8 (GT8) family [30-33]. The family name, GAUT, originated with the identification of Arabidopsis galacturonosyltransferase 1 (*GAUT1*). GAUT1 is a pectin biosynthetic homogalacturonan (HG):α-1,4-galacturonosyltransferase (GalAT) that functions in an HG:GalAT protein complex with GAUT7 [34-36]. Arabidopsis *GAUT12/irx8* has highest expression in cells with secondary walls, and the encoded protein has 61% amino acid sequence similarity with GAUT1. GAUT12 is predicted to be a type II membrane protein targeted to the Golgi. The *irx8/gaut12* mutation leads to a reduction in GX; however, microsomes from *irx8* mutant stems did not show any reduction in xylan XylT activity [7,10] or xylan GlcAT (glucuronosyltransferase) activity [7] compared to microsomes from wild type (WT). Structural analysis of cell walls from *irx8 Arabidopsis* mutant plants identified a dramatic reduction in GX and in a tetrasaccharide sequence β-D-Xyl*p*-(1-3)-α-L-Rha*p*-(1-2)-α-D-Gal*p*A-(1-4)-D-Xyl*p* located at the reducing end of xylan [8,29]. A concomitant reduction in a subfraction of the pectin HG was also identified in walls of *Arabidopsis irx8* mutants compared to WT leading to the hypotheses that GAUT12 is involved in either xylan or HG synthesis [29].

Populus has two orthologs of the *Arabidopsis GAUT12* gene: *PtGAUT12.1* (*POPTR_0001s44250*, Genemodel V2.0, Phytozome 8.0, http://www.phytozome.net; *Potri.001G416800*, Genemodel V3.0, Phytozome 10.0, http://phytozome.jgi.doe.gov) and *PtGAUT12.2* (*POPTR_0011s13600*, Genemodel V2.0, Phytozome 8.0; *Potri.011G132600*, Genemodel V3.0, Phytozome 10.0). Downregulation of both genes simultaneously in transgenic *P. trichocarpa* RNA interference (RNAi) lines (*PtrGT8D1/POPTR_0001s44250/Potri.001G416800* and *PtrGT8D2/POPTR_0011s13600/Potri.011G132600*) led to a major reduction in xylan and an increase in lignin in the stem, along with reduced wall thickness in fiber cells [37]. Based on these results, a role for *PtrGT8D* in GX biosynthesis in *Populus* wood has been suggested [37]. The irregular xylem and dwarf phenotype observed in the *Arabidopsis irx8* mutant, however, was not observed in these *Populus* transgenic RNAi lines. In another study, it was reported that overexpression of full-length *PoGT8D*, the *Populus alba x tremula* homolog of *POPTR_0001s44250/Potri.001G416800* (Phytozome 8.0/Phytozome 10.0), did not complement the *Arabidopsis irx8* mutant [38], although RNAi downregulation of *PoGT8D* suggested a slight reduction in the amount of xylan reducing end sequence [39]. The PoGT8D protein was shown to be targeted to the Golgi, matching its predicted type II membrane topology and in agreement with its involvement in the biosynthesis of non-cellulosic polysaccharides in *Populus* wood. Since the biochemical activity of the GAUT12 protein remains to be determined, it is not clear how this protein is involved in GX biosynthesis in *Populus*. Both of the *GAUT12 Populus trichocarpa* genes are expressed in primary xylem, differentiating xylem, secondary xylem, and phloem fibers in the woody stem. The expression of *PtrGT8D1* has been reported to be seven times greater than *PtrGT8D2* [37,38].

The recalcitrance of *Populus* biomass to deconstruction is the major obstacle to the bioconversion of this lignocellulosic feedstock into biofuels [40]. Heavily acetylated xylan cannot be efficiently hydrolyzed [41] and xylose, xylan, and xylooligomers, when produced during saccharification, inhibit hydrolysis of cellulose by cellulase as well as conversion rates and yields [42]. Reducing xylan levels in *Populus* is, therefore, one strategy to reduce recalcitrance. Similarly, pectin degradation in crop plants as well as in energy plants can enhance saccharification and enable the remaining polymers to be better substrates for the production of biofuels [43-45]. The genetic manipulation of xylan and pectin biosynthetic genes is a potential means to develop *Populus* plants with reduced xylan and pectin content and biomass that yields increased sugar release for biofuel production.

Here, we report that targeted RNAi construct-driven downregulation of one of the *Populus deltoides GAUT12* orthologs (*POPTR_0001s44250/Potri.001G416800*, *GAUT12.1*-KD) results in plants with reduced pectin and xylan content; increased plant height, stem expansion, and biomass yield; and greater glucose and xylose release and lignin syringyl-to-guaiacyl (S/G) ratio. No change in the amount of total lignin was observed in these transgenic lines nor was an irregular xylem cell phenotype observed in secondary *Populus* wood as has been reported in the *Arabidopsis irx8* mutant. There was, however, a significant increase in phloem fiber cell length and width and in xylem fiber and vessel cell size in *PdGAUT12.1* knockdown (*PdGAUT12.1*-KD) wood. *PdGAUT12.1*-KD lines also exhibited larger leaf area, larger palisade, and spongy parenchyma cell size and greater relative water content compared to controls.

Results and discussion

GAUT12 is highly expressed in xylem in *P. deltoides*

The *GAUT1*-related gene family consists of 15 GAUTs in *Arabidopsis* and 20 GAUTs in *P. trichocarpa* [46], suggesting that one or two poplar orthologs exist for each *Arabidopsis GAUT* gene. The *Arabidopsis* GAUT family falls into three broad clades in which clade A is subdivided into clades A1-4, clade B into clades B 1-2, and clade C is undivided [46]. A phylogenetic analysis of the 15 *Arabidopsis* GAUT proteins (TAIR10) and the 20 *P. trichocarpa* GAUTs (Phytozome 8.0) using MEGA5 [47] confirmed the family structure as previously described [46]. There are two *Populus* orthologs of *Arabidopsis* GAUT12 (IRX8) in clade C (Figure 1) [46]. For the purposes of this paper, we named them GAUT12.1 (encoded by POPTR_0001s44250, Genemodel V2.0, Phytozome 8.0; *Potri.001G416800*, Genemodel V3.0, Phytozome 10.0) and GAUT12.2 (encoded by POPTR_0011s13600, Genemodel V2.0, Phytozome 8.0; *Potri.011G132600*, Genemodel V3.0, Phytozome 10.0). PtGAUT12.1 has 82% amino acid identity (Figure 2A) and 75% nucleotide sequence identity to AtGAUT12. The two *Populus* GAUT12 proteins, PtGAUT12.1 and PtGAUT12.2, share 91% amino acid sequence identity and 90% nucleotide sequence identity with each other.

Quantitative real-time (RT)-PCR was used to measure the transcript level of *PdGAUT12.1* and *PdGAUT12.2* in *P. deltoides*. Both genes were most highly expressed in differentiating xylem; however, *PdGAUT12.1* expression was approximately eight times greater than *PdGAUT12.2* (Figure 2B,C) [37,38]. *PdGAUT12.1* transcript expression in leaf, stem, and petiole was lower than in stem cambium in *P. deltoides* (Figure 2B). No *PdGAUT12.2* transcript was detected in leaf and there was low expression in the petiole and stem (Figure 2C). Due to the abundant expression of *PdGAUT12.1* transcript in differentiating *Populus* xylem, we characterized the effect of reduced expression of this gene on plant and wood development and sugar release. Several prior reports described aspects of the function of this gene during wood formation in *Populus* [37,39]. However, this is the first report of a knockdown (KD) targeted against the *PdGAUT12.1* gene (*PdGAUT12.1*-KD) alone and of the resulting phenotypes.

Downregulation of the expression of GAUT12.1 in *P. deltoides* by RNAi-silencing

An RNAi construct (Figure 3B) containing *PtGAUT12.1*-specific sequence to trigger RNAi-silencing of *PdGAUT12.1*, but not *PdGAUT12.2* (Figure 3A), was generated to study the function of GAUT12 during wood development. The construct was transferred into eastern cottonwood (*P. deltoides*) clone WV94. Integration of the *PdGAUT12.1* RNAi construct into the

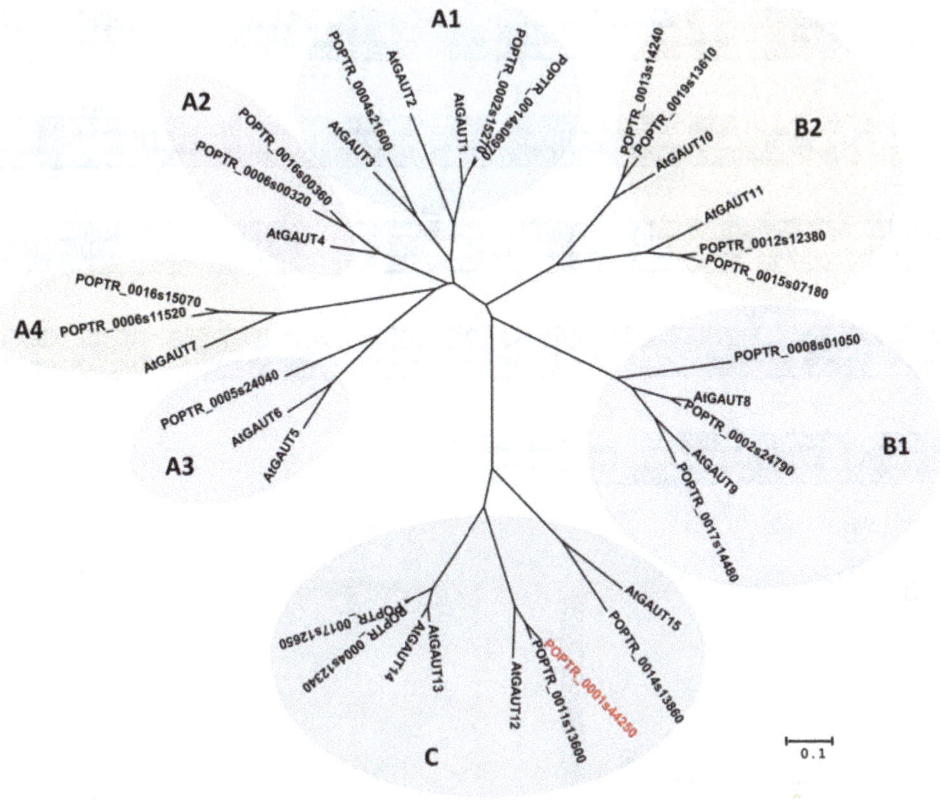

Figure 1 Phylogenetic tree of GAUTs in *Arabidopsis thaliana* and *Populus trichocarpa*. Phylogenetic tree showing the relationship between amino acid sequences of the GAUT Protein Family of *Arabidopsis thaliana* (TAIR10) and *Populus trichocarpa* (Phytozome 8.0). The tree was constructed by the Neighbor-Joining method using MEGA5 [44]. POPTR_0001s44250 is named *GAUT12.1* in this paper.

genome was confirmed by gene-specific PCR of leaf DNA from the 11 *PdGAUT12.1*-KD transgenic lines obtained, AB30.1 through AB30.11 (data not shown). The WT non-transformed cottonwood (*P. deltoides*) clone WV94 and eight plants transformed with the vector served as controls. Vector control lines were verified by PCR to contain the control construct (data not shown). Transgenic plants were produced by regeneration of shoots from transformed callus grown on shoot elongation medium, transfer of 3- to 5-cm-tall shoots from callus to BTM medium for rooting [48], and propagation of rooted seedlings to amplify the clonal lines. WT WV94 (25 plants), 10 to 15 clones of each vector control line, and all 11 *PdGAUT12.1*-KD transgenic lines were analyzed for this study.

Transcript expression of *PdGAUT12.1* was measured by quantitative RT-PCR. The transcript level of *PdGAUT12.1* was reduced by 50 to 67% in *GAUT12.1*-KD lines compared to the controls (Figure 3C). Two patterns of *GAUT12.1* transcript reduction were observed. Transgenic lines AB30.1, AB30.2, AB30.3, AB30.5, AB30.8, AB30.10, and AB30.11 had 60 to 67% reduced *GAUT12.1* transcript levels and lines AB30.4, AB30.6, AB30.7, and AB30.9 were reduced by 50 to 54%. We also

quantified the transcript level of *GAUT12.2* to determine whether the sequence-specific RNAi construct for *PdGAUT12.1* had any effect on *PdGAUT12.2* transcript expression in the lines. No significant change in *PdGAUT12.2* transcript level was observed in any of the transgenic lines in relation to control plants (Figure 3C).

RNAi downregulation of *P. deltoides* GAUT12.1 expression leads to improved saccharification without changing the amount of total lignin

To determine the effect of downregulation of *PdGAUT12.1* on the recalcitrance of *Populus* to enzymatic deconstruction, 9-month-old WT, vector control, and *PdGAUT12.1*-KD transgenic lines were subjected to pretreatment and enzymatic hydrolysis. Glucose release per gram dry biomass was significantly increased by 4 to 8% across ten of the eleven *PdGAUT12.1*-KD lines in comparison to the controls (Figure 4A). Xylose release per gram dry biomass was increased 3 to 7% in ten of the eleven transgenic lines (Figure 4B). Total sugar release in all 11 *PdGAUT12.1*-KD lines increased 4 to 6% compared to WT and vector controls (Figure 4C).

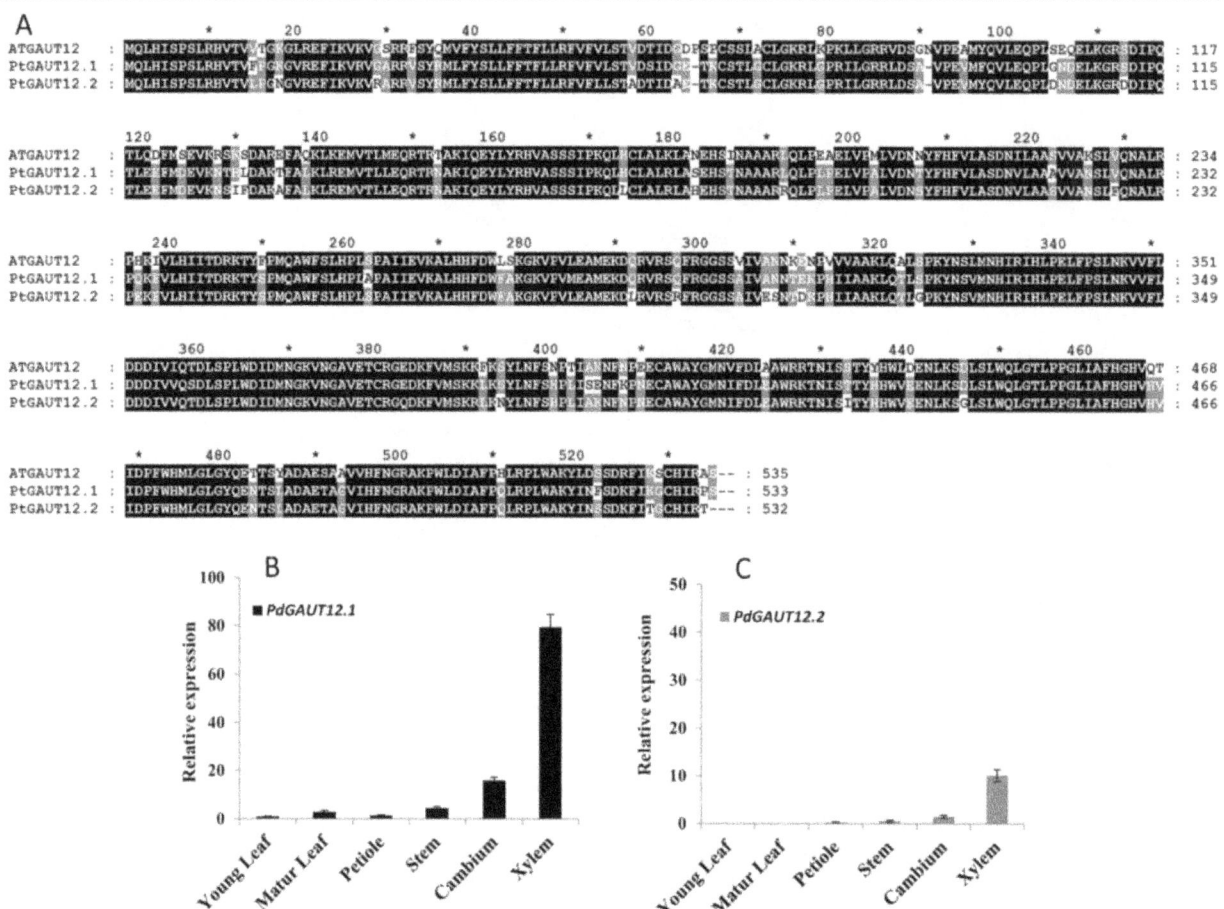

Figure 2 Sequence alignment of *Populus deltoides* and Arabidopsis *GAUT12* homologs and expression of *Populus GAUT12* genes. (A) Amino acid sequence alignment between *Arabidopsis* GAUT12 and *Populus* GAUT12. Numbers at the right of each sequence are the positions of amino acid residues in the corresponding proteins. Identical amino acid residues across all three genes are shaded with black and identical residues between two genes are shaded with gray. (B) *PdGAUT12.1* and (C) *PdGAUT12.2* expression in different tissues of *P. deltoides* as analyzed by quantitative Real-Time PCR. A *Populus 18S ribosomal RNA* (*18S rRNA*) gene was used as internal standard for normalization and the expression of *PdGAUT12.1* in young leaf was set to 1. Data are means of three biological replicates ± SE.

Arabidopsis irx8/gaut12 mutants have been reported to have reduced levels of lignin [30,49]. To determine if the reduced recalcitrance in the *Populus PdGAUT12.1*-KD lines was associated with reduced lignin content, total lignin and lignin S/G levels were measured by pyrolysis molecular beam mass spectrometry. There was no significant difference in the total lignin content of the *PdGAUT12.1*-KD (26.25% dry biomass weight) compared to WT (26.13%) and vector controls (26.19%) (Table 1). This is in contrast to the report on double RNAi-silenced *PtrGT8D1* and *PtGT8D2 P. trichocarpa* lines which had 11 to 25% increased lignin content [37].

All 11 *PdGAUT12.1*-KD lines analyzed showed a significant increase in lignin S/G ratio levels (Table 1), similar to the *Arabidopsis irx8* mutant [30]. There was a positive correlation between S/G ratio and glucose, xylose, and total sugar release (R^2 0.77, 0.81, and 0.81, respectively). No correlation was observed between

sugar release and total lignin content in *PdGAUT12.1*-KD lines. The results indicate that knockdown expression of *GAUT12.1* in *P. deltoides* results in increased sugar release and lignin S/G ratios.

A small reduction in S lignin was previously reported in *P. alba x tremula PoGT8D* RNAi lines [39]. However, in the present study with *P. deltoides* the S/G ratio was significantly increased (3 to 7%) in the *PdGAUT12.1*-KD lines compared to WT and vector control plants (Table 1). Total sugar release was previously shown to increase with increasing S/G ratio [50]. However, in that study, glucose and total sugar release were negatively correlated with lignin content. In the present study, although total lignin levels did not differ between the control and transgenic lines, there was still a positive correlation between high S/G ratios and increased sugar release in the *PdGAUT12.1*-KD lines, suggesting that *GAUT12.1* expression affects the lignin composition in

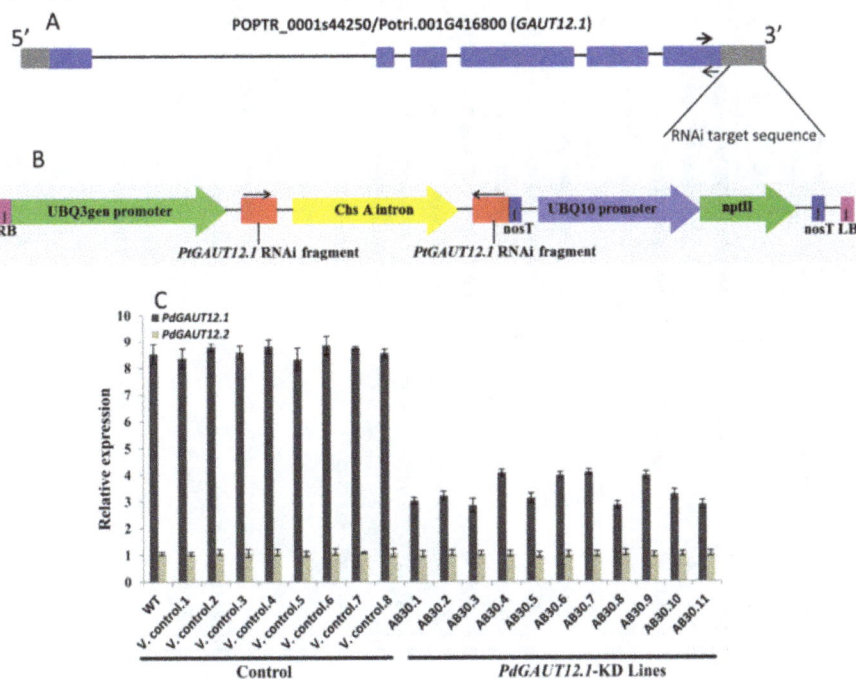

Figure 3 Knockdown of *PdGAUT12.1* in *Populus deltoides*. **(A)** Positions of the RNAi target sequence in the *GAUT12.1* (*POPTR_0001s44250/ Potri.001G416800*) gene. Blue boxes indicate exons, lines indicate introns, and gray boxes are the 5′ and 3′ untranslated regions (UTRs). The RNAi target sequence was 168 bp in the 3′UTR. The arrows indicate the targeted sequences used for quantitative real-time PCR. **(B)** Schematic presentation of *PdGAUT12.1* RNAi silencing construct. **(C)** Quantitative real-time PCR analysis of controls and *PdGAUT12.1*-KD lines of 3-month-old plants. The relative expression level of each gene was normalized using *Populus 18S ribosomal RNA* (*18S rRNA*) as the reference gene and the expression of *PdGAUT12.2* in AB30.5 was set to 1. *N* = 5. Error bars represent SE.

wood and that the composition of the lignin is a factor affecting recalcitrance. It is also noteworthy that, contrary to the results obtained here, simultaneous knockdown expression of both *Populus GAUT12* homologs in *P. trichocarpa* resulted in 11 to 25% increased lignin content [37]. As there was no measurement of biomass recalcitrance in that report, it is not known how the transgenic plants were affected in sugar release. Other studies, however, have shown that increased lignin content is detrimental to effective sugar release [50-52]. The knockdown of both *GAUT12* homologs in *P. trichocarpa* transgenics also resulted in more brittle wood, 17 to 29% reduced stem modulus of elasticity and 16 to 23% reduced modulus of rupture [37], suggesting that reduced expression of both homologs may be deleterious to wood quality.

RNAi downregulation of *PdGAUT12.1* results in increased plant growth in Poplar

Substantial morphological differences were observed between the controls and *PdGAUT12.1*-KD transgenic lines. Measurement of plant height and stem diameter at 3 months showed that *PdGAUT12.1* RNAi-silenced transgenic lines had increased vegetative growth compared to controls (Figure 5A). Nine of the eleven *PdGAUT12.1*-KD transgenic lines had 12 to 52%

increased plant height compared to WT and vector control plants (Table 2), and nine lines had 12 to 44% greater stem diameter (Table 2). We selected four transgenic lines (AB30.1, AB30.3, AB30.8, AB30.11) for further phenotype analysis based on GAUT12.1 transcript expression and growth phenotype, lignin S/G ratio, and the saccharification data. Measurement of plant height and stem diameter over a 9-month growth period showed that the increased height and stem diameter of the *PdGAUT12.1*-KD transgenic lines continued throughout the growth period (Figure 5B,C). Indeed, the total aerial biomass of the *PdGAUT12.1*-KD plants was 17% to 38% greater than controls (Figure 5D) by the end of the 9-month growth period. The increase in plant height and stem diameter were negatively correlated (R^2 of 0.66 and 0.75, respectively) with *GAUT12.1* transcript levels (Additional file 1), which were reduced by 50 to 67% compared to WT and vector control plants. These results suggest that systematic manipulation of the *GAUT12.1* gene can lead to greater plant growth.

The *PdGAUT12.1*-KD plants had increased vegetative growth compared to controls. This is in contrast to *Arabidopsis irx8* mutant plants which have stunted growth [8,10,29]. This difference may be due to the 50 to 67% reduction in *GAUT12.1* transcript level in the RNAi-

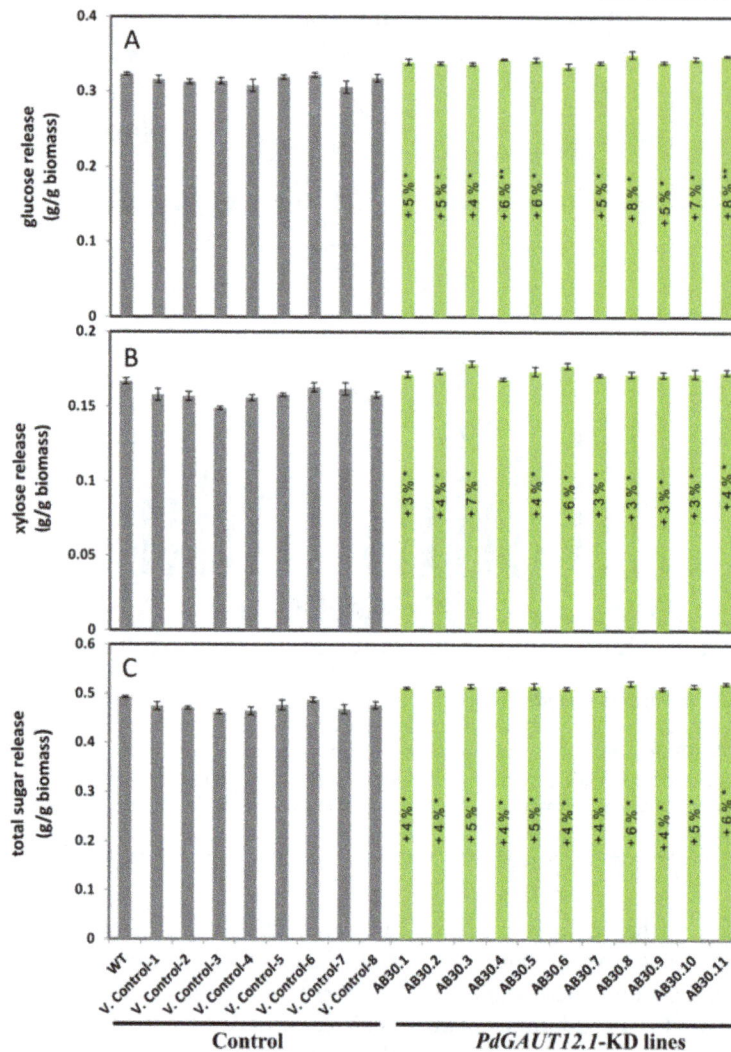

Figure 4 RNAi downregulation of *P. delotides* *PdGAUT12.1* expression leads to improved saccharification. Glucose (**A**), xylose (**B**) and total sugar (**C**) release from controls and *PdGAUT12.1*-KD lines. Significance *P* values are expressed as *$P < 0.05$, **$P < 0.001$ calculated by statistical analysis using Statistica 5.0 with one-way analysis of variance (ANOVA) followed by Tukey's multiple comparison test. $n = 25$ for wild-type *P. deltoides*, $n = 10$ to 15 for vector control and *PdGAUT12.1*-KD lines.

silenced *P. deltoides* transgenic plants compared to the total knock out of *GAUT12/IRX8* in the *Arabidopsis irx8* mutants [8,29]. However, the previously reported simultaneous knockdown expression of both *PtrGT8D1* and *PtGT8D2* in *P. trichocarpa*, similar to the work described here, also did not yield a dwarf phenotype in *Populus* even though those plants had 85 to 94% reduced *GAUT12* transcript levels [37]. The *PtrGT8D* RNAi-silenced transgenic plants [37] also did not have collapsed vessels as observed in the *Arabidopsis irx8* mutant [8,29]. Based on those results, it was proposed that growth of the *PtrGT8D* RNAi-silenced transgenic plants was not affected because the vessels in the transgenic lines retained their normal transport function [37]. The simultaneous knockdown expression of both

Populus GAUT12 homologs in *P. trichocarpa* [33], however, did not result in the increased growth observed in the current study where only *GAUT12.1* expression was reduced. The increased plant size obtained upon targeted downregulation of a specific *GAUT12* homolog as reported here suggests that a controlled manipulation of this gene to yield greater plant growth could be used to improve feedstock for the biofuel industry.

Altered *PdGAUT12.1* expression in leaves causes increased leaf cell size and number and increased relative water content

To determine how leaf growth and size was affected in the *PdGAUT12.1*-KD lines, every fifth successive leaf from the apex down to leaf 40 (Additional file 2C-D)

Table 1 Total lignin content and S/G ratio of controls and *PdGAUT12.1*-KD lines

Plant types	Lignin content (%)	S/G ratio
WT	26.13 ± 0.15	1.90 ± 0.02
V Control-1	25.82 ± 0.17	1.68 ± 0.01
V Control-2	26.68 ± 0.35	1.70 ± 0.03
V Control-3	26.00 ± 0.34	1.62 ± 0.02
V Control-4	26.45 ± 0.21	1.73 ± 0.03
V Control-5	26.43 ± 0.42	1.85 ± 0.03
V Control-6	26.24 ± 0.16	1.85 ± 0.04
V Control-7	25.61 ± 0.40	1.62 ± 0.07
V Control-8	26.30 ± 0.25	1.78 ± 0.05
AB30.1	25.53 ± 0.50	2.00 ± 0.03*
AB30.2	25.99 ± 0.39	1.96 ± 0.02*
AB30.3	26.37 ± 0.15	1.98 ± 0.02 *
AB30.4	26.61 ± 0.26	1.98 ± 0.03*
AB30.5	26.84 ± 0.21	1.97 ± 0.03*
AB30.6	26.11 ± 0.41	2.02 ± 0.04*
AB30.7	26.43 ± 0.38	1.97 ± 0.02*
AB30.8	25.78 ± 0.37	1.97 ± 0.02*
AB30.9	26.44 ± 0.63	2.04 ± 0.02*
AB30.10	26.38 ± 0.24	1.98 ± 0.02*
AB30.11	26.28 ± 0.17	2.03 ± 0.02*

Lignin content is expressed as percentage (wt/wt). S/G ratios were determined by summing the intensity of the syringyl peaks at 154, 167, 168, 182, 194, 208, and 210 and dividing by the sum of intensity of guaiacyl peaks 124, 137, 138, 150, 164, and 178 (see methods for details, Additional file 9). Mean ± SE, Significant P values are expressed as *$P < 0.05$, **$P < 0.001$ as calculated by statistical analysis using Statistica 5.0 with one-way analysis of variance (ANOVA) followed by Tukey's multiple comparison test. $n = 25$ for wild-type *P. deltoides*, $n = 10$ to 15 for vector control (v control-1 to v control-8) and *PdGAUT12.1*-KD lines (AB30.1-AB30.11).

was examined. Both leaf length and width was significantly ($P = 0.05$) larger for each of the *PdGAUT12.1*-KD lines examined compared to the controls (Additional file 2C-D). To compare the growth of developing leaves, every fifth, sixth, and seventh leaf from the apex was collected (Additional file 2B, E) and leaf area was measured. All *PdGAUT12.1*-KD lines examined had 52 to 117% significantly greater leaf area compared to control plants (Additional file 2E). Simultaneously, we collected the 20th to 40th leaf from the apex of five transgenic and control lines and compared the fully expanded leaf area. All four transgenic lines had 22 to 76% greater leaf area than control plants (Additional file 2 F). Therefore, both the leaf length and width was significantly greater in both the developing and expanded leaves of *PdGAUT12.1*-KD lines resulting in a larger leaf area. The *PdGAUT12.1*-KD lines also had accelerated growth in the top part of the *PdGAUT12.1*-KD plant (Additional file 2A) and 2.8 times greater biomass yield compared to WT (Additional file 2H).

To establish whether the larger leaf size in the *PdGAUT12.1*-KD lines was a result of increased cell number or cell expansion, we analyzed cross sections of leaf blades of the sixth leaf from the apex of the plant and hand cut cross sections of the 20th leaf. The knockdown lines had increased numbers of palisade parenchyma cells (17%, AB30.1; 13%, AB30.8) compared to WT (Additional file 3A-C, M). The knockdown lines also had larger bundle-sheath cells (Additional file 3D-G, I, K). Although there were more palisade cells in the young leaves, by the time these leaves matured, there was no overall major change in cross sectional total leaf area occupied by palisade cells of *PdGAUT12.1*-KD lines compared to WT, except for a small 9% increase in line AB30.1 (Additional file 4A-F). As shown in Additional file 2; however, there was a significant increase in the total leaf surface area in the mature *PdGAUT12.1*-KD leaves compared to the control (Additional file 2C-F). Also, the palisade and spongy parenchyma cells in mature leaves of *PdGAUT12.1*-KD lines were 13 to 56% and 23 to 62% larger, respectively, than in the controls (Additional file 4G-H). Thus, the increased leaf size of the *PdGAUT12.1*-KD lines is due to both increased cell number and cell size.

Leaf water content is a useful indicator of plant water balance and expresses the relative amount of water present in the plant tissues. The measurement of relative water content (RWC) in leaf tissues is commonly used to assess the water status of plants. Measurements of water content in a tissue on a fresh or dry mass basis can be correlated with the maximum amount of water a tissue can hold. Leaf water status is positively correlated with several leaf physiological variables, such as leaf turgor, transpiration, stomatal conductance, and growth [53]. We thus measured the RWC of *PdGAUT12.1*-KD lines to determine if it was correlated with the positive growth effect observed in the lines. All four knockdown lines examined had 5 to 10% increased RWC compared to the WT at 72 h (Additional file 2G). After 72 h, there was a 7% increase in RWC of *PdGAUT12.1*-KD plants compared to the WT (Additional file 2G). In summary, the results show that the increased leaf size in the *PdGAUT12.1*-KD lines could be explained by enhanced cell number, enhanced cell size, and enhanced RWC.

Downregulation of *PdGAUT12.1* in secondary cell wall-containing tissues leads to increased phloem and vessel size

To examine the role of downregulation of *GAUT12.1* in tissues containing secondary walls, stems from transgenic lines and WT were analyzed by sectioning internode numbers 10 and 20 from the top of the stem. A comparison of the internode 10 stem cross sections from *PdGAUT12.1*-KD lines versus WT revealed a significant

Figure 5 RNAi downregulation of *PdGAUT12.1* results in increased plant growth. Phenotype and development of *PdGAUT12.1*-KD transgenic lines. **(A)** Plant phenotypes of 3-month-old wild-type *P. deltoides* (left two plants), vector control (middle two plants), and *PdGAUT12.1*-KD lines (right two plants). **(B)** Height and **(C)** radial growth of *PdGAUT12.1*-KD lines in comparison to the controls. **(D)** Dry aerial biomass of WT and *PdGAUT12.1*-KD lines of three-month-old plants. $n = 7$ for plant height and stem diameter, $n = 6$ for dry weight measurement. Asterisks indicate values significantly different from WT by ANOVA followed by Tukey's multiple comparison test, $*P < 0.05$, $**P < 0.001$. Error bars represent SE.

increase in the *PdGAUT12.1*-KD lines of the amount of phloem tissue (63%, AB30.1; 69%, AB30.8) compared to WT (Figure 6A,B,D,E,G,H,S). We did not observe the collapsed xylem cell phenotype reported for the *irx8 Arabidopsis* mutants [8,29]. However, there was a significant increase in xylem vessel cell size in the *PdGAUT12.1*-KD lines compared to WT. The 10th internode from the transgenic lines had 40% and 31% larger vessel diameter than WT (Figure 6A,C,D,F,G,I,T). Our data corroborate similar vessel phenotypes previously reported in *P. trichocarpa* when both *PtGAUT12.1* and *PtGAUT12.2* were simultaneously knocked down through RNAi silencing [37]. We also observed an increased amount of phloem (50%, AB30.1; 55%, AB30.8) and increased xylem vessel cell lumen diameter (48%, AB30.1; 39%, AB30.8) in the 20th internodes from the transgenic lines compared to the WT (Figure 6J to R,U,V).

Since we observed increased stem diameter in the *PdGAUT12.1*-KD lines, we hypothesized that increased wood diameter was a result of an increased radial expansion of the xylem cells. To test this hypothesis, wood cells from internode 40 of the transgenic and WT plants were separated by maceration. The measurement of fiber cell size revealed that all knockdown lines had greater fiber cell length (19% to 26%) and two transgenic lines had wider fiber cells (6%, 8%) than the WT (Figure 7E,F). Vessel cell length and diameter were also measured in the WT and

PdGAUT12.1-KD lines (Figure 7D). All transgenic lines examined had significantly larger vessel cell total length (12% to 21%) and vessel cell lumen length (7% to 9%) and wider vessel cell diameter (12% to 29%) than WT vessels (Figure 7A to C). Thus, at least a portion of the change in stem length and diameter in *PdGAUT12.1*-KD lines is likely due to increased xylem and vessel cell size. The larger vessel size may also support the larger leaf size and greater water content of the *pdGAUT12*-KD lines.

Populus *PdGAUT12.1* complements and *PdGAUT12.2* partially complements the Arabidopsis *irx8* (*gaut12*) mutant
To examine whether *PdGAUT12.1* and *PdGAUT12.2* are functional homologs of the Arabidopsis *GAUT12*, we transformed T-DNA constructs harboring the full-length coding sequences of *Arabidopsis thaliana GAUT12* as well as *P. deltoides GAUT12.1* and *GAUT12.2* (Additional file 5) under the control of a CaMV 35S promoter into *irx8-5/+* plants. For each construct, approximately 12 primary transgenics homozygous for the *irx8* allele were selected and their phenotypes examined. Both the *AtGAUT12* and *PdGAUT12.1* constructs showed similar activity in complementing *irx8*: about 20% of the primary transgenics did not show any of the visible defects associated with the mutation; these plants produced WT-like rosette leaves and intact vessels (Figure 8A to C), resulting in significantly increased rosette size and overall height (Figure 8D,E). In

Table 2 Height and diameter of *PdGAUT12.1*-KD and control lines

Plant types	Plant height (inches)	Plant diameter (mm)
WT	48.31 ± 3.1	6.3 ± 0.24
V Control-1	46.67 ± 2.5	6.1 ± 0.19
V Control-2	47.18 ± 0.9	6.2 ± 0.13
V Control-3	47.95 ± 1.2	6.3 ± 0.26
V Control-4	47.62 ± 1.9	6.1 ± 0.31
V Control-5	48.43 ± 2.0	6.3 ± 0.37
V Control-6	48.51 ± 1.5	6.1 ± 0.36
V Control-7	48.81 ± 2.0	6.3 ± 0.22
V Control-8	48.81 ± 1.0	6.1 ± 0.23
AB30.1	74.17 ± 2.9**	8.2 ± 0.46**
AB30.2	61.81 ± 3.3**	7.7 ± 0.34**
AB30.3	60.29 ± 2.3**	8.7 ± 0.32**
AB30.4	49.22 ± 2.2	6.8 ± 0.29
AB30.5	60.86 ± 2.0**	9.1 ± 0.53**
AB30.6	54.61 ± 3.8**	6.6 ± 0.41
AB30.7	49.86 ± 2.4	7.1 ± 0.28**
AB30.8	64.75 ± 2.4**	7.9 ± 0.37**
AB30.9	54.82 ± 1.0**	7.2 ± 0.37**
AB30.10	61.72 ± 2.3**	8.4 ± 0.30**
AB30.11	68.07 ± 2.0**	8.5 ± 0.33**

Plant height and plant diameter of three-month-old wild type *P. deltoides*, vector control and *PdGAUT12.1*-KD lines. Means ± SE, Significance *P* values are expressed as **$P < 0.05$, **$P < 0.001$ as calculated by statistical analysis using Statistica 5.0 with one-way analysis of variance (ANOVA) followed by Tukey's multiple comparison test. $n = 25$ for wild type *P. deltoides*, $n = 10$ to 15 for vector control (v control-1 to v control-8) and *PdGAUT12.1*-KD lines (AB30.1-AB30.11).

contrast, the *PdGAUT12.2* construct only partially rescued the phenotypes associated with *irx8*, as primary transgenics had only marginally larger rosette leaves and slightly taller stature (Figure 8D,E). The *PdGAUT12.2* lines displaying the strongest rescue effect still had many collapsed xylem vessels (Figure 8C). We conclude that the Poplar *PdGAUT12.1* gene is as effective in complementing Arabidopsis *irx8* mutants as the native *GAUT12* gene.

RNAi downregulation of *GAUT12.1* expression leads to reductions in pectin and xylan content and increased extractability of *P. deltoides* wood

Sugar composition analyses of total cell walls from *PdGAUT12.1*-KD plants

We investigated the effects of RNAi downregulation of *P. deltoides GAUT12.1* expression on the content of cell wall polysaccharides in *P. deltoides* wood. Cell walls were isolated as alcohol-insoluble residues (AIRs) from cored, debarked wood from the bottom 6 cm of stems. All four *PdGAUT12.1*-KD lines had 17 to 30% significantly reduced xylose content in AIR compared to control lines (Table 3), indicating that RNAi downregulation of *P. deltoides*

GAUT12.1 expression causes a defect in xylan formation. This result corroborated two earlier findings in *Populus* that *PtGAUT12.1* has a role in xylan formation [37,39]. Simultaneously, the galacturonic acid (GalA) content was significantly reduced by 25% to 47% in all four transgenic lines compared to the WT. This result indicates that *PdGAUT12.1*-KD lines have defects in pectin formation in addition to the defect in xylan formation. A substantial increase in galactose and mannose content was also observed in *PdGAUT12.1*-KD walls compared to controls. The AIR composition data indicated a clear alteration in the composition of the *PdGAUT12.1*-KD walls compared to controls and indicated that at least two types of polymers, xylan and pectin, were affected.

Glycosyl residue composition analyses of fractionated cell walls from PdGAUT12.1-KD plants

The AIR (that is, cell walls) of control and transgenic lines were subjected to sequential extraction with increasingly harsh solvents to fractionate the walls to determine if cell wall changes caused by the *GAUT12.1*-KD could be identified in a specific subfraction of cell wall polymers. The total amount of material recovered in each sequential extract of the starting AIR for each of the six solvents was greater for the *PdGAUT12.1*-KD RNAi lines than for the controls. There was a significant 9 to 21% increase in the amount of material recovered in the ammonium oxalate extract, 7 to 15% increase in the carbonate extract, 17 to 34% increase in the 1 M KOH extract, 12 to 30% increase in the 4 M KOH extract, 8 to 10% increase in the chlorite extract, and 7 to 15% increase in the post-chlorite 4 M KOH extract, yielding an 11 to 17% overall increase in the total amount of extractable material from the starting AIR of the *PdGAUT12.1*-KD lines compared to WT (Additional file 6A-H). Since the total amount of AIR isolated from the WT versus the transgenic lines was comparable (Additional file 6A), these results suggested that the remaining insoluble wall material should have been reduced in the transgenic lines. Indeed, analysis of the mass of the final AIR pellet remaining after all extractions of AIR from the transgenic lines revealed that there was 6 to 13% less final pellet from the *PdGAUT12.1*-KD lines compared to control lines (Additional file 6I). These results suggest that GAUT12 functions in the synthesis of a structure that is highly integrated into the basic cell wall architecture, and that when GAUT12 function is reduced, the plants synthesize more easily extractable walls.

To determine if there were any differences in the type or mol% amounts of each type of sugar in the different extracts from the *PdGAUT12.1*-KD lines compared to controls, the individual extracts were analyzed by glycosyl residue composition. Wall fractions extracted with mild reagents such as ammonium oxalate and sodium carbonate generally contain large amounts of GalA along with rhamnose (Rha), arabinose (Ara), galactose (Gal), glucose (Glc), xylose (Xyl),

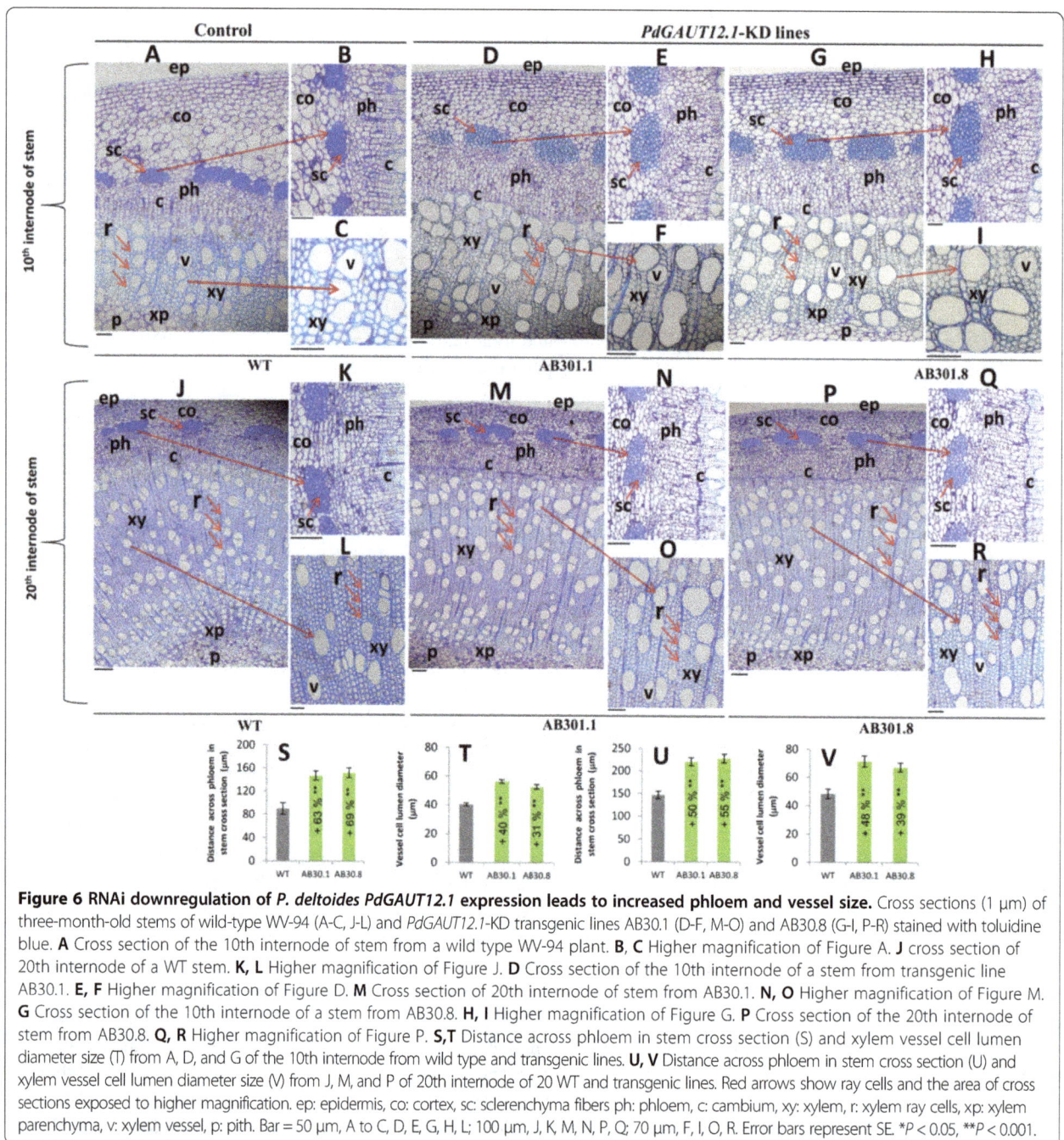

Figure 6 RNAi downregulation of *P. deltoides PdGAUT12.1* expression leads to increased phloem and vessel size. Cross sections (1 µm) of three-month-old stems of wild-type WV-94 (A-C, J-L) and *PdGAUT12.1*-KD transgenic lines AB30.1 (D-F, M-O) and AB30.8 (G-I, P-R) stained with toluidine blue. **A** Cross section of the 10th internode of stem from a wild type WV-94 plant. **B, C** Higher magnification of Figure A. **J** cross section of 20th internode of a WT stem. **K, L** Higher magnification of Figure J. **D** Cross section of the 10th internode of a stem from transgenic line AB30.1. **E, F** Higher magnification of Figure D. **M** Cross section of 20th internode of stem from AB30.1. **N, O** Higher magnification of Figure M. **G** Cross section of the 10th internode of a stem from AB30.8. **H, I** Higher magnification of Figure G. **P** Cross section of the 20th internode of stem from AB30.8. **Q, R** Higher magnification of Figure P. **S,T** Distance across phloem in stem cross section (S) and xylem vessel cell lumen diameter size (T) from A, D, and G of the 10th internode from wild type and transgenic lines. **U, V** Distance across phloem in stem cross section (U) and xylem vessel cell lumen diameter size (V) from J, M, and P of 20th internode of 20 WT and transgenic lines. Red arrows show ray cells and the area of cross sections exposed to higher magnification. ep: epidermis, co: cortex, sc: sclerenchyma fibers ph: phloem, c: cambium, xy: xylem, r: xylem ray cells, xp: xylem parenchyma, v: xylem vessel, p: pith. Bar = 50 µm, A to C, D, E, G, H, L; 100 µm, J, K, M, N, P, Q; 70 µm, F, I, O, R. Error bars represent SE. *$P < 0.05$, **$P < 0.001$.

and mannose (Man) and, thus, are considered to be enriched in pectin. The oxalate extracts of AIR from the *PdGAUT12.1*-KD RNAi line had 39 to 78% decreased mol % GalA compared to the controls, and the carbonate extracts had 17 to 42% decreased mol% GalA (Table 4). The mol% rhamnose in both the oxalate and carbonate fractions was also significantly reduced by 44 to 76% and 39 to 55%, respectively, in the *PdGAUT12.1*-KD lines compared to the controls (Table 4). These results show that *GAUT12.1* downregulation results in reduced amounts of easily extractable HG and RG. Simultaneously, the xylose, mannose,

galactose, and glucose contents in the oxalate extracts were significantly increased in the *PdGAUT12.1*-KD lines compared to the WT (Table 4), while in the carbonate fractions, the glucose content was increased, suggesting that reduced levels of easily extractable pectin were accompanied by increased mol% hemicellulose in these wall fractions.

As expected, extraction of the remaining insoluble wall residues with harsher alkaline solvents such as 1 M and 4 M KOH (conditions which typically enrich for hemicelluloses) released primarily xylose with lesser amounts of glucose, mannose, GlcA, and galactose as well as

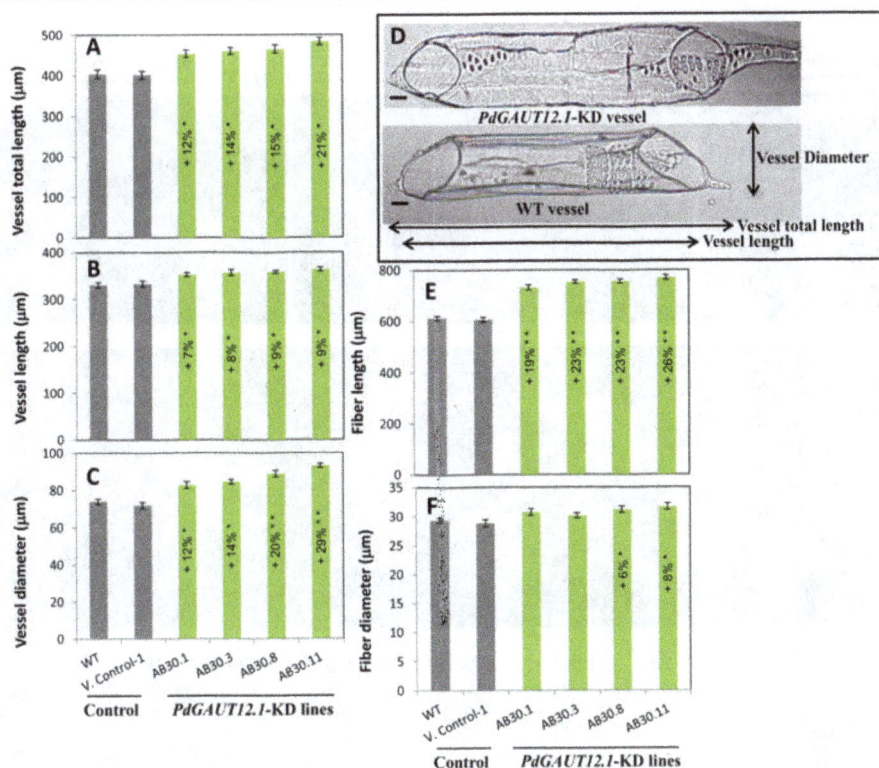

Figure 7 Xylem (fiber and vessel) cell size in WT and *PdGAUT12.1*-KD line. Downregulation of expression of *GAUT12.1* results in increased xylem total vessel length **(A)**, xylem lumen length **(B)**, xylem vessel lumen diameter **(C)**, xylem fiber cell length **(E)**, xylem fiber cell diameter **(F)**. Both xylem fibers and vessels were separated by maceration and approximately 80 fiber cells and 60 vessel elements from three independent *PdGAUT12.1*-KD transgenic lines were analyzed. Images were captured with a Nikon DS-Ri1 camera (Nikon, Melville, NY) using NIS-Elements Basic Research software. **(D)** Representative individual vessels of WT and *PdGAUT12.1*-KD lines illustrating the parameters measured. Asterisks indicate values significantly different from wild type based on ANOVA followed by Tukey's multiple comparison test, $*P < 0.05$, $**P < 0.001$. $n = 240$ fibers and 180 vessels cells. The vertical bar represents standard error. Bar = 20 μm, D.

small amounts of rhamnose, GalA, and arabinonse. The *PdGAUT12.1*-KD 1 M KOH extract contained significantly reduced (35 to 45%) mol% xylose compared to the controls (Table 4). There was also a 35 to 45% reduction in the mol% GlcA and a 60 to 76% reduction in the mol% GalA in these extracts from the *PdGAUT12.1*-KD lines. Similarly, we found a 31 to 41% reduction in xylose in the 4 M KOH extracts from *PdGAUT12.1*-KD RNAi lines. There was also a noticeable increase in mannose, galactose and glucose content in both the 1 M and 4 M KOH extracts. The reduced 35 to 45% mol% xylose and GlcA in the transgenic *Populus* plants (Table 4) agrees with the reduced xylan content in *Arabidopsis irx8/gaut12* mutant plants [8,29]. The even larger reduction (60 to 76%) of GalA in the 1 M KOH extracts of the *PdGAUT12.1*-KD RNAi *Populus* lines suggests that reduced expression of this *GAUT12.1* homolog may also impact a type of pectin that is released from the walls by this extraction.

More tightly bound lignin-associated components in the AIR can be removed by extraction with 100 mM sodium chlorite. There was a 34 to 47% and 24 to 38% reduction of xylose and GalA, respectively, in the chlorite extract from the *PdGAUT12.1*-KD lines compared to the WT controls (Table 4). Similarly, a substantial decrease in arabinose (23 to 39%) and rhamnose (14 to 37%) was observed in this extract from the transgenic lines compared with controls.

After removal of lignin, the residual pellet following the chlorite extractions was treated with a second treatment of 4 M KOH to solubilize additional carbohydrates. Treatment with 4 M KOH after chlorite released primarily xylose. There was a 27 to 45% reduction in the amount of xylose in the post chlorite 4 M KOH extracts from *PdGAUT12.1*-KD lines compared to the controls (Table 4). Similarly, the post chlorite 4 M KOH extract from the *PdGAUT12.1*-KD lines had 27 to 45% reduced amounts of GlcA. No major change in GalA content was observed in this fraction (Table 4).

Taken together, the results indicate that downregulation of *GAUT12.1* in *Populus* wood leads to reductions in a population of pectin and xylan. To compare the mass yield of each sugar from the starting AIR in the *PdGAUT12.1*-KD versus WT cell walls, the data were analyzed as micrograms sugar/milligram AIR. Analyses

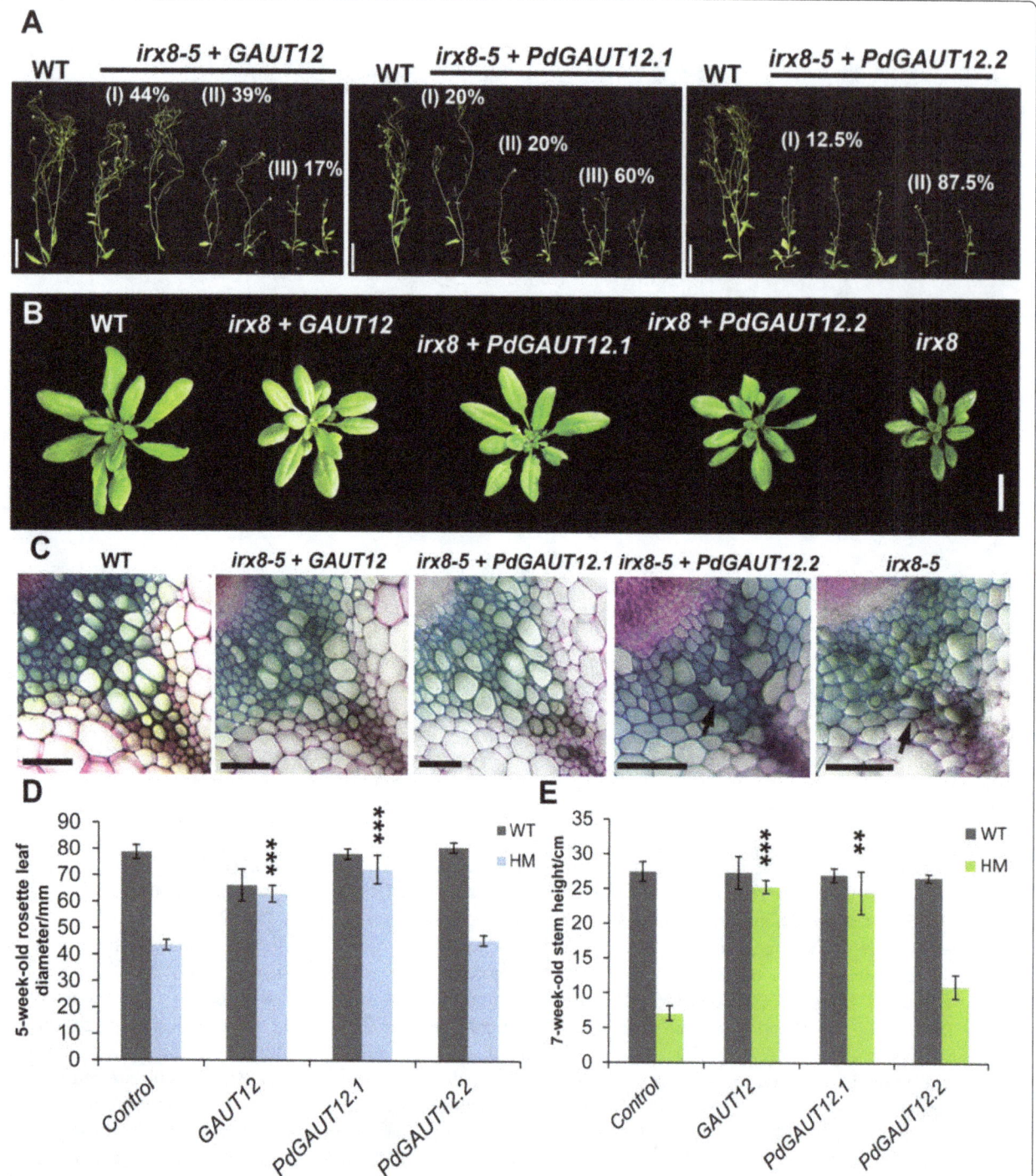

Figure 8 Characterization of *irx8* mutants complemented by CaMV35S driven *GAUT12*, *PdGAUT12.1*, and *PdGAUT12.2*. (A) Pictures of 7-week-old wild-type (WT), and *irx8* + *GAUT12*, *irx8* + *PdGAUT12.1*, and *irx8* + *PdGAUT12.2* complemented lines. The complemented lines were classified into groups according to the phenotypes and labeled by percentage. Bar = 5 cm. **(B)** Images of 5-week-old rosette leaves of WT, *irx8* + *GAUT12*, *irx8* + *PdGAUT12.1*, *irx8* + *PdGAUT12.2*, and *irx8*. Bar = 2 cm. **(C)** Hand-cut transverse sections of the basal stems stained with Toluidine blue O, arrows indicate collapsed xylem vessels. Bar = 50 μm. Measurements of **(D)** 5-week-old rosette leaf diameter and **(E)** 7-week-old stem height. Value = mean ± standard error ($n \geq 3$), Value = mean ± standard deviation (n = 3), ***$P < 0.001$, significant increase of rosette leaf size or stem height compared to the *irx8* mutant determined by one-way ANOVA and *post hoc* Bonferroni corrected *t*-test.

Table 3 Monosaccharide sugar composition of total cell walls (AIR) from nine-month-old *Populus* plants

	Sugar composition of wood AIR stems walls (mol%; ± SE standard error)								
	Ara	Rha	Fuc	Xyl	GlcA	GalA	Man	Gal	Glc
WT	3.11 ± 0.11	2.01 ± 0.05	0.08 ± 0.01	46.27 ± 1.02	0.23 ± 0.01	6.31 ± 0.13	6.29 ± 0.41	3.07 ± 0.32	33.21 ± 0.85
V Control-1	3.00 ± 0.12	1.87 ± 0.11	0.09 ± 0.01	47.01 ± 1.11	0.22 ± 0.02	6.35 ± 0.11	6.61 ± 0.42	3.11 ± 0.47	32.25 ± 0.99
AB30.1	3.01 ± 0.09	2.31 ± 0.12	0.10 ± 0.01	*38.33 ± 0.56**	0.21 ± 0.02	*4.70 ± 0.14**	*10.09 ± 0.36**	*7.32 ± 0.33**	33.82 ± 0.76
AB30.3	2.95 ± 0.08	2.01 ± 0.13	0.06 ± 0.01	*36.14 ± 1.01**	0.21 ± 0.01	*4.30 ± 0.09**	*11.58 ± 0.38**	*8.53 ± 0.53**	34.19 ± 1.01
AB30.8	2.76 ± 0.01	1.76 ± 0.21	0.04 ± 0.01	*32.30 ± 0.85***	0.20 ± 0.01	*3.34 ± 0.08***	*14.59 ± 0.53**	*10.99 ± 0.58**	33.98 ± 0.89
AB30.11	2.86 ± 0.08	1.85 ± 0.07	0.17 ± 0.02	*34.44 ± 1.12***	0.19 ± 0.01	*3.79 ± 0.11***	*11.72 ± 0.39**	*10.95 ± 0.61**	33.96 ± 0.97

Glycosyl residue composition of stem AIR as measured by GC-MS of tetramethylsilane (TMS) derivatives from controls and *PdGAUT12.1*-KD lines. The amount of sugar is represented as average mol% of AIR (alcohol insoluble residue) isolated from stem walls. Means ± SE of three biological replicates with two technical replicates each. Ital values with asterisks represent significant difference between WT and *PdGAUT12.1*-KD lines at *$^*P \leq 0.05$, *$^{**}P \leq 0.001$ significant level (one-way ANOVA followed by Tukey's multiple comparison test).

of the total amount of each sugar in each extract of the starting cell wall (that is, AIR) (Additional file 7) show that only the GalA and rhamnose (Rha) were decreased in each *PdGAUT12.1*-KD cell wall extract compared to WT. These results could be consistent with HG and/or RG-I as the glycan directly affected in the *PdGAUT12.1*-KD lines. However, significant reductions in Ara, Xyl, and GlcA were also observed, with the amount of Ara decreased in five of six extracts and Xyl and GlcA decreased in the four fractions released under the harshest extraction conditions, fractions known to be enriched for hemicelluloses. Thus, the data are also consistent with a role for GAUT12 in xylan synthesis.

Glycosyl linkage analyses of fractionated cell walls from PdGAUT12.1-KD RNAi plants

Glycosyl residue linkage analyses were carried out to test if the reductions in xylose, GalA, and Rha were associated with xylan, HG, and RG-I, respectively. Specifically, linkage analyses were performed on the ammonium oxalate, sodium carbonate, and 1 M KOH extracts from *PdGAUT12.1*-KD lines AB30.1 and AB30.8 and compared with WT *Populus* (Table 5). The *PdGAUT12.1*-KD plants had 42 to 82% reduced levels of 4-GalAp and 55 to 64% reduced amounts of terminal-GalAp in the oxalate fractions compared with WT. These same *PdGAUT12.1*-KD fractions, compared to WT, had a 29 to 43% reduction in 2-linked Rhap, a linkage found in RG-I. Since the mole percent of 4-linked GalA to 2-linked Rha was approximately 7 (Table 5), taken together, these linkage data indicate that the *PdGAUT12.1*-KD oxalate fraction extracts had reduced amounts of HG and lesser reductions in RG-I. There was also an increase in 4-Xylp, T-Manp, 4-Manp, 4-Galp, T-Glcp, and 4-Glcp contents in *PdGAUT12.1*-KD plants compared with WT.

In the carbonate fractions, we observed 21 to 44% reduced amounts of 4-GalAp and 30 to 50% reduced levels of terminal-GalAp in the *PdGAUT12.1*-KD lines compared to WT (Table 5). The 2-linked Rhap content was

also reduced by 32 to 48% in the carbonate extracts of *PdGAUT12.1*-KD lines compared to WT, although again, the 2-linked Rha was only 8 mole% of the GalA content. There was a corresponding increase in the amount of 4-Xylp and 4-Glcp in the *PdGAUT12.1*-KD carbonate fraction. These data are consistent with the hypothesis that the *Populus GAUT12.1* gene encodes a glycosyltransferase that makes a subfraction of HG.

A significant reduction of 34 to 46% in 4-Xylp content in the 1 M KOH extract of the *PdGAUT12.1*-KD line corroborates the results from the glycosyl residue composition analysis (Table 5) and indicates that the 1 M KOH fraction of the *PdGAUT12.1*-KD plants had reduced amounts of xylan (4-Xylp). A 73 to 93% reduction in 4-GalAp content in the KD lines indicates a reduction in the amount of HG compared to WT. Also, a 50 to 67% reduction of 2-GalAp and a 44 to 50% reduction in 3-Rhap suggests a simultaneous reduction in the xylan reducing end sequence [39]. The reduced content of 4-Xylp in this 1 M KOH extract along with reduced amounts of 2-GalAp and 3-Rhap would be consistent with a role for GAUT12 in the addition of GalA into xylan reducing end oligosaccharide sequence in *Populus*. The greater mole percent reduction in 4-GalAp compared to 2-GalAp would be consistent with a role for GAUT12 in synthesizing a GalA-containing moiety, possibly HG, required for the formation of xylan containing the reducing end oligosaccharide or linking the xylan to another structure in the wall. These overall results show that GAUT12 knockdown leads to defects in pectin and xylan production in *Populus*.

Glycome profiling of fractionated cell walls from PdGAUT12.1-KD plants

In an effort to better understand the specific defects in xylan and pectin, glycome profiling of sequential extracts from WT and *PdGAUT12.1*-KD walls was performed. Glycome profiling is an ELISA-based technique in which monoclonal antibodies directed against distinct cell wall carbohydrate epitopes are used to identify differences in

Table 4 Glycosyl residue composition of cell wall (AIR) fractions from stem of controls and *PdGAUT12.1-KD* plants

Sugar composition of cell wall fractions (mol %; ± SE standard error)

	Ara	Rha	Fuc	Xyl	GlcA	GalA	Man	Gal	Glc
Ammonium oxalate									
WT	20.9 ± 1.0	4.1 ± 0.2	0.7 ± 0.02	8.9 ± 0.2	0.6 ± 0.01	23.1 ± 1.0	11.4 ± 0.9	3.7 ± 0.3	26.8 ± 1.2
V Control-1	21.0 ± 0.6	4.9 ± 0.3	0.8 ± 0.02	8.3 ± 0.4	0.5 ± 0.02	23.2 ± 1.1	11.6 ± 1.0	3.6 ± 0.2	26.9 ± 1.0
AB30.1	18.5 ± 0.4	1.0 ± 0.3**	0.2 ± 0.01	10.2 ± 0.5	0.7 ± 0.02	14.2 ± 1.3*	19.5 ± 1.1*	5.6 ± 0.3*	30.6 ± 0.9*
AB30.3	17.7 ± 0.6*	2.1 ± 0.2*	0.4 ± 0.01	10.5 ± 0.2*	0.7 ± 0.01	12.9 ± 0.9**	18.2 ± 1.3*	6.0 ± 0.1*	31.4 ± 1.0*
AB30.8	19.8 ± 0.8	2.1 ± 0.1*	0.4 ± 0.01	11.1 ± 0.3*	0.8 ± 0.01	5.0 ± 0.5**	21.9 ± 1.2*	6.2 ± 0.3*	32.8 ± 1.1*
AB30.11	19.3 ± 0.6	2.3 ± 0.2*	0.4 ± 0.02	11.4 ± 0.3*	0.8 ± 0.01	8.9 ± 0.8**	20.2 ± 1.3*	7.1 ± 0.2*	29.4 ± 1.2
Sodium carbonate									
WT	10.7 ± 0.6	6.0 ± 0.2	0.3 ± 0.02	23.0 ± 0.2	1.6 ± 0.01	19.3 ± 0.5	14.0 ± 0.9	7.0 ± 0.3	18.0 ± 0.9
V Control-1	11.1 ± 0.4	5.9 ± 0.1	0.5 ± 0.02	22.5 ± 0.6	1.5 ± 0.02	19.5 ± 0.8	15.0 ± 0.8	6.7 ± 0.2	17.9 ± 0.7
AB30.1	4.8 ± 0.4*	2.7 ± 0.2*	0.4 ± 0.01	24.2 ± 0.5	1.7 ± 0.02	16.0 ± 0.3*	18.0 ± 1.0	5.8 ± 0.3	26.9 ± 0.9*
AB30.3	7.3 ± 0.3	2.7 ± 0.1*	0.3 ± 0.01	24.7 ± 0.2	1.7 ± 0.01	14.0 ± 0.4*	17.2 ± 0.6	6.8 ± 0.3	25.0 ± 1.0*
AB30.8	7.9 ± 0.3	3.7 ± 0.2*	0.8 ± 0.01	26.2 ± 0.3	2.6 ± 0.01	11.2 ± 0.5*	19.3 ± 0.9	6.7 ± 0.4	21.9 ± 1.1
AB30.11	7.3 ± 0.4	3.4 ± 0.3*	0.4 ± 0.02	25.7 ± 0.3	1.9 ± 0.01	13.5 ± 0.4*	16.3 ± 0.7	7.4 ± 0.2	24.2 ± 1.2*
1 M KOH									
WT	0.5 ± 0.04	4.2 ± 0.2	0.2 ± 0.02	69.3 ± 1.9	8.5 ± 0.3	3.3 ± 0.1	6.4 ± 0.6	1.7 ± 0.2	7.4 ± 1.2
V Control-1	0.6 ± 0.03	4.1 ± 0.1	0.3 ± 0.01	70.1 ± 2.3	8.7 ± 0.2	3.4 ± 0.2	6.0 ± 0.5	1.6 ± 0.1	7.1 ± 1.4
AB30.1	1.1 ± 0.05	3.1 ± 0.2	0.4 ± 0.01	44.8 ± 2.3*	5.5 ± 0.2*	1.3 ± 0.09*	10.7 ± 0.5*	4.1 ± 0.3**	29.3 ± 2.1**
AB30.3	1.6 ± 0.06	3.1 ± 0.1	0.4 ± 0.01	42.4 ± 2.4*	5.3 ± 0.3*	1.2 ± 0.1*	11.1 ± 1.0*	3.7 ± 0.1**	31.9 ± 2.5**
AB30.8	1.8 ± 0.1	2.6 ± 0.1*	0.5 ± 0.01	37.8 ± 2.6**	4.7 ± 0.4*	0.8 ± 0.08**	14.6 ± 0.8*	4.8 ± 0.2**	33.2 ± 2.1**
AB30.11	1.6 ± 0.06	3.3 ± 0.2	0.4 ± 0.02	38.4 ± 3.2**	4.8 ± 0.3*	1.0 ± 0.1**	11.2 ± 0.4*	3.8 ± 0.2**	35.9 ± 2.4**
4 M KOH									
WT	2.3 ± 0.08	3.0 ± 0.1	0.5 ± 0.02	37.8 ± 1.3	4.7 ± 0.2	0.8 ± 0.04	11.9 ± 1.2	7.8 ± 0.3	31.2 ± 1.2
V Control-1	2.4 ± 0.06	3.2 ± 0.2	0.3 ± 0.01	38.1 ± 1.5	4.9 ± 0.2	0.9 ± 0.03	11.4 ± 0.5	7.9 ± 0.3	30.8 ± 1.5
AB30.1	1.6 ± 0.09	2.4 ± 0.1	0.4 ± 0.01	26.2 ± 1.7*	3.4 ± 0.2*	0.5 ± 0.03*	18.2 ± 0.5**	8.4 ± 0.3	38.9 ± 1.8*
AB30.3	2.1 ± 0.07	2.2 ± 0.2	0.3 ± 0.01	25.2 ± 1.1*	3.3 ± 0.3*	0.4 ± 0.02*	15.6 ± 1.0*	10.9 ± 0.5**	39.9 ± 1.5*
AB30.8	1.7 ± 0.05	1.9 ± 0.2*	0.3 ± 0.01	22.3 ± 1.3**	2.7 ± 0.4*	0.3 ± 0.03**	18.9 ± 0.8**	8.6 ± 0.6*	43.2 ± 1.3**
AB30.11	1.9 ± 0.09	2.3 ± 0.1	0.4 ± 0.02	23.2 ± 1.4**	2.9 ± 0.3*	0.3 ± 0.05**	15.9 ± 0.4*	8.1 ± 0.2	44.9 ± 1.9**
Chlorite									
WT	7.2 ± 0.5	4.3 ± 0.1	-	10.9 ± 0.5	0	6.4 ± 0.2	2.1 ± 0.1	11.2 ± 0.4	57.9 ± 1.1
V Control-1	7.0 ± 0.3	4.4 ± 0.1	0.2 ± 0.02	11.1 ± 0.7	0	6.5 ± 0.3	2.2 ± 0.09	10.9 ± 0.5	57.6 ± 1.0
AB30.1	5.5 ± 0.3*	2.7 ± 0.2*	-	7.2 ± 0.4*	0	4.9 ± 0.1*	3.0 ± 0.1**	13.6 ± 0.6*	63.1 ± 0.9*

Table 4 Glycosyl residue composition of cell wall (AIR) fractions from stem of controls and PdGAUT12.1-KD plants (Continued)

AB30.3	5.1 ± 0.3*	2.7 ± 0.3*	–	7.0 ± 0.5*	0	4.6 ± 0.2*	3.2 ± 0.08**	14.5 ± 0.3*	63.6 ± 1.2*
AB30.8	4.5 ± 0.4*	3.7 ± 0.2*	–	5.8 ± 0.6**	0	3.4 ± 0.3*	3.1 ± 0.09**	14.2 ± 0.4*	64.9 ± 1.4*
AB30.11	4.4 ± 0.3*	3.4 ± 0.1*	0.4 ± 0.01	6.2 ± 0.7*	0	4.3 ± 0.2*	3.6 ± 0.1**	14.1 ± 0.5*	63.7 ± 0.9*
4 M KOH PC									
WT	1.1 ± 0.05	2.5 ± 0.10	–	65.1 ± 2.1	7.9 ± 0.2	2.4 ± 0.06	6.2 ± 1.2	5.0 ± 0.8	10.1 ± 0.9
V Control-1	1.3 ± 0.06	2.4 ± 0.09	–	64.6 ± 2.3	7.8 ± 0.2	2.5 ± 0.07	6.3 ± 1.0	5.2 ± 0.6	10.3 ± 1.0
AB30.1	1.1 ± 0.04	2.4 ± 0.08	–	47.5 ± 1.9*	5.9 ± 0.2*	2.2 ± 0.06	13.2 ± 1.3*	9.1 ± 0.7*	18.8 ± 1.1**
AB30.3	1.0 ± 0.03	2.2 ± 0.06	–	42.3 ± 1.5*	5.2 ± 0.3*	2.1 ± 0.06	14.6 ± 1.4**	10.7 ± 0.9**	21.8 ± 1.2**
AB30.8	0.9 ± 0.05	2.1 ± 0.07	–	35.7 ± 1.8**	4.3 ± 0.2**	2.0 ± 0.08	17.9 ± 1.7**	12.4 ± 1.0**	24.6 ± 1.3**
AB30.11	1.0 ± 0.05	2.2 ± 0.09	–	37.8 ± 1.6**	4.7 ± 0.3**	2.0 ± 0.09	18.5 ± 0.8**	11.8 ± 0.5**	22.2 ± 0.9**

The amount of sugar is represented as average mol% in each wall extract. AIR (alcohol insoluble residue) was sequentially extracted using 50 mM ammonium oxalate, 50 mM Na_2CO_3, 1 M KOH, 100 mM sodium chlorite (chlorite) and 4 M KOH PC (post chlorite), hydrolyzed with methanolic HCl, and analyzed by GC. Data are mean ± SE of three biological and two technical replicates. Ital values represent significant difference between WT and PdGAUT12.1-KD plants at *$P \leq 0.05$, **$P \leq 0.001$ significance level (one-way ANOVA followed by Tukey's multiple comparison test). A dash designates that the amount of sugar was below detection limits.

Table 5 Glycosyl linkage analysis of fractionated cell wall fractions from wild type *P. deltoides* and *PdGAUT12.1*-KD lines

Fractions	Ammonium oxalate soluble			Na$_2$CO$_3$ soluble			1 M KOH soluble		
	WT	AB30.1	AB30.8	WT	AB30.1	AB30.8	WT	AB30.1	AB30.8
t-Ara*f*	3.6	1.5	3.3	3.9	1.2	2.6	0.2	0.6	0.9
t-Ara*p*	0.2	0.2	0.3	0.2	0.2	0.2	0.1	0.1	0.1
3-Ara*f*	0.5	0.5	1.0	0.6	0.3	0.5	-	0.2	0.2
3-Ara*p* or 5-Ara*f*	3.2	2.4	2.5	2.2	0.8	1.3	0.2	0.2	0.6
3,4-Ara*p* or 3,5-Ara*f*	1.5	1.3	2.1	-	-	-	-	-	-
t-Rha*p*	0.9	0.6	1.4	1.9	1.0	1.4	0.2	0.4	0.6
2-Rha*p*	2.1	1.2	1.5	3.1	1.6	2.1	0.7	0.5	0.4
3-Rha*p*	-	-	-	-	-	-	1.6	0.9	0.8
2,4-Rha*p*	1.1	1.0	1.4	0.8	0.7	0.8	0.5	1.0	1.0
t-Fuc*p*	0.3	0.2	0.4	0.6	0.4	0.8	0.2	0.5	0.5
t-Xyl*p*	0.4	0.4	0.9	1.6	1.3	1.8	0.9	0.6	0.8
4-Xyl*p*	6.5	9.0	12.5	20.5	21.8	22.1	59.3	38.9	32.3
2,4-Xyl*p*	0.9	1.3	1.5	1.5	1.6	1.6	9.2	5.3	4.1
t-Man*p*	1.2	1.2	2.2	1.8	1.8	2.0	-	-	-
4-Man*p*	17.8	22.9	19.2	19.6	20.1	20.9	6.0	11.0	15.0
4,6-Man*p*	0.7	0.9	0.5	-	-	-	-	-	-
t-GlcA*p*	-	-	2.0	1.6	1.7	2.3	8.5	6.8	3.7
2-GlcA*p*	0.1	0.2	0.2	0.2	0.2	0.2	-	-	-
t-GalA*p*	1.1	0.4	0.5	2.0	1.0	1.4	-	-	-
2-GalA*p*	-	-	0.2	0.9	0.7	1.0	1.8	0.9	0.6
4-GalA*p*	23.8	13.7	4.2	15.5	12.3	8.7	1.5	0.4	0.1
3,4-GalA*p*	-	-	-	0.4	-	0.7	-	-	-
2,4-GalA*p*	0.1	0.1	0.1	0.1	-	-	-	-	0.1
t-Gal*p*	2.9	2.1	2.7	3.2	3.1	3.0	1.2	2.9	3.1
2-Gal*p*	0.4	0.2	0.3	0.6	-	-	-	0.4	0.6
4-Gal*p*	-	-	3.3	-	-	2.8	-	-	0.1
6-Gal*p*	0.4	1.1	0.8	-	-	0.8	0.5	0.8	1.0
3,4-Gal*p*	-	1.2	1.0	0.3	-	-	-	-	-
2,4-Gal*p*	0.2	0.3	0.3	0.2	-	-	-	0.1	0.2
t-Glc*p*	3.3	3.4	4.3	3.6	2.9	3.2	-	2.6	2.8
2-Glc*p*	0.3	0.3	0.5	0.1	0.1	0.2	-	-	-
4-Glc*p*	26.2	31.6	28.3	12.9	25.5	17.7	7.4	24.4	29.8
4,6-Glc*p*	0.4	0.7	0.6	-	-	-	-	0.4	0.7

All numbers are mole percentages. A dash designates that the amount of sugar was below detection limits. Glycosyl linkage analyses were provided by the CCRC Analytical Services.

specific carbohydrate structures present in sequential cell wall extracts [54-56]. Glycome profiling using a panel of 155 monoclonal antibodies directed against diverse cell wall matrix polysaccharide epitopes (Additional file 8) revealed changes in the extractability of three types of wall polysaccharides (Figure 9). In general, a marginal increase in the amounts of 1 M KOH extractable material (as noted in the bar graphs above the heat maps) was observed for all *PdGAUT12.1*-KD lines, suggesting an enhanced extractability of some hemicelluloses in these lines compared to the WT. A marginally enhanced abundance of xylan epitopes (recognized by the xylan-6 and 7 mAb groups) was observed in carbonate extracts from two GAUT12-KD lines (AB30.8 and AB30.11), suggesting that a portion of cell wall xylan is more easily extracted in the GAUT12-KD lines. Concomitantly, these two lines exhibited reduced abundance of these xylan epitopes in

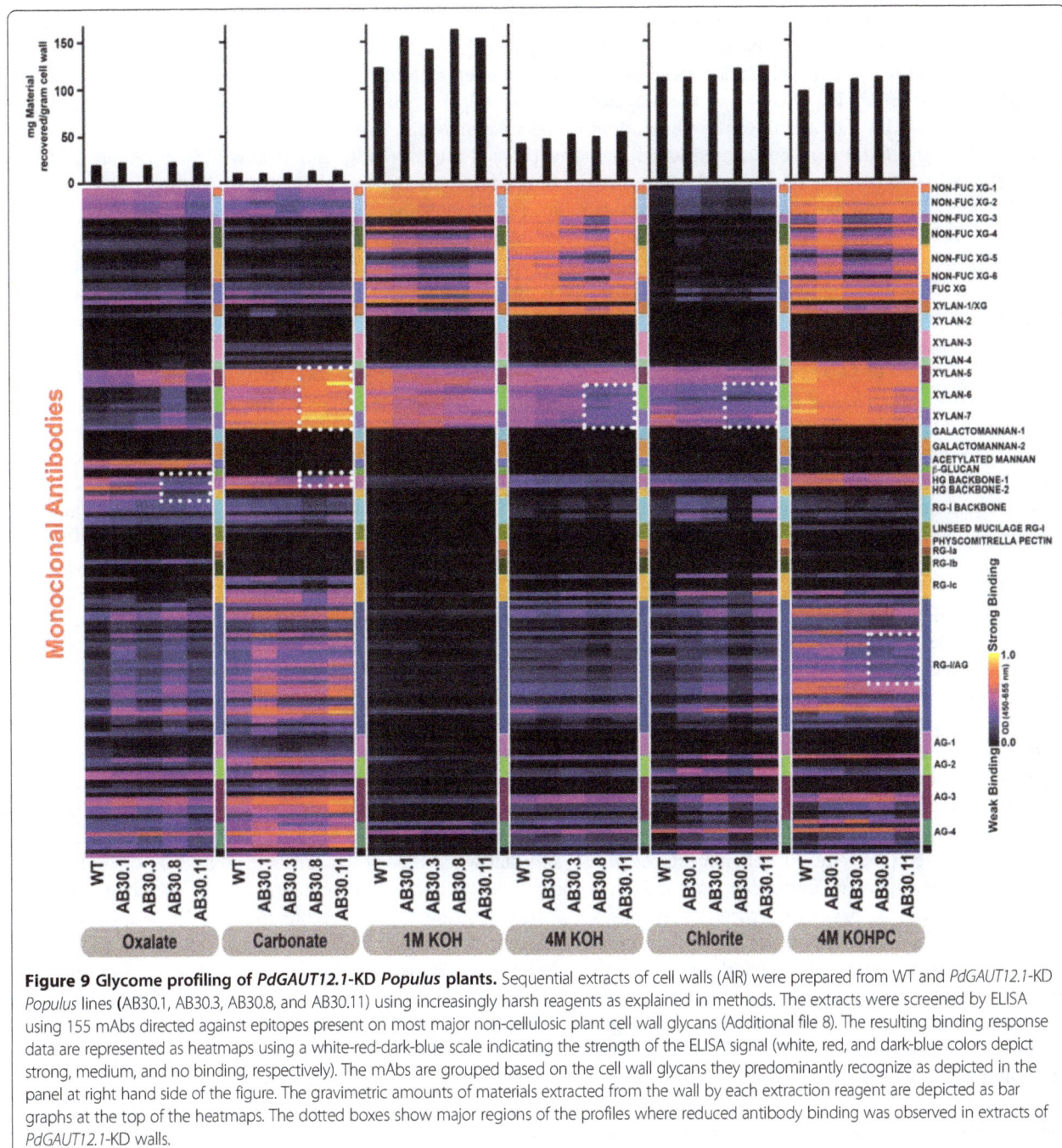

Figure 9 Glycome profiling of *PdGAUT12.1*-KD *Populus* plants. Sequential extracts of cell walls (AIR) were prepared from WT and *PdGAUT12.1*-KD *Populus* lines (AB30.1, AB30.3, AB30.8, and AB30.11) using increasingly harsh reagents as explained in methods. The extracts were screened by ELISA using 155 mAbs directed against epitopes present on most major non-cellulosic plant cell wall glycans (Additional file 8). The resulting binding response data are represented as heatmaps using a white-red-dark-blue scale indicating the strength of the ELISA signal (white, red, and dark-blue colors depict strong, medium, and no binding, respectively). The mAbs are grouped based on the cell wall glycans they predominantly recognize as depicted in the panel at right hand side of the figure. The gravimetric amounts of materials extracted from the wall by each extraction reagent are depicted as bar graphs at the top of the heatmaps. The dotted boxes show major regions of the profiles where reduced antibody binding was observed in extracts of *PdGAUT12.1*-KD walls.

the 4 M KOH and chlorite extracts and a lesser reduction in the 1 M KOH extract. These results suggest that a portion of xylan in the *PdGAUT12.1*-KD lines (AB30.8 and AB30.11) was held less tightly in the walls and was recovered in the carbonate fraction. In these *PdGAUT12.1*-KD lines, there was also a reduction in HG epitopes in the oxalate extract, carbonate, and 4 M KOH postchlorite fractions, as well as a slight reduction of some RG-I/AG mAb binding in the 4 M KOH postchlorite extract.

The combination of results obtained from glycosyl residue and linkage composition analyses of total AIR together with sugar composition analyses, glycome profiling, and total yield of sequential wall extracts indicate the following. (1) The only sugars consistently and statistically reduced in total AIR from *PdGAUT12.1*-KD lines were xylose and GalA, suggesting that GAUT12 may function in the addition of one or both of these into one or more wall polymers. (2) The only sugars consistently and statistically reduced in all the wall fractions, both the pectin- and xylan-enriched fractions, were GalA and rhamnose. Combined with linkage data, these results suggest that a pectin-like HG or HG/RG-I structure is either synthesized by GAUT12, or is part of the polymer synthesized by GAUT12. The significant reduction in xylose content in the four hemicellulose-enriched wall extracts suggests that xylan or a xylan-containing moiety may also be either the, or a part of, the polymer synthesized by GAUT12. (3) GAUT12.1-KD walls are more easily extracted as evidenced by the reduced yield of insoluble AIR pellet and the recovery of increased amounts of material in the sequential cell wall extracts. These results strongly support the hypothesis that the structure(s) synthesized by GAUT12 are part of the basic wall infrastructure. (4) Glycome profiling confirms that both xylan and HG along with some rhamnogalacturonan/arabinogalactan epitopes are affected in the PdGAUT12.1-KD lines, with losses in the HG and RG/AG epitopes and an increased ease of extraction of the xylan epitope.

Conclusions

We applied a genetic engineering approach to modify GAUT12.1 expression in Populus. Eleven transgenic lines were obtained that had reduced PdGAUT12.1 transcript expression and an accompanying 25 to 47% reduction in GalA, 17 to 30% reduction in xylan content, and 3 to 7% increase in sugar release across the lines. We also found a significant increase in plant height (12 to 52%) and stem diameter (12 to 44%) and leaf size (52 to 117%).

It has been reported that removal of hemicellulose, especially xylose, xylo-oligomers, and xylan from lignocellulosic materials prior to enzymatic hydrolysis increases enzyme accessibility, conversion rates, and hydrolysis yield of plant biomass [41,42,57]. Similarly, pectin degradation has been shown to enhance saccharification [43]. Thus, a possible explanation for the greater sugar release in the P. deltoides PdGAUT12.1-KD lines is that the reduction of 17 to 30% xylose and 25 to 47% GalA associated with xylan and HG in the cell wall leads to an increased accessibility of the biomass to enzymes and results in greater conversion and sugar release. These data support our hypothesis that downregulation of GAUT12.1 reduces recalcitrance in Populus.

There was also a significant increase in phloem tissue, xylem fiber cell size, and xylem vessel diameter size in PdGAUT12.1-KD wood. The larger leaf area in both developing and expanded leaves, the larger individual palisade and spongy parenchyma cells, the higher RWC, and the increased phloem and wider vessels in the PdGAUT12.1-KD lines may explain the greater growth and biomass yield in PdGAUT12.1-KD RNAi Populus lines.

We also demonstrated that PdGAUT12.1, which is highly expressed in developing xylem (Figure 2B), can complement the phenotype of Arabidopsis irx8 mutants (Figure 8), indicating that it is a functional homolog of the Arabidopsis GAUT12. PdGAUT12.2, which is expressed more broadly and at lower levels (Figure 2C), showed only partial complementation in the same experiments, suggesting that it might have functionally diverged from PdGAUT12.1.

The substantial decrease in GalA and xylose content in the PdGAUT12.1-KD cell walls and the content of fractionated walls as determined by glycosyl residue composition, linkage analyses, and glycome profiling support the hypothesis that reduced GAUT12.1 transcript levels affect both pectin and xylan biosynthesis in Populus. Sequential extraction of cell walls with increasingly harsh solvents in an effort to identify the specific polymers affected revealed an 11 to 17% increase in extractable material in walls of the PdGAUT12.1-KD lines compared to the controls and a concomitant 8 to 13% reduced mass of the final insoluble pellet, indicating a loosened wall structure in the PdGAUT12.1-KD lines that enables more facile extraction. Analyses of sequential cell wall extracts confirmed a reduction of both GX and the pectins HG and RG-I in extracts from PdGAUT12.1-KD RNAi lines. These transgenic lines had a higher S/G ratio but no major change in the total lignin content. Since no major change in the total lignin was observed, the data support the hypothesis that GAUT12.1 synthesizes a wall structure that is either required for deposition of xylan and pectin or that contains xylan and pectin and that when this structure is not made, the wall is held together less tightly, leading to increased sugar release and increased growth. These results indicate that a systematic biotechnological manipulation of pectin and xylan biosynthetic genes in woody Populus can be used to develop less recalcitrant woody biomass with increased growth potential that can be more easily deconstructed and fermented into biofuels. The exact structure synthesized by GAUT12.1 remains to be determined, but the data presented here suggest it is a core architectural structure in Populus wood cell walls.

Methods
Plant materials and growth conditions

P. deltoides plants were grown in Fafards 3B Soil mix (GroSouth Inc, Atlanta, GA, USA) with osmocote (250 ml/bag 79 L 3B soil), bone meal (84 ml/bag 79 L 3B soil), gypsum (84 ml/bag 79 L 3B soil), and dolomite/limestone (42 ml/1bag 3B soil) in the greenhouse under a 16-h-light/8-h-dark cycle at 25 to 32°C, depending on the season. All WT, vector control, and transgenic plants were grown for 9 months and fertilized weekly with Peters 20-10-20 (nitrogen-phosphorus-potassium; GroSouth Inc,

Atlanta, GA, USA). The bottom 6 cm of the stem from WT, vector control, and transgenic 9-month-old plants was harvested and the bark removed by peeling with a razor. The peeled stem samples were air-dried and the pith was removed using a hand drill. The remaining stem material was milled to a particle size of 20 mesh (0.85 mm) using a Wiley Mini-Mill (model number 3383 L10, Thomas Scientific). Ground material was used for analytical pyrolysis, saccharification, and cell wall composition analyses.

Arabidopsis plants were grown on soil in a controlled-environment chamber (Conviron, Pembina, ND, USA) under a 14-h-light/10-h-dark cycle at 19°C during the light and 15°C during the dark. The light intensity was 150 μEm^{-2}/s and relative humidity was maintained at 50%. T-DNA insertions were confirmed using primers from genomic regions flanking the T-DNA and the general T-DNA left border primer (Additional file 5). Arabidopsis Col-0 and heterozygous irx8-5 (SALK_044387) plants were transformed via the floral dip method [58], and transgenic plants harboring BASTA-resistant plasmids were selected on MS media plates containing 10 mg/l DL-phosphinothricin (PhytoTechnology Laboratories G523) and 100 mg/l timentin (PhytoTechnology Laboratories T869). Timentin inhibits residual Agrobacterium growth resulting from the dipping transformation. T1 transgenic plants harboring the construct in Col-0 and irx8-5 homozygote mutant backgrounds were selected by genotyping PCR (Additional file 5) and used for complementation analyses.

Vector construction and transformation of Populus

The PtGAUT12.1 RNAi construct used for transformation of P. deltoides was generated by cloning a 168-bp fragment from the 3′-untranslated region of POPTR_0001s44250/Potri.001G416800, which was amplified via PCR from a P. trichocarpa cDNA library using the forward primer 5′-CACCCCCGGGGGAGAGTTCTTCACAAATCCA-3′ and the reverse primer 5′- TCTAGAAAATCCATAACATGCT TAATTTCTC-3′. The integrity of the fragment was verified by DNA sequencing (ACGT, Inc., Wheeling, IL, USA) after cloning into the Gateway® entry vector, pENTR/D-TOPO (Life Technologies, Grand Island, NY, USA). The fragment was transferred to a binary Gateway® destination plasmid, pAGSM552 (GenBank accession KP259613) using LR Clonase II recombination (Life Technologies). The resulting binary transformation vector, pAGW641 (GenBank accession KP259612), was transformed into A. tumefaciens strain EHA105 via electroporation. The RNAi cassette comprised the Arabidopsis UBIQUITIN3 promoter, inverted repeats of the PtGAUT12.1 fragment flanking a spacer, and the nopaline synthase terminator. The spacer was a 1.3-kb fragment of the Petunia hybrida chsA intron. P. deltoides genotype WV94 was transformed using

a modified Agrobacterium-based method [59,60]. Shoots regenerated from isolated calli were tested using PCR to verify the presence of the GAUT12.1 construct.

Generation of constructs for complementation of Arabidopsis irx8 mutant

Arabidopsis GAUT12 coding sequence (CDS) was amplified from Arabidopsis cDNA using primers whose sequences are provided in Additional file 5. Poplar GAUT12 cDNAs were generated by reverse transcription-PCR as follows: approximately 1 μg of RNA harvested from WT P. deltoides stems was reverse transcribed into first-strand cDNA with oligo-dT primers. This library was diluted tenfold and used as a template for amplifying the predicted coding sequences of PdGAUT12.1(POPTR_0001s44250, Phytozome 8.0/Potri.001G416800, Phytozome 10.0) and PdGAUT12.2 (POPTR_0011s13600, Phytozome 8.0/Potri.011G132600, Phytozome 10.0) by a nested PCR with specific primers designed on the basis of the P. trichocarpa genome sequence available at Phytozome (http://www.phytozome.net/, http://phytozome.jgi.doe.gov/). PCR products were gel-purified and ligated into pGEM®-T Easy vector (Promega, Madison, WI, USA) and confirmed by sequencing. BamHI and ApaI restriction sites included in the second-round primers were used for lacing the coding sequences downstream of the 35S promoter of a T-DNA derived from pCAMBIA3300. T-DNAs were moved into Agrobacterium tumefaciens strain GV3101 by electroporation and the cultures used for plant transformation as described above.

RNA isolation and quantitative real-time PCR

Plant tissues used to isolate RNA for transcript analysis were P. deltoides young leaves, mature leaves, stems of 6 to 8 internodes, petioles pooled from internodes 1 and 2, cambium samples lightly scraped from frozen peeled bark, and xylem samples scraped from debarked frozen stem. All of the tissues were ground to a fine powder in liquid nitrogen. Total RNA was isolated from 100 to 150 mg frozen powder from three individual tissue samples using hexadecyl-trimethylammonium bromide (CTAB; 2% CTAB, 2% polyvinylpyrolidone, 100 mM Tris-HCl, 25 mM EDTA, 2 M NaCl, 0.5 g L21 spermidine, and 2% β-mercaptoethanol, pH 8.0) [61] and purified using columns from an RNeasy Plant Mini Kit (Qiagen, Venlo, Netherlands; 74903). Genomic DNA was removed by treatment with DNase (Ambion) and first-strand cDNA was synthesized from 1 μg total RNA using SuperScript III First-Strand Synthesis Super Mix (Invitrogen). Primers for real-time PCR were designed using Beacon Designer (v 2.1, Premier Biosoft International, Palo Alto, CA, USA) and tested for efficiency and specificity. The least conserved region of the

PtGAUT12.1 and PtGAUT12.2 genes in the poplar genome (Phytozome 8.0/Phytozome 10.0) was used for primer design. The primers used for this analysis were as follows: *GAUT12.1* F 5'-GGTCGAGCAAAGCCTTGGCT AGATATAGC-3' and *GAUT12.1* R 5'-AGATGGCCTAAT ATGACAGCCCTTTA-3' yielding a PCR product of 111 bp and *GAUT12.2* F 5'- CATTTCAATGGTCGAGCA AAGCCTTGGC-3' and *GAUT12.2* R 5'- GACAGCCCGT AATGAACTTGTCAGA-3' yielding a PCR product of 106 bp. The *18S rRNA* reference gene was used to amplify a product of 74 bp (*18S* F 5'-AATTGTTGGTCTTCAACG AGGAA-3' and *18S* R 5'-AAAGGGCAGGGACGTAGTC AA-3'). The PCR reactions were performed in triplicate in a 96-well plate format using iQ TM SYBR Green Supermix (Bio-Rad, Hercules, CA, USA; 170-8882) in a CFX96 TM Real-Time PCR Detection System (Bio-Rad) according to the manufacturer's instructions. The relative transcript levels were calculated by normalizing target gene expression to control gene expression, where expression of the control gene was set to 1 [62,63].

Plant growth analysis, water status, and measurement of dry weight

After the WT, vector control and transgenic plants developed proper roots in BTM media, all plants were transferred to soil in the greenhouse. Twelve to 15 plants per line were measured for plant height and radial stem growth. All plants were grown for nine months and initial plant stem growth was measured at three months. To determine the sizes of leaves from controls and transgenic lines at different developmental stages, the length and width of every fifth successive leaf from the apex was measured from three-month-old plants. Every 5th, 6th and 7th leaf from the apex of six WT and six transgenic plants was collected and measured for leaf area to compare developing leaf growth. Similarly, the 20th to 40th leaves from the apex of five transgenic lines were collected and measured to compare fully expanded leaf areas.

    Plant water status was measured as RWC of WT and *PdGAUT12.1*-KD plants [64]. The RWC is expressed as RWC (%) = [(FM − DM)/(TM − DM)] × 100, where, FM, DM, and TM are the fresh, dry, and turgid mass, respectively. Leaves were detached from the plants and used to measure FW. To obtain the turgid mass (TM), the detached leaves were rehydrated at 4°C for 24 h. The rehydrated leaves were placed in a pre-heated oven at 70°C for 72 h to measure the dry mass (DM). For the dry matter measurements, two sets of three individual three month old plants (above ground plant parts, that is, entire shoots) were harvested and each set was dried separately at 70°C for 5 days, weighed and mass values averaged ($N = 6$).

Preparation of AIR cell wall and cell wall fractionations

The ground biomass was sequentially extracted with 80% (*v*/*v*) ethanol, 100% ethanol, and chloroform/methanol (1:1[*v*/*v*]), and after centrifugation, the resulting cell wall residue (AIR) was vacuum-dried for 24 h at room temperature. Sequential fractionations were carried out at 10 mg/ml based on the starting dry weight of AIR in order to isolate fractions enriched for different types of cell wall components. AIR wall was suspended at 25°C in 50 mM ammonium oxalate, pH 5.0 for 24 h with constant shaking at 100 rpm. After incubation, the mixture was centrifuged at $4,000\,g$ for 15 min at room temperature and the supernatant saved as the ammonium oxalate extract. The residual pellet was subsequently washed three times with the same volume of deionized water, centrifuged, and the supernatant discarded. The pellet was treated at 25°C with 50 mM Na_2CO_3, pH 10.0 (containing 0.5% (*w*/*v*) sodium borohydride) with constant shaking (100 rpm). After centrifugation, the supernatant was saved as the sodium carbonate extract. The washed residual pellet was re-suspended in 1 M KOH with 1% (*w*/*v*) sodium borohydride and incubated at 25°C for 24 h with constant shaking at 100 rpm. After incubation, the suspension was centrifuged at $4,000\,g$ for 15 min, the supernatant collected and labeled as 1 M KOH extract, and stored at 4°C. The residual pellet from the 1 M KOH fraction was re-suspended in 4 M KOH with 1% (*w*/*v*) sodium borohydride and incubated at 25°C for 24 h with constant shaking at 100 rpm. After incubation, the suspension was centrifuged at $4,000\,g$ for 15 min and the supernatant collected, labeled as 4 M KOH extract and stored at 4°C. The residual pellet from the 4 M KOH fraction was treated with 100 mM sodium chlorite at 70°C for 3 h to break down lignin polymers. After centrifugation, the supernatant was collected and labeled as chlorite extract. The residual pellet left after the sodium chlorite treatment was treated with 4 M KOH containing 1% (*w*/*v*) sodium borohydride extraction at 25°C for 24 h with constant shaking (100 rpm) and the supernatant collected and labeled as 4MKOH PC. The 1MKOH, 4MKOH, and 4MKOH PC fractions were neutralized using glacial acetic acid, dialyzed against six changes of de-ionized water at room temperature for 72 h, and lyophilized [55].

Glycosyl residue composition of AIR and fractionation of *Populus* stem cell walls

Glycosyl residue composition analysis of stems from 9-month-old WT, vector control, and *PdGAUT12.1*-KD plants was carried out by preparation and analysis of trimethylsilyl (TMS) derivatives of the monosaccharide methyl glycosides produced from the samples by acidic methanolysis [65,66]. The stem AIR (approximately

2 mg) and fractionated cell walls (100 to 300 μg) were aliquoted into individual tubes, and 20 μg of inositol was added as an internal standard. The samples were lyophilized and dry samples hydrolyzed for 18 h at 80°C in 1 M methanolic-HCl. The samples were cooled, evaporated under a stream of dry air, and further dried two additional times with anhydrous methanol. The walls were derivatized with 200 μl of TriSil Reagent (Pierce-Endogen, Rockford, IL, USA) and heated to 80°C for 20 min. The cooled samples were evaporated under a stream of dry air, resuspended in 3 ml hexane, and filtered through packed glass wool. The dried samples were resuspended in 150 μl hexane and 1 μl of sample was injected into the gas chromatograph-mass spectrometer (GC-MS). GC/MS analysis of the TMS methyl glycosides was performed on an Agilent 7890A GC interfaced to a 5975C MSD using a Supelco EC-1 fused silica capillary column (30 m × 0.25 mm ID) and helium as carrier gas.

Glycosyl residue linkage analysis
For glycosyl residue linkage analysis, the samples were permethylated, reduced, re-permethylated, depolymerized, reduced, and acetylated. The resulting partially methylated alditol acetates (PMAAs) were analyzed by gas chromatography-mass spectrometry (GC-MS) as previously described [66].

The fractionated wall samples from WT and *PdGAUT12.1*-KD lines were first suspended in 200 ul dimethyl sulfoxide and the samples stirred for 2 days at 25°C. The samples were permethylated using potassium dimsyl anion and iodomethane. The permethylated uronic acids were reduced using lithium borodeuteride and then permethylated again by the method described earlier [67] with sodium hydroxide and methyl iodide in dry DMSO. This additional permethylation was to insure complete methylation of the polymer. Following sample workup, the permethylated material was hydrolyzed using 2 M trifluoroacetic acid (2 h in sealed tube at 121°C), reduced with $NaBD_4$, and acetylated using acetic anhydride/TFA. The resulting PMAAs were analyzed on an Agilent 7890A GC interfaced to a 5975C MSD (mass selective detector, electron impact ionization mode). Separation was performed on a Supelco SP-2380 fused silica capillary column (30 m × 0.25 mm ID).

Analytical pyrolysis for lignin analysis
All *Populus* biomass samples were analyzed by analytical pyrolysis for lignin content and S/G ratio. A commercially available molecular beam mass spectrometer (MBMS) designed specifically for biomass analysis was used for pyrolysis vapor analysis [68-70]. Approximately 4 mg of air-dried 20-mesh biomass was introduced into the quartz pyrolysis reactor via 80 μl deactivated

stainless steel Eco-Cups provided with the autosampler. Mass spectral data from m/z 30 to 450 were acquired on a Merlin Automation data system version 3.0 using 17 eV electron impact ionization. S/G ratios were determined by summing the syringyl peaks 154, 167, 168, 182, 194, 208, and 210 and dividing by the sum of guaiacyl peaks 124, 137, 138, 150, 164, and 178 (Additional file 9). Several lignin peaks were omitted in the syringyl or guaiacyl summations due to individual peaks having associations with both S and G precursors [68]. Lignin estimates were determined by summing the intensities of the major lignin precursors listed in Additional file 9.

Saccharification assay
High-throughput thermochemical pretreatment and enzymatic hydrolysis of biomass was carried out as previously described [71,72]. Briefly, ground *Populus* biomass was extracted with glucoamylase (Spirizyme Ultra - 0.25%) and alpha-amylase (Liquozyme SC DS - 1.5%) in 0.1 M sodium acetate (24 h, 55°C, pH 5.0) to remove possible starch content (16 ml enzyme solution per 1 g biomass) [73]. This was followed by an ethanol Soxhlet extraction for an additional 24 h to remove extractives. After drying overnight, 5 mg (±0.5 mg) was weighed in triplicate into individual wells of custom-made 96-well format Hastelloy reactor plates. Water was added (250 μl), the samples sealed with silicone adhesive-backed Teflon tape, clamped tightly, and treated at 180°C for 17.5 min in a customized steam reactor. Once cooled, 40 μl of buffer-enzyme stock was added. The buffer-enzyme stock was 8% CTec2 (Novozymes) in 1.0 M sodium citrate buffer. The samples were gently mixed and left to statically incubate at 50°C for 70 h. After 70 h of incubation, an aliquot of the saccharified hydrolysate was diluted and analyzed using Megazyme's GOPOD (glucose oxidase/peroxidase) and XDH assays (xylose dehydrogenase), using mixed glucose/xylose standards.

Tissue fixation, embedding and microscopy
Both leaf (6th leaf from the apex) and stem (10th and 20th internode from the apex) of *P. deltoides* tissues were cut into small pieces (~3 mm) with a razor blade and immediately fixed in 25 mM sodium phosphate buffer pH (7.1) with 4% (*v/v*) paraformaldehyde, 0.5% (*v/v*) glutaraldehyde, and 0.02% (*v/v*) Triton X-100 for 24 h at 4°C. Tissues were washed three times for 30 min each with 25 mM sodium phosphate buffer, pH 7.1, followed by three washes with water. In the next step, the samples underwent a graded ethanol series (35%, 50%, 70%, 95%, and 100% [*v/v*]) wash for 25 min at each step (one time) to dehydrate the tissue. The 100% ethanol step was repeated two times to remove as much water as possible from the tissue. The samples were infiltrated with LR

White embedding resin (Ted Pella) as follows: 1:3 (LR White: 100% ethanol) for 24 h (one time); 1:1 (LR White: 100% ethanol) for 24 h (one time); 3:1 (LR White: 100% ethanol) for 24 h (one time); and LR White (100%, no ethanol) for 24 h (three times). After the last resin change, the samples were transferred into gelatin capsules with fresh LR White. The gelatin capsules filled with resin were polymerized under 365-nm UV light for 48 h at 4°C. Sections, approximately 1 um in thickness, were cut with a diamond Histo-knife on a Reichert-Jung Ultracut S ultramicrotome (Leica Microsystems, Wetzlar, Germany) and mounted onto pre-coated slides (Color-frost/Plus, Fisher Scientific, Waltham, MA, USA). The mounted samples were stained with 1% toluidine blue for light microscopy. Images were captured using a Nikon DS-Ri1 camera (Nikon, Melville, NY, USA) and NIS-Elements Basic Research software.

Maceration of xylem

The bottom part of the stem (approximately 6 cm) from 9-month-old WT, vector control, and *PdGAUT12.1*-KD lines was harvested and debarked. Stem segments approximately 10 mm in length and 3 mm in diameter were used for maceration. The debarked stem segments were macerated with little modification in 50% glacial acetate acid and 3% hydrogen peroxide at 100°C for 12 h [74]. The samples were washed three times in water and the pH neutralized. The samples were suspended in 70% ethanol and vigorously shaken to release individual xylem fibers and vessel cells. The length and width of approximately 80 fibers and 60 vessel elements were measured from images captured using a Nikon DS-Ri1 camera (Nikon, Melville, NY) and NIS-Elements Basic Research software.

Glycome profiling

Glycome profiling of the sequential extracts prepared above was carried out as described previously [55]. Plant glycan-directed monoclonal antibodies [54-56] were from laboratory stocks (CCRC, JIM, and MAC series) at the Complex Carbohydrate Research Center (available through CarboSource Services; http://www.carbosource.net) or were obtained from BioSupplies (Australia) (BG1, LAMP). A description of the mAbs used in this study can be found in Additional file 8, which includes links to a web database, Wall*MAb*DB (http://www.wallmabdb.net) that provides detailed information about each antibody.

Statistical analysis

Statistical analysis was performed using Statistica 5.0. The significance of differences between control and transgenic was analyzed using a one-way ANOVA followed by Tukey's multiple comparison test or *post hoc* Bonferroni corrected *t*-test.

Additional files

Additional file 1: Relationship of GAUT12.1 transcript expression to plant height and diameter.

Additional file 2: Physiological measurements on *PdGAUT12.1*-KD lines.

Additional file 3: Downregulation of *GAUT12.1* leads to increased numbers of palisade parenchyma cells.

Additional file 4: Leaf anatomy of *PdGAUT12.1*-KD plants.

Additional file 5: List of primers used for complementation study.

Additional file 6: Mass of extracts recovered upon extraction of *PdGAUT12.1*-KD AIR.

Additional file 7: Monosaccharide composition and total carbohydrate of cell wall fractions from *PdGAUT12.1*-KD and control plants.

Additional file 8: List of cell wall glycan-directed monoclonal antibodies (mAbs) used for glycome profiling analyses.

Additional file 9: Lignin peak and precursor assignments of analytical pyrolysis mass spectra.

Abbreviations

AIR: Alcohol-insoluble residue; GalA: Galacturonic acid; GAUT1: Galacturonosyltransferase1; GAUT12: Galacturonosyltransferase12; GlcA: Glucuronic acid; GT8: Glycosyltransferase 8; GX: Glucuronoxylan; HG: Homogalacturonan; MBMS: Molecular beam mass spectrometer; MeGlcA: Methyl glucuronic acid; RG: Rhamnogalacturonan; TMS: Trimethylsilyl; WT: Wild type.

Competing interests

The strategy to produce improved biomass as described in this paper has been included in a patent application. Cassandra Collins and Kim Winkeler are employees of ArborGen Inc., a global provider of conventional and next generation plantation tree seedling products for the forestry industry.

Authors' contributions

AB is responsible for all aspects of the study, oversaw plant growth, line selection, tissue handling, and distribution, identified analyses to be done, carried out plant molecular, physiological, and cell wall analyses, and wrote the manuscript. ZH designed and performed the *irx8* complementation experiments, interpreted the results, and contributed to qRT-PCR analyses and writing of the manuscript. SP performed the glycome profiling. XY identified *Populus* genes based on provided Arabidopsis homolog and designed the DNA construct. KW performed the molecular cloning and produced the RNAi plasmid. CC performed *Populus* transformation and propagated transgenic plants. SSM and DR aided in growth and analysis of the plants. EAR performed microtome sectioning of leaf and stem resin-embedded tissues. IG-A and K.H. contributed to the identification of genes for transgenesis and growth of plants. RWS conducted saccharification assays. GBT carried out sample preparation and operation of the high-throughput recalcitrance pipeline, including data acquisition and analysis for sugar release values. AZ and EG coordinated analysis of samples through the BioEnergy Science Center (BESC) high-throughput molecular beam mass spectrometry (MBMS) and saccharification pipelines. WL helped design, carry out, and interpret the *irx8* complementation of the study. MFD developed and provided leadership for the BESC MBMS pipeline. SRD provided leadership and oversight in developing the standardized saccharification high-throughput pipeline and provided data analysis. MGH helped design and interpret the glycome profiling study. DM conceived of the study, coordinated research, contributed to interpretation of results and experimental design, and helped draft and finalize the manuscript. All authors read and approved the final manuscript.

Acknowledgements

We thank Lee Gunter for validation of constructs containing amplified gene targets from *P. trichocarpa* leaf cDNA libraries, Rick Nelson for directing the BESC transformation pipeline and critical review of the manuscript, Will Rottmann for overseeing the *Populus* transformation pipeline, Sheilah Dixon Huckabee for administrative assistance, and the CCRC Analytical Services for glycosyl residue linkage analysis. The authors also thank Crissa Doeppke, Melissa Glenn, Kimberly Mazza, Logan Schuster, and Kevin Cowley for preparation of samples for the HTP biomass recalcitrance pipeline and Breeanna R. Urbanowicz for providing the poplar cDNA. The generation of the CCRC series of plant cell wall glycan-directed monoclonal antibodies used in this work was supported by the US National Science Foundation Plant Genome Program (DBI-0421683 and IOS-0923992). The research was funded by The BioEnergy Science Center (BESC) Grant DE-PS02-06ER64304 and partially by the Department of Energy Center Grant DE-FG02-93ER20097. The BioEnergy Science Center is a U.S. Department of Energy Bioenergy Research Center supported by the Office of Biological and Environmental Research in the DOE Office of Science.

Author details

[1]Department of Biochemistry and Molecular Biology, University of Georgia, B122 Life Sciences Bldg., Athens, GA 30602, USA. [2]Department of Plant Biology, University of Georgia, 2502 Miller Plant Sciences, Athens, GA 30602, USA. [3]Complex Carbohydrate Research Center, University of Georgia, 315 Riverbend Road, Athens, GA 30602, USA. [4]DOE-BioEnergy Science Center (BESC), Oak Ridge, USA. [5]Bioscience Division, Oak Ridge National Laboratory, Oak Ridge, TN 37831, USA. [6]ArborGen Inc., 2011 Broadbank Ct, Ridgeville, SC 29472, USA. [7]National Renewable Energy Laboratory, 15013 Denver West Parkway, Golden, CO 80401-3305, USA.

References

1. Himmel ME, Ding SY, Johnson DK, Adney WS, Nimlos MR, Brady JW, et al. Biomass recalcitrance: engineering plants and enzymes for biofuels production. Science. 2007;315:804–7.
2. Somerville C, Youngs H, Taylor C, Davis SC, Long SP. Feedstocks for lignocellulosic biofuels. Science. 2010;329:790–2.
3. Pauly M, Keegstra K. Plant cell wall polymers as precursors for biofuels. Curr Opin Plant Biol. 2010;13:305–12.
4. Mellerowicz EJ, Baucher M, Sundberg B, Bojeran W. Unraveling cell wall formation in the woody dicot stem. Plant Mol Biol. 2001;47:239–74.
5. Hao Z, Mohnen D. A review of xylan and lignin biosynthesis: foundation for studying Arabidopsis irregular xylem mutants with pleiotropic phenotypes. Crit Rev Biochem Mol Biol. 2014;49:212–41.
6. Brown DM, Zeef LA, Ellis J, Goodacre R, Turner SR. Identification of novel genes in Arabidopsis involved in secondary cell wall formation using expression profiling and reverse genetics. Plant Cell. 2005;17:2281–95.
7. Lee C, O'Neill MA, Tsumuraya Y, Darvill AG, Ye ZH. The *irregular xylem9* mutant is deficient in xylan xylosyltransferase activity. Plant Cell Physiol. 2007;48:1624–34.
8. Pena MJ, Zhong RQ, Zhou GK, Richardson EA, O'Neill MA, Darvill AG, et al. Arabidopsis *irregular xylem8* and *irregular xylem9*: Implications for the complexiy of glucuronoxylan biosynthesis. Plant Cell. 2007;19:549–63.
9. Wu AM, Hornblad E, Voxeur A, Gerber L, Rihouey C, Lerouge P, et al. Analysis of the Arabidopsis *IRX9/IRX9-L* and *IRX14/IRX14-L* pairs of glycosyltransferase genes reveals critical contributions to biosynthesis of the hemicellulose glucuronoxylan. Plant Physiol. 2010;153:542–54.
10. Brown DM, Goubet F, Vicky WWA, Goodacre R, Stephens E, Dupree P, et al. Comparison of five xylan synthesis mutants reveals new insight into the mechanisms of xylan synthesis. Plant J. 2007;52:1154–68.
11. Keppler BD, Showalter AM. IRX14 and IRX14-LIKE, two glycosyl transferases involved in glucuronoxylan biosynthesis and drought tolerance in Arabidopsis. Mol Plant. 2010;3:834–41.
12. Brown DM, Zhang ZN, Stephens E, Dupree P, Turner SR. Characterization of IRX10 and IRX10-like reveals an essential role in glucuronoxylan biosynthesis in Arabidopsis. Plant J. 2009;57:732–46.
13. Wu AM, Rihouey C, Seveno M, Hornblad E, Singh SK, Matsunaga T, et al. The Arabidopsis IRX10 and IRX10-LIKE glycosyltransferases are critical for

glucuronoxylan biosynthesis during secondary cell wall formation. Plant J. 2009;57:718–31.
14. Brown D, Wightman R, Zhang Z, Gomez LD, Atanassov I, Bukowski JP, et al. Arabidopsis genes *IRREGULAR XYLEM* (IRX15) and IRX15L encode DUF579-containing proteins that are essential for normal xylan deposition in the secondary cell wall. Plant J. 2011;66:401–13.
15. Jensen JK, Kim H, Cocuron JC, Orler R, Ralph J, Wilkerson CG. The DUF579 domain containing proteins IRX15 and IRX15-L affect xylan synthesis in Arabidopsis. Plant J. 2011;66:387–400.
16. Urbanowicz BR, Peña MJ, Moniz HA, Moremen KW, York WS. Two Arabidopsis proteins synthesize acetylated xylan *in vitro*. Plant J. 2014;80:197–206.
17. Jensen JK, Johnson NR, Wilkerson CG. *Arabidopsis thaliana* IRX10 and two related proteins from *psyllium* and *Physcomitrella patens* are xylan xylosyltransferases. Plant J. 2014;80:207–15.
18. Rennie EA, Hansen SF, Baidoo EE, Hadi MZ, Keasling JD, Scheller HV. Three members of the Arabidopsis glycosyltransferase family 8 are xylan glucuronosyltransferases. Plant Physiol. 2012;159:1408–17.
19. Mortimer JC, Miles GP, Brown DM, Zhang ZN, Segura MP, Weimar T, et al. Absence of branches from xylan in Arabidopsis *gux* mutants reveals potential for simplification of lignocellulosic biomass. Proc Natl Acad Sci U S A. 2010;107:17409–14.
20. Urbanowicz BR, Pena MJ, Ratnaparkhe S, Avci U, Backe J, Steet HF, et al. 4-O-Methylation of glucuronic acid in Arabidopsis glucuronoxylan is catalyzed by a Domain of Unknown Function family 579 protein. Proc Natl Acad Sci U S A. 2012;109:14253–8.
21. Zhong RQ, Pena MJ, Zhou GK, Nairn CJ, Wood-Jones A, Richardson EA, et al. Arabidopsis *fragile fiber8*, which encodes a putative glucuronyltransferase, is essential for normal secondary wall synthesis. Plant Cell. 2005;17:3390–408.
22. Lee C, Teng Q, Huang WL, Zhong RQ, Ye ZH. The Poplar GT8E and GT8F glycosyltransferases are functional orthologs of Arabidopsis PARVUS involved in glucuronoxylan biosynthesis. Plant Cell Physiol. 2009;50:1982–7.
23. Bouton S, Leboeuf E, Mouille G, Leydecker MT, Talbotec J, Granier F, et al. QUASIMODO1 encodes a putative membrane-bound glycosyltransferase required for normal pectin synthesis and cell adhesion in Arabidopsis. Plant Cell. 2002;14:2577–90.
24. Orfila C, Sørensen SO, Harholt J, Geshi N, Crombie H, Truong HN, et al. QUASIMODO1 is expressed in vascular tissue of Arabidopsis thaliana inflorescence stems, and affects homogalacturonan and xylan biosynthesis. Planta. 2005;222:613–22.
25. Leboeuf E, Guillon F, Thoiron S, Lahaye M. Biochemical and immunohistochemical analysis of pectic polysaccharides in the cell walls of Arabidopsis mutant QUASIMODO 1 suspension-cultured cells: implications for cell adhesion. J Exp Bot. 2005;56:3171–82.
26. Shao M, Zheng H, Hu Y, Liu D, Jang JC, Ma H, et al. The *GAOLAOZHUANGREN1* gene encodes a putative glycosyltransferase that is critical for normal development and carbohydrate metabolism. Plant Cell Physiol. 2004;45:1453–60.
27. Lee C, Zhong R, Richardson E, Himmelsbach D, McPhail B, Ye ZH. The *PARVUS* gene is expressed in cells undergoing secondary wall thickening and is essential for glucuronoxylan biosynthesis. Plant Cell Physiol. 2007;48:1659–72.
28. Kong Y, Zhou G, Avci U, Gu X, Jones C, Yin Y, et al. Two poplar glycosyltransferase genes, PdGATL1.1 and PdGATL1.2, are functional orthologs to PARVUS/AtGATL1 in Arabidopsis. Mol Plant. 2009;2:1040–50.
29. Persson S, Caffall KH, Freshour G, Hilley MT, Bauer S, Poindexter P, et al. The Arabidopsis *irregular xylem8* mutant is deficient in glucuronoxylan and homogalacturonan, which are essential for secondary cell wall integrity. Plant Cell. 2007;19:237–55.
30. Hao Z, Avci U, Tan L, Zhu X, Glushka J, Pattathil S, et al. Loss of Arabidopsis GAUT12/IRX8 causes anther indehiscence and leads to reduced G lignin associated with altered matrix polysaccharide deposition. Frontiers Plant Sci. 2014;5:357.
31. Campbell JA, Davies GJ, Bulone V, Henrissat B. A classification of nucleotide-diphospho-sugar glycosyltransferases based on amino acid sequence similarities. Biochem J. 1997;326:929–39.
32. Coutinho PM, Deleury E, Davies GJ, Henrissat B. An evolving hierarchical family classification for glycosyltransferases. J Mol Biol. 2003;328:307–17.
33. Yin Y, Chen H, Hahn MG, Mohnen D, Xu Y. Evolution and function of the plant cell wall synthesis-related Glycosyltransferase Family 8. Plant Physiol. 2010;153:1729–46.

34. Sterling JD, Atmodjo MA, Inwood SE, Kumar Kolli VS, Quigley HF, Hahn MG, et al. Functional identification of an Arabidopsis pectin biosynthetic homogalacturonan galacturonosyltransferase. Proc Natl Acad Sci U S A. 2006;103:5236–41.

35. Atmodjo MA, Hao Z, Mohnen D. Evolving views of pectin biosynthesis. Annu Rev Plant Biol. 2013;64:747–79.

36. Atmodjo MA, Sakuragi Y, Zhu X, Burrell JA, Mohanty SS, Atwood III JA, et al. GAUT1:GAUT7 are the core of a plant cell wall pectin biosynthetic homogalacturonan:galacturonosyltransferase complex. Proc Natl Acad Sci U S A. 2011;108:20225–30.

37. Li Q, Min D, Wang JP-Y, Peszlen I, Horvath L, Horvath B, et al. Down-regulation of glycosyltransferase 8D genes in Populus trichocarpa caused reduced mechanical strength and xylan content in wood. Tree Physiol. 2011;31:226–36.

38. Zhou GK, Zhong R, Himmelsbach DS, McPhail BT, Ye ZH. Molecular characterization of PoGT8D and PoGT43B, two secondary wall-associated glycosyltransferases in poplar. Plant Cell Physiol. 2007;48:689–99.

39. Lee C, Teng Q, Zhong R, Ye ZH. Molecular dissection of xylan biosynthesis during wood formation in poplar. Mol Plant. 2011;4:730–47.

40. Olson DG, McBride JE, Shaw AJ, Lynd LR. Recent progress in consolidated bioprocessing. Curr Opin Biotechnol. 2012;23:396–405.

41. Shin HD, McClendon S, Vo T, Chen RR. Escherichia coli binary culture engineered for direct fermentation of hemicellulose to a biofuel. Appl Environ Microbiol. 2010;76:8150–9.

42. Qing Q, Yang B, Wyman CE. Xylooligomers are strong inhibitors of cellulose hydrolysis by enzymes. Bioresour Technol. 2010;101:9624–30.

43. Lionetti V, Francocci F, Ferrari S, Volpi C, Bellincampi D, Galletti R, et al. Engineering the cell wall by reducing de-methyl-esterified homogalacturonan improves saccharification of plant tissues for bioconversion. Proc Natl Acad Sci U S A. 2010;107:616–21.

44. Biswal AK, Soeno K, Gandla ML, Immerzeel P, Pattathil S, Lucenius J, et al. Aspen pectate lyase PtxtPL1-27 mobilizes matrix polysaccharides from woody tissues and improves saccharification yield. Biotechnol Biofuels. 2014;7:11.

45. Mohnen DA, Biswal AK, Hao Z, Kataeva I, Adams MW, Hunt KD, et al. Plants with altered cell wall biosynthesis and methods of use. PCT Application # PCT/US20011/032733, filed on 4/15/2011, published on 10/20/2011 as WO 2011130666, and claiming priority to 4/16/2010.

46. Caffall KH, Pattathil S, Phillips SE, Hahn MG, Mohnen D. Arabidopsis thaliana T-DNA mutants implicate GAUT genes in the biosynthesis of pectin and xylan in cell walls and seed testa. Mol Plant. 2009;2:1000–14.

47. Tamura K, Peterson D, Peterson N, Stecher G, Nei M, Kumar S. MEGA5: molecular evolutionary genetics analysis using maximum likelihood, evolutionary distance, and maximum parsimony methods. Mol Biol Evol. 2011;28:2731–9.

48. Chalupa V. Clonal propagation of broad-leaved forest trees in vitro. Commun Inst Forest Cechosl. 1981;12:255–71.

49. Petersen PD, Lau J, Ebert B, Yang F, Verhertbruggen Y, Kim JS, et al. Engineering of plants with improved properties as biofuels feedstocks by vessel-specific complementation of xylan biosynthesis mutants. Biotechnol Biofuels. 2012;5:84.

50. Studer MH, DeMartini JD, Davis MF, Sykes RW, Davison B, Keller M, et al. Lignin content in natural Populus variants affects sugar release. Proc Natl Acad Sci U S A. 2011;108:6300–5.

51. Van Acker R, Leplé JC, Aerts D, Storme V, Goeminne G, Ivens B, et al. Improved saccharification and ethanol yield from field-grown transgenic poplar deficient in cinnamoyl-CoA reductase. Proc Natl Acad Sci U S A. 2014;111:845–50.

52. Voelker SL, Lachenbruch B, Meinzer FC, Jourdes M, Ki C, Patten AM, et al. Antisense down-regulation of 4CL expression alters lignification, tree growth, and saccharification potential of field-grown poplar. Plant Physiol. 2010;154:874–86.

53. Kramer PJ, Boyer JS. Water relations of plants and soils. San Diego: Academic Press; 1995. p. 495.

54. Pattathil S, Avci U, Baldwin D, Swennes AG, McGill JA, Popper Z, et al. A comprehensive toolkit of plant cell wall glycan-directed monoclonal antibodies. Plant Physiol. 2010;153:514–25.

55. Pattathil S, Avci U, Miller JS, Hahn MG. Immunological approaches to plant cell wall and biomass characterization: glycome profiling. In: Himmel ME, editor. Biomass conversion. Methods and protocols (Springer), methods in molecular biology, vol, vol. 908. New York: Humana Press; 2012. p. 61–72.

56. DeMartini JD, Pattathil S, Avci U, Szekalski K, Mazumder K, Hahn MG, et al. Application of monoclonal antibodies to investigate plant cell wall deconstruction for biofuels production. Energy Environ Sci. 2011;4:4332–9.

57. Liao W, Wen ZY, Hurley S, Liu Y, Liu CB, Chen SL. Effects of hemicellulose and lignin on enzymatic hydrolysis of cellulose from dairy manure. Appl Biochem Biotechnol. 2005;2005(121):1017–30.

58. Clough SJ, Bent AF. Floral dip: a simplified method for Agrobacterium-mediated transformation of Arabidopsis thaliana. Plant J. 1998;16:735–43.

59. Tsai CJ, Podila GK, Chiang VL. Agrobacterium-mediated transformation of quaking aspen (Populus tremuloides) and regeneration of transgenic plants. Plant Cell Rep. 1994;14:94–7.

60. Mingozzi M, Montello P, Merkle S. Adventitious shoot regeneration from leaf explants of eastern cottonwood (Populus deltoides) cultured under photoautotrophic conditions. Tree Physiol. 2009;29:333–43.

61. Chang SJ, Puryear J, Cairney J. A simple and efficient method for isolating RNA from pine trees. Plant Mol Biol Rep. 1993;11:113–6.

62. Livak KJ, Schmittgen TD. Analysis of relative gene expression data using real-time quantitative PCR and the $2^{-\Delta\Delta CT}$ method. Methods. 2001;25:402–8.

63. Brunner AM, Yakovlev IA, Strauss SH. Validating internal controls for quantitative plant gene expression studies. BMC Plant Biol. 2004;4:14.

64. Boyer JS. Measurement of the water status of plants. Annu Rev Plant Physiol. 1968;9:351–63.

65. Tan L, Eberhard S, Pattathil S, Warder C, Glushka J, Yuan C, et al. Arabidopsis cell wall proteoglycan consists of pectin and arabinoxylan covalently linked to an arabinogalactan protein. Plant Cell. 2013;25:270–87.

66. York W, Darvill AG, McNeil M, Stevenson TT, Albersheim P. Isolation and characterization of plant cell walls and cell wall components. Methods Enzymol. 1985;118:3–40.

67. Ciucanu I, Kerek F. A simple and rapid method for the permethylation of carbohydrates. Carbohydr Res. 1984;131:209–17.

68. Evans RJ, Milne TA. Molecular characterization of the pyrolysis of biomass. Energy Fuels. 1987;1:123–37.

69. Sykes R, Yung M, Novaes E, Kirst M, Peter G, Davis M. High-throughput screening of plant cell-wall composition using pyrolysis molecular beam mass spectroscopy. In: Mielenz JR, editor. Biofuels:methods and protocols, methods in molecular biology. New York: Humana Press; 2009. p. 169–83.

70. Tuskan G, West D, Bradshaw HD, Neale D, Sewell M, Wheeler N, et al. Two high-throughput techniques for determining wood properties as part of a molecular genetics analysis of hybrid poplar and loblolly pine. Appl Biochem Biotechnol. 1999;77:55–65.

71. Selig MJ, Tucker MP, Sykes RW, Reichel KL, Brunecky R, Himmel ME, et al. Biomass recalcitrance screening by integrated high throughput hydrothermal pretreatment and enzymatic saccharification. Ind Biotechnol. 2010;6:104–11.

72. Decker SR, Brunecky R, Tucker MP, Himmel ME, Selig MJ. High throughput screening techniques for biomass conversion. Bioenergy Res. 2009;2:79–192.

73. Decker SR, Carlile M, Selig MJ, Doeppke C, Davis M, Sykes R, et al. Reducing the effect of variable starch levels in biomass recalcitrance screening. In: Himmel ME, editor. Biomass conversion. Methods in molecular biology, vol, vol. 908. New York: Humana Press; 2012. p. 181–95.

74. Gray-Mitsumune M, Blomquist K, McQueen-Mason S, Teeri TT, Sundberg B, Mellerowicz EJ. Ectopic expression of a wood-abundant expansin PttEXPA1 promotes cell expansion in primary and secondary tissues in aspen. Plant Biotechnol J. 2008;6:62–72.

4

Trichoderma reesei meiosis generates segmentally aneuploid progeny with higher xylanase-producing capability

Yu-Chien Chuang[1,2,3], Wan-Chen Li[3,4], Chia-Ling Chen[3], Paul Wei-Che Hsu[3], Shu-Yun Tung[3], Hsiao-Che Kuo[3,6], Monika Schmoll[5] and Ting-Fang Wang[1,3*]

Abstract

Background: *Hypocrea jecorina* is the sexual form of the industrial workhorse fungus *Trichoderma reesei* that secretes cellulases and hemicellulases to degrade lignocellulosic biomass into simple sugars, such as glucose and xylose. *H. jecorina* CBS999.97 is the only *T. reesei* wild isolate strain that is sexually competent in laboratory conditions. It undergoes a heterothallic reproductive cycle and generates CBS999.97(1-1) and CBS999.97(1-2) haploids with *MAT1-1* and *MAT1-2* mating-type loci, respectively. *T. reesei* QM6a and its derivatives (RUT-C30 and QM9414) all have a *MAT1-2* mating type locus, but they are female sterile. Sexual crossing of CBS999.97(1-1) with either CBS999.97(1-2) or QM6a produces fruiting bodies containing asci with 16 linearly arranged ascospores (the sexual spores specific to ascomycetes). This sexual crossing approach has created new opportunities for these biotechnologically important fungi.

Results: Through genetic and genomic analyses, we show that the 16 ascospores are generated via meiosis followed by two rounds of postmeiotic mitosis. We also found that the haploid genomes of CBS999.97(1-2) and QM6a are similar to that of the ancestral *T. reesei* strain, whereas the CBS999.97(1-1) haploid genome contains a reciprocal arrangement between two scaffolds of the CBS999.97(1-2) genome. Due to sequence heterozygosity, most 16-spore asci (>90%) contain four or eight inviable ascospores and an equal number of segmentally aneuploid (SAN) ascospores. The viable SAN progeny produced higher levels of xylanases and white conidia due to segmental duplication and deletion, respectively. Moreover, they readily lost the duplicated segment approximately two weeks after germination. With better lignocellulosic biomass degradation capability, these SAN progeny gain adaptive advantages to the natural environment, especially in the early phase of colonization.

Conclusions: Our results have not only further elucidated *T. reesei* evolution and sexual development, but also provided new perspectives for improving *T. reesei* industrial strains.

Keywords: *Trichoderma reesei*, *Hypocrea jecorina*, Genome evolution, Aneuploidy, Sexual development, Meiosis, Xylanase, Conidia pigmentation, Lignocellulosic biomass

* Correspondence: tfwang@gate.sinica.edu.tw
[1]Taiwan International Graduate Program in Molecular and Cellular Biology, Academia Sinica, Taipei 115, Taiwan
[3]Institute of Molecular Biology, Academia Sinica, Taipei 115, Taiwan
Full list of author information is available at the end of the article

Background

Meiosis is a special type of cell division that gives rise to genetic diversity in sexually reproductive organisms. Programmed DNA double-strand breaks (DSBs) are spontaneously generated throughout the genome by the meiosis-specific Spo11 endonucleases in many organisms, such as yeast and mice [1]. In some fungi (for example, *Neurospora crassa* and *Coprinus cinereus*) meiotic DSBs are also induced via Spo11-independent mechanisms [2,3]. Both Spo11-dependent and Spo11-independent DSBs are repaired robustly by error-free homologous recombination to ensure genomic stability and accurate segregation of homologous chromosomes.

Previous studies also revealed that infertile or abnormal meiotic products are generated in some fungi [4]. For example, several fungi carry spore-killing meiotic drive elements [5], including *Neurospora sitophila*, *Neurospora intermedia*, *Podospora anserina*, and *Cochliobolus heterostrophus*. Recently, two Spore killer elements (*Sk-2* and *Sk-3*) were reported to be located near a chromosome rearrangement site in *Neurospora crassa* [6]. It is still unclear whether chromosome rearrangement can cause meiotic drive. The molecular mechanism of the Spore killer in *Neurospora crassa* is still a mystery. Plant pathogenic fungi, for example, *Mycosphaerella graminicola* (anamorph *Septoria tritici*) and *Nectria haematococca* mating population VI (anamorph *Fusarium solani*), generate ascospores containing "extra" chromosomes, termed "conditionally dispensable" chromosomes [7-11]. In the human fungal pathogen *Cryptococcus neoformans*, a large segmental duplication occurs during meiosis via telomere-telomere fusion and chromosomal translocation between two different chromosomes [12,13]. The hybrid infertility of two closely related *Schizosaccharomyces* strains (*S. pombe* and *S. kambucha*) resulted from the chromosome rearrangement of their genomes [14]. Therefore, genome heterozygosity may be a cause of the production of segmental aneuploidy (SAN) and whole chromosome aneuploidy (WCA) ascospores in fungal meiosis.

SAN and WCA are often deleterious for survival due to the altered gene dosage [15]. However, they sometimes provide benefits in fitness to cells under stress; for instance, the *SUL1* gene is amplified in budding yeast cells cultured in sulfate-depleted medium for generations [16]. Budding yeast cells also exhibit SAN or WCA after being cultured at high temperature for over 490 generations [17] or when continuously cultured in copper-containing medium [18]. Notably, these strains readily exhibit return to euploidy (RTU) after the relief of stress-inducing conditions, indicating that genome plasticity is an adaptive strategy to gain transient advantages [15-18].

Trichoderma is a genus of ascomycete fungi that is present in soils and in other diverse habitats [19]. These fungi are beneficial symbiotic partners for plants, particularly crops. *Trichoderma* spp. secrete cellulases and hemicellulases that degrade β-glucan and xylan, the key structural components of lignocellulosic biomass, to produce glucose and xylose, respectively. Several cellulase-overproducing mutants (such as RUT-C30 and QM9414) derived from the *Trichoderma reesei* QM6a isolate have been widely used in industrial applications [20-23]. Due to multiple rounds of chemical and/or physical mutagenesis, the genomes of these hypersecretion mutants have numerous mutations, deletions, and DNA rearrangements [20,24,25]. Currently, the major bottleneck of lignocellulosic biomass degradation is enzyme cost, and these industrial strains secrete less xylan-degrading hemicellulases than β-glucan-degrading cellulases. *T. reesei* is the anamorph of the pantropical ascomycete *Hypocrea jecorina* [26]. The *H. jecorina* CBS999.97 wild isolate undergoes a heterothallic reproductive cycle, generating CBS999.97(1-1) and CBS999.97 (1-2) haploids with *MAT1-1* and *MAT1-2* mating-type loci, respectively. QM6a has a *MAT1-2* mating type locus, but it is female sterile. Sexual crossing of the CBS999.97(1-1) haploid with the CBS999.97(1-2) haploid or QM6a(1-2) yields fruiting bodies that contain asci with 16-part ascospores [27]. This sexual crossing approach has opened up new perspectives for these biotechnologically important fungi.

Here, we report that there is a reciprocal exchange (*re*) between two scaffolds in CBS999.97(1-1) compared with other *Trichoderma* wild isolates. Herein, the two CBS999.97 haploid strains and QM6a are referred as to as CBS999.97(1-1, *re*), CBS999.97(1-2, *wt* [*wild type*]), and QM6a(1-2, *wt*), respectively. Due to genome heterozygosity, sexual crossing of CBS999.97(1-1, *re*), CBS999.97(1-2, *wt*), or QM6a(1-2, *wt*) frequently generated equal numbers of inviable and viable SAN ascospores.

Results

The 16 ascospores in *T. reesei* asci are generated via meiosis followed by two rounds of postmeiotic mitosis

CBS999.97 is the only *T. reesei* wild isolate strain that is sexually competent in laboratory conditions [27]. Sexual crossing of CBS999.97(1-1, *re*) with CBS999.97(1-2, *wt*) or QM6a (1-2, *wt*) yields fruiting bodies containing asci with 16 ascospores (that is, hexadecads). To reveal the molecular mechanism of *T. reesei* sexual development, we applied the yeast tetrad dissection technique to sequentially separate the 16 ascospores from an ascus (Figure 1A). Each ascospore was individually cultured, and the spore viability, colony morphology, and colony color were determined (Figure 1B and 1C). Genomic

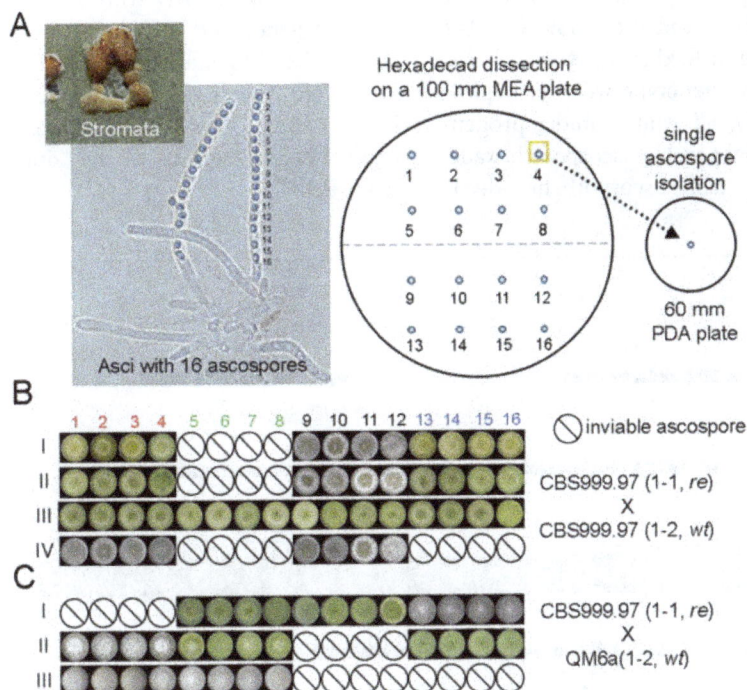

Figure 1 Hexadecad dissection. (A) Upper panel, stromata; lower panel, developing asci of which two contain 16 ascospores. Each ascospore is numbered according to its order in the ascus. Sixteen ascospores from a hexadecad were sequentially separated and grown on individual 100-mm malt extract agar (MEA) plates. A single colony from one ascospore was isolated and transferred individually to a 60-mm potato dextrose agar (PDA) plate to determine the spore viability, spore color, and colony morphology. **(B-C)** Ascospore phenotype of hexadecads from the sexual crossing of the wild isolate CBS999.97(1-1, re) with CBS999.97(1-2, wt) (n ≥ 20) or QM6a(1-2, wt) (n ≥ 10). Hexadecads were grown in constant darkness. Sixteen single-ascospore colonies were aligned sequentially according to the ascospore order. The inviable ascospores are indicated by a black circle with a cross.

PCR genotyping (Additional file 1: Figure S1 and Additional file 2: Table S1) of all 16 ascospores from one hexadecad (Figure 1B, III) further revealed that each hexadecad was classified into four linearly arranged groups, and each group contained four genetically indistinguishable ascospores. Finally, by staining the developing asci with 4',6-diamidino-2-phenylindole (DAPI), we also visualized two meiotic divisions and two further mitotic divisions (Additional file 3: Figure S2). Together, these results indicate that the 16 ascospores are generated via meiosis followed by two rounds of postmeiotic mitosis.

The CBS999.97 wild isolate frequently generates abnormal sexual progeny

QM6a(1-2, *wt*), CBS999.97(1-2, *wt*), and CBS999.97(1-1, *re*) all propagate mycelia and produce green conidia (that is, asexual spores). We found that sexual crossing of CBS999.97(1-1, *re*) with CBS999.97(1-2, *wt*) (Figure 1B) or QM6a(1-2, *wt*) (Figure 1C) often yielded hexadecads (>90%) with abnormal ascospores. The majority of the hexadecads (>80%, n ≥ 30) contained only 12 viable ascospores: eight of them germinated and then produced green conidia, whereas the other four

produced white conidia (Figure 1B, I and II). The inviable ascospores were unable to germinate. Only approximately 10% of hexadecads could generate 16 viable ascospores, and these ascospores germinated and produced green conidia (Figure 1B, III). Finally, the remaining approximately 9% of the hexadecads produced eight inviable ascospores as well as an equal number of viable, white-conidia ascospores (Figure 1B, IV).

The *T. reesei* v2.0 database (Department of Energy, Joint Genome Institute, USA) of the QM6a(1-2, *wt*) genome comprises 89 scaffolds and 97 contigs [25]. Using several commercially available cell-wall digesting enzymes (such as β-glucanase, Driselase, and lyticase), we were unable to prepare intact CBS999.97 and QM6a chromosomes for clamped homogeneous electric field (CHEF) gel electrophoresis. Novozyme 234 was previously used to prepare *Trichoderma* protoplasts [28] and intact chromosomes for CHEF analysis [29,30], but it has not been commercially available since 2000 [31]. Instead, we applied the microarray-based comparative genomic hybridization (aCGH) technique to identify genome-wide gene copy-number variation. Comparing the aCGH data between the two CBS999.97 parental haploid strains, their genome-wide gene copy numbers seemed to be equivalent, suggesting that both strains are euploid (Figure 2A). We then used the CBS999.97(1-2, *wt*) as the reference genome in further analyses. The ascospores with green-conidia phenotype were all euploid (Figure 2B and D). Notably, all white-conidia progeny generated from either the eight viable ascospore hexadecad (Figure 2C) or the 12 viable ascospore hexadecad

(Figure 2D) were SAN with a copy gain within scaffold 27 (130 genes and about 431 kb), the 3′ terminus of scaffold 28 (12 genes, about 40 kb), and the 5′ terminus of scaffold 36 (15 genes, about 52 kb) (Additional file 4: Figure S3A). This approximately 523-kb duplicated region was referred to as the "duplicated (D) segment" (Figure 3).

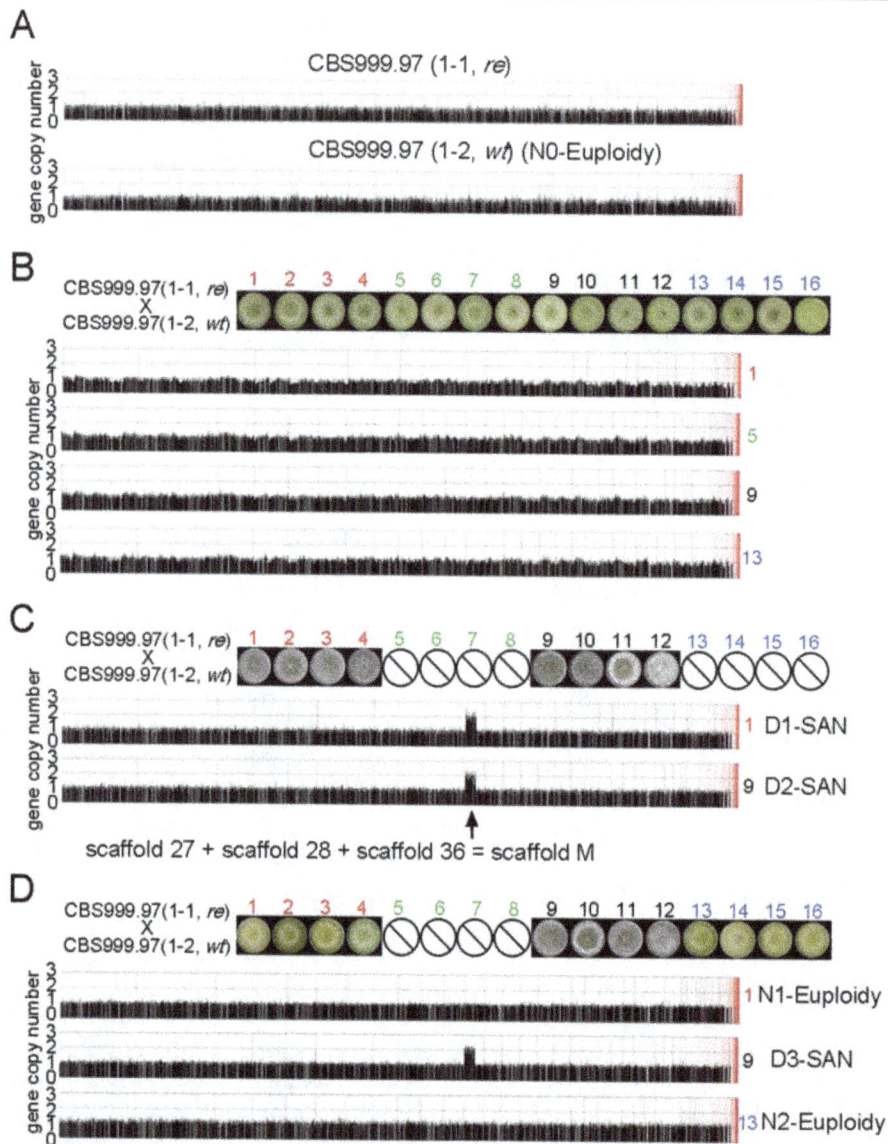

Figure 2 aCGH. (A) Gene copy number of the two parental wild isolate haploids. CBS999.97(1-2, *wt*) genomic DNA was used as a reference to measure DNA copy number changes of CBS999.97(1-1, *re*) and *vice versa*. Each line in the histogram represents one oligonucleotide and its position in the CBS999.97(1-2, *re*) and CBS999.97(1-2, *wt*) genomic sequence. The CBS999.97(1-2, *wt*) genome, such as that of QM6a(1-2, *wt*) [25], comprises 89 scaffolds. Moreover, three contiguous scaffolds (27, 28, 36) were reassembled into a much larger scaffold, scaffold M. The length of scaffold M is slightly shorter than that of scaffold 11. Normalized means of gene copy number for the oligonucleotides covering the 87 scaffolds are shown. The 87 scaffolds are ordered from left to right according to their length. **(B)** Gene copy number of four representative ascospores (numbers 1, 5, 9, and 13) of asci III (Figure 1B) with 16 viable ascospores. **(C)** Gene copy number of two representative ascospores (numbers 1 and 9) of asci IV (Figure 1B) with eight inviable ascospores. **(D)** Gene copy number of two representative ascospores (numbers 1, 9, and 13) of asci I (Figure 1B) with 12 inviable ascospores. The three strains with duplicated segments (D1-SAN, D2-SAN, and D3-SAN) and the three euploid control strains (N0-Euploidy, N1-Euploidy, and N2-Euploidy) are also indicated.

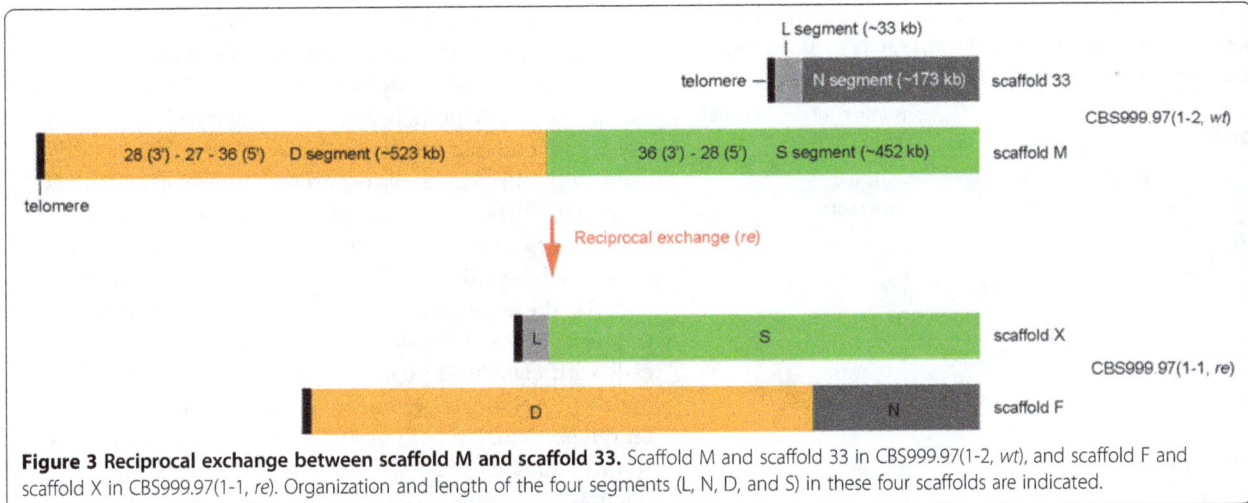

Figure 3 Reciprocal exchange between scaffold M and scaffold 33. Scaffold M and scaffold 33 in CBS999.97(1-2, *wt*), and scaffold F and scaffold X in CBS999.97(1-1, *re*). Organization and length of the four segments (L, N, D, and S) in these four scaffolds are indicated.

Deep sequencing of the CBS999.97(1-2, *wt*) genome (http://bc.imb.sinica.edu.tw/~lab229/Text_file_T1-4.rar) and genomic PCR analyses (Additional file 4: Figures S3B and S3C) revealed that three scaffolds (27, 28, and 36) were contiguous segments and were together referred to as scaffold "M" (Figure 2C and Figure 3). Accordingly, we further referred to the about 452-kb unduplicated region of scaffold M as the "single (S) segment" (Figure 3). Genomic PCR and sequencing results also revealed that a repeated hexanucleotide sequence TTAGGG, which is the telomeric repeat of QM6a(1-2, *wt*) [25], is connected to the 3′ terminus of scaffold 28. This finding provides evidence that the telomere is located near the 5′ terminus of scaffold M (Figure 3).

Identification of the chromosome rearrangement region in CBS999.97 wild isolate strains

The ascospore phenotype of our study seemed to be similar to that of the hybrid infertility of *S. pombe* and *S. kambucha* [14]. Zanders *et al.* indicated that chromosome rearrangement leads to genome heterozygosity of the two closely related haploid strains, and the meiotic recombination between the two homoeologous chromosomes generates both inviable and viable SAN progeny [14]. Our aCGH results also supported this conclusion. The duplication of the D segment in the viable SAN ascospores apparently resulted from a chromosome rearrangement region located in the second intron of a novel gene (ID 112288; scaffold 36:54323-54324 bp) (Additional file 5: Figure S4A). Additional PCR and sequencing analysis showed that, compared to QM6a(1-2, *wt*) [25], the first intron of 112288 was deleted in both CBS999.97(1-2, *wt*) (Additional file 5: Figure S4A) and CBS999.97(1-1, *re*) (Additional file 5: Figure S4B). Moreover, compared with the genomes of CBS999.97(1-2, *wt*) and QM6a(1-2, *wt*), a large chromosome translocation was found at this region of the CBS999.97(1-1, *re*)

genome: (1) the 5′ terminus of scaffold 33 (1-33,249 bp; referred to as the "L" segment) links the S segment of scaffold M to form a new scaffold "X" in CBS999.97(1-1, *re*); and (2) the 3′ terminus of scaffold 33 (33,250-207,997 bp, referred to as the "N" segment) and the D segment of scaffold M physically link and form a new scaffold "F" in CBS999.97(1-1, *re*) (Figure 3, (http://bc.imb.sinica.edu.tw/~lab229/Text_file_T1-4.rar) and Additional file 5: Figure S4B).

Genome heterozygosity is responsible for the production of SAN ascospores

To confirm whether this reciprocal exchange (*re*) allele is the cause of meiotic drive segmental duplication, we identified two new haploids, CBS999.97(1-1, *wt*) and CBS999.97(1-2, *re*), from the offspring of the two parental haploids, CBS999.97(1-1, *re*) and CBS999.97(1-2, *wt*) (Additional file 1: Figure S1, Additional file 5: Figure S4C and S4D). We found that all 16 ascospores generated from sexually crossing CBS999.97(1-1, *re*) with CBS999.97(1-2, *re*) or CBS999.97(1-1, *wt*) with CBS999.97(1-2, *wt*) were viable. In addition, they all germinated to form mycelia with green conidia (Table 1). These results indicate that the genome heterozygosity is the primary cause for production of the inviable ascospores as well as the viable, white-conidia ascospores.

Next, we examined whether the non-homologous end joining (NHEJ) DNA repair pathway is responsible for the production of inviable ascospores. Two NHEJ proteins, Ku70 and Ku80, form a heterodimer and function as a molecular scaffold at DSB ends to which other NHEJ proteins (such as the DNA ligase IV, Lig4) can bind [32]. The *T. reesei* genes that encode Ku70, Ku80, and Lig4 were previously referred to as *tku70*, *tku80*, and *tmus53*, respectively [33,34]. We found that deletion of either *tku70* or *tmus53* did not affect meiosis, ascospore number, or ascospore viability in any of the

Table 1 The chromosome reciprocal exchange (*re*) allele, but not NHEJ genes (*tku70* and *tmus53*), is responsible for the formation of meiotic drive SAN progeny

Strain background	Sexual crossing	Number of asci with 4 or 8 inviable ascospores	Number of asci dissected
CBS999.97	(1-1, *re*) x (1-2, *wt*)	19	20
	(1-1, *wt*) x (1-2, *wt*)	0	10
	(1-1, *re*) x (1-2, *re*)	0	10
CBS999.97 *tku70*Δ	(1-1, *re*) x (1-2, *wt*)	8	10
	(1-1, *wt*) x (1-2, *wt*)	0	10
	(1-1, *re*) x (1-2, *re*)	0	10
CBS999.97 *tmus53*Δ	(1-1, *re*) x (1-2, *wt*)	10	10
	(1-1, *wt*) x (1-2, *wt*)	0	10
	(1-1, *re*) x (1-2, *re*)	0	10

relevant strains (Table 1 and Additional file 5: Figure S4). These results suggest that homologous recombination (or crossover) during meiotic prophase or random chromosome segregation during the meiotic nuclear division (MI) may be responsible for the production of SAN ascospores.

Interhomolog recombination between two homoeologous chromosomes accounts for most, if not all, SAN ascospores during meiosis

We propose a hypothetical model (Figure 4) to explain the sexual crossing results in Figure 1. First, the hexadecads with 16 viable ascospores were referred to as "parental ditype (PD)"; all of these euploid ascospores could germinate and produce green conidia. Eight of them, such as CBS999.97(*wt*), have scaffold M and scaffold 33; the other eight ascospores, such as CBS999.97 (*re*), have scaffold F and scaffold X. Second, the hexadecads with eight white-conidia SAN ascospores and eight inviable SAN ascospores were referred to as "non-parental ditype (NPD)"; these viable, white-conidia SAN ascospores contain two D segments but no L segment (Figures 2C and 5). Accordingly, we inferred that the inviable SAN ascospores have two L segments but no D segment. Finally, the hexadecads with four inviable ascospores and 12 viable ascospores were referred to as "tetratype (TT)", containing four euploid ascospores with scaffold M and scaffold 33, four euploid ascospores with scaffold F and scaffold X, four viable SAN ascospores with two D segments but no L segment, and four

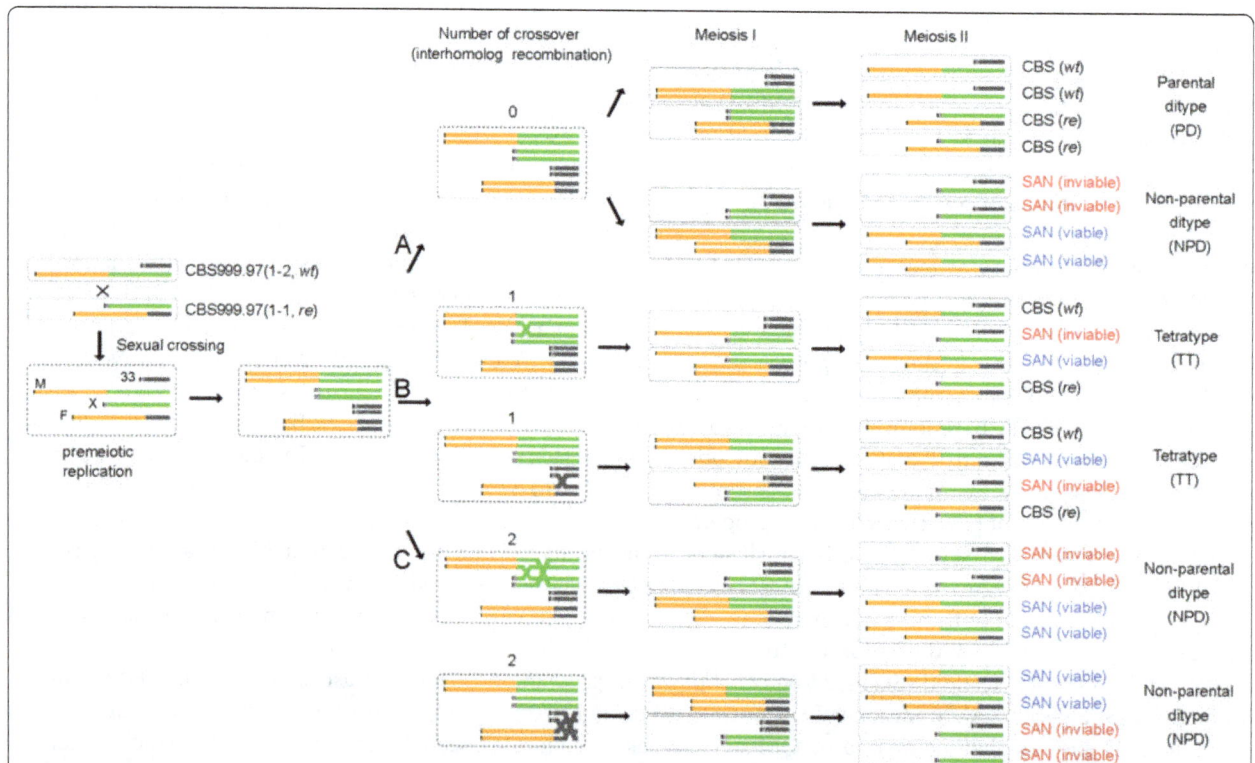

Figure 4 Meiotic crossover and chromosome segregation are responsible for the formation of segmental aneuploidy. The hypothetical model of meiotic drive viable and inviable ascospores. **(A)** In the absence of crossover (interhomolog recombination), the parental ditype (PD) hexadecad asci and the non-parental ditype (NPD) hexadecad asci can be generated by random segregation of homologous chromosomes during the first meiotic nuclear division (MI). **(B)** A single crossover between scaffold M and scaffold F or scaffold X and scaffold 33 results in the tetratype (TT) hexadecad asci. **(C)** The NPD hexadecad asci can also be generated if two crossovers exist between scaffold M and F or scaffold X and 33.

Trichoderma reesei meiosis generates segmentally aneuploid progeny with higher xylanase-producing...

65

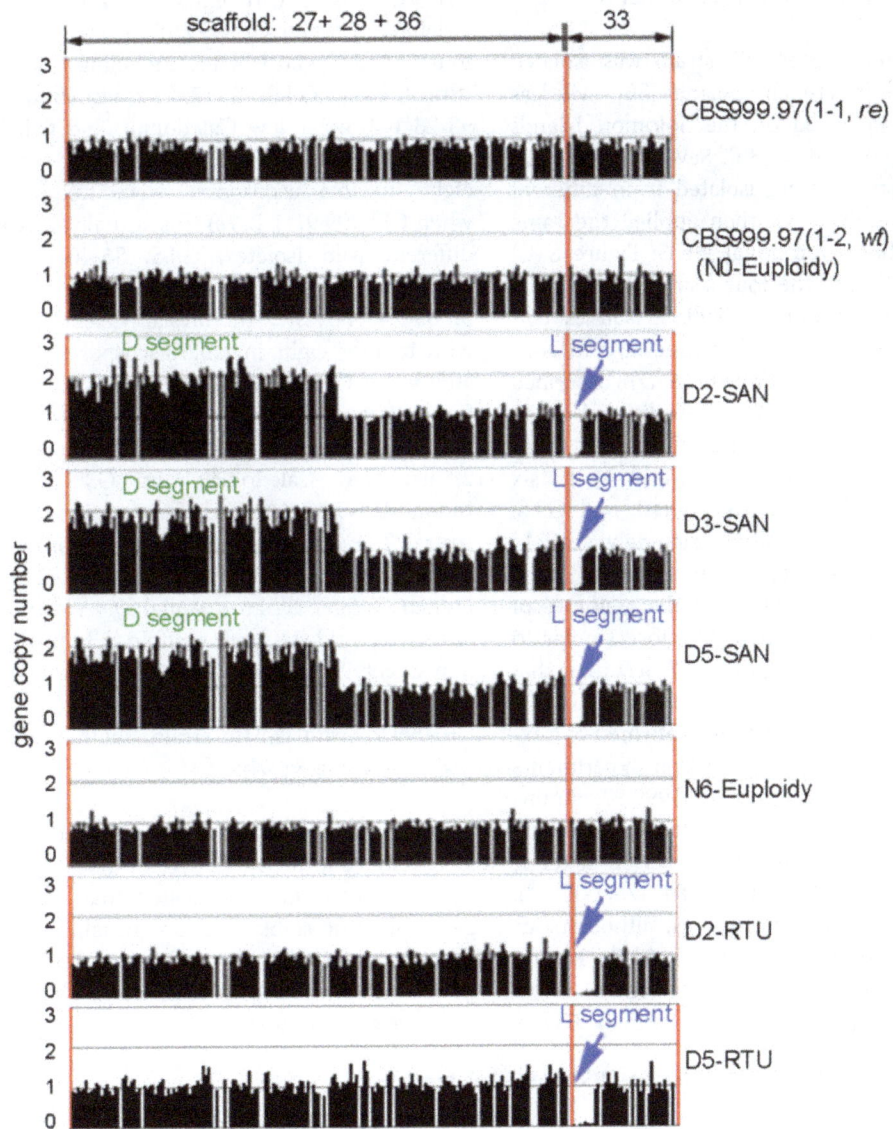

Figure 5 The SAN mutants contain a duplicated D segment but no L segment. Shown are aCGH results in scaffold M (27 + 28 + 36) and scaffold 33 of the two parental strains (upper two panels), three viable SAN progeny (D2, D3, and D5), and one viable euploid progeny (N6) (middle four panels), and the return-to-euploid (RTU) strains of D2-SAN and D5-SAN (D2-RTU and D5-RTU, lower two panels). The three SAN progeny contain two D segments and lose the L segment, whereas the D2-RTU and D5-RTU strains have a D segment but no L segment.

inviable SAN ascospores with two L segments but no D segment (Figures 2D and 5).

When no interhomolog recombination (or crossover) occurs between the four scaffolds (M, 33, X, and F) during meiotic prophase, PD and NPD can be produced simply by random chromosome segregation during MI (Figure 4A). In contrast, TT is likely generated via a single crossover between scaffold M and scaffold X or between scaffold 33 and scaffold F (Figure 4B). NPD may also be generated when two crossovers occur between two of these four scaffolds (Figure 4C). Because our single ascospore isolation experiments (Figure 1) revealed that there were more TT (>80%) than PD (about 10%) and NPD (about 9%), we suggest that meiotic recombination occurs at a high frequency between scaffold M and scaffold X or between scaffold 33 and scaffold F.

Notably, the hypothesis described here is consistent with our aCGH and genetic results: first, all three viable SAN ascospores we examined (D2, D3, and D5) had two D segments but no L segment (Figure 5); second, we inferred that the inviable SAN ascospores have two L segments but no D segment. The D segment is about 523 kb in length and contains at least 113 genes (Additional file 6: Table S2), including several putative essential genes such as an actin-like protein (ID 111468). Due to the lack of D segments, these inviable SAN ascospores failed to germinate.

The ancestral *T. reesei* genome is similar to that of CBS999.97(1-2, *wt*)

The sexually competent CBS999.97 strain was isolated from a storage lake in French Guiana [35], whereas QM6a(1-2, *wt*) was collected on the Solomon Islands during the Second World War [36]. Several other non-CBS999.97 haploid strains were isolated from different geographical locations [27]. We then applied the same diagnostic PCR method (Additional file 5: Figure S4C) to study the distribution of the four scaffolds (M, 33, F, X) in nine representative non-CBS999.97 isolates collected from French Guiana, Brazil, Indonesia, and New Caledonia. Using the four primers (A-D) described above (Additional file 2: Table S1 and Additional file 5: Figure S4C), we were able to detect scaffold 33 in eight isolates (except G.J.S. 84-473) and scaffold M in six isolates (except G.J.S. 84-473, G.J.S. 86-410, and G.J.S. 93-23) (Additional file 7: Table S3). Intriguingly, scaffold F and scaffold X were not detected in any of the nine non-CBS999.97 isolates, indicating that the genomes of these non-CBS999.97 isolates might be more similar to those of CBS999.97(1-2, *wt*) or QM6a(1-2, *wt*) than that of CBS999.97(1-1, *re*).

This hypothetical model was further confirmed by sexual crossing and single ascospore isolation experiments between those wild isolates and CBS999.97 strains (Table 2). CBS999.97(1-2, *wt*) sexually crossed with the three French Guiana *MAT1-1* isolates, G.J.S. 86-404(1-1, *wt*), G.J.S. 86-410(1-1, *wt*), and G.J.S. 84-473(1-1, *wt*), generated asci with 16 viable ascospores, although our diagnostic PCR method failed to detect scaffold M and scaffold 33 in G.J.S. 84-473(1-1, *wt*). Asci with 16 viable ascospores were also generated when CBS999.97(1-2,

wt) was crossed with G.J.S. 85-249(1-1, *wt*) (Indonesia, Celebes) or when CBS999.97(1-1, *wt*) was crossed with four *MAT1-1* wild isolates, including G.J.S. 89-7(1-1, *wt*) (Brazil, Para), G.J.S. 97-178(1-2, *wt*) (Brazil, Para), G.J.S. 93-23(1-2, *wt*) (New Caledonia), and G.J.S. 85-236(1-2, *wt*) (Indonesia, Celebes). In contrast, asci with four or eight inviable ascospores were frequently generated when CBS999.97(1-2, *re*) was sexually crossed with three different wild isolates, G.J.S. 86-404(1-1, *wt*; French Guiana), G.J.S. 86-410(1-1, *wt*; French Guiana), and G.J.S. 85-249(1-1, *wt*; Indonesia, Celebes). Similarly, asci with four or eight inviable ascospores were also generated when CBS999.97(1-1, *re*) was sexually crossed with four different wild isolates, G.J.S. 89-7(1-2, *wt*; Brazil, Para), G.J.S. 97-178(1-2, *wt*; Brazil, Para), G.J.S. 93-23(1-2, *wt*; New Caledonia), and G.J.S. 85-236(1-2, *wt*; Indonesia, Celebes). Therefore, we inferred that the ancestral *T. reesei* genomes likely contain scaffold M and scaffold 33, and scaffold F and scaffold X evolved later in French Guiana via an unequal DNA rearrangement between scaffold M and scaffold 33. Intriguingly, these non-CBS999.97 isolates examined here could only sexually cross with the haploid progeny generated from the CBS999.97 wild isolate strain but not with each other (Additional file 8: Table S4).

The viable SAN progeny exhibit return to euploidy (RTU) in vegetative growth

Several studies have suggested that the duplicated regions are not stable and are usually lost after several generations [15-18]. To test the stability of our SANs, they were continuously cultured in rich MEA medium and the genome-wide gene copy number was analyzed

Table 2 Sexual crossing of CBS999.97 with non-CBS999.97 isolates

Sexual crossing	Number of asci dissected	Number of asci with 4 or 8 inviable ascospores
CBS999.97(1-2, *wt*; French Guiana) & G.J.S. 86-404(1-1, *wt*; French Guiana)	7	0
CBS999.97(1-2, *wt*; French Guiana) & G.J.S. 86-410(1-1, *wt*; French Guiana)	8	0
CBS999.97(1-2, *wt*; French Guiana) & G.J.S. 84-473(1-1, #; French Guiana)	7	0
CBS999.97(1-1, *wt*; French Guiana) & G.J.S. 89-7(1-2, *wt*; Brazil, Para)	7	0
CBS999.97(1-1, *wt*; French Guiana) & G.J.S. 97-178(1-2, *wt*; Brazil, Para)	9	0
CBS999.97(1-1, *wt*; French Guiana) & G.J.S. 93-23(1-2, #; New Caledonia)	9	0
CBS999.97(1-1, *wt*; French Guiana) & G.J.S. 85-236(1-2, *wt*; Indonesia Celebes)	9	0
CBS999.97(1-1, *wt*; French Guiana) & G.J.S. 85-249(1-1, *wt*; Indonesia Celebes)	9	7
CBS999.97(1-2, *re*; French Guiana) & G.J.S. 86-404(1-1, *wt*; French Guiana)	9	8
CBS999.97(1-2, *re*; French Guiana) & G.J.S. 86-410(1-1, *wt*; French Guiana)	7	5
CBS999.97(1-1, *re*; French Guiana) & G.J.S. 89-7(1-2, *wt*; Brazil, Para)	9	8
CBS999.97(1-1, *re*; French Guiana) & G.J.S. 97-178(1-2 *wt*; Brazil, Para)	8	6
CBS999.97(1-1, *re*; French Guiana) & G.J.S. 93-23(1-2, #; New Caledonia)	8	4
CBS999.97(1-1, *re*; French Guiana) & G.J.S. 85-236(1-2, *wt*; New Caledonia)	7	6

#: Diagnostic PCR failed to detect M, F, 33 or X in G.J.S. 84-473 and G.J.S. 93-23

every ten days. The aCGH data showed that only one D segment was detected after 34 days of vegetative growth (Figure 6A). The results of physical analysis (n = 10) also confirmed that the D segment from scaffold F was lost during vegetative growth (Figure 6B), suggesting that the D segment in scaffold F is highly unstable compared to the one in scaffold M. Therefore, these RTU strains have genomes similar to CBS999.97 (1-2, *wt*) but lack the L segments (Figure 5).

Loss of the L segment resulted in a white-conidia phenotype

The viable SAN progeny exhibited a white-conidia phenotype before (Figure 1) and after RTU (Figure 6A). The *T. reesei* polyketide synthase 4 gene (*tpks4*) has been reported to be responsible for the green conidial pigmentation, and the *tpks4Δ* mutant produces white conidia [37]. Notably,

tpks4 is one of the nine annotated genes in the L segment (Figure 6C). We inferred that deletion of the L segment, but not D segment duplication, may lead to the white-conidia phenotype. To further determine whether the loss of the L segment is the only factor responsible for the white-conidia phenotype, we backcrossed the RTU strains with the parental strain CBS999.97(1-2, *wt*). The 16 offspring of the backcross showed an eight green-conidia and eight white-conidia phenotype (Figure 6D), supporting our hypothesis. Genomic PCR genotyping experiments further showed that only the green-conidia progeny but not the white-conidia progeny have the *tpks4* gene (Figure 6D). Finally, the aCGH results of the representing progeny confirmed that the green-conidia ascospore but not the white-conidia ascospore had the L segment (Figure 6E). These data imply that lack of *tpks4* is responsible for the white-conidia phenotype.

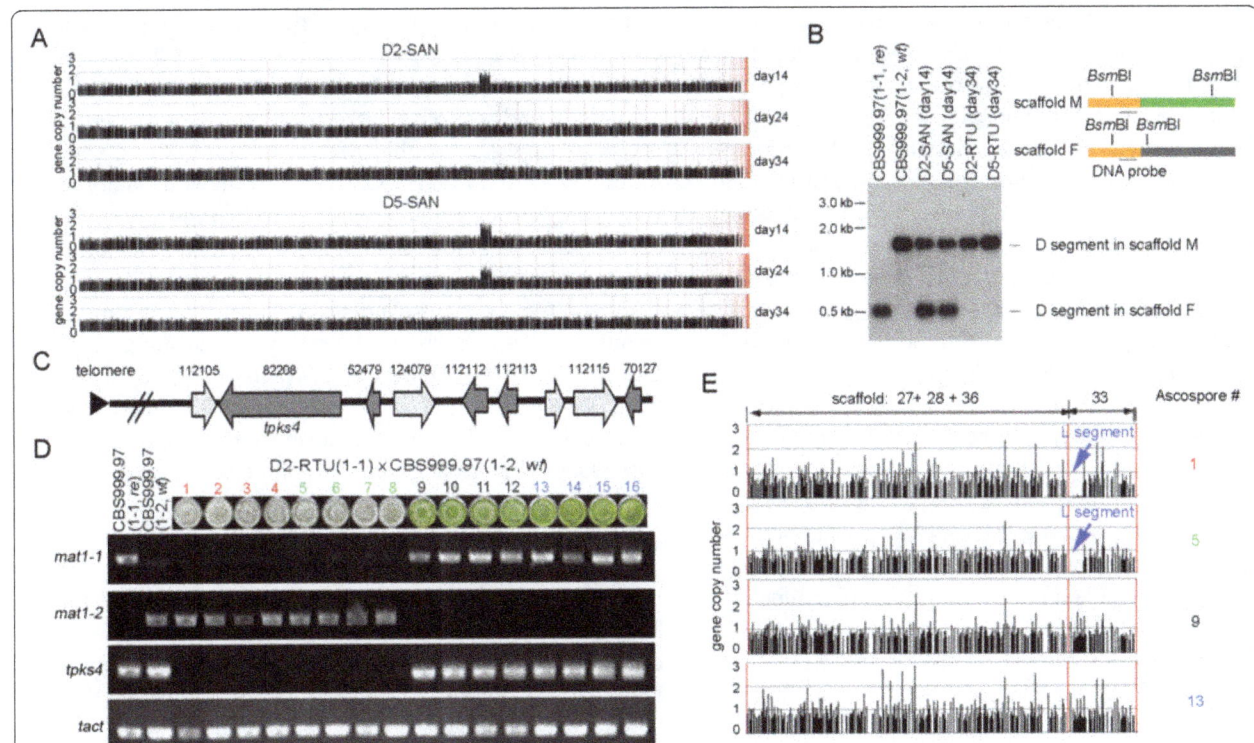

Figure 6 Loss of the L segment leads to a white colony of ascospores. (A) The genome stability of two different SAN strains (D2-SAN and D5-SAN) during vegetative growth was examined. The number of days post ascospore germination in a dextrose-containing malt extract agar (MEA) medium is indicated on the right. The Gene Expression Omnibus accession number is GSE-42359. **(B)** Southern blot analysis of the two D segments before and after RTU. The locations of the *BsmB*I restriction enzyme sites and the DNA probe used for Southern blot analysis are indicated in the right panel. The restriction fragments of the D segments in scaffold F and scaffold M are about 400 bp and 1,500 bp in length, respectively. The two parental haploid strains, CBS999.97(1-2, *re*) and CBS999.97(1-2, *wt*), were used as positive controls. **(C)** The organization and protein ID numbers of the nine annotated genes in the L segment. *T. reesei* polyketide synthase (*tpks4*) gene has been shown to be responsible for the formation of green conidia [37]. **(D)** Loss of the L segment or *tpks4* led to the white-conidia phenotype. Twenty hexadecads generated from sexual crossing of CBS999.97(1-2, *wt*) with D2-RTU or D5-RTU were dissected and isolated as described in Figure 1A. These 20 hexadecads all generated 16 ascospores. After germination, eight produced white conidia and the other eight produced green conidia. Shown are the conidia-color phenotype and genotyping data (by genomic PCR) of a representative hexadecad with 16 viable ascospores. **(E)** The aCGH results of four representative ascospores (numbers 1, 5, 9, and 16) in (D). The two white-conidia progeny (numbers 1 and 5) have no L segment, whereas the two green-conidia progeny (numbers 9 and 16) have the L segment.

The viable SAN progeny showed high hemicellulase-producing capability

A hallmark of the QM6a(1-2, *wt*) genome is that many genes encoding the carbohydrate-active enzymes (CAZymes) are non-randomly distributed in several gene clusters [25]. The CAZymes can cleave, build, and rearrange oligo- and polysaccharides [38]. The majority of the CAZyme genes in these clusters encode glycoside hydrolases that contribute to the degradation of lignocelluloses and plant cell walls. Previous transcriptomic studies also indicated that adjacent or nearly adjacent genes were coexpressed in four CAZyme gene clusters located in scaffolds 1, 6, 28, and 29 [25]. Notably, the CAZyme gene cluster in scaffold 28 is located at the D segment (Additional file 6: Table S2), including an endo-β-1,4-xylanase gene (ID 69276), a β-mannosidase (ID 69245), and the *cip2* glucuronoyl esterase gene (ID 123940). The D segment also contains the *xyn2* xylanase II gene in scaffold 27 (ID 123818). These four genes all encode enzymes with hemicellulase activity [38]. Indeed, the SAN progeny with two D segments exhibited a higher xylanase activity than the euploid progeny or their parental haploid strains (Figure 7A).

T. reesei is well known as an industrial workhorse for synthesizing cellulases and hemicellulases [25], and we therefore wanted to know if the SANs also showed enhancement in cellulase activity. The cellulase activity assay showed that compared with the wild-type control, the SANs did not produce higher levels of cellulases (Figure 7B). These data suggest that the production of cellulase and xylanase is differentially regulated in these SAN progeny.

The viable SAN progeny showed growth advantage

Although variations in chromosome copy number are often detrimental to organisms [15], studies in several microorganisms also indicate that DNA copy-number alterations can be beneficial, increasing survival under selective pressure [16-18]. When grown on a xylan-based medium, the SAN strains produced more biomass than either the parental CBS999.97 euploid strains or the RTU strains (Figure 7C). Given the lack of L segment in the RTU strains, we conclude that D segment duplication, but not L segment deletion, is responsible for the growth advantage in xylan-based media. Intriguingly, the SAN strain also grew better than the CBS999.97 euploid haploid on rice straw (Figure 7D). Taken together, these results suggest that meiotic drive segmental duplication apparently provides an advantage to transiently enhance the efficiency for degrading and utilizing lignocellulosic biomass.

Applying the sexual crossing strategy to improve industrial strains

To further improve the xylanase production activity of two cellulase-overproducing mutants, RUT-C30(1-2, *wt*) and QM9414(1-2, *wt*), we tried to generate SANs under both backgrounds. However, sexual crossing of CBS999.97 (1-1, *re*) with either RUT-C30(1-2, *wt*) or QM9414(1-2, *wt*) usually produced asci or hexadecads with no or only four viable ascospores (Additional file 9: Figure S5). Because the genomes of these two cellulase-overproducing mutants acquired numerous mutations, deletions, and chromosomal rearrangements via multiple rounds of physical and chemical mutagenesis [20,24], genome heterozygosity could also account for the meiotic drive lethality we observed here. Our results here suggest that sexual crossing should be more cautiously used to improve these industrially used hypersecretion mutants.

Discussion

The current study reveals several novel characteristics of *T. reesei* sexual development and genome plasticity. First, *T. reesei* generates asci with 16-part ascospores via meiosis and two further rounds of mitotic nuclear divisions. Second, our data suggest that the genome of ancestral *T. reesei*, such as that of CBS999.97(1-2, *wt*) and QM6a(1-2, *wt*), contains scaffold M and scaffold 33. Scaffold X and scaffold F were generated via a reciprocal exchange between scaffold M and scaffold 33 and were found only in the CBS999.97(1-1, *re*) haploid genome. Accordingly, the *re* allele was likely to be evolved later in French Guiana. How such a chromosomal rearrangement occurred remains an open question. Genomic plasticity is often found in filamentous fungal species, and is thought to be a widely used strategy during fungal evolution [39]. In fact, different *Trichoderma* isolates also show variation in the karyotype [29,30]. Our data had identified a specific DNA rearrangement locus in the CBS999.97 wild isolate, and this information may provide a better picture in the study of the evolution of *Trichoderma* spp. Notably, the *re* locus may not be the only DNA rearrangement site within the *T. reesei* wild isolates. The sexual crossing experiment showed that the non-CBS999.97 wild isolates could not mate with each other, including those isolated from the same geographic areas [27] (Additional file 8: Table S4). However, they could all undergo sexual reproduction with the CBS999.97 strains (Table 2). Within these wild isolates, novel diversity may be found and linked to the evolution of sexual development genes.

Third, fungi have been reported to utilize different mechanisms to generate SAN progeny via meiosis, including hybrid infertility in *S. pombe* and *S. kambucha* [14] and telomere-telomere fusion and chromosomal translocation between two different chromosomes in *C. neoformans* [12,13]. Here, we showed that, due to a reciprocal DNA rearrangement between two scaffolds, the *T. reesei* CBS999.97 wild isolate produces both viable and inviable SAN progeny via meiosis. Our genetic

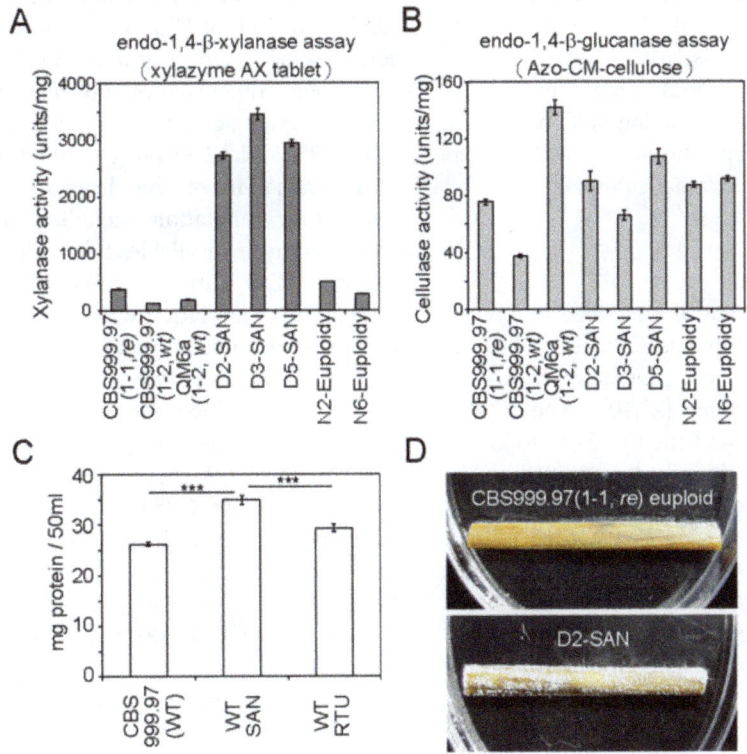

Figure 7 The SAN progeny produce higher levels of xylanases. (A) Xylanase and **(B)** cellulase specific activities (U/mg of mycelium) of the indicated strains were measured using xylazyme AX tablet and Azo-CM-cellulose as substrates, respectively. Experiments were conducted in triplicate and are presented with standard deviations. **(C)** The SAN strains produced more biomass than the two CBS999.97 euploid strains or the RTU strains in a xylan-based Mandels-Andreotti medium. Experiments were conducted with two different colonies, each in triplicate, and are presented with mean values ± SEM (error bars). **(D)** The D2-SAN grew better than the parental euploid strain CBS999.97(1-1, re) on rice straw.

results also revealed that interhomolog recombination or crossover occurred at a high frequency (>80%) to produce the "tetratype" asci (Figure 4B). Further investigation is needed to identify whether these crossover products are generated via a Spo11-induced or Spo11-independent DSB, where the DSB hotspot may be located, and if the DSB hotspot is created by a reciprocal exchange between scaffold M and scaffold 33.

Many *Trichoderma* species are found as anamorphs present in soils, where they act as plant beneficial fungi. In contrast, the teleomorphic *Hypocrea* species are most frequently found on decorticated wood or wood-rotting fungi (for example, wood ear fungi, shelf fungi, or agarics). With better lignocellulosic biomass degradation capability, the SAN sexual progeny we described in this report can likely provide adaptive advantages to the natural environments, especially in the early phase of colonization (the first two weeks of growth). Further study of the molecular mechanism leading to high hemicellulase production is thus of high interest. The D segment contains not only a few hemicellulase or xylanase encoding genes but also several novel transcription factors, including Gal4-like genes (ID: 70414, 111446, 111466, 36913) and fungal specific transcription factors

(ID: 123860, 5664, 111515, and 69077) (Additional file 6: Table S2). Further investigation will reveal whether these transcription factors control the genome-wide expression of hemicellulase or xylanase encoding genes. Finally, for economical application, finding a way to prevent the high hemicellulase-producing SAN strains from RTU or at least to prolong the stability of these SAN strains is important. The very first step is to determine the molecular mechanism of RTU.

Conclusion

Trichoderma reesei QM6a and its derivatives are industrial workhorse fungi that secrete cellulases and hemicellulases to degrade lignocellulosic biomass into glucose and xylose. CBS999.97 is the only *T. reesei* wild isolate strain that is sexually competent under laboratory conditions. Here we show that CBS999.97 sexual reproduction undergoes meiosis and two rounds of postmeiotic mitosis to yield asci with 16 linearly arranged ascospores. Notably, the two haploid genomes of the CBS999.97 wild isolate comprise a reciprocal arrangement between two chromosomal scaffolds. Due to sequence heterozygosity, most 16-spore asci contain four or eight inviable ascospores and an equal number of segmentally aneuploid

(SAN) ascospores. The meiotic driven gene copy number change readily allows these viable SAN progeny to display new phenotypes, that is, white conidia, higher levels of hemicellulases, and genome instability. Our results have revealed a new understanding of *T. reesei* evolution and sexual development and also provided novel perspectives for improving industrial strains.

Materials and methods

Strains and sexual crossing

Sexual crossing was carried out as described previously [27,40]. The two parental haploids CBS999.97(1-1, *re*) and CBS999.97(1-2, *wt*) were generated from the CBS999.97 wild isolate strain [27,40]. The QM6a *tku70Δ*(1-2, *wt*) mutant [34] and the QM6a *tmus53Δ*(1-2, *wt*) mutant [33] were described previously. The *tku70Δ* and *tmus53Δ* mutants in the CBS999.97 background were generated by crossing each QM6a mutant with the wild isolate CBS999.97(1-1, *re*), respectively. The corresponding offspring mutants were backcrossed at least twice with the CBS999.97(1-2, *wt*) strains. All the strains used in this study are listed in Additional file 10: Table S5.

Hexadecad dissection

Mature hexadecads were manually isolated from stromata and transferred onto the center of a 10-cm malt extract agar (MEA) plate. Yeast tetrad dissection using a micromanipulator was applied to sequentially separate and isolate each ascospore in a hexadecad (Figure 1A). The fiberglass needle could readily break the fragile ascus wall and separate each ascospore, leaving the remaining part intact.

Deep sequencing and *de novo* assembly of the wild-type CBS999.97(1-2, *wt*) genome

The shotgun library for 454 Sequencing was prepared with 0.5 μg of genomic DNA from the wild isolate CBS999.97(1-2, *wt*) haploid using the GS Rapid Library Prep Kit following the manufacturer's protocol (Roche 454; 454 Life Sciences, Branford, CT, USA). The resulting library was examined by the BioAnalyzer DNA Chip assay (Agilent Technologies; Santa Clara, CA, USA), and FAM fluorescence was quantified using a Modulus fluorometer (Turner Biosystems, Sunnyvale, CA, USA). Sequencing was performed on a GS FLX Titanium system in the High Throughput Sequencing Core Facility at the Biodiversity Research Center at Academia Sinica, Taiwan. Raw reads were obtained from 2.5 sequencing runs totaling 873 Mb, with median read lengths ranging from 351 to 454 nt among the five datasets. *De novo* assembly was performed using Newbler v.2.5.3 (Roche 454) on a single CPU. The draft genome assembly consisted of 1,087 contigs with sizes ranging from 500 bp to

404,555 bp, with an average contig and N50 size of 29,833 bp and 66,873 bp, respectively. Additional *de novo* assembly and gene annotation were conducted by an in-house computational core. The assembled CBS999.97 (1-2, *wt*) genomic sequences are available online (http://140.109.32.39/~lab229/contig_info/index.php).

For evaluation of the deep sequencing results, the community annotation including the Gene Ontology classification is available from the *T. reesei* genome database v.2.0 (http://genome.jgi-psf.org/Trire2/Trire2.home.html). Annotation was performed using BLAST to search for orthologous genes of *Trichoderma reesei* v.2.0 and *Trichoderma virens* Gv29-8 v2.0. Gene sequences were downloaded from the Joint Genome Institute (JGI) database (http://genome.jgi-psf.org/) [25,41]; there are 9,143 genes in QM6a(1-2, *wt*) and 12,427 in *T. virens*. We identified 8,106 CBS999.97(1-2, *wt*) and QM6a(1-2, *wt*) orthologous genes with high sequence similarity (≥90%). Of the remaining genes (similarity <90%), we found 49 that existed only in QM6a(1-2, *wt*), without any similarity to genes in the CBS999.97(1-2, *wt*) genome.

Array-based comparative genomic hybridization (aCGH) and data analysis

Genomic DNA was isolated using standard techniques and fragmented by a Bioruptor Sonicator (Diagenode, UCD-200) using repeated cycles of 75 seconds on (high) and then 75 seconds off for a total of 15 minutes, producing a median DNA size of 500 bp (range 200 to 1,000 bp). The fragmented DNA was then quantified using a NanoDrop ND-1000 UV-Vis Spectrophotometer to assess the genomic DNA concentration and purity. Fragmented genomic DNA samples were labeled with Cy5 or Cy3 using a NimbleGen Dual-Color DNA Labeling Kit (Roche NimbleGen, Madison, WI, USA). Test sample genomic DNA was end-labeled with Cy3, whereas CBS999.97(1-2, *wt*) genomic DNA was labeled with Cy5 and used as a reference to measure DNA copy number changes. The Cy5- and Cy3-labeled genomic DNA samples were hybridized to custom-designed oligonucleotide arrays (4 × 72,000 formation) by Roche-NimbleGen based on the CBS999.97(1-2, *wt*) genome sequence and *T. reesei* v2.0 genome sequence, respectively [40,42].

DNA end-labeling, hybridization, and scanning were performed by the Academia Sinica Institute of Molecular Biology Microarray Core using the NimbleGen Systems technique (NG_CGH&CNV_Guide_v7p0), following the vendor's standard operating protocol. Image data were processed using NimbleScan software version 2.6.3 (Roche NimbleGen) to obtain the raw intensity data NimbleGen(.pair file). Data analysis and normalization were performed using Agilent GeneSpring GX 11.5.1 by an in-house bioinformatics core. Raw intensity scales were transformed by quantile normalization which was used to correct array biases and make all distributions uniform.

Shake flask cultures

Conidia from three-day-old plates (10-cm diameter dishes) were harvested with 1 mL of sterile spore solution [0.8% NaCl, 0.05% Tween 20 (Sigma)], vortexed and filtered through glass wool. The volume was then adjusted to OD_{600nm} of 0.3, and 1 mL of the spore suspension was transferred to 250-mL Erlenmeyer flasks containing 50 mL of Mandels-Andreotti basal medium prepared in 0.1 M citrate-phosphate buffer (pH 5.0) and supplemented with 1 g L^{-1} peptone and 0.33 g L^{-1} urea. As a carbon source, 10 g L^{-1} Solka Floc 200 (International Fiber Corporation, North Tonawanda, NY, USA) or birchwood xylan (Sigma Aldrich, USA) was added to induce cellulase or xylanase expression, respectively. After three days of cultivation in constant darkness at 25°C on a rotary shaker (200 rpm), the flask culture was used for biomass determination and enzymatic assays.

Biomass determination

For biomass determination, 50 mL shake flask culture was filtered onto a dry Miracloth (Calbiochem, Darmstadt, Germany), washed with distilled water, and dried with paper towels. The fungal mycelia were collected, frozen with liquid nitrogen, and stored at -80°C until use. For total protein quantification, the fungal sample was dissolved in 5 mL of 0.1 N NaOH. The solution was sonicated by a digital sonifier (Branson) at a duty cycle setting of 30% for six minutes total with 30 seconds on and 30 seconds off. The sample was then incubated at room temperature for three hours and centrifuged at 5251 × g for 10 minutes. The protein concentration of the supernatant was determined by modified Lowry protein assay, using bovine serum albumin as a standard.

Enzymatic assays

For endo-1,4-β-glucanase (cellulase) activity measurements, supernatant diluted from 1:2 to 1:10 was added to an Azo-CM-cellulose solution (S-ACMCL; Megazyme International, Wicklow, Ireland). The procedure was carried out according to the manufacturer's instructions. Azo-CM-cellulose is a dyed polysaccharide containing Remazol Brilliant Blue R at a concentration of approximately one dye molecule per 20 sugar residues (Megazyme International). This assay is much more sensitive (50- to 100-fold) than filter paper assays (such as the DNS method).

To measure endo-1,4-β-D-xylanase (xylanase) activity, supernatant diluted from 1:2 to 1:10 was prewarmed at 40°C for five minutes. A xylazyme AX test tablet substrate

(T-XAX200; Megazyme International) was added to the supernatant (0.5 mL) and incubated at 40°C for 10 minutes. The reaction was stopped by adding 10 mL stop solution (2% Trizma Base, pH about 9). The sample was centrifuged at 2000 × g for 10 minutes, and the absorbance was measured at 590 nm. *Aspergillus niger* xylanase (about 300 mU/mL) was used as a control following the procedure specified by the manufacturer.

Southern blot analysis

The Southern blot analysis procedure was followed using standard techniques. Genomic DNA was first digested by the indicated restriction enzyme, and 0.8% agarose gel was used for the gel electrophoresis. The DNA probes were amplified with indicated primers by PCR. The Southern blots were exposed to an X-ray film (GE Healthcare, USA).

Rice straw growth assay

The conidia of the indicated strains were collected with Mandels-Andreotti basal medium (without carbohydrate supplement) and purified through glass wool. 2×10^6 conidia were cultured on autoclaved 70 mm rice straw at 25°C, under conditions of 12 hours light/dark for five days. 1 mL of Mandels-Andreotti basal medium (without carbohydrate supplement) was added to the rice straw before the incubation.

Microarray data access

The microarray data and the related protocols are available at the GEO site (http://www.ncbi.nlm.nih.gov/geo/) under accession numbers GSE-41965, GSE-42359, GSE-59350.

Additional files

Additional file 1: Figure S1. Genotyping. Genomic PCR analysis of *mat1-1*, *mat1-2*, scaffold F, scaffold X, scaffold 33, scaffold M, and *tact* (actin) genes in all 16 viable ascospores of asci IV in Figure 1B. The parental wild isolate haploid strains, CBS999.97(1-1, *re*) and CBS999.97(1-2, *wt*), were used as controls. The nucleotide sequences of the PCR primers are listed in Additional file 2: Table S1.

Additional file 2: Table S1. Oligonucleotide primers used in this study.

Additional file 3: Figure S2. Visualization of four rounds of nuclear division during *T. reesei* ascospore formation/maturation. **(A-G)** Developing asci were manually dissected, stained with DAPI, and then visualized by fluorescent microscopy. DIC and DAPI fluorescent images are shown. Nuclei (N) are marked by white arrows. **(H)** A DIC image of developing asci showing synchronous division of eight nuclei (8 N) into 16 nuclei (16 N).

Additional file 4: Figure S3. Scaffold M is a contiguous segment that includes the scaffolds 27, 28, and 36. **(A)** A representative aCGH result of viable SAN progeny. Normalized means of gene copy number for the oligonucleotides covering scaffold 25 to scaffold 37 are shown. These 13 scaffolds are ordered from left to right according to their length. **(B)** A schema illustrates the order of scaffold 28(3'), scaffold 27, scaffold 36, and scaffold 28(5') in the scaffold M. The locations of six PCR primers (E, F, G, H, I, and J; see Additional file 2: Table S1) in the scaffold M are indicated.

(C) Genomic PCR. The genomic DNA of two parental haploids strains, CBS999.97(1-1, *re*) (1) and CBS999.97(1-2, *wt*) (2), were amplified using three pairs of PCR primers (E/F, G/H, and I/J). The expected PCR products are 7,168 bp, 1,000 bp, and 1,272 bp in length, respectively. The *tact* (actin) gene was used as positive control for genomic PCR.

Additional file 5: Figure S4. Differential chromosomal organization in the genomes of QM6a(1-2, *wt*), CBS999.97(1-1, *re*), and CBS999.97(1-2, *wt*), respectively. **(A)** Scaffold 33 and scaffold M in QM6a(1-2, *wt*) and CBS999.97(1-2, *wt*). Scaffold 27, scaffold 28, scaffold 33, and scaffold 36 are indicated. The 5′ terminus of scaffold M has a telomere (in black), because the corresponding end in scaffold 28(3′) is connected to multiple copies of a repeated hexanucleotide sequence, TTAGGG, which is the telomeric repeat of QM6a(1-2, *wt*) [25]. The D segment and the S segment of scaffold M are indicated in orange and green, respectively. The L segment, located at the 5′ terminus of scaffold 33, is indicated in light gray, and the remaining portion of scaffold 33, the N segment, is indicated in dark gray. The three exons (E1, E2, and E3) of a novel gene (ID: 112288) in scaffold 36 are indicated. The first exon (E1) only exists in the QM6a(1-2, *wt*) genome. **(B)** Scaffold F and scaffold X in CBS999.97 (1-1, *re*). The nucleotide sequences of scaffold M, scaffold 33, scaffold F, and scaffold X are available online (http://bc.imb.sinica.edu.tw/~lab229/Text_file_T1-4.rar). **(C)** Schema illustrating the location of PCR primers (A, B, C, and D) used for the genotyping scaffold M, scaffold F, scaffold 33, and scaffold X, respectively. The nucleotide sequences of these four primers are listed in Additional file 2: Table S1. **(D)** PCR genotyping of indicated haploid strains. The two parental haploid strains, CBS999.97 (1-1, *re*) and CBS999.97(1-2, *wt*), were used as positive controls. CBS999.97 (1-1, *wt*) and CBS999.97(1-2, *re*) are progeny generated by sexually crossing CBS999.97(1-1, *re*) with CBS999.97(1-2, *wt*). The corresponding progeny in *tku70*Δ and *tmus53*Δ were also generated by sexually crossing, respectively (Additional file 10: Table S5).

Additional file 6: Table S2. Genes on the D segment.

Additional file 7: Table S3. Genomic PCR genotyping the four scaffolds (M, F, 33, X) in *T. reesei* wild isolates and industrial strains.

Additional file 8: Table S4. Sexual crossing of the non-CBS999.97 isolates.

Additional file 9: Figure S5. Hexadecad dissection of asci generated from sexual crossing CBS999.97(1-1, *re*) with RUT-C30(1-2, *wt*) or QM9414(1-2, *wt*). Hexadecads from sexual crossing of CBS999.97(1-1, *re*) with RUT-C30(1-2, *wt*) (n ≥ 10) **(A)** or QM9414(1-2, *wt*) (n ≥ 10) **(B)**. Sixteen ascospores from each hexadecad were sequentially separated and aligned as described in Figure 1A. The inviable ascospores are indicated by a black circle with a cross.

Additional file 10: Table S5. Strains used in this study.

Abbreviations
aCGH: array-based comparative genomic hybridization; DAPI: 4′,6-diamidino-2-phenylindole; DIC: differential interference contrast; DSB: double-strand break; RTU: return to euploidy; SAN: segmentally aneuploid.

Competing interests
The authors declare that they have no competing interests.

Authors' contributions
YCC, WCL, CLC, and HCK designed and performed the experiments and analyzed the data. PWCH and SYT analyzed the deep-sequencing and aCGH data. MS provided the *T. reesei* wild isolate strains. TFW conceived and designed the experiments and analyzed the data. YCC and TFW wrote the paper. All authors read and approved the final manuscript.

Acknowledgements
This work received support by Academia Sinica to TFW. We thank Shu-Yu Liang for research assistance and Astrid R. Mach-Aigner (Vienna University of Technology, Austria) for the QM6a *tmus53*Δ mutant.

Author details
[1]Taiwan International Graduate Program in Molecular and Cellular Biology, Academia Sinica, Taipei 115, Taiwan. [2]Institute of Life Sciences, National Defense Medical Center, Taipei 115, Taiwan. [3]Institute of Molecular Biology, Academia Sinica, Taipei 115, Taiwan. [4]Institute of Genome Sciences, National Yang-Ming University, Taipei 112, Taiwan. [5]Austrian Institute of Technology, Health and Environment Department, Bioresources, University and Research Center, UFT Campus Tulln, Tulln/Donau 3430, Austria. [6]Present address: Department of Forest Sciences, University of Helsinki, Helsinki, Finland.

References

1. Cole F, Keeney S, Jasin M. Evolutionary conservation of meiotic DSB proteins: more than just Spo11. Genes Dev. 2010;24:1201–7.
2. Crown KN, Savytskyy OP, Malik SB, Logsdon J, Williams RS, Tainer JA, et al. A mutation in the FHA domain of *Coprinus cinereus* Nbs1 leads to Spo11-independent meiotic recombination and chromosome segregation. G3 (Bethesda). 2013;3:1927–43.
3. Bowring FJ, Yeadon PJ, Catcheside DE. Residual recombination in *Neurospora crassa spo11* deletion homozygotes occurs during meiosis. Mol Genet Genomics. 2013;288:437–44.
4. Turner BC, Perkins DD. Spore killer, a chromosomal factor in *Neurospora* that kills meiotic products not containing it. Genetics. 1979;93:587–606.
5. Raju NB. Spore killers: meiotic drive elements that distort genetic ratios. In: Osiewacz HD, editor. Molecular biology of fungal development. New York: Marcel Dekker; 2002. p. 275.
6. Harvey AM, Rehard DG, Groskreutz KM, Kuntz DR, Sharp KJ, Shiu PK, et al. A critical component of meiotic drive in *Neurospora* is located near a chromosome rearrangement. Genetics. 2014;197:1165–74.
7. Goodwin SB, M'Barek SB, Dhillon B, Wittenberg AH, Crane CF, Hane JK, et al. Finished genome of the fungal wheat pathogen *Mycosphaerella graminicola* reveals dispensome structure, chromosome plasticity, and stealth pathogenesis. PLoS Genet. 2011;7:e1002070.
8. Wittenberg AH, van der Lee TA, Ben M'barek S, Ware SB, Goodwin SB, Kilian A, et al. Meiosis drives extraordinary genome plasticity in the haploid fungal plant pathogen *Mycosphaerella graminicola*. PLoS One. 2009;4:e5863.
9. Coleman JJ, Rounsley SD, Rodriguez-Carres M, Kuo A, Wasmann CC, Grimwood J, et al. The genome of *Nectria haematococca*: contribution of supernumerary chromosomes to gene expansion. PLoS Genet. 2009;5:e1000618.
10. Taga M, Murata M, VanEtten HD. Visualization of a conditionally dispensable chromosome in the filamentous ascomycete *Nectria haematococca* by fluorescence in situ hybridization. Fungal Genet Biol. 1999;26:169–77.
11. Tegtmeier KJ, VanEtten H. Genetic studies on selected traits of *Nectria haematococca*. Phytopathology. 1982;72:604–7.
12. Ni M, Feretzaki M, Li W, Floyd-Averette A, Mieczkowski P, Dietrich FS, et al. Unisexual and heterosexual meiotic reproduction generate aneuploidy and phenotypic diversity de novo in the yeast *Cryptococcus neoformans*. PLoS Biol. 2013;11:e1001653.
13. Fraser JA, Huang JC, Pukkila-Worley R, Alspaugh JA, Mitchell TG, Heitman J. Chromosomal translocation and segmental duplication in *Cryptococcus neoformans*. Eukaryot Cell. 2005;4:401–6.
14. Zanders SE, Eickbush MT, Yu JS, Kang JW, Fowler KR, Smith GR, et al. Genome rearrangements and pervasive meiotic drive cause hybrid infertility in fission yeast. Elife. 2014;3:e02630.
15. Tang YC, Amon A. Gene copy-number alterations: a cost-benefit analysis. Cell. 2013;152:394–405.
16. Gresham D, Desai MM, Tucker CM, Jenq HT, Pai DA, Ward A, et al. The repertoire and dynamics of evolutionary adaptations to controlled nutrient-limited environments in yeast. PLoS Genet. 2008;4:e1000303.
17. Yona AH, Manor YS, Herbst RH, Romano GH, Mitchell A, Kupiec M, et al. Chromosomal duplication is a transient evolutionary solution to stress. Proc Natl Acad Sci U S A. 2012;109:21010–5.
18. Chang SL, Lai HY, Tung SY, Leu JY. Dynamic large-scale chromosomal rearrangements fuel rapid adaptation in yeast populations. PLoS Genet. 2013;9:e1003232.
19. Schuster A, Schmoll M. Biology and biotechnology of *Trichoderma*. Appl Microbiol Biotechnol. 2010;87:787–99.
20. Peterson R, Nevalainen H. *Trichoderma reesei* RUT-C30 - thirty years of strain improvement. Microbiology. 2012;158:58–68.
21. Montenecourt Bland S, Eveleigh Douglas E. Selective screening methods for the isolation of high yielding cellulase mutants of *Trichoderma reesei*. In: Hydrolysis of

Trichoderma reesei meiosis generates segmentally aneuploid progeny with higher xylanase-producing...

73

cellulose: mechanisms of enzymatic and acid catalysis. Washington, DC: ACS Publications; 1979. p. 289–301. Advances in Chemistry, Vol. 181.

22. Montenecourt BS, Eveleigh DE. Preparation of mutants of *Trichoderma reesei* with enhanced cellulase production. Appl Environ Microbiol. 1977;34:777–82.

23. Mandels M, Weber J, Parizek R. Enhanced cellulase production by a mutant of *Trichoderma viride*. Applied Microbiol. 1971;21:152–4.

24. Vitikainen M, Arvas M, Pakula T, Oja M, Penttila M, Saloheimo M. Array comparative genomic hybridization analysis of *Trichoderma reesei* strains with enhanced cellulase production properties. BMC Genomics. 2010;11:441.

25. Martinez D, Berka RM, Henrissat B, Saloheimo M, Arvas M, Baker SE, et al. Genome sequencing and analysis of the biomass-degrading fungus *Trichoderma reesei* (syn. *Hypocrea jecorina*). Nat Biotechnol. 2008;26:553–60.

26. Samuels GJ, Petrini O, Manguin S. Morphological and macromolecular characterization of *Hypocrea schweinitzii* and its *Trichoderma* anamorph. Micologia. 1994;86:421–35.

27. Seidl V, Seibel C, Kubicek CP, Schmoll M. Sexual development in the industrial workhorse *Trichoderma reesei*. Proc Natl Acad Sci U S A. 2009;106:13909–14.

28. Penttila M, Nevalainen H, Ratto M, Salminen E, Knowles J. A versatile transformation system for the cellulolytic filamentous fungus *Trichoderma reesei*. Gene. 1987;61:155–64.

29. Herrera-Estrella A, Goldman GH, van Montagu M, Geremia RA. Electrophoretic karyotype and gene assignment to resolved chromosomes of *Trichoderma* spp. Mol Microbiol. 1993;7:515–21.

30. Mantyla AL, Rossi KH, Vanhanen SA, Penttila ME, Suominen PL, Nevalainen KM. Electrophoretic karyotyping of wild-type and mutant *Trichoderma longibrachiatum* (*reesei*) strains. Curr Genet. 1992;21:471–7.

31. Jung MK, Ovechkina Y, Prigozhina N, Oakley CE, Oakley BR. The use of beta-D-glucanase as a substitute for Novozym 234 for immunofluorescence and protoplasting. Fungal Genet Newsletter. 2000;47:65–6.

32. Deriano L, Roth DB. Modernizing the nonhomologous end-joining repertoire: alternative and classical NHEJ share the stage. Annu Rev Gen. 2013;47:433–55.

33. Steiger MG, Vitikainen M, Uskonen P, Brunner K, Adam G, Pakula T, et al. Transformation system for *Hypocrea jecorina* (*Trichoderma reesei*) that favors homologous integration and employs reusable bidirectionally selectable markers. Appl Environ Microbiol. 2011;77:114–21.

34. Guangtao Z, Hartl L, Schuster A, Polak S, Schmoll M, Wang T, et al. Gene targeting in a nonhomologous end joining deficient *Hypocrea jecorina*. J Biotechnol. 2009;139:146–51.

35. Lieckfeldt E, Kullnig C, Samuels GJ, Kubicek CP. Sexually competent, sucrose- and nitrate-assimilating strains of *Hypocrea jecorina* (*Trichoderma reesei*) from South American soils. Mycologia. 2000;92:374–80.

36. Mandels M, Reese ET. Induction of cellulase in *Trichoderma viride* as influenced by carbon sources and metals. J Bacteriol. 1957;73:269–78.

37. Atanasova L, Knox BP, Kubicek CP, Druzhinina IS, Baker SE. The polyketide synthase gene *pks4* of *Trichoderma reesei* provides pigmentation and stress resistance. Eukaryot Cell. 2013;12:1499–508.

38. Hakkinen M, Arvas M, Oja M, Aro N, Penttila M, Saloheimo M, et al. Re-annotation of the CAZy genes of *Trichoderma reesei* and transcription in the presence of lignocellulosic substrates. Microb Cell Fact. 2012;11:134.

39. Stukenbrock EH. Evolution, selection and isolation: a genomic view of speciation in fungal plant pathogens. New Phytologist. 2013;199:895–907.

40. Chen CL, Kuo HC, Tung SY, Hsu PW, Wang CL, Seibel C, et al. Blue light acts as a double-edged sword in regulating sexual development of *Hypocrea jecorina* (*Trichoderma reesei*). PLoS One. 2012;7:e44969.

41. Nordberg H, Cantor M, Dusheyko S, Hua S, Poliakov A, Shabalov I, et al. The genome portal of the Department of Energy Joint Genome Institute: 2014 updates. Nucleic Acid Res. 2014;42:D26–31.

42. Tisch D, Kubicek CP, Schmoll M. The phosducin-like protein PhLP1 impacts regulation of glycoside hydrolases and light response in *Trichoderma reesei*. BMC Genomics. 2011;12:613.

5

Detoxification of 5-hydroxymethylfurfural by the *Pleurotus ostreatus* lignolytic enzymes aryl alcohol oxidase and dehydrogenase

Daria Feldman[1], David J Kowbel[2], N Louise Glass[2], Oded Yarden[1] and Yitzhak Hadar[1*]

Abstract

Background: Current large-scale pretreatment processes for lignocellulosic biomass are generally accompanied by the formation of toxic degradation products, such as 5-hydroxymethylfurfural (HMF), which inhibit cellulolytic enzymes and fermentation by ethanol-producing yeast. Overcoming these toxic effects is a key technical barrier in the biochemical conversion of plant biomass to biofuels. *Pleurotus ostreatus*, a white-rot fungus, can efficiently degrade lignocellulose. In this study, we analyzed the ability of *P. ostreatus* to tolerate and metabolize HMF and investigated relevant molecular pathways associated with these processes.

Results: *P. ostreatus* was capable to metabolize and detoxify HMF 30 mM within 48 h, converting it into 2,5-bis-hydroxymethylfuran (HMF alcohol) and 2,5-furandicarboxylic acid (FDCA), which subsequently allowed the normal yeast growth in amended media. We show that two enzymes groups, which belong to the ligninolytic system, aryl-alcohol oxidases and a dehydrogenase, are involved in this process. HMF induced the transcription and production of these enzymes and was accompanied by an increase in activity levels. We also demonstrate that following the induction of these enzymes, HMF could be metabolized *in vitro*.

Conclusions: Aryl-alcohol oxidase and dehydrogenase gene family members are part of the transcriptional and subsequent translational response to HMF exposure in *P. ostreatus* and are involved in HMF transformation. Based on our data, we propose that these enzymatic capacities of *P. ostreatus* either be integrated in biomass pretreatment or the genes encoding these enzymes may function to detoxify HMF via heterologous expression in fermentation organisms, such as *Saccharomyces cerevisiae*.

Keywords: *Pleurotus ostreatus*, 5-hydroxymethylfurfural (HMF), Aryl-alcohol oxidase, Aryl-alcohol dehydrogenase

Background

Ethanol biofuel derived from lignocellulosic biomass is a viable alternative to fossil fuel-based transportation fuels [1]. Unlike first-generation biofuels, which are produced from corn, lignocellulosic biofuels do not compete with food-derived ethanol and can be made from abundant and renewable plant biomass sources [2]. Complex carbohydrates in the plant cell wall, such as cellulose and hemicellulose, are closely associated with lignin, and pretreatment using thermo and/or chemical processes is necessary to increase their availability for enzymatic hydrolysis and fermentation

[1,2]. In spite of the necessity for pretreatment of plant biomass, it can be a rate-limiting step in fermentation due to the production of inhibitory compounds, particularly furans, such as 5-hydroxymethylfurfural (HMF) and furfural and other phenolic compounds [3-5]. The level of furans produced during pre-treatment varies according to the type of raw material and the pretreatment procedure [6-9].

Furfural and HMF are the most potent inhibitors generated via pretreatment [10-13]. The effect on the growth rate on *Saccharomyces cerevisiae* and the subsequent decrease in the fermentation rate is higher for furfural than for HMF, but the effect of HMF lasts longer [14]. Several mechanisms may explain the inhibition effects on yeast growth and ethanol fermentation by exposure to furans. *In vitro* experiments and crude cell extract measurements showed that HMF directly inhibited alcohol dehydrogenase, pyruvate

* Correspondence: yitzhak.hadar@mail.huji.ac.il
[1]Department of Plant Pathology and Microbiology, The R.H. Smith Faculty of Agriculture, Food and Environment, The Hebrew University of Jerusalem, POB 12, Rehovot 76100, Israel
Full list of author information is available at the end of the article

dehydrogenase, and aldehyde dehydrogenase. This inhibition of enzyme activity occurs along with the re-direction of yeast energy to repair the damage caused by furans and by reduced intracellular ATP and NAD(P)H levels, either by enzymatic inhibition or consumption/regeneration of co-factors [15].

Microarray-based expression studies in S. cerevisiae identified more than 300 genes that were expressed at significantly higher levels after exposure to furans. Based on these results, it was concluded that furan degradation is catalyzed by multiple aldehyde reductases and tolerance to these compounds can be conferred by enhanced expression of members of pleiotropic drug resistance genes [16,17]. An HMF metabolic conversion product was isolated and identified as 2,5-bis-hydroxymethylfuran (HMF alcohol) [18,19], which is catalyzed by various aldehyde reductases in the presence of NAD(P)H as a co-factor [16]. The bacterium Cupriavidus basilensis was shown to grow on HMF as a sole carbon source and harbors a gene cluster involved in HMF metabolism. In C. basilensis, HMF oxidation activity [20], was catalyzed by HMF oxidase (HMFO) encoded by HmfH [20,21]. The corresponding homologue was cloned from a Methylovorus sp. strain MP688, and an HMFO enzyme was shown to oxidize HMF to 5-(hydroxymethyl)furoic acid (HMF acid) and to 2,5-furandicarboxylic acid (FDCA), during which H_2O_2 was generated [21]. The fungus Amorphotheca resinae ZN1 was isolated from pretreated corn stover and was shown to also degrade HMF, both to HMF alcohol and HMF acid, under aerobic conditions [22].

Overcoming the effects of pretreatment toxicity in biofuel-producing organisms, such as yeast, is a key technical challenge in the biochemical conversion of biomass feed-stocks to biofuels. The basidiomycete Pleurotus ostreatus, a white-rot fungus, can efficiently degrade lignin, cellulose, and hemicellulose [23-25], thus providing potential tools for biological pretreatment in biofuel production [26]. Furthermore, P. ostreatus has been shown to degrade a wide variety of phenolic compounds including those that are inhibitory to S. cerevisiae [27]. Hence, we hypothesized that P. ostreatus may metabolize HMF by enzymatic pathways that are specific and/or abundant in white rot fungi.

In this study, we demonstrate, for the first time, that P. ostreatus can bio-convert HMF to HMF alcohol and FDCA, thus detoxifying the compound. We show that exposure to HMF increases the expression, translation, and activity of enzymes involved in the ligninolytic system, including aryl-alcohol oxidases and a dehydrogenase. Both enzyme families can specifically bio-convert HMF and contribute to the tolerance of P. ostreatus to HMF.

Results

HMF is bio-converted by P. ostreatus

In order to explore the effect of HMF on P. ostreatus, we first examined the tolerance of the fungus toward the compound. The growth of P. ostreatus PC9 on a solid glucose-peptone (GP) medium supplemented with different concentrations of HMF was measured. Under these conditions, the IC_{50} of HMF to P. ostreatus was 12.5 mM (Figure 1), which is significantly higher than the value reported for S. cerevisiae (viability percent was $log_{10} = 10$ on YPD) [28]. The fact that P. ostreatus is more tolerant than S. cerevisiae to the compound suggests that it may harbor more efficient mechanisms to metabolize HMF or otherwise avoid the toxic effects of this compounds.

To determine whether P. ostreatus can metabolize HMF, we conducted experiments in liquid GP medium, in which the fungus was cultured for 5 days to accumulate biomass prior to the addition of HMF 30 mM. Control treatments were identical, excluding the HMF amendment. The amount of HMF and metabolites were monitored colorimetrically and verified by gas chromatography–mass spectrometry (GC-MS) analyses using standards. After 8 h, the extracellular concentration of HMF was reduced by approximately 10%, 24 h marked the point of 50% reduction, and complete transformation occurred after 48 h. HMF alcohol was detected after 8 h and remained in the media for 48 h (Additional file 1). From the oxidation derivatives of HMF, we only detected FDCA after 24 h, but not after 48 h (Additional file 1).

To determine if the bio-transformation of HMF by P. ostreatus also results in reducing its toxic effects on yeast, we preformed experiments in which HMF-amended medium was subjected to P. ostreatus detoxification prior to cultivation of S. cerevisiae on the spent medium. Yeast grown in the presence of 30 mM HMF for 30 h accumulated only 30% of the biomass as compared to control cultures lacking HMF. By contrast, when S. cerevisiae was inoculated into spent medium from P. ostreatus cultures

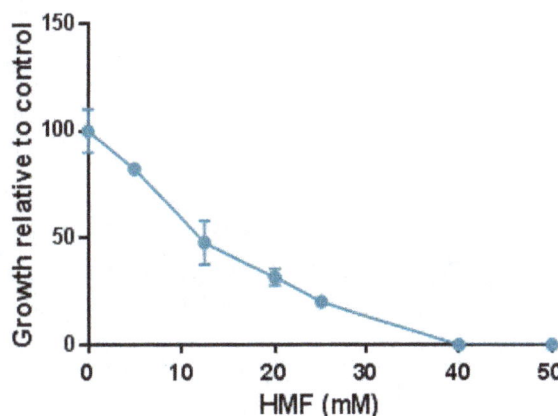

Figure 1 Relative growth of *P. ostreatus* in the presence of different concentrations of HMF. Different concentrations of HMF were added to GP solid media and linear growth of *P. ostreatus* was measured relative to a control lacking HMF. Bars indicate standard errors.

grown for 8 or 24 h in the presence of HMF, growth of the culture was elevated to 50% and to 85%, respectively. No inhibition of yeast growth was observed in spent HMF-amended medium after 48 h of *P. ostreatus* growth (Additional file 2). These results are in agreement with the level of transformation of HMF by *P. ostreatus* at those time points and suggest that the fungus has the potential to be used for reducing the toxic effects of HMF and may serve as a safe pretreatment amendment.

Abundance of aryl-alcohol oxidases and dehydrogenases increases in the presence of HMF

To investigate the enzymatic processes involved in HMF transformation by *P. ostreatus*, a time course experiment in which changes in the profiles of secreted and intracellular proteins as a result of HMF exposure was monitored. Proteins were concentrated from the extracellular fraction at different stages of HMF transformation (8, 24, and 48 h after HMF addition to the media, corresponding to approximately 5%, 50%, and 100% transformation, respectively). Surprisingly, a significant increase in the abundance of an approximately 75-kDa secreted protein was evident from samples at the beginning of the transformation process (only 8 h after HMF addition). Its abundance peaked at a time corresponding to 50% HMF transformation and diminished once all the HMF was detoxified (Figure 2). The approximately 75-kDa band was sequenced and identified as having a high coverage and score of 6 aryl-alcohol oxidase proteins (AAO; EC 1.1.3.7): 69649, 82653, 93955, 114510, 116309, and 121882, designated *aao1-6*, respectively. AAOs are secreted peroxide-producing flavoenzymes and members of the glucose-methanol-choline oxidase (GMC) oxidoreductase super-family [29-31]. *P. ostreatus* has a large AAO, a gene family

consisting of at least 36 genes [32]. Phylogenetic analysis of the AAOs in *P. ostreatous* revealed that they cluster into two groups (Additional file 3).

Among the intracellular proteins obtained from the same cultures, a significant increase in an approximately 45-kDa protein band was observed, whose abundance also correlated with the extent of HMF detoxification. An increase in the abundance of the approximately 45-kDa protein was observed at the early stage of transformation, peaking during the mid-phase; in contrast to the AAOs, the levels of this protein remained constant even when HMF transformation was completed (Figure 2). The sequence of the approximately 45-kDa band identified it as an aryl alcohol dehydrogenase (AAD; EC 1.1.1.90): 75413, which we designated *aad1*. AADs have been shown to reduce alcohols and acids to aldehydes and alcohols, respectively, using nicotinamide adenine dinucleotide phosphate (NADPH) as a co-factor [33]. As HMF is both an aryl alcohol and an aldehyde (Additional file 1), the increase in the abundance of these extracellular and intracellular proteins suggests that there is an involvement of both AAOs and AADs in the response of *P. ostreatus* to HMF.

HMF induces elevated expression levels of AAO and AAD

In order to verify whether the observed AAO and AAD protein accumulation is a consequence of alteration in the expression of the corresponding genes, we constructed specific primers to monitor their expression levels by quantitative real-time polymerase chain reaction (RT-PCR). RNA was extracted from *P. ostreatus* hyphae at different time points following exposure to HMF. Following the kinetics of expression, we observed a significant and substantial increase in all AAO and AAD mRNA levels (Figure 3), with three patterns of

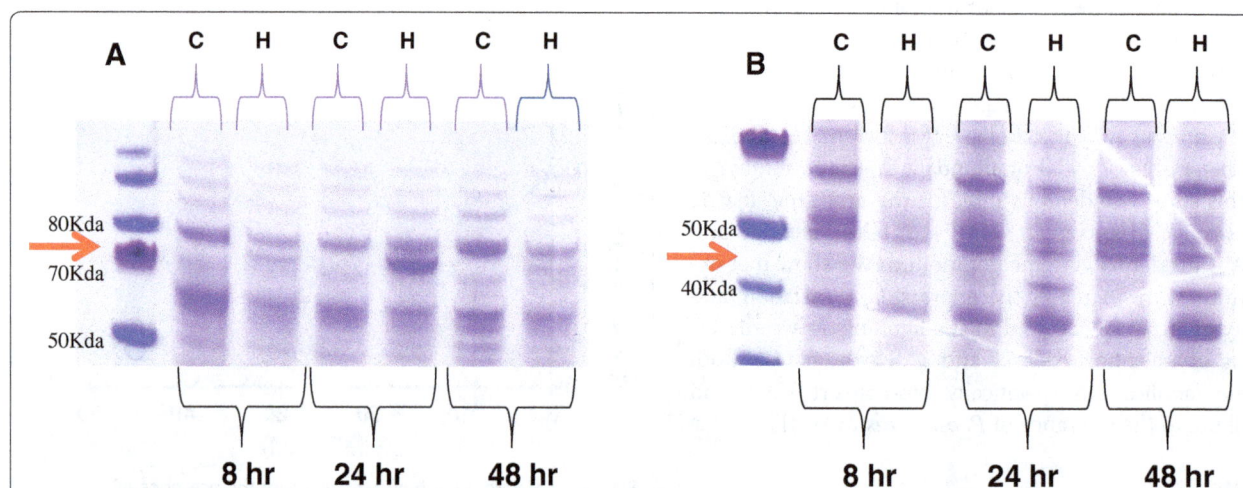

Figure 2 Profiles of secreted and cellular *P. ostreatus* proteins obtained from cultures grown in the presence of HMF. Secreted **(A)** and cellular **(B)** proteins were extracted from *P. ostreatus* 8, 24, and 48 h after addition of 30 mM of HMF to the media. The proteins were resolved by SDS-PAGE 4% to 12%. C: control cultures without HMF, H: cultures exposed to HMF. Arrows point to major visible difference in the profiles.

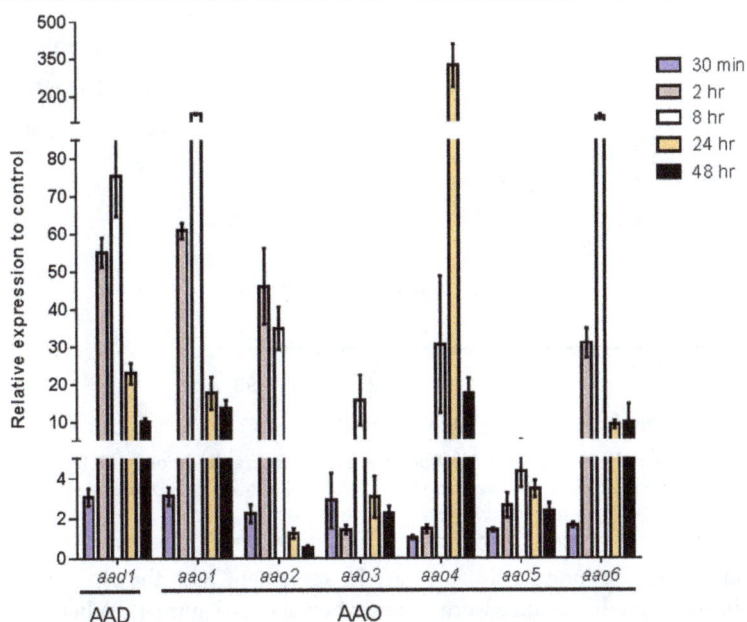

Figure 3 Time course expression of *P. ostreatus aao1-6* and *aad1* genes following the addition of HMF. The expression levels of *aao1-6*, *aad1*, and *vp1* were monitored by real-time RT-PCR (for primers information see Additional file 4). RNA was extracted from *P. ostreatus* at different time points (0.5, 2, 8, 14, and 48 h) after addition of 30 mM HMF to the media. The expression levels calculated relative to β-tubulin, as the endogenous control, and represent the expression relative to control without HMF addition. Bars indicate standard errors.

expression. The first group (*aao1-3* and *aad1*) was rapidly transcribed following exposure to HMF and expression levels increased at least twofold relative to the control as early as 30 min after exposure to the furan. The second group (*aao5-6*) showed increased expression levels after 2 h of exposure to HMF. Both these groups, the maximum expression change ranged from 4- to 150-fold, which was observed following 8 or 24 h of exposure to HMF. After this time period, transcript abundance of both of these groups of genes declined. *aao4* exhibited a third expression pattern characterized by a slower increase in expression, starting only 8 h after exposure to HMF. Although the observed increase in the rate of *aao4* expression was slower than the other groups of genes, it exhibited the most dramatic change (approximately 300-fold) in its expression level, at 24 h after HMF exposure.

The differential expression pattern of the *aaos* and *aad* genes suggest that they may have different roles in the molecular response of *P. ostreatus* to HMF, as the expression kinetics corroborate with the observed AAO and AAD protein levels (Figure 2).

HMF is a substrate of both AAOs and AADs

The enzymatic systems of AAO and AAD families are not substrate-specific, and these proteins are responsible for the degradation and detoxification of a variety of organic compounds [33,34]. Since HMF has both aldehyde and alcohol functional groups (Additional file 1), we

assumed that, in addition of being inducer, HMF could be a potential substrate of these enzymes. To explore whether these enzyme families could directly reduce or oxidize HMF to its derivatives, we further monitored their enzymatic activities *in vitro*. AAOs oxidize alcohols and aldehydes to acids and alcohols, respectively [29,33]. Enzyme activity was measured using a specific well-studied substrate, veratryl alcohol (3,4-dimethoxybenzyl alcohol) [35] in samples prepared from *P. ostreatus* culture media at different time points of HMF exposure. A significant increase in the activity of AAO was observed 24 h after the addition of 30 mM HMF and remained at a level of approximately 5 mU/ml even 4 days after HMF was added (Figure 4). The enzyme activity level was found to be dependent on HMF concentration in the culture. In fact the enzyme activity in the presence of HMF 20 Mm was only 2 mU/ml. In the *P. ostreatus* control cultures that were not exposed to HMF, activity was detected 7 days later (12 days after inoculation) and reached a higher level (1,250 mU/ml), which represents the anticipated increase in enzyme activity at that late growth phase [30,33]. Thus, an increase in AAO protein abundance was correlated with an increase in corresponding enzymatic activities (as determined colorimetrically).

Secreted AAOs oxidize alcohols and aldehydes while generating H_2O_2. We therefore also monitored peroxide production as another indication for AAO activity [21,36]. Indeed, an increase in H_2O_2 concentration was observed in the samples containing veratryl alcohol as a

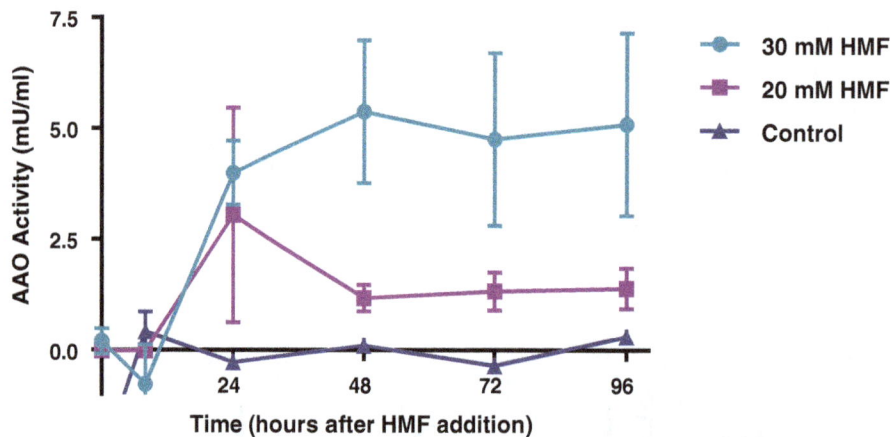

Figure 4 *In vitro* AAO activity is increased in the extracellular fraction of *P. ostreatus* after addition of HMF to the medium. The activity of AAO following 1 mM veratryl alcohol addition as a substrate was monitored in free cell extracts of *P. ostreatus*, at different time points after addition of 30 or 20 mM HMF to the media. Bars indicate standard errors.

substrate, 24 h after HMF addition to the fungal culture. Furthermore, peroxide levels in the reaction mixture were maintained at approximately 700 nM/µl, even 4 days after exposure. By contrast, at the same time points, H_2O_2 was not detected in control cultures (Figure 5A). These results are in agreement with the changes in AAO activity levels described above (Figure 4). When HMF was introduced as a substrate to the assay mixture, peroxide levels increased to approximately 70 nM/µl as soon as 8 h after exposure (Figure 5B). The maximum level of H_2O_2 detected

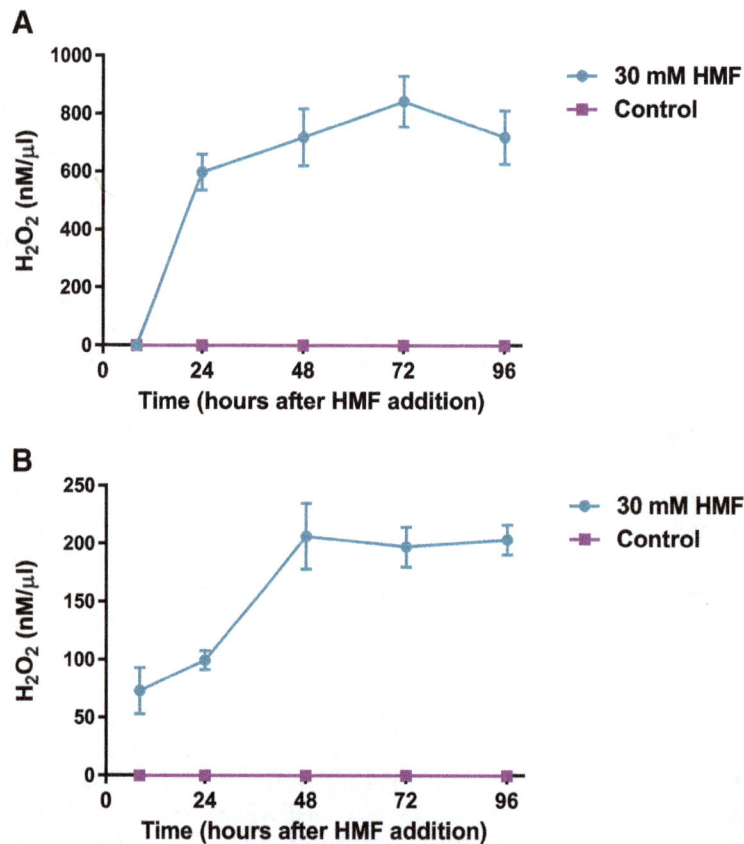

Figure 5 *In vitro* generation of peroxide is increased in the extracellular fraction of *P. ostreatus* after addition of HMF to the medium. Concentration of H_2O_2 generated during activity *in vitro* with 1 mM veratryl alcohol **(A)** or 10 mM HMF **(B)** over time in free cell extracts of *P. ostreatus*. The measurements were performed at different time points after addition of 30 mM HMF to the medium. Bars indicate standard errors.

(approximately 200 nM/μl) occurred 2 to 4 days after HMF addition to the fungal culture. These results could be explained by changes in preferential expression of the different AAOs (Figure 3), suggesting that AAOs whose expression is induced shortly after exposure to HMF may have a higher affinity for HMF than for veratryl alcohol.

HMF can be reduced in the intracellular fraction in *S. cerevisiae* by either NADPH or nicotinamide-adenine dinucleotide (NADH) as co-factors in the reaction [16,19,37]. In *P. ostreatus*, we observed an increase in intracellular AAD expression and protein accumulation (Figures 2 and 3), as well as an accumulation of the AAD reaction product, HMF-alcohol (Additional file 1). We therefore examined whether *P. ostreatus* cell free extracts can directly reduce HMF, by monitoring NAD(P)H depletion in cell extracts of *P. ostreatus* exposed to HMF. Specific activity coupled with NADPH doubled in samples extracted from *P. ostreatus* 8 h after addition of HMF and was maintained over time at a steady level of 400 to 500 mU/mg protein (Figure 6A). Specific activity coupled with NADH did not alter as a response to exposure to HMF and remained at the level of approximately 250 mU/mg protein. However, after 4 days, activity increased by 80% (Figure 6B), which may suggest an additional secondary response to the compound. We

conclude that AAD can directly reduce HMF while using NADPH as a preferred co-factor.

Discussion

P. ostreatus is a commercially important edible, lignino-lytic white-rot filamentous basidiomycete species. It can be easily cultivated on a variety of organic substrates, including agricultural wastes. Hence, it has potential to be harnessed as a pretreatment process for recycling of lignocellulosic substrates, in view of biofuel production [24-26,38]. In this study, we probed the ability of *P. ostreatus* to tolerate and detoxify HMF, a compound that imposes a rate-limiting step in thermo-chemical pretreatment [5] via inhibition of efficient biomass utilization [3,4].

This is the first time that the ability of a white-rot fungus to tolerate HMF was explored. We found that *P. ostreatus* is more tolerant to HMF than *S. cerevisiae*. Furthermore, *P. ostreatus* has the capability in metabolizing HMF to HMF alcohol and FDCA. While HMF alcohol remained in the media and was not metabolized, FDCA was detected only at one time point, suggesting that the fungus can process it further. As *S. cerevisiae* has been shown to degrade HMF to HMF alcohol

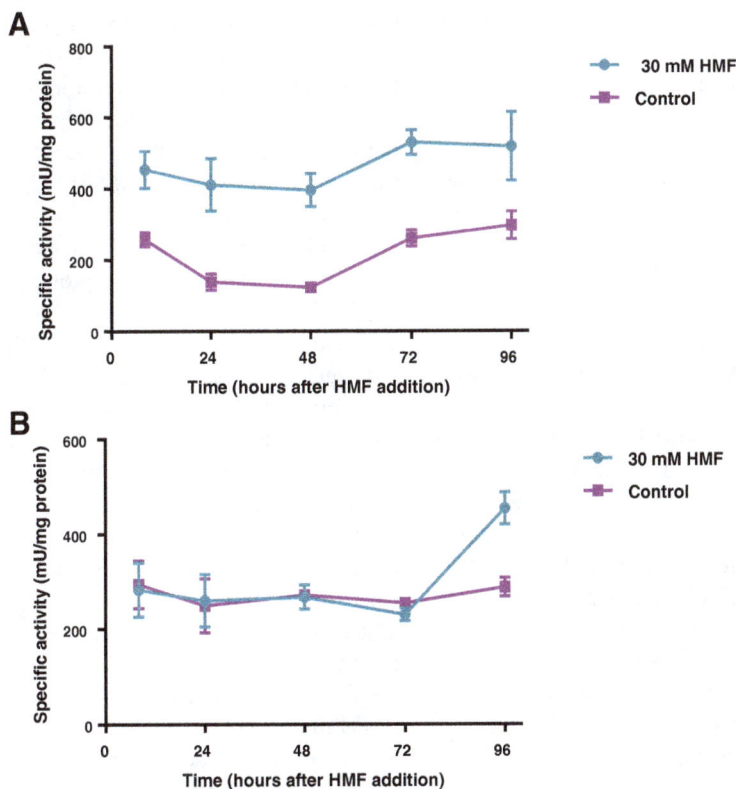

Figure 6 Specific activity coupled with NAD(P)H in free cell extracts of *P. ostreatus* after addition of HMF to the medium. Depletion of NADPH **(A)** or NADH **(B)** was monitored over time *in vitro* with cell extracts of *P. ostreatus* with 10 mM HMF as a substrate. The measurements were performed at different time points after addition of 30 mM HMF to the culture.

[18,19], but not to HMF acid or FDCA, we hypothesized that other, perhaps unique, oxidative pathways may be involved in *P. ostreatus'* capability to do so.

The secreted AAOs, which are peroxide-producing flavoenzymes that belong to GMC oxidoreductase superfamily [29], oxidized HMF and generated H_2O_2 during the process (Figure 5). We speculate that it may catalyze a reaction similar to HMFO from *C. basilensis* [20,21]. HMFO also belongs to the GMC superfamily and can oxidize HMF to HMF acid and FDCA [21]. The changes in AAO expression patterns (Figure 3), combined with HMF being the preferred substrate over veratryl alcohol (Figure 5), along with the large number of AAOs in the *P. ostreatus* genome [32], lead us to suspect that the different AAOs, induced at different times after exposure, have different substrate specificities. Since no AAO homologues have been found in *S. cerevisiae*, heterologous expression of the AAO-encoding genes may be a means to improve yeast tolerance to HMF. Such an approach has already proved feasible by the expression of a laccase gene from the white-rot fungus *Trametes versicolor* in *S. cerevisiae*, to produce a strain that exhibited increased resistance to phenolic inhibitors present in lignocellulose hydrolysates [5,39]. Flavin-based redox enzymes, such as AAOs, have gained enormous interest and importance in the development of biosensors and production of industrially useful carbonyl compounds [40], suggesting another biotechnological potential for the described AAOs from *P. ostreatus*.

Within the cell, transformation of HMF was catalyzed by AAD, an intracellular reducing enzyme that uses NADPH as a co-factor [33] (Figure 6). The pathway involved is probably similar to the conversion of HMF to HMF alcohol which was shown to be catalyzed by various aldehyde reductases in *S. cerevisiae* [16]. The link between AAD and HMF was described in yeast, where expression of two AADs (AAD4, AAD14) was increased in HMF-grown cells [41]. We suggest that the *P. ostreatus aad1* can be another potential candidate whose heterologous expression may enhance HMF degradation in yeast, while improving tolerance to this furan during pretreatment. The preference of yeast enzymes toward co-factors varies between NADPH to NADH [16]. As predicted, the reaction catalyzed by AAD was increased with NADPH (Figure 6A). Since basal levels of HMF transformation was coupled with NADH as well (Figure 6B), we suspect that hydrolases and reductases other than AAD1 are also involved in the process. Surprisingly, a significant increase in the transformation with NADH occurred 4 days after HMF was added to the media (Figure 6B), which supports the involvement of an additional, yet unknown, transformation reaction or pathway for HMF in *P. ostreatus*.

Conclusion

Most of the research on *P. ostreatus* has focused on its ability to degrade lignin [25,38], as a means for providing an alternative biological pretreatment of biomass in biofuel production [26]. Here, we have described and subsequently analyzed the ability of *P. ostreatus* to bio-convert a pre-treatment toxic byproduct, HMF, by enzymes usually associated with the lignin degradation complex (Figure 7). Based on our findings, we suggest that *P. ostreatus* can be potentially integrated as part of the physical and chemical pretreatment process, either by direct use of the fungus or its enzymes or by mining the genetic pool of this white-rot fungus for genes to be heterologously expressed in yeast or other biofuel-producing microorganisms. Such a strategy has already been employed to facilitate direct fermentation of cellodextrins by yeast, where *Neurospora crassa* was used as a gene pool and its two cellodextrin transporters were introduced into *S. cerevisiae* along with an intracellular β-glucosidase, which subsequently improved cellobiose fermentation [42,43]. Expanding and/or combining the resources available from various fungal gene pools may well prove beneficial in engineering yeast strains tailored to challenge rate limiting steps in biofuel production.

Methods
Fungal growth and experimental conditions

P. ostreatus monokaryon strain PC9 (Spanish Type Culture Collection accession number CECT20311), which is a protoclone derived by de-dikaryotization of the commercial dikaryon strain N001 (Spanish Type Culture Collection accession number CECT20600) [44], was used throughout this study.

Fungal strains were grown and maintained in GP medium (2% glucose, 0.5% peptone (Difco, Franklin Lakes, NJ, USA), 0.2% yeast extract (Difco, Franklin Lakes, NJ, USA), 0.1% K_2HPO_4, and 0.05% $MgSO_4 \cdot 7H_2O$). When required, 1.5% agar was added. The gene and protein expression as well as activity assays were conducted in samples of fungal biomass or cell free extracellular extracts taken from liquid cultures that were maintained in stationary 100 ml Erlenmeyer flasks containing 10 ml of media. The fungus was grown for 5 days to accumulate biomass, after which HMF (Sigma-Aldrich, St. Louis, MO, USA) was added to the media to obtain concentrations of 20 to 30 mM. All experiments were accompanied by controls that lacked the chemical amendment. The inoculum for all growth conditions was one disk (5 mm diameter) of mycelium obtained from the edge of a young colony grown on solid medium and positioned at the center of the Petri dish or a flask. Cultures were incubated at 28°C in the dark.

S. cerevisiae strain CBS8066 was maintained on 10 g/l yeast extract, 20 g/l peptone, 20 g/l glucose, and 20 g/l

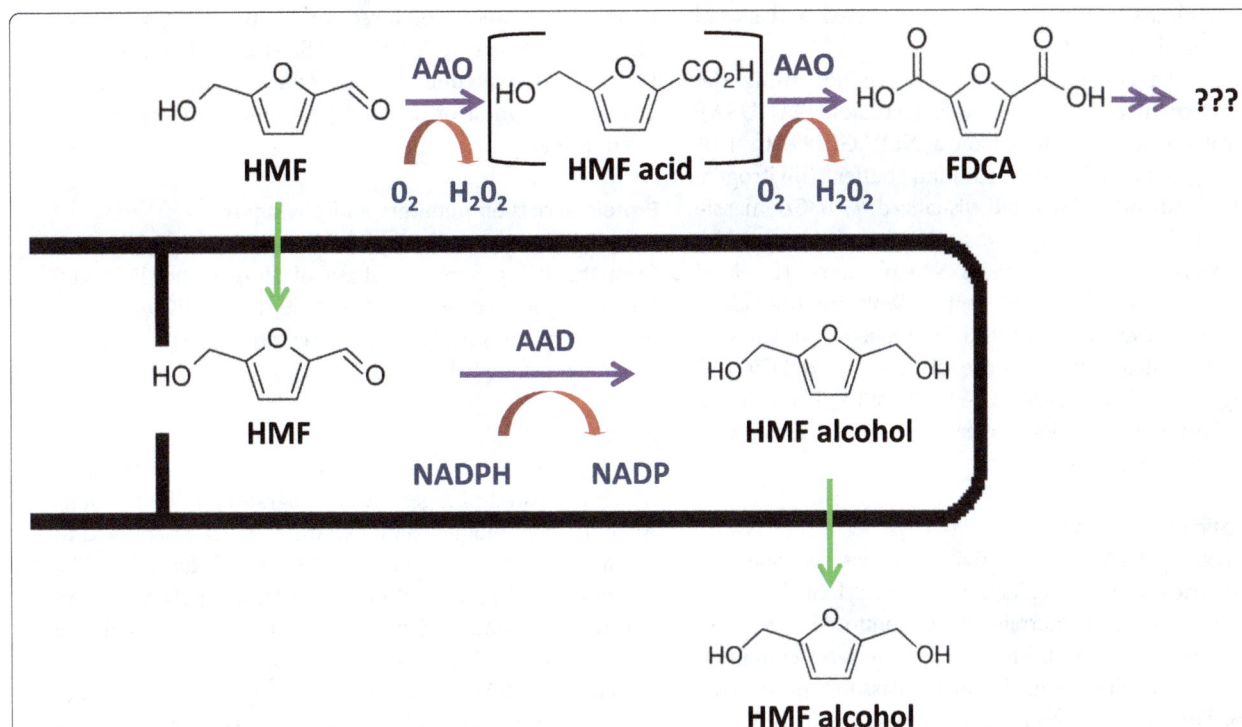

Figure 7 Scheme for the enzymatic degradation of HMF by *P. ostreatus*. HMF is metabolized extracellularly by AAO to FDCA, which generates H$_2$O$_2$. These reactions probably involve an intermediate conversion of HMF to HMF acid and further conversion to unknown products. When HMF enters the cell, it is reduced by AAD with NADPH as a co-factor and metabolized to HMF alcohol. HMF alcohol can be secreted and accumulates extracellularly.

agar (YPD agar) or on 10 g/l yeast extract, 20 g/l peptone, and 20 g/l glucose (YPD). For the spent medium experiments, the cultures were grown on spent *Pleurotus* medium (GP) supplemented initially with 30 mM HMF. The cultures were grown in an orbital shaker (180 rpm) at 30°C under aerobic conditions. The growth was monitored at 600 nm using the Synergy 2 Multi-Mode Microplate Reader (BioTek, Winooski, VT, USA). The assay was performed in triplicate, and an average reading was plotted.

Gene expression analyses

Total RNA was extracted from culture biomass, first ground under liquid nitrogen with mortar and pestle, then homogenized with QIA shredder spin columns (Qiagen, Hilden, Germany) and purified from the lysate using the RNeasy Plus Mini Kit (Qiagen, Hilden, Germany). cDNA was synthesized using the High Capacity cDNA Reverse Transcription Kit (Applied Biosystems, Carlsbad, CA, USA). Gene expression analyses were performed on an ABI StepOne Real-Time PCR Sequence Detection System and software (Applied Biosystems, Foster City, CA, USA), using Power SYBR Green PCR Master Mix (Applied Biosystems, Foster City, CA, USA). The PCR volume was 10 µl, using 20 ng of total cDNA and 300 nM oligonucleotide primers (Additional file 4). The thermal cycling conditions were as follows: an initial step at 95°C for 20 s and 40 cycles at 95°C for 5 s, 60°C for 30 s, followed by a denaturation step to

verify the absence of unspecific products or primer dimmers. The *β-tubulin* (ID: 117235) gene was used as the endogenous control. The primer efficiency levels of the genes were with the range of 90% to 110%. Amplification data were compared on the basis of the $\Delta\Delta$CT method and presented as $2^{-\Delta\Delta CT}$. Data was normalized with respect to *β-tubulin* and calculated where ΔCT = CT *target gene* – CT *β-tubulin*, and then $\Delta\Delta$CT = ΔCT *treatment with 30 mM HMF* – ΔCT *control without HMF*.

Protein expression profiles

For extracellular protein analyses, culture fluids were filtered through Whatman No. 1 filter paper followed by 0.45-µm mixed cellulose ester filter paper (Whatman, Buckinghamshire, UK). The sample was then concentrated using a 10-kDa cutoff PM-10 membrane (Millipore, Amicon Division, Billerica, MA, USA) and treated with cOmplete (Roche Applied Science, Mannheim, Germany), after concentration. For intracellular protein extraction, mycelial samples were frozen in liquid nitrogen, pulverized, and suspended in lysis buffer (1 M sorbitol, 10 mM HEPES (pH 7.5), 5 mM EDTA, 5 mM EGTA, 5 mM NaF, 0.1 M KCl, 0.2% Triton X-100, cOmplete (Roche Applied Science, Mannheim, Germany). The samples were homogenized by ten strokes of pestle A in a Dounce homogenizer. The homogenates were centrifuged for 40 min at $10,000 \times g$

at 4°C, and the supernatants were recovered and stored at −70°C until analyzed.

The protein concentration was determined using the BioRad protein assay kit (BioRad, Hercules, CA, USA). The proteins were separated on a NuPAGE 4% to 12% Bis-Tris gel in MES-SDS running buffer (Invitrogen, Grand Island, NY, USA) and visualized with Coomassie R-250 (0.125%). The sample was subsequently analyzed by HPLC/mass spectrometry/mass spectrometry (LC-MS/MS) in an Orbitrap (Thermo Scientific, Waltham, MA, USA) mass spectrometer and identified by Sequest 3.31 software against the JGI genome database of *P. ostreatus* PC9 v1.0 (http://genome.jgi-psf.org/PleosPC9_1/PleosPC9_1.home.html) at The Smoler Proteomics Center of The Israel Institute of Technology (Technion).

Enzymatic activity assays

AAO activity (AAO): the activity was assayed spectrophotometrically, as the oxidation of veratryl alcohol (3,4-dimethoxybenzyl) to veratraldehyde, monitored at 310 nm ($\varepsilon_{310} = 9,300$ M^{-1} cm^{-1}). The reaction mixtures contained 1 mM veratryl alcohol in 50 mM potassium phosphate, pH = 6. The assay was conducted in a volume of 200 μl in microtiter plates at 30°C, using the Synergy 2 Multi-Mode Microplate Reader (BioTek, Winooski, VT, USA). An enzyme unit was defined as the amount enzyme producing 1 μmol of product per minute.

AAO activity coupled with H_2O_2: This assay was based on a highly sensitive and stable probe for H_2O_2, 10-acetyl-3,7-dihydroxyphenoxazine (Amplex Red reagent). The Amplex Red™ kit assay (Invitrogen, Carlsbad, CA, USA) was performed on each sample, according to the manufacturer's instructions. The filtrate was added to 50 mM potassium phosphate, pH = 6 with 1 mM of veratryl alcohol or 10 mM HMF, to a total volume of 50 μl. The samples were placed in microtiter plates, and Amplex Red reaction mixture (50 μl) was added to each well. The reaction was incubated for 3.5 h at 30°C, after which fluorescence (conversion of the reagent to resorufin) was measured using the Synergy 2 Multi-Mode Microplate Reader (BioTek, Winooski, VT, USA). Excitation and emission were at 540 ± 20 and 590 ± 25 nm, respectively.

AAD activity: The mycelial samples were disrupted using a Bead Beater (BioSpec Products, Inc, Bartlesville, OK, USA) in 500 μl of 100 mM potassium phosphate, pH = 7. The homogenates were centrifuged for 1 min at 4,000 × g at 4°C. The protein concentration of the clear lysate was determined using the BioRad protein assay kit (BioRad, Hercules, CA, USA), and 30 μl was used per each reaction. The activity was assayed spectrophotometrically, as NAD(P)H was added to a concentration of 30 μM and monitored at 340 nm ($\varepsilon_{340} = 6,220$ M^{-1} cm^{-1}) and 10 mM HMF with 100 mM potassium phosphate, pH = 7. The assay was conducted in a total volume of 600 μl at 30°C, and changes

in absorption were monitored for 15 min using a spectrophotometer (Biomate 3, Thermo Scientific, Waltham, MA, USA). An enzyme unit was defined as the amount enzyme producing 1 μmol of product per minute and divided by mg of protein.

Protein accession numbers and phylogeny

For phylogenetic analysis, protein sequences were obtained from the JGI genome database of *P. ostreatus* PC9 v1.0 (http://genome.jgi-psf.org/PleosPC9_1/PleosPC9_1.home.html) using blastp algorithm. The phylogenetic tree was generated using phylogeny.fr [45].

Chemical analysis

The samples were analyzed using a GC-MS apparatus which consisted of a gas chromatograph (Agilent 7890A, Agilent Technologies, Santa Clara, CA, USA) coupled to the mass selective (Agilent 5975C MSD) detector. The compounds were separated on a BPX-5 capillary column (30 m × 0.25 mm, 0.25 μm, SGE). Helium was used as a carrier gas at a 1.3 ml/min flow rate. Prior to analysis, the samples (150 μl) were evaporated upon dry nitrogen at 50°C and derivatized with 100 μl trimethylsilylation reagent which consisted of pyridine, BSA, and TMCS (20:20:1). Analytical equipment was controlled, and data was analyzed using MassHunter Acquisition and Data Analysis software (Agilent). Analytical standards of HMF and FDCA were purchased from Sigma-Aldrich (St. Louis, MO, USA). HMF acid and HMF alcohol were purchased from Toronto Research Chemicals Inc (North York, Canada).

Additional files

Additional file 1: Summary of the GC-MS analysis of *P. ostreatus* spent medium after growth in the presence of 30 mM HMF. The molecules were identified using standards. ND (not detected), +++ (high concentration), ++ (medium concentration), and + (low concentration).

Additional file 2: Growth of *S. cerevisiae* cultures on spent medium of *P. ostreatus* grown in the presence of HMF. Yeast cultures were inoculated into spent *P. ostreatus* media, initially supplemented with 30 mM of HMF. Yeast growth was monitored for 30 h at 600 nm. Control (**A**) or after an addition of 30 mM of HMF (**B**).

Additional file 3: Phylogenetic analysis of the AAO family in *P. ostreatus*. Phylogenetic analysis based on the protein sequences of AAOs from *P. ostreatus*. Green represents proteins whose abundance increased after HMF addition to the media. Blue represents additional genes whose expression was induced (as determined by real-time PCR; data not shown) 24 h after HMF was added to the medium.

Additional file 4: Primer information.

Abbreviations

AAD: aryl-alcohol dehydrogenase; AAO: aryl-alcohol oxidase; FDCA: 2,5-furandicarboxylic acid; GP: glucose-peptone; HMF: 5-hydroxymethylfurfural; HMF acid: 5-(hydroxymethyl)furoic acid; HMF alcohol: 2,5-bis-hydroxymethylfuran; NADH: nicotinamide-adenine dinucleotide; NADPH: nicotinamide adenine dinucleotide phosphate.

Competing interests

The authors declare that they have no competing interests.

Authors' contributions

DF carried out the experimental studies and drafted the manuscript. DJK participated in genomic analyses. YH, OY, and NLG conceived the study, and participated in its design and coordination, and helped to draft the manuscript. All authors read and approved the final manuscript.

Acknowledgements

This study was supported by grants from US-Israel Binational Fund (BSF) and the Israel Science Foundation (ISF). Daria Feldman was supported by a fellowship from the President of Israel fund granted by the Estates Committee.

Author details

[1]Department of Plant Pathology and Microbiology, The R.H. Smith Faculty of Agriculture, Food and Environment, The Hebrew University of Jerusalem, POB 12, Rehovot 76100, Israel. [2]Department of Plant and Microbial Biology, University of California at Berkeley, 111 Koshland Hall, Berkeley, California 94720, USA.

References

1. Carroll A, Somerville C. Cellulosic biofuels. Annu Rev Plant Biol. 2009;60:165–82.
2. Almeida JRM, Bertilsson M, Gorwa-Grauslund MF, Gorsich S, Liden G. Metabolic effects of furaldehydes and impacts on biotechnological processes. Appl Microbiol Biotechnol. 2009;82:625–38.
3. Klinke HB, Thomsen AB, Ahring BK. Inhibition of ethanol-producing yeast and bacteria by degradation products produced during pre-treatment of biomass. Appl Microbiol Biotechnol. 2004;66:10–26.
4. Palmqvist E, Hahn-Hagerdal B. Fermentation of lignocellulosic hydrolysates. II: inhibitors and mechanisms of inhibition. Bioresour Technol. 2000;74:25–33.
5. Parawira W, Tekere M. Biotechnological strategies to overcome inhibitors in lignocellulose hydrolysates for ethanol production: review. Crit Rev Biotechnol. 2011;31:20–31.
6. Cantarella M, Cantarella L, Gallifuoco A, Spera A, Alfani F. Effect of inhibitors released during steam-explosion treatment of poplar wood on subsequent enzymatic hydrolysis and SSF. Biotechnol Progr. 2004;20:200–6.
7. Klinke HB, Olsson L, Thomsen AB, Ahring BK. Potential inhibitors from wet oxidation of wheat straw and their effect on ethanol production of Saccharomyces cerevisiae: Wet oxidation and fermentation by yeast. Biotechnol Bioeng. 2003;81:738–47.
8. Taherzadeh MJ, Karimi K. Pretreatment of lignocellulosic wastes to improve ethanol and biogas production: a review. Int J Mol Sci. 2008;9:1621–51.
9. Zhang J, Zhang W-X, Wu Z-Y, Yang J, Liu Y-H, Zhong X, et al. A comparison of different dilute solution explosions pretreatment for conversion of distillers' grains into ethanol. Prep Biochem Biotechnol. 2013;43:1–21.
10. Dunlop AP. Furfural formation and behavior. Ind Eng Chem. 1948;2:204–9.
11. Antal MJJ, Leesomboon T, Mok WS, Richards GN. Mechanism of formation of 2-furaldehyde from D-xylose. Carbohydr Res. 1991;217:71–86.
12. Larsson S, Palmqvist E, Hahn-Hagerdal B, Tengborg C, Stenberg K, Zacchi G, et al. The generation of fermentation inhibitors during dilute acid hydrolysis of softwood. Enzyme Microb Technol. 1999;24:151–9.
13. Lewkowski J. Synthesis, chemistry and applications of 5-hydroxymethylfurfural and its derivatives. Arkivoc. 2001;2001:17–54.
14. Taherzadeh MJ, Gustafsson L, Niklasson C, Liden G. Physiological effects of 5-hydroxymethylfurfural on Saccharomyces cerevisiae. Appl Microbiol Biotechnol. 2000;53:701–8.
15. Almeida JRM, Modig T, Petersson A, Hahn-Hagerdal B, Liden G, Gorwa-Grauslund MF. Increased tolerance and conversion of inhibitors in lignocellulosic hydrolysates by Saccharomyces cerevisiae. J Chem Technol Biotechnol. 2007;82:340–9.
16. Liu ZL, Moon J, Andersh BJ, Slininger PJ, Weber S. Multiple gene-mediated NAD(P)H-dependent aldehyde reduction is a mechanism of in situ detoxification of furfural and 5-hydroxymethylfurfural by Saccharomyces cerevisiae. Appl Microbiol Biotechnol. 2008;81:743–53.
17. Almario MP, Reyes LH, Kao KC. Evolutionary engineering of Saccharomyces cerevisiae for enhanced tolerance to hydrolysates of lignocellulosic biomass. Biotechnol Bioeng. 2013;110:2616–23.
18. Liu ZL, Slininger PJ, Dien BS, Berhow MA, Kurtzman CP, Gorsich SW. Adaptive response of yeasts to furfural and 5-hydroxymethylfurfural and

new chemical evidence for HMF conversion to 2,5-bis-hydroxymethlfuran. J Ind Microbiol Biot. 2004;31:345–52.
19. Liu ZL. Genomic adaptation of ethanologenic yeast to biomass conversion inhibitors. Appl Microbiol Biotechnol. 2006;73:27–36.
20. Koopman F, Wierckx N, de Winde JH, Ruijssenaars HJ. Identification and characterization of the furfural and 5-(hydroxymethyl)furfural degradation pathways of Cupriavidus basilensis HMF14. Proc Natl Acad Sci USA. 2010;107:4919–24.
21. Dijkman WP, Fraaije MW. Discovery and characterization of a 5-hydroxymethylfurfural oxidase from Methylovorus sp strain MP688. Appl Environ Microbiol. 2014;80:1082–90.
22. Ran H, Zhang J, Gao Q, Lin Z, Bao J. Analysis of biodegradation performance of furfural and 5-hydroxymethylfurfural by Amorphotheca resinae ZN1. Biotechnol Biofuels. 2014;7:51.
23. Cohen R, Persky L, Hadar Y. Biotechnological applications and potential of wood-degrading mushrooms of the genus Pleurotus. Appl Microbiol Biotechnol. 2002;58:582–94.
24. Sanchez C. Cultivation of Pleurotus ostreatus and other edible mushrooms. Appl Microbiol Biotechnol. 2010;85:1321–37.
25. Stajic M, Vukojevic J, Duletic-Lausevic S. Biology of Pleurotus eryngii and role in biotechnological processes: a review. Crit Rev Biotechnol. 2009;29:55–66.
26. Wan C, Li Y. Fungal pretreatment of lignocellulosic biomass. Biotechnol Adv. 2012;30:1447–57.
27. Giardina P, Faraco V, Pezzella C, Piscitelli A, Vanhulle S, Sannia G. Laccases: a never-ending story. Cell Mol Life Sci. 2010;67:369–85.
28. Greetham D, Wimalasena T, Kerruish DWM, Brindley S, Ibbett RN, Linforth RL, et al. Development of a phenotypic assay for characterisation of ethanologenic yeast strain sensitivity to inhibitors released from lignocellulosic feedstocks. J Ind Microbiol Biot. 2014;41:931–45.
29. Hernandez-Ortega A, Ferreira P, Martinez AT. Fungal aryl-alcohol oxidase: a peroxide-producing flavoenzyme involved in lignin degradation. Appl Microbiol Biotechnol. 2012;93:1395–410.
30. Kumar VV, Rapheal VS. Induction and purification by three-phase partitioning of aryl alcohol oxidase (AAO) from Pleurotus ostreatus. Appl Biochem Biotechnol. 2011;163:423–32.
31. Sanchez C. Lignocellulosic residues: Biodegradation and bioconversion by fungi. Biotechnol Adv. 2009;27:185–94.
32. Riley R, Salamov AA, Brown DW, Nagy LG, Floudas D, Held BW, et al. Extensive sampling of basidiomycete genomes demonstrates inadequacy of the white-rot/brown-rot paradigm for wood decay fungi. Proc Natl Acad Sci USA. 2014;111:9923–8.
33. Gutiérrez A, Caramelo L, Prieto A, Martinez MJ, Martinez AT. Anisaldehyde production and aryl-alcohol oxidase and dehydrogenase activities in ligninolytic fungi from the genus Pleurotus. Appl Environ Microbiol. 1994;60:1783–8.
34. Yang D-D, Francois JM, de Billerbeck GM. Cloning, expression and characterization of an aryl-alcohol dehydrogenase from the white-rot fungus Phanerochaete chrysosporium strain BKM-F-1767. BMC Microbiol. 2012;12:126.
35. Guillen F, Martinez AT, Martinez MJ. Substrate-specificity and properties of the aryl-alcohol oxidase from the ligninolytic fungus Pleurotus-eryngii. Eur J Biochem. 1992;209:603–11.
36. Jimenez DJ, Korenblum E, van Elsas JD. Novel multispecies microbial consortia involved in lignocellulose and 5-hydroxymethylfurfural bioconversion. Appl Microbiol Biotechnol. 2014;98:2789–803.
37. Ma M, Wang X, Zhang X, Zhao X. Alcohol dehydrogenases from Scheffersomyces stipitis involved in the detoxification of aldehyde inhibitors derived from lignocellulosic biomass conversion. Appl Microbiol Biotechnol. 2013;97:8411–25.
38. Knop D, Yarden O, Hadar Y. The ligninolytic peroxidases in the genus Pleurotus: divergence in activities, expression, and potential applications. Appl Microbiol Biotechnol. 2015;99:1025–38.
39. Larsson S, Cassland P, Jonsson LJ. Development of a Saccharomyces cerevisiae strain with enhanced resistance to phenolic fermentation inhibitors in lignocellulose hydrolysates by heterologous expression of laccase. Appl Environ Microbiol. 2001;67:1163–70.
40. Goswami P, Chinnadayyala SSR, Chakraborty M, Kumar AK, Kakoti A. An overview on alcohol oxidases and their potential applications. Appl Microbiol Biotechnol. 2013;97:4259–75.
41. Bajwa PK, Ho C-Y, Chan C-K, Martin VJJ, Trevors JT, Lee H. Transcriptional profiling of Saccharomyces cerevisiae T2 cells upon exposure to hardwood spent sulphite liquor: comparison to acetic acid, furfural and hydroxymethylfurfural. Antonie Van Leeuwenhoek. 2013;103:1281–95.

42.   Kim H, Lee W-H, Galazka JM, Cate JHD, Jin Y-S. Analysis of cellodextrin transporters from *Neurospora crassa* in *Saccharomyces cerevisiae* for cellobiose fermentation. Appl Microbiol Biotechnol. 2014;98:1087–94.

43.   Galazka JM, Tian C, Beeson WT, Martinez B, Glass NL, Cate JHD. Cellodextrin transport in yeast for improved biofuel production. Science. 2010;330:84–6.

44.   Larraya LM, Perez G, Penas MM, Baars JJP, Mikosch TSP, Pisabarro AG, et al. Molecular karyotype of the white rot fungus *Pleurotus ostreatus*. Appl Environ Microbiol. 1999;65:3413–7.

45.   Dereeper A, Guignon V, Blanc G, Audic S, Buffet S, Chevenet F, et al. Phylogeny.fr: robust phylogenetic analysis for the non-specialist. Nucleic Acids Res. 2008;36:465–9.

6

Impact of the supramolecular structure of cellulose on the efficiency of enzymatic hydrolysis

Ausra Peciulyte[1], Katarina Karlström[2], Per Tomas Larsson[2,3] and Lisbeth Olsson[1,4*]

Abstract

Background: The efficiency of enzymatic hydrolysis is reduced by the structural properties of cellulose. Although efforts have been made to explain the mechanism of enzymatic hydrolysis of cellulose by considering the interaction of cellulolytic enzymes with cellulose or the changes in the structure of cellulose during enzymatic hydrolysis, the process of cellulose hydrolysis is not yet fully understood. We have analysed the characteristics of the complex supramolecular structure of cellulose on the nanometre scale in terms of the spatial distribution of fibrils and fibril aggregates, the accessible surface area and the crystallinity during enzymatic hydrolysis. Influence of the porosity of the substrates and the hydrolysability was also investigated. All cellulosic substrates used in this study contained more than 96% cellulose.

Results: Conversion yields of six cellulosic substrates were as follows, in descending order: nano-crystalline cellulose produced from never-dried soda pulp (NCC-OPHS-ND) > never-dried soda pulp (OPHS-ND) > dried soda pulp (OPHS-D) > Avicel > cotton treated with sodium hydroxide (cotton + NaOH) > cotton.

Conclusions: No significant correlations were observed between the yield of conversion and supramolecular characteristics, such as specific surface area (SSA) and lateral fibril dimensions (LFD). A strong correlation was found between the average pore size of the starting material and the enzymatic conversion yield. The degree of crystallinity was maintained during enzymatic hydrolysis of the cellulosic substrates, contradicting previous explanations of the increasing crystallinity of cellulose during enzymatic hydrolysis. Both acid and enzymatic hydrolysis can increase the LFD, but no plausible mechanisms could be identified. The sample with the highest initial degree of crystallinity, NCC-OPHS-ND, exhibited the highest conversion yield, but this was not accompanied by any change in LFD, indicating that the hydrolysis mechanism is not based on lateral erosion.

Keywords: Cellulose I, Enzymatic hydrolysis, Cellulose supramolecular structure, Solid-state cross-polarization magic angle spinning carbon-13 nuclear magnetic resonance (CP/MAS ^{13}C-NMR), Porosity, Crystallinity

Background

Cellulose, one of the most abundant organic materials on Earth, is found in raw materials such as forestry and agricultural residues and various kinds of waste. The enzymatic hydrolysis of cellulose is generally considered to be a sustainable means of obtaining monosaccharides that can be converted into a number of products via microbial fermentation [1]. Bioethanol is the prime example of the conversion of monosaccharides into renewable transportation fuels employing fermentation [2]. However, the enzymatic hydrolysis of cellulose is often incomplete [3] and we do not yet have a full understanding of the process.

The types of enzymes required for the enzymatic hydrolysis of cellulose include endocellulases, exocellulases and β-glucosidase [3-5]. A new enzyme family, AA9 (formerly GH61), harbouring fungal enzymes and which functions in the same way as lytic polysaccharide monooxygenases has recently been introduced [6,7]. Filamentous fungi, such as the thoroughly investigated *Trichoderma reesei*, and certain bacteria secrete cellulases (non-complexed) extracellularly. Also, multienzyme complexes (cellulosomes) are known to be produced by other bacteria. It has been shown that non-complexed

---
* Correspondence: lisbeth.olsson@chalmers.se
[1]Department of Biology and Biological Engineering, Division of Industrial Biotechnology, Chalmers University of Technology, Kemivägen 10, Gothenburg SE-412 96, Sweden
[4]Wallenberg Wood Science Center, Chalmers University of Technology, Kemigården 4, Gothenburg SE-412 96, Sweden
Full list of author information is available at the end of the article

fungal cellulases deconstruct plant cell walls using a different mechanism from the one used by cellulosomes [8]. The rate of enzymatic hydrolysis usually decreases in time. Various enzyme-related factors as possible causes have been discussed in the literature, such as: product inhibition [9]; enzyme deactivation due to heat, exposure to the air-liquid interface and/or mechanical factors [9]; overcrowding of bound cellulases on cellulose surface (called as traffic jams) causes interference between cellulases and reduces their hydrolytic efficiency [10]; non-productively bound cellulases [9]; loss of enzyme synergy [11] and competition between enzymes [12]. Among the substrate-related factors discussed are available surface area of cellulose which can be covered by lignin and thereby become unaccessible for cellulases [13], cellulose crystallinity, degree of polymerization and pore volume [14].

Native cellulose I exists in the form of fibrils, which are bundles of β-(1,4)-D-glucan polymer chains. Cellulose has a simple chemical structure but the arrangement of these chains into a supramolecular structure makes cellulose surprisingly complex. In the solid state, cellulose polymers are packed together and are described in terms of the spatial distribution of fibrils and fibril aggregates, the specific surface area (SSA), degree of crystallinity (DCr) and porosity. The supramolecular characteristics of these fibrils, such as lateral dimensions, are dependent on the species and isolation process [15]. The arrangement of the individual polymers in a fibril is highly dependent on the location within the fibril. Molecular dynamic simulations have shown that the cellulose structure at the surface is different from the crystalline bulk structure [16]. Analysis of cellulose-water interfaces has shown that the C4 atoms in cellulose chains located on top of different crystallographic planes have different mobilities [15].

In most cases, cellulose fibrils are embedded in a lignocellulose matrix, where the cellulose content ranges from approximately 40 to 50% of the plant (dry weight) [4]. In case of cotton, cellulose is the major constituent (over 94%). Cellulose is also synthesized by bacteria and prokaryotes [17]. The cellulose I fibrils can assemble into larger supramolecular structures or fibril aggregates that constitute the cellulose network. During the isolation procedure of cellulose I-rich plant fibres, non-cellulose I components are removed and therefore, additional pores are formed in the fibre wall which create a porous cellulose network [18-20]. This network of cellulose has a complex orientation in space [20] (Figure 1).

Different fibril models have been proposed in the literature. In one of them, the fibril aggregates have a rectangular cross-section [21,22]. In the other, the fibril is composed of 36 elementary chains, in which it is assumed that the fibril aggregates have a hexagonal

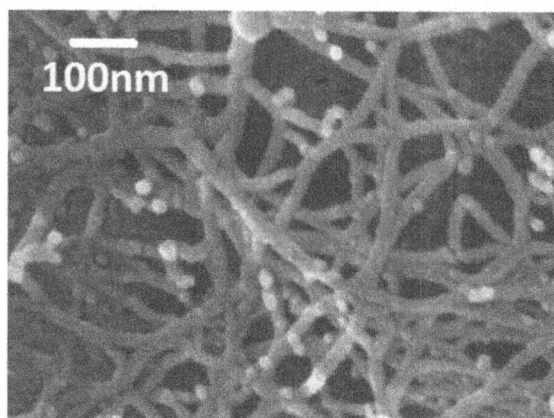

Figure 1 A field emission scanning electron microscopy image of a deliginified kraft pulp fibre adapted from [20] (reprinted with modifications with permission). After removal of non-cellulose I material during the kraft cooking process, it has resulted in a highly porous fibre wall morphology where fibril aggregates form a network.

cross-section [23]. It has been shown that endo-β-1,4-glucanase and its isolated carbohydrate-binding domain adsorb preferentially on certain edges or faces of *Valonia* cellulose microcrystals [24]. It has been suggested that the faces preferred for the binding of enzymes have hydrophobic character [8,25,26]. These findings demonstrate the complexity of cellulose and the necessity to understand the mechanisms of cellulose hydrolysis in greater detail. The solubilized sugars released from cellulose are usually measured during the course of enzymatic hydrolysis, and the non-hydrolysed fraction is attributed to the recalcitrant nature of cellulose. This approach is very common in research on the production of bioethanol. In contrast, in research on forestry products, a great deal of attention is paid to the overall structure of cellulose fibres and the understanding of their reactivity in thermochemical processes [27].

Various imaging techniques have been used to investigate cellulose structure and cellulose-cellulase interactions, such as fluorescence microscopy, transmission electron microscopy, atomic force microscopy and scanning electron microscopy [10,28,29]. The drawback of these microscopic techniques is that only a very small part of the substrate can be visualized, from which conclusions are drawn regarding the whole substrate.

An interdisciplinary approach was adopted in the current study, combining knowledge from the fields of biotechnology, chemical pulping and cellulose structure, employing solid-state cross-polarization magic angle spinning carbon-13 nuclear magnetic resonance (CP/MAS ^{13}C-NMR) with spectral fitting in order to characterize the supramolecular structure of cellulosic substrates during enzymatic hydrolysis. We investigated whether the supramolecular structure of cellulose, in terms of the spatial distribution of fibrils and fibril aggregates, the accessible surface area, crystallinity and porosity, can influence its hydrolysability. We believe that investigations of cellulose structures with nanometre resolution will provide a more precise picture of the process as both enzymes and the building blocks of cellulose have these dimensions.

Results and discussion

The model used for the supramolecular structure of cellulose

Collecting the knowledge from the literature, we built a model, emphasizing the complexity of supramolecular structure of cellulose, depicted in Figure 2. This model was used for interpreting our experimental data. Earlier results which have shown that during the isolation of the pulp fibres, i.e. removal of non-cellulosic material, fibril aggregates form a highly porous fibre wall morphology, were incorporated in the present model [20]. In our discussion of the results, we used the term fibre wall morphology which describes the network formed by the fibril aggregates inside the fibre wall remaining after the isolation procedure. Along the fibril axis, there are ordered (crystalline) and disordered (non-crystalline) regions [21,30,31]. However, it is not clear exactly how these regions are distributed within the fibril nor to which extent they occur. CP/MAS ^{13}C-NMR measurements with

spectral fitting [19,27,31,32] (Figure 2B) were done on the cellulosic substrates based on which the supramolecular structure of cellulose was represented in terms of the spatial distribution of fibrils and fibril aggregates, the accessible surface area, crystallinity and porosity. A representative CP/MAS ^{13}C-NMR spectrum from cellulose, together with a typical spectral fitting result, is shown in Figure 3. It serves to emphasize that for efficient enzymatic hydrolysis, the surface of cellulosic substrates need to be accessible to the enzymes, implying that the typical fibre wall pore sizes need to be larger than the typical sizes of the enzyme molecules which are around 10 nm [26].

Cellulosic substrates

Six cellulosic substrates with different supramolecular and physical properties were studied. Avicel PH-101 (Avicel) and cotton are commercially available celluloses. Both are non-lignin-containing materials and were delivered as dry materials. During the manufacture of Avicel, the cellulose pulp is subjected to acid hydrolysis and spray drying [33], resulting in a powder containing particles with remnants of the fibre wall morphology. Avicel is used extensively in enzymatic studies as a model substrate for cellulose. Cotton is a fibrous material with essentially intact fibre wall morphology. In an attempt to increase the SSA of cotton, it was treated with sodium

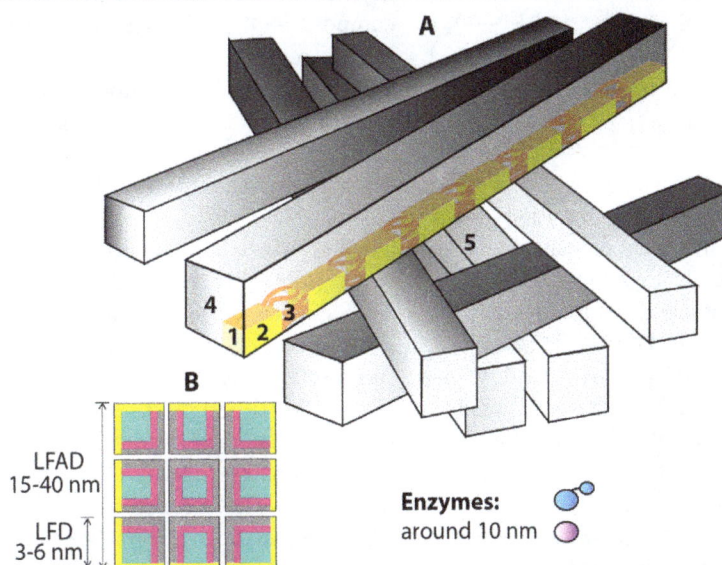

Figure 2 Complexity of supramolecular structure of cellulose. A Schematic representation of the model used in the present study to represent the supramolecular structure of cellulose, where fibril aggregates are shown as having square cross-sections with the following key elements: (1) fibril, a bundle of β-(1,4)-D-glucan polymers which is a mixture of the structures showing a high degree of three-dimensional order (crystalline) (2) and disordered (non-crystalline) (3) domains; (4) fibril aggregate, a structural element of cellulose composed of a bundle of fibrils; (5) pore, a cavity between fibril aggregates. **B** Front view of the model of the aggregated cellulose I fibrils used in calculations of lateral fibril aggregate dimensions (LFAD) and lateral fibril dimensions (LFD). The model differentiates between crystalline regions (turquoise), para-crystalline regions (magenta), accessible surface area (yellow) and inaccessible surface area that results from the close proximity of fibrils in a fibril aggregate (grey). Different enzymes are involved in the enzymatic hydrolysis of cellulose; some have a carbohydrate-binding module and a catalytic module, while others have only a catalytic module.

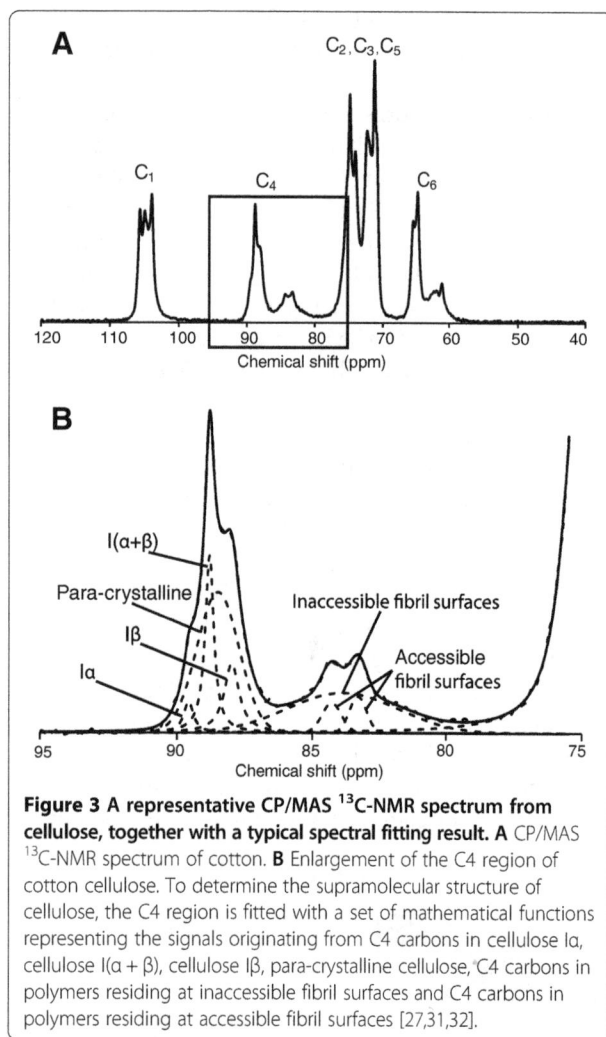

Figure 3 A representative CP/MAS ^{13}C-NMR spectrum from cellulose, together with a typical spectral fitting result. A CP/MAS ^{13}C-NMR spectrum of cotton. **B** Enlargement of the C4 region of cotton cellulose. To determine the supramolecular structure of cellulose, the C4 region is fitted with a set of mathematical functions representing the signals originating from C4 carbons in cellulose Iα, cellulose I(α + β), cellulose Iβ, para-crystalline cellulose, C4 carbons in polymers residing at inaccessible fibril surfaces and C4 carbons in polymers residing at accessible fibril surfaces [27,31,32].

Table 1 Chemical composition (in percentage) of the substrates studied

Substrate	Avicel	OPHS-ND and OPHS-D	NCC-OPHS-ND	Cotton and cotton + NaOH[a]
Acid-insoluble lignin	n.a.	0.5	n.d.	n.a.
Acid-soluble lignin	n.a.	0.5	n.d.	n.a.
Extractives	n.d.	<0.25	n.d.	0.7
Ash content	n.d.	<0.1	n.d.	0.2
Xylose	2.2	1.5	0.8	0.5
Mannose	0.9	0.8	0.8	0.1
Arabinose	<0.1	<0.1	<0.1	0.1
Galactose	<0.1	<0.1	<0.1	0.2
Glucose	96.8	96.7	98.4	98.2

Major hemicelluloses in softwood (OPHS-ND, OPHS-D and NCC-OPHS-ND substrates) are (galacto)glucomannan, galactoglucomannan and arabinoglucuronoxylan and major hemicelluloses in hardwood are glucuronoxylan and glucomannan [35]. [a]Carbohydrate composition analysis was done on cotton substrate. Since cotton has such a high degree of purity, NaOH treatment is not expected to change its chemical composition. n.a., not applicable; n.d., not determined.

(OPHS-D). This is known to induce changes in the supramolecular structure of cellulose I, partly responsible for the phenomenon known as hornification, traditionally measured as a reduction in the water-binding capacity of pulp [27]. It should be noted that drying induces little or no change in the chemical composition [34]. The chemical compositions of the substrates, representing wood [35] and cotton, are given in Table 1. It should be emphasized that all the substrates used in this study were cellulose-rich, all having cellulose content typically above 96%, and the main differences between the substrates were of a structural nature.

Initial cellulose supramolecular structure and the hydrolysability of the cellulosic substrates

The enzymatic hydrolysability, measured as the conversion yield of cellulosic substrates to glucose after 2 days, was quite different for the six substrates investigated (Table 2). Although the experimental conditions were identical, the conversion yield of the cellulosic substrates ranged from 2% to 91%. This was interpreted as an indication that the structural properties of cellulose may affect enzymatic hydrolysability. NMR results given in Table 2 provide average values of the supramolecular properties of the sample before enzymatic hydrolysis and after 2 days of hydrolysis. For example, after 2 days of hydrolysis, 31% of the cellulose in Avicel was hydrolysed which means that 69% residue remained, on which a supramolecular structure was measured.

Only weak correlations were found between initial SSA and enzymatic conversion yield of the cellulose substrates. No clear correlations were seen between the

hydroxide (cotton + NaOH), under conditions not leading to any significant conversion of cellulose I to cellulose II nor changes in chemical composition. Never-dried, oxygen-delignified, softwood pulp from pre-hydrolysis soda cooking technique (OPHS-ND) was used to represent cellulose isolated from wood. The OPHS-ND substrate can in a sense represent the best possible case of commercial pulps specifically designed for high enzymatic reactivity. This was achieved by combining high cellulose content with a suitable balance between fibre wall SSA and fibre wall average pore size. In order to reduce the possible influence of fibre wall morphology on enzymatic reactivity, cellulose nano-particles, sometimes referred to as nano-crystalline cellulose (NCC), were manufactured from OPHS-ND by prolonged hydrochloric acid hydrolysis at elevated temperature (NCC-OPHS-ND). The result was a colloidal suspension of cellulose nano-particles resulting from the removal of the fibre wall morphology. Part of the OPHS-ND material was also subjected to drying and rewetting

Table 2 Structural characteristics of the substrates before and after 2 days of enzymatic hydrolysis

Substrate	Physical state	Size	Before hydrolysis				After 2 days of hydrolysis				
			LFD (nm)	LFAD (nm)	SSA (m^2 g^{-1})	DCr (%)	LFD (nm)	LFAD (nm)	SSA (m^2 g^{-1})	DCr (%)	Conversion yield (%)
Avicel	Dried powder	Micro (round particles)	4.5 ± 0.1	23.7 ± 1.1	113 ± 5	56 ± 3	4.5 ± 0.1	20.4 ± 0.6	131 ± 4	56 ± 1	30.5 ± 0.2
Cotton	Dried fibres	Milli (in length)	5.8 ± 0.1	29.7 ± 1.2	90 ± 4	65 ± 2	n.d.	n.d.	n.d.	n.d.	2.1 ± 0.4
Cotton + NaOH	Dried fibres	Milli (in length)	6.1 ± 0.1	26.2 ± 0.9	102 ± 4	66 ± 2	n.d.	n.d.	n.d.	n.d.	6.3 ± 1.1
OPHS-ND	Never-dried fibres	Milli (in length)	4.7 ± 0.1	17.5 ± 0.8	153 ± 7	57 ± 1	5.8 ± 0.2	20.1 ± 1.1	132 ± 7	65 ± 3	67.7 ± 0.2
OPHS-D	Dried fibres	Milli (in length)	4.7 ± 0.1	28.3 ± 0.7	94 ± 2	57 ± 1	5.0 ± 0.1	29.5 ± 1.1	90 ± 3	59 ± 2	46.2 ± 5.2
NCC-OPHS-ND	Never-dried particles	Nano (rod-like particles)	5.6 ± 0.2	28.9 ± 1.7	92 ± 5	63 ± 2	5.5 ± 0.2	20.6 ± 1.1	130 ± 7	63 ± 3	90.7 ± 2.1

LFD, lateral fibril dimensions (or fibril thickness); LFAD, lateral fibril aggregate dimensions (or aggregate thickness); SSA, specific surface area; DCr, degree of crystallinity estimated from CP/MAS ^{13}C-NMR measurements; n.d., not determined.

DCr before hydrolysis and conversion yield. The sample with the highest conversion yield (NCC-OPHS-ND) had one of the highest DCr before hydrolysis.

In the model for supramolecular cellulose I used here (Figure 2), the SSA accessible to enzymes is assumed to be the envelope area of cellulose fibril aggregates computed from the LFAD, and the lateral dimensions of the individual fibrils are therefore assumed to be of less importance. Ultimately, the fibril aggregate envelope surface is composed of fibril surfaces, but aggregation prevents access to the whole fibril surface area within a fibril aggregate. It is therefore assumed that the fibril aggregate envelope surface constitutes an upper limit of the SSA accessible to the enzymes during the initial stage of hydrolysis. This is consistent with the data in Table 2, since no strong correlations were found between the LFD of the starting materials and the ability of the enzymes to hydrolyse them.

Furthermore, in the cellulose model used here (Figure 2), it is assumed that there is a relation between the DCr and LFD. Since the largest structural element of cellulose capable of maintaining crystalline cellulose I lattice is the cellulose fibrils, the DCr is ultimately limited by the LFD. The CP/MAS ^{13}C-NMR spectrum recorded on cellulose I shows separate signals originating from C4 carbons in polymers residing on fibril surfaces and C4 carbons in polymers residing in the fibril interior (Figure 3). The difference in carbon-13 chemical shift is in the range of 4 to 6 ppm, indicating a significant difference in polymer conformation between surface and interior polymers. This is further supported by previous measurements of laboratory frame carbon-13 spin–lattice relaxation times [15,32]. The C4 carbons on the fibril surfaces show a significantly shorter carbon-13 spin–lattice relaxation time than any other structural domain detected, and the difference in relaxation time can be ten or more times that for the interior structures. This means that the polymers on fibril surfaces are quite mobile compared to interior polymers. For these

reasons, it is assumed in the NMR model that surface polymers, on both accessible and inaccessible surfaces, are not identical to polymers in the cellulose I crystal lattice and that they therefore do not contribute to the DCr. This might have two consequences related to the cellulose structure. Firstly, even in the hypothetical case of perfect longitudinal polymer ordering in a fibril, the small lateral dimensions of a cellulose I fibril prevent the occurrence of completely crystalline cellulose I. Secondly, since the polymers at the fibril surface are not considered to be in a crystalline state, enzymes attach to fibril surfaces where the polymers are in a conformational and dynamical state quite distinct from that of the polymers in the crystalline fibril interior. This hypothesis is the basis of our suggested cellulose model, and it is supported by the lack of any strong correlations between the LFD or the DCr, which is inversely proportional to LFD and the conversion yields (Table 2).

In a previous study by our group, enzyme production during the filamentous fungus *T. reesei* growth on cellulose-rich pulps isolated from softwood and NCC-OPHS-ND was compared with growth of the fungus on Avicel. It was found that enzyme production by the filamentous fungus *T. reesei* showed the highest protein production when Avicel was used as a carbon and energy source for the fungus and that protein production levels were significantly lower when cellulose-rich pulps or NCC-OPHS-ND were used as carbon sources [36]. This observation led us to hypothesize that there may exist a relation between the structural properties of the substrate used for the fungal growth and the capacity of the produced enzymes to hydrolyse the same substrate. In the present study, a commercial mixture of cellulases from *T. reesei* and excess of β-glucosidase to prevent cellobiose inhibition were used. Interestingly, in the present study, Avicel showed a lower conversion yield than the cellulose-rich pulp samples. This is opposite to the trends observed for protein production among our samples and suggests that cellulosic substrates that are more recalcitrant to enzymatic hydrolysis may result in higher enzyme production when they are used as carbon and energy sources for fungi.

Changes in the supramolecular structure of cellulose resulting from enzymatic hydrolysis

The characteristics of the Avicel, OPHS-ND, OPHS-D and NCC-OPHS-ND substrates were measured before and after enzymatic hydrolysis. No significant change in LFD was observed for Avicel or NCC-OPHS-ND as a result of hydrolysis. This could be because these substrates had been subjected to acid hydrolysis prior to enzymatic hydrolysis.

The LFD for the OPHS-ND and OPHS-D substrates increased as a result of enzymatic hydrolysis, significantly in the case of OPHS-ND and slightly in the case of OPHS-D. Interestingly, the observed increase in LFD for OPHS-ND after enzymatic hydrolysis (from 4.7 to 5.8 nm) was about the same as the increase in LFD observed between OPHS-ND and the NCC-OPHS-ND (from 4.7 to 5.6 nm) as the result of acid hydrolysis of OPHS-ND during the manufacturing of NCC-OPHS-ND. It appears that the observed increase in LFD in this case is independent of whether acid or enzymatic hydrolysis is used. The mathematical model used estimates the DCr based on the LFD. Therefore, the observed increase in the DCr as the result of enzymatic hydrolysis observed for OPHS-ND, OPHS-D and NCC-OPHS-ND is simply a mathematical reflection of the observed increase in LFD.

If the longitudinal structure of cellulose I fibrils is as illustrated in Figure 2, with recurring regions of non-crystalline polymers along the fibril axis, preferential removal of these zones could result in an apparent increase in LFD. The calculations of LFD and LFAD are based on ratios of relative signal intensities obtained by spectral fitting of the NMR data. However, the mathematical model does not accommodate longitudinally recurring non-crystalline regions, and if these are in fact present, they will constitute a source of error in the estimates of LFD and LFAD. The signal intensity affected by the presence of a signal intensity from such non-crystalline regions will be that from the C4 atoms in polymers present on the surface of inaccessible fibrils, i.e. the broad signal centred at about 83 to 84 ppm (Figure 3). If the longitudinally recurring non-crystalline zones are preferentially attacked by hydrolysis without any other significant structural effects, the mathematical model will predict an increase in LFD accompanied by a (small) decrease in LFAD. No such observations were made for any of the substrates in this study (Table 2), as the result of acid hydrolysis or enzymatic hydrolysis. Although the treatment conditions used

elsewhere are harsher, both NMR and wide-angle X-ray scattering method (WAXS) have shown the same trends, increasing LFD/crystallite size as a result of treatment [37]. As shown by both, NMR and X-ray, the increase in LFD is not a measurement artefact. Something does happen with the LFD as the result of acid and enzymatic hydrolysis, though we cannot explain the mechanism. We therefore conclude that longitudinally recurring regions of non-crystalline polymers may be present along the axis of the fibril, which may be more susceptible to hydrolysis, but the amount of material in these regions constitutes only a small fraction of the fibril. Therefore, another mechanism, or mechanisms, must be responsible for the observed increase in LFD resulting from hydrolysis.

Interestingly, the substrate showing the highest DCr before hydrolysis was the NCC-OPHS-ND sample, which showed no change in LFD (or DCr) after enzymatic hydrolysis, but still showed the highest conversion yield.

The two samples exhibiting the greatest changes in LFAD during enzymatic hydrolysis were Avicel and NCC-OPHS-ND. The mechanism behind these observations is not known, but both substrates had been subjected to acid hydrolysis prior to enzymatic hydrolysis. The structural changes induced by acid hydrolysis may make it easier for the enzymes to split the fibril aggregates apart during enzymatic hydrolysis, possibly due to fibril or fibril aggregate shortening which might occur during the acid hydrolysis step. Furthermore, it can be that the lack of change in LFD following enzymatic hydrolysis in the substrates having been subjected to acid hydrolysis could indicate that no significant lateral erosion took place during enzymatic hydrolysis. It is also worth noting that the sample showing the greatest decrease in LFAD as the result of enzymatic hydrolysis was NCC-OPHS-ND (29%), which was never dried, while the decrease in LFAD for Avicel, which was dried, was less (14%).

In the mathematical model used to interpret the NMR data, LFAD is assumed to be inversely proportional to SSA. In contrast to LFD, LFAD is known to vary depending on the isolation procedure used [18]. The OPHS-D substrate was made by drying OPHS-ND. During the drying of the OPHS pulp, there was an irreversible increase in LFAD (Table 2). This could be explained as an irreversible increase in the degree of aggregation with an associated decrease in SSA [18,27].

The effect of enzymatic hydrolysis on degree of crystallinity

One common explanation of the slowing down or cessation of the enzymatic hydrolysis of cellulose is that after the more 'easily accessible' cellulose for enzymes has been converted to sugars, the remaining more 'difficult,' i.e. crystalline, cellulose is recalcitrant to enzymatic hydrolysis [30]. However, spectral fitting of the CP/MAS ^{13}C-NMR spectra revealed no direct correlations between the DCr before enzymatic hydrolysis and the conversion yield. Furthermore, in some substrates the DCr did not change as a result of enzymatic hydrolysis. In the model of the cellulose supramolecular structure, most cellulose fibrils are part of aggregates, which means that a high proportion of the fibril surface area and crystalline regions are in the interior of the aggregates (Figure 2), preventing direct enzymatic attack during enzymatic hydrolysis, at least during the initial stages of hydrolysis. It has been observed in recent studies that there were no changes in the DCr of Avicel during the course of hydrolysis [9,38], in agreement with our findings. It has been proposed that the DCr remains constant during enzymatic hydrolysis due to the simultaneous hydrolysis of crystalline and amorphous domains [39].

The NCC-OPHS-ND substrate was made by acid hydrolysis of the OPHS-ND pulp using concentrated (2.5 M) hydrochloric acid at elevated temperature (95°C to 100°C) for 17 h. The NCC-OPHS-ND sample was the material remaining after acid hydrolysis. Traditionally, it is assumed that the so-called amorphous domains in cellulose are removed during acid hydrolysis. For this reason, the explanation according to Puls and Wood [39] seems less likely for explaining the unchanging DCr observed during the subsequent enzymatic hydrolysis of the NCC-OPS-ND sample since little or no 'amorphous' cellulose remained in the starting material for the enzymatic hydrolysis. This may also be the case with Avicel since acid hydrolysis is employed during its manufacture [33].

No change in crystallinity was observed for Avicel, OPHS-D or NCC-OPHS-ND substrates after 2 days of enzymatic hydrolysis. The OPHS-ND substrate showed a significant increase in the DCr after enzymatic hydrolysis but, as discussed above, without any accompanying decrease in LFAD (Table 2). Based on the model of cellulose I as a supramolecular structure and the findings presented in Table 2, we conclude that the degree of cellulose I crystallinity is not a major factor controlling enzymatic hydrolysability.

Hydrolysability is improved when the pore size of the substrate is larger than the size of the enzyme molecules

The efficiency of the enzymatic hydrolysis of cellulose depends on the surface area accessible to the enzymes. An important parameter affecting this is the pore size of the material in relation to the molecular size of the enzymes [14]. The projection of a bound cellobiohydrolase I molecule onto a cellulose surface has been estimated to be around 10 nm [26]. In the present work, we

combined SSA measurements obtained from CP/MAS ^{13}C-NMR spectra with results from fibre saturation point (FSP) measurements to estimate the average pore sizes in water-swollen samples. It should be noted that the FSP method used is not applicable to substrates lacking fibre cell wall morphology, such as the NCC-OPHS-ND substrate.

In cellulose I-rich substrates with remaining fibre wall morphology, a pore system exists in a water-swollen state of the samples. Fibre wall pore sizes are in the range of 1 to 100 nm [40], covering the range of sizes of typical enzyme molecules. Pore size could thus influence the penetration of enzymes into porous fibre walls and their subsequent attack on the cellulose surfaces available inside the water-swollen fibre wall. Measuring pore size distributions, or average pore sizes, in a water-swollen fibre wall or a fragment of such a wall is quite complicated and prone to errors, partly due to the soft nature of the material when in a water-swollen state. Stone and Scallan have developed a method based on solute exclusion that can be used to estimate the total water content of the fibre wall pore system when in a water-swollen state, which they call the FSP [41]. When the results of FSP measurements are combined with NMR estimates of the SSA, also measured when the sample is in a water-swollen state, the average pore size of the fibre wall can be estimated [19]. One advantage of combining FSP and NMR results is that estimates of the average pore size can be obtained without the need for sample drying, which often causes structural changes. Larsson et al. have shown that for cellulose-rich samples, the often employed nitrogen- or krypton-adsorption Brunauer-Emmett-Teller (BET) method [42] underestimates the SSA by as much as a factor of two, resulting in a similar overestimation of the pore size by a factor of two [19]. Köhnke et al. have recently published estimates of FSP and SSA for Avicel PH-105, the latter using nitrogen BET [43], and it can be expected that values of the average pore size based on these measurements will be overestimated by a factor of about two, compared with those made using the method described by Larsson et al. [19].

During enzymatic hydrolysis, the NCC-OPHS-ND sample is a sol (solid particles in a liquid) without any well-defined pore size. In order to assess if the cellulose concentration used could result in a cellulose nano-particle density so high that it might exert restrictions on the accessibility of the enzyme to the cellulose particles, an order-of-magnitude estimate of the average inter-particle distance in the reaction medium was made (See Additional file 1). The estimate showed that at the cellulose nano-particle concentration used, the average inter-particle distance in a well-dispersed system, during sufficient agitation, was an order of magnitude larger

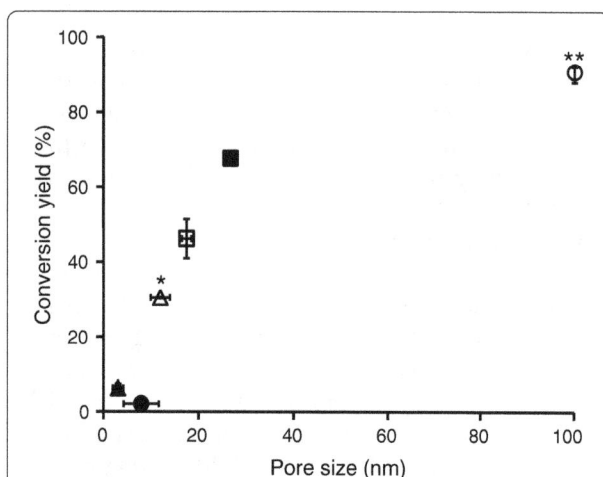

Figure 4 Average pore size in the substrates before enzymatic hydrolysis, together with conversion yields after 2 days of enzymatic hydrolysis: cotton + NaOH (black triangle), cotton (black circle), Avicel (white triangle), OPHS-D (white square), OPHS-ND (black square) and NCC-OPHS-ND (white circle). Errors in the conversion yields (%) are represented by the minimum and maximum values of two replicates. *The value for the average pore size in Avicel was calculated from the data given by Köhnke et al. 20 (±1) nm (FSP = 0.47 ± 0.02 g g^{-1}, SSA(BET) = 47 ± 1 m^2 g^{-1}) [43]. Taking into account the underestimation of the SSA by the BET method, the estimated pore size of Avicel is probably in the range 10 to 14 nm, calculated according to Larsson et al. [19]. ** Average pore size is >100 nm (see Additional file 1).

than the typical enzyme molecules. The interpretation was that no significant restrictions in accessibility of enzymes to cellulose particles were imposed by the cellulose nano-particle density in the reaction medium. A possible limiting factor for enzymatic conversion in this system could be the amount of cellulose nano-particle SSA.

Investigating the relationship between the average pore size and the hydrolysability of the substrates showed that conversion yield increased with increasing pore size of the substrate (Figure 4). A significant reduction in conversion yield was observed when the average pore size was in the order of, or smaller than, the size of typical enzyme molecules. Pore size is affected by drying conditions. It has been shown that LFAD are related to the rate at which water is removed from the cellulose fibres during drying. Pore size also depends on the presence of spacers. Spatial distribution of spacers such as hemicellulose and lignin can limit the degree of irreversibility of the structural changes caused by drying [18,27,44]. The OPHS-D substrate was prepared by drying OPHS-ND substrate, and the average pore size decreased from about 27 to 18 nm, but was still larger than the enzyme molecules (Figure 4). These results suggest that the average pore size may be an important factor governing the enzymatic hydrolysability of the substrates.

The NCC-OPHS-ND and OPHS-D substrates were made from the same material, OPHS-ND, and thus had similar SSA before hydrolysis, 92 and 94 m^2 g^{-1} (Table 2), respectively, which suggests that the observed difference in conversion between these two substrates is dominated by effects related to pore size/enzyme accessibility.

We investigated whether there were any correlations between substrate characteristics before enzymatic hydrolysis and the conversion yield measured after hydrolysis. During hydrolysis, the action of the enzymes may alter both the SSA and average pore size. Since identical reaction conditions were used throughout the experiments, the effects on conversion yield will be cumulative, making kinetic factors during the early stages of hydrolysis important for the conversion yield.

It should be noted that average pore sizes were used in this study. Pore size distributions and the detailed structural characteristics of the pore system, such as bottlenecks and dead-end pores may considerably reduce the SSA accessible to the enzymes. This means that the correlation between average pore size and conversion yield may not be perfect but, as the data in Figure 4 suggests, the impact of such pore system characteristics on the conversion yield is probably most pronounced in substrates where the average pore size is in the order of, or smaller than, the typical size of enzymes molecules.

It is also interesting to note that the substrate resulting in the highest conversion yield was NCC-OPHS-ND, which was a sol of cellulose nano-particles without any well-defined pore size, exhibiting a 91% conversion yield and the highest DCr. Its effective pore size during hydrolysis is the average inter-particle distance in the suspension, which is significantly larger than any enzyme molecule. Prolonged sedimentation or centrifugation gives an apparently dense pellet of cellulose nano-particles, but little is known about the average pore size in such pellets. When carrying out the enzymatic hydrolysis experiments, agitation was performed to prevent the formation of dense non-porous regions of nano-particles in the reaction tube. It was thus assumed that during enzymatic hydrolysis of NCC-OPHS-ND substrate the average distance between cellulose nano-particles in the reaction tube was much greater than the typical dimensions of enzyme molecules, imposing no significant additional hindrance on the enzyme molecules.

Possible mechanisms in enzymatic hydrolysis of cellulose
Several aspects of the supramolecular structure of cellulose may be affected during enzymatic hydrolysis, such as scission of fibrils and fibril aggregates, preferential hydrolysis of non-crystalline zones along the fibril axis, lateral erosion and cleavage of liberated fibril aggregates into individual fibrils. The interpretation of the data

in Table 2 is, to a large extent, based on CP/MAS ^{13}C-NMR analysis. It would be useful to evaluate the changes in NMR spectra resulting from changes in the supramolecular structure of cellulose that may occur during the time course of enzymatic hydrolysis. Since NMR spectral features mainly results from the lateral properties of the supramolecular cellulose I structure, little can be concluded about changes in the longitudinal properties. It is not until the LFD becomes so small that the signal from reducing end groups appears at about 96 ppm that any longitudinal characteristics affect spectra. The signal-to-noise ratio of the NMR spectra is typically such that the signal from reducing end groups becomes detectable when the degree of polymerization is below 150 to 200 anhydroglucose units. Theoretically, lateral erosion should be easily detected in these spectra. Successive erosion of the fibrils comprising the fibril aggregate surface is expected to be detected as a decrease in the average LFD. Similarly, the cleavage of fibril aggregates into separate fibrils is expected to result in a decrease in LFAD, possibly accompanied by a decrease in LFD, as some of the liberated fibrils start to become laterally eroded. When there are significant amounts of non-crystalline material along the fibril axis, preferential hydrolysis of these regions would lead to an apparent increase in LFD (and a corresponding increase in the DCr), accompanied by a slight decrease in LFAD, as discussed above. Other mechanisms may also alter the supramolecular structure of cellulose I. Several mechanisms may be working simultaneously, enhancing or counteracting the expected spectral changes, making it difficult to relate the observed spectral changes to specific mechanisms. Despite the complexity of the situation, a few cases in Table 2 stand out, and these will be discussed below.

The increase in LFD (or DCr) observed as the result of acid hydrolysis of the OPHS-ND substrate (NCC-OPHS-ND) and as the result of enzymatic hydrolysis of the OPHS-ND substrate was significant, about 20% in both cases, without any significant decrease in the LFAD. The magnitude of the change in LFD, together with the lack of decrease in LFAD, indicates that the hydrolysis of longitudinally distributed non-crystalline regions is not a likely explanation. Also, the less pronounced increase in LFD resulting from enzymatic hydrolysis of the OPHS-D substrate is difficult to explain, unless drying and rewetting would decrease the extent of longitudinally distributed non-crystalline regions. If the removal of longitudinally distributed non-crystalline regions is the explanation for the increase in LFD observed during hydrolysis, this would suggest that the actual fibril width of isolated wood-based cellulose I is 5 to 6 nm, rather than the 3 to 4 nm normally quoted. It is therefore likely that some other mechanism is responsible for the observed increase in LFD.

Enzymatic hydrolysis of NCC-OPHS-ND did not result in any change in the LFD or DCr, while the LFAD changed from 29 to 21 nm due to the hydrolysis of about 91% of the material. It is difficult to explain these observations by any other mechanism than successive shortening of the structural elements during enzymatic hydrolysis. It is also plausible that as the fibril aggregates become shorter, the cohesion between individual fibrils may became weaker, resulting in a decrease in LFAD due to cleavage of the fibril aggregates.

Conclusions

In the present study, we used a set of substrates with high cellulose content to investigate if the supramolecular structure of cellulose could influence the enzymatic hydrolysability. Initial cellulose supramolecular properties were used to interpret conversion yield measured after 2 days of enzymatic hydrolysis. Larger pore sizes in the cellulosic substrates resulted in higher conversion yield. We found no pronounced correlations between conversion yield and the SSA and LFD. Both acid and enzymatic hydrolysis can increase the LFD, but no plausible mechanism leading to the observed increase in LFD could be given. The substrate with the highest initial DCr, NCC-OPHS-ND, showed the highest conversion yield (91%), without any accompanying change in LFD, seemingly in conflict with a mechanism based on lateral erosion.

Methods
Substrates
Six different cellulosic substrates were used in this study (Table 3). Avicel® PH-101 (Fluka BioChemika, Cork, Ireland) has been used in many studies and therefore served as a reference. Two cotton (Original Topz®, Gunry AB, Kungsbacka, Sweden) substrates were used: one as delivered and in other case 1% of cotton (w/v) was treated with 5% NaOH (w/v) for 1 h at room temperature without stirring, in an attempt to increase the SSA. After treatment with NaOH, the cotton was washed with deionized water, until the pH was the same as that of the washing water. The sample material was filtered through a Miracloth (Calbiochem, La Jolla, CA, USA) and stored at 4°C in airtight plastic bags until it was used in enzymatic hydrolysis experiments. OPHS was produced from softwood biomass consisting of an industrially chipped and screened mixture of 40% Scots pine (Pinus sylvestris) and 60% Norway spruce (Picea abies). After screening, chips with bark and knots were removed by hand from the accept fraction, chip thickness 2 to 8 mm. The chips were subjected to pre-hydrolysis followed by an alkaline soda cooking according to the procedure described elsewhere [36,45]. The chemical composition of the softwood biomass was 28.3% acid-insoluble

Table 3 Description of the substrates used in this study

Abbreviation	Full name	Treatment	Origin
Avicel® PH-101	Commercial microcrystalline cellulose	Acid hydrolysis of wood pulp	Purchased
Cotton	Pharmaceutical grade cellulose	None	Purchased
Cotton + NaOH		NaOH treatment of cotton	This study
OPHS-ND	Oxygen-delignified pre-hydrolysis soda pulp-never-dried	Pre-hydrolysis soda cooking and oxygen delignification of prehydrolysis soda pulp from softwood	This study
OPHS-D	Oxygen-delignified pre-hydrolysis soda pulp-dried	Drying of OPHS-ND pulp	This study
NCC-OPHS-ND	Nano-crystalline cellulose	Acid hydrolysis of OPHS-ND pulp	This study

(Klason) lignin, 0.3% acid-soluble lignin, 41.1% cellulose, 9.5% xylan, 17.7% (galacto)glucomannan, 2.8% extractives and 0.3% ash. Never-dried (ND) and dried (D) OPHS substrates were used in the study. To produce OPHS-D substrate, OPHS-ND was gently stirred in deionized water until it became homogeneous. It was then filtered through a Miracloth, and the solid material was manually separated into small pieces, which were left at room temperature for 2 h and then dried overnight in an oven at 85°C. This material was stirred in deionized water until the mixture became homogeneous, filtered through a Miracloth and stored at 4°C in airtight plastic bags until it was used in the enzymatic hydrolysis experiments. NCC-OPHS-ND substrate was produced from OPHS-ND according to the procedure described elsewhere and stored in excess of deionized water in a bottle with a lid in room temperature [36].

Analysis of the chemical composition of the substrates

The ash content of biomass and OPHS was determined according to ISO 1762. The biomass and OPHS were extracted by acetone prior to analysis of the carbohydrates and lignin to determine the amount of acetone-extractable matter (extractives), according to SCAN-CM 49:03 [46]. To analyse the carbohydrates, samples were hydrolysed at 121°C in an autoclave with 0.4 M H_2SO_4, according to SCAN-CM 71:09 [47]. The solubilized monosaccharides were quantified using high-performance anion exchange chromatography with a Dionex ISC-5000 system coupled to a CarboPac PA1 (250 mm × 4 mm i.d.) column (Dionex, Stockholm, Sweden) and a pulsed amperometric detector (HPAEC-PAD). To determine the lignin content, the samples were hydrolysed with sulphuric acid and then filtered, and the acid-insoluble lignin (Klason lignin) was determined gravimetrically according to TAPPI T 222 om-02 [48]. The acid-soluble lignin was measured by UV spectrophotometry at 205 nm, according to TAPPI UM 250 [49]. MilliQ water was used as a blank and for dilution of the hydrolysate. The acid-soluble fraction was calculated using an absorption coefficient of 110 L g^{-1} cm^{-1}. The total content of lignin was assumed to be the sum of the acid-soluble and acid-insoluble fractions. Samples were analysed in duplicates.

Characterization of the substrates by CP/MAS ^{13}C-NMR

All samples were wetted with deionized water in the range of 40 to 60% water content and packed uniformly in a zirconium oxide rotor. Recording spectra from wet rather than dry samples give a higher apparent resolution. The CP/MAS ^{13}C-NMR spectra were recorded in a Bruker Avance III AQS 400 SB instrument operating at 9.4 T. All measurements were carried out at 295 (±1) K with a magic angle spinning (MAS) rate of 10 kHz. A 4-mm double air-bearing probe was used. Data acquisition was performed using a cross-polarization (CP) pulse sequence, i.e. a 2.95 μs proton 90 degree pulse and an 800-μs ramped (100 to 50%) falling contact pulse, with a 2.5 s delay between repetitions. A SPINAL64 pulse sequence was used for 1H decoupling. The Hartmann-Hahn matching procedure was based on glycine. The chemical shift scale was calibrated to tetramethylsilane (TMS $((CH_3)_4Si)$) scale by assigning the data point of maximum intensity in the alpha-glycine carbonyl signal to a shift of 176.03 ppm. A total of 4,096 transients were recorded from each sample, leading to an acquisition time of approximately 3 h. The software for spectral fitting was developed at Innventia AB and is based on a Levenberg-Marquardt algorithm [31]. All computations were based on integrated signal intensities obtained from spectral fitting [32]. To determine the supramolecular structure of cellulose, the C4 region is fitted with a set of mathematical functions representing the C4 signal intensity originating from C4 carbons in cellulose Iα, cellulose I(α + β), cellulose Iβ, para-crystalline cellulose, C4 carbons in polymers residing on inaccessible fibril surfaces and C4 carbons in polymers residing on accessible fibril surfaces. The integrated signal intensities obtained from the spectral fitting procedure are subsequently used for the calculation of the supramolecular characteristics of cellulose I, based on a cellulose model assuming square cross-sections of both fibrils and fibril aggregates [31]. More specifically, the LFAD, LFD, SSA and DCr were determined. The SSA for cellulose I was calculated from LFAD by assigning a density of 1,500 kg m^{-3} to cellulose I [27]. The errors given for LFD, LFAD, SSA and DCr are the standard error of the mean with respect to the quality of the fit. After convergence of the fitting procedure (using a chi^2 objective function with respect to spectral noise) the parameter standard errors (asymptotic standard errors) are computed as the square root of the sum squared residuals multiplied by the corresponding diagonal element from the variance-covariance matrix, divided by the number of degrees of freedom. The number of degrees of freedom is set to the number of data points reduced by the number of fitted parameters. The variance-covariance matrix is the inverse of the transpose Jacobian matrix multiplied with the Jacobian matrix using unit weights. The Jacobian matrix is estimated from analytical partial parameter derivatives of all adjusted parameters calculated at each spectral point. The aim of this procedure is to estimate an uncertainty in the parameter mean value, not to establish a measure representative of the variability of an underlying population of fibril aggregates widths.

Fibre saturation point measurements

FSP measurements were conducted according to Stone and Scallan [41]. Water-swollen sample material with a known solid content was mixed with a dextran solution of known concentration (approximately 1%, dextran mass/solution mass) (CAS No. 9004-54-0, Dextran 2000, from Leuconostoc spp., molecular mass approx. 2,000 kDa; Sigma-Aldrich, Stockholm, Sweden) in deionized water, approximately 1 mass unit of wet sample mass being mixed with 3 mass units of dextran solution. After mixing, the sample was stored in a sealed vessel at room temperature for 3 days to equilibrate. A liquid sample was subsequently taken and filtered through a Puradisc syringe filter (Whatman, Maidstone, UK) equipped with a 0.45-μm polytetrafluoroethylene (PTFE) membrane in a polypropylene housing (VWR International AB, Stockholm, Sweden). The concentration of dextran in the sample was determined using a calibration curve established for the optical rotation of polarized light measured using a Polartronic NH8 polarimeter (Schmidt + Haensch, Berlin, Germany) operating at 589 nm, with a resolution of 0.005°. The calibration curve was computed using three dextran concentrations: approximately 0.5%, 1.0% and 1.5% (dextran mass/solution mass), covering the range of all measurements. Dynamic light scattering was used to determine the hydrodynamic diameter of the dextran molecules at high dilution in deionized water (Zetasizer ZEN3600; Malvern Instruments Ltd., Malvern, UK), using a He-Ne 4.0 mW, 633 nm laser and a detector angle of 178°. The hydrodynamic diameter was found to be 101 ± 2 nm with a polydispersity index of 0.2, measured at a dextran concentration of 0.15 g dextran per litre solution. Based on the determined size of the dextran, the results obtained for the FSP were interpreted as representing liquid contained in pores smaller than approximately 100 nm in diameter. The FSP value is expressed as the dimensionless ratio of the mass of pore water to the mass of dry solids (g g^{-1}).

Pore size determination of the substrates

Average pore sizes were computed by combining estimates of the SSA from NMR spectra with results from FSP measurements. Advantages of this approach is that average pore sizes can be estimated without the need for sample drying and no pore geometry needs to be assumed [19].

Pore size was calculated from the following relation:

$$2t = \frac{2(\text{FSP})}{\sigma \varphi_L}$$

where

$2t$
  average pore size
FSP
  fibre saturation point
σ
  cellulose SSA
φ_L
  density of water.

The reported average pore size ($2t$) is comparable to the average pore diameter.

Enzymatic hydrolysis experiments

Enzymatic hydrolysis experiments were carried out in 50 mL tubes (TPP Techno Plastic Products AG, Trasadingen, Switzerland) with a 40 mL working volume at 50°C and 30 rpm on an adjustable angle mixing rotator (SB3, Stuart®, Bibby Scientific Limited, Staffordshire, UK). The cellulose concentration was 15 mg mL^{-1}, and 50 mM sodium citrate buffer (pH = 4.8) was added. The enzyme mixture was Celluclast 1.5 L and Novozyme 188 (both from Novozymes A/S Bagsvaerd, Denmark) at 5 mg protein per gram cellulose and 500 nkat g^{-1} cellulose, respectively. Novozyme 188 was added to the hydrolysis mixture to compensate for low β-glucosidase activity in Celluclast 1.5 L mixture. Two reaction tubes were withdrawn after 2 days of enzymatic hydrolysis and centrifuged (10 min, 3,000 rcf). Errors in the conversion yields (%) are represented by the minimum and maximum values of two replicates. The supernatant was used to determine the sugar concentration, and the solid residue after two days of hydrolysis was characterized by CP/MAS ^{13}C-NMR. The solubilized sugars were quantified using a Dionex ISC-3000 high-performance anion exchange chromatography system coupled to a CarboPac PA1 (250 mm × 4 mm i.d.) column (Dionex, Sweden) and a pulsed amperometric detector (HPAEC-PAD). The conversion of cellulose to glucose was calculated assuming that the theoretical yield of glucose is 1.11 times higher than the weight of cellulose (1.11 g glucose per gram cellulose) due to the addition of water during hydrolysis.

Additional file

Additional file 1: Estimation of inter-particle distance in a sol consisting of cellulose nano-particles in a liquid.

Abbreviations

DCr: Degree of crystallinity; FSP: Fibre saturation point; LFAD: Lateral fibril aggregate dimensions; LFD: Lateral fibril dimensions; NCC-OPHS-ND: Nano-crystalline cellulose produced from OPHS-ND pulp substrate; SSA: Specific surface area; OPHS-ND: Oxygen-delignified pre-hydrolysis soda pulp, never-dried; OPHS-D: Oxygen-delignified pre-hydrolysis soda pulp, dried.

Competing interests

The authors declare that they have no competing interests.

Authors' contributions

AP, PTL and LO designed the experiments. AP performed the experiments. KK produced OPHS-ND substrate. PTL did CP/MAS ^{13}C-NMR measurements. AP wrote the manuscript. PTL, KK and LO participated in data analysis and revised the manuscript. All authors read and approved the final manuscript.

Acknowledgements

This work was funded by the Swedish Research Council (VR) under the programme for strategic energy research (no. 621-2010-3788).

Author details

[1]Department of Biology and Biological Engineering, Division of Industrial Biotechnology, Chalmers University of Technology, Kemivägen 10, Gothenburg SE-412 96, Sweden. [2]Innventia AB, Drottning Kristinas väg 61, Stockholm SE-114 86, Sweden. [3]Wallenberg Wood Science Center, KTH Royal Institute of Technology, Teknikringen 56-58, Stockholm SE-100 44, Sweden. [4]Wallenberg Wood Science Center, Chalmers University of Technology, Kemigården 4, Gothenburg SE-412 96, Sweden.

References

1. FitzPatrick M, Champagne P, Cunningham MF, Whitney RA. A biorefinery processing perspective: treatment of lignocellulosic materials for the production of value-added products. Bioresour Technol. 2010;101:8915–22.
2. Otero J, Panagiotou G, Olsson L. Fueling industrial biotechnology growth with bioethanol. In: Olsson L, editor. Biofuels, vol. 108. Berlin Heidelberg: Springer; 2007. p. 1–40.
3. Mansfield SD, Mooney C, Saddler JN. Substrate and enzyme characteristics that limit cellulose hydrolysis. Biotechnol Prog. 1999;15:804–16.
4. Pauly M, Keegstra K. Cell-wall carbohydrates and their modification as a resource for biofuels. Plant J. 2008;54:559–68.
5. Yang B, Dai Z, Ding S-Y, Wyman CE. Enzymatic hydrolysis of cellulosic biomass. Biofuels. 2011;2:421–50.
6. Hemsworth GR, Henrissat B, Davies GJ, Walton PH. Discovery and characterization of a new family of lytic polysaccharide monooxygenases. Nat Chem Biol. 2014;10:122–6.
7. Horn SJ, Vaaje-Kolstad G, Westereng B, Eijsink VG. Novel enzymes for the degradation of cellulose. Biotechnol Biofuels. 2012;5:45.
8. Ding S-Y, Liu Y-S, Zeng Y, Himmel ME, Baker JO, Bayer EA. How does plant cell wall nanoscale architecture correlate with enzymatic digestibility? Science. 2012;338:1055–60.
9. Yu ZY, Jameel H, Chang HM, Philips R, Park S. Evaluation of the factors affecting Avicel reactivity using multi-stage enzymatic hydrolysis. Biotechnol Bioeng. 2012;109:1131–9.
10. Igarashi K, Uchihashi T, Koivula A, Wada M, Kimura S, Okamoto T, et al. Traffic jams reduce hydrolytic efficiency of cellulase on cellulose surface. Science. 2011;333:1279–82.
11. Ooshima H, Kurakake M, Kato J, Harano Y. Enzymatic activity of cellulase adsorbed on cellulose and its change during hydrolysis. Appl Biochem Biotechnol. 1991;31:253–66.
12. Andersen N, Johansen KS, Michelsen M, Stenby EH, Krogh KBRM, Olsson L. Hydrolysis of cellulose using mono-component enzymes shows synergy during hydrolysis of phosphoric acid swollen cellulose (PASC), but competition on Avicel. Enzyme Microb Tech. 2008;42:362–70.
13. Zeng Y, Zhao S, Yang S, Ding S-Y. Lignin plays a negative role in the biochemical process for producing lignocellulosic biofuels. Curr Opin Biotechnol. 2014;27:38–45.
14. Chandra RP, Bura R, Mabee WE, Berlin A, Pan X, Saddler JN. Substrate pretreatment: the key to effective enzymatic hydrolysis of lignocellulosics? Adv Biochem Eng Biotechnol. 2007;108:67–93.
15. Bergenstrahle M, Wohlert J, Larsson PT, Mazeau K, Berglund LA. Dynamics of cellulose-water interfaces: NMR spin–lattice relaxation times calculated from atomistic computer simulations. J Phys Chem B. 2008;112:2590–5.
16. Heiner AP, Kuutti L, Teleman O. Comparison of the interface between water and four surfaces of native crystalline cellulose by molecular dynamics simulations. Carbohyd Res. 1998;306:205–20.
17. Brown RM. Cellulose structure and biosynthesis: what is in store for the 21st century? J Polym Sci, Part A Polym Chem. 2004;42:487–95.
18. Hult E-L, Larsson P, Iversen T. Cellulose fibril aggregation - an inherent property of kraft pulps. Polymer. 2001;42:3309–14.
19. Larsson PT, Svensson A, Wagberg L. A new, robust method for measuring average fibre wall pore sizes in cellulose I rich plant fibre walls. Cellulose. 2013;20:623–31.
20. Duchesne I, Daniel G. Changes in surface ultrastructure of Norway spruce fibres during kraft pulping - visualisation by field emission-SEM. Nord Pulp Pap Res J. 2000;15:54–61.
21. Preston RD, Cronshaw J. Constitution of the fibrillar and non-fibrillar components of the walls of Valonia ventricosa. Nature. 1958;181:248–50.
22. Ohad I, Danon IO, Hestrin S. Synthesis of cellulose by Acetobacter xylinum. V Ultrastructure of polymer J Cell Biol. 1962;12:31–46.
23. Ding SY, Himmel ME. The maize primary cell wall microfibril: a new model derived from direct visualization. J Agric Food Chem. 2006;54:597–606.
24. Gilkes NR, Kilburn DG, Miller Jr RC, Warren RA, Sugiyama J, Chanzy H, et al. Visualization of the adsorption of a bacterial endo-beta-1,4-glucanase and its isolated cellulose-binding domain to crystalline cellulose. Int J Biol Macromol. 1993;15:347–51.
25. Tormo J, Lamed R, Chirino AJ, Morag E, Bayer EA, Shoham Y, et al. Crystal structure of a bacterial family-III cellulose-binding domain: a general mechanism for attachment to cellulose. EMBO J. 1996;15:5739–51.
26. Liu YS, Baker JO, Zeng Y, Himmel ME, Haas T, Ding SY. Cellobiohydrolase hydrolyzes crystalline cellulose on hydrophobic faces. J Biol Chem. 2011;286:11195–201.
27. Chunilall V, Bush T, Larsson PT, Iversen T, Kindness A. A CP/MAS 13C-NMR study of cellulose fibril aggregation in eucalyptus dissolving pulps during drying and the correlation between aggregate dimensions and chemical reactivity. Holzforschung. 2010;64:693–8.
28. Ding SY, Xu Q, Ali MK, Baker JO, Bayer EA, Barak Y, et al. Versatile derivatives of carbohydrate-binding modules for imaging of complex carbohydrates approaching the molecular level of resolution. Biotechniques. 2006;41:435–42.
29. Zhu P, Moran-Mirabal JM, Luterbacher JS, Walker LP, Craighead HG. Observing Thermobifida fusca cellulase binding to pretreated wood particles using time-lapse confocal laser scanning microscopy. Cellulose. 2011;18:749–58.
30. Park S, Baker JO, Himmel ME, Parilla PA, Johnson DK. Cellulose crystallinity index: measurement techniques and their impact on interpreting cellulase performance. Biotechnol Biofuels. 2010;3:10.
31. Larsson PT, Wickholm K, Iversen T. A CP/MAS 13C NMR investigation of molecular ordering in celluloses. Carbohydr Res. 1997;302:19–25.
32. Wickholm K, Larsson PT, Iversen T. Assignment of non-crystalline forms in cellulose I by CP/MAS 13C NMR spectroscopy. Carbohydr Res. 1998;312:123–9.
33. Reier EG. Avicel PH microcrystalline cellulose, NF, Ph Eur., JP, BP. FMC Corporation. 2000;11:1–27.
34. Larsson PT, Salmén L. Influence of cellulose supramolecular structure on strength properties of chemical pulp. Holzforschung. 2014;68:861–6.

35. Pettersen RC. The chemical-composition of wood. Adv Chem Ser. 1984;207:57–126.

36. Peciulyte A, Anasontzis GE, Karlström K, Larsson PT, Olsson L. Morphology and enzyme production of *Trichoderma reesei* Rut C-30 are affected by the physical and structural characteristics of cellulosic substrates. Fungal Genet Biol. 2014;72:64–72.

37. Testova L, Borrega M, Tolonen LK, Penttilä PA, Serimaa R, Larsson PT, et al. Dissolving-grade birch pulps produced under various prehydrolysis intensities: quality, structure and applications. Cellulose. 2014;21:2007–21.

38. Hall M, Bansal P, Lee JH, Realff MJ, Bommarius AS. Cellulose crystallinity - a key predictor of the enzymatic hydrolysis rate. FEBS J. 2010;277:1571–82.

39. Puls J, Wood TM. The degradation pattern of cellulose by extracellular cellulases of aerobic and anaerobic microorganisms. Bioresour Technol. 1991;36:15–9.

40. Li T, Henriksson U, Ödberg L. Determination of pore sizes in wood cellulose fibers by ^{2}H and ^{1}H NMR. Nord Pulp Pap Res J (Sweden). 1993;8:326–30.

41. Stone J, Scallan A. The effect of component removal upon the porous structure of the cell wall of wood. II. Swelling in water and the fiber saturation point. Tappi. 1967;50:496–501.

42. Brunauer S, Emmett PH, Teller E. Adsorption of gases in multimolecular layers. J Am Chem Soc. 1938;60:309–19.

43. Kohnke T, Ostlund A, Brelid H. Adsorption of arabinoxylan on cellulosic surfaces: influence of degree of substitution and substitution pattern on adsorption characteristics. Biomacromolecules. 2011;12:2633–41.

44. Chunilall V, Bush T, Larsson PT. Supra-molecular structure and chemical reactivity of cellulose I studied using CP/MAS ^{13}C-NMR. In: van de Ven T, Godbout L, editors. Cellulose-fundamental aspects. 2013.

45. Rydholm SA. Pulping processes. New York, US: Interscience Pub; 1965.

46. SCAN-CM standard 49:03. Content of acetone-soluble matter. Scandinavian pulp, paper and board testing committee; 2003.

47. SCAN-CM 71:09. Carbohydrate composition. Scandinavian pulp, paper and board testing committee; 2009.

48. TAPPI. Test Method T 222 om-02. Acid-insoluble lignin in wood and pulp. Atlanta: TAPPI Test Methods; 2002.

49. TAPPI. Useful Method UM-250. Acid-soluble lignin in wood and pulp. Atlanta: Tappi Press; 1991.

7

A constitutive expression system for glycosyl hydrolase family 7 cellobiohydrolases in *Hypocrea jecorina*

Jeffrey G Linger[2], Larry E Taylor II[1], John O Baker[1], Todd Vander Wall[1], Sarah E Hobdey[1], Kara Podkaminer[1], Michael E Himmel[1] and Stephen R Decker[1]*

Abstract

Background: One of the primary industrial-scale cellulase producers is the ascomycete fungus, *Hypocrea jecorina*, which produces and secretes large quantities of diverse cellulolytic enzymes. Perhaps the single most important biomass degrading enzyme is cellobiohydrolase I (*cbh1*or Cel7A) due to its enzymatic proficiency in cellulose depolymerization. However, production of Cel7A with native-like properties from heterologous expression systems has proven difficult. In this study, we develop a protein expression system in *H. jecorina* (*Trichoderma reesei*) useful for production and secretion of heterologous cellobiohydrolases from glycosyl hydrolase family 7. Building upon previous work in heterologous protein expression in filamentous fungi, we have integrated a native constitutive enolase promoter with the native *cbh1* signal sequence.

Results: The constitutive *eno* promoter driving the expression of Cel7A allows growth on glucose and results in repression of the native cellulase system, severely reducing background endo- and other cellulase activity and greatly simplifying purification of the recombinant protein. Coupling this system to a Δ*cbh1* strain of *H. jecorina* ensures that only the recombinant Cel7A protein is produced. Two distinct transformant colony morphologies were observed and correlated with high and null protein production. Production levels in 'fast' transformants are roughly equivalent to those in the native QM6a strain of *H. jecorina*, typically in the range of 10 to 30 mg/L when grown in continuous stirred-tank fermenters. 'Slow' transformants showed no evidence of Cel7A production. Specific activity of the purified recombinant Cel7A protein is equivalent to that of native protein when assayed on pretreated corn stover, as is the thermal stability and glycosylation level. Purified Cel7A produced from growth on glucose demonstrated remarkably consistent specific activity. Purified Cel7A from the same strain grown on lactose demonstrated significantly higher variability in activity.

Conclusions: The elimination of background cellulase induction provides much more consistent measured specific activity compared to a traditional *cbh1* promoter system induced with lactose. This expression system provides a powerful tool for the expression and comparison of mutant and/or phylogenetically diverse cellobiohydrolases in the industrially relevant cellulase production host *H. jecorina*.

Keywords: *Hypocrea jecorina*, *Trichoderma reesei*, Cellobiohydrolase, Cellulase expression, Fungal molecular biology, Biomass hydrolysis

* Correspondence: steve.decker@nrel.gov
[1]Biosciences Center, National Renewable Energy Laboratory, 16253 Denver West Parkway, Golden, CO 80401, USA
Full list of author information is available at the end of the article

Background

Enzymatic deconstruction of biomass to liberate monomeric sugars for the biological production of fuels and chemicals has been a research direction of global importance over the last several decades. One of the primary industrial-scale cellulase producers is the ascomycete fungus, *Hypocrea jecorina*, which produces and secretes large quantities of diverse cellulolytic enzymes. *H. jecorina* is not as genetically malleable as many other microorganisms, making it a challenging organism to use as a tool for the manipulation and expression of heterologous enzymes. However, recent work has expanded the tools available for genetically manipulating *H. jecorina*, including enhanced homology-based gene targeting via disruption of the non-homologous end joining (NHEJ) pathway [1,2], reusable genetic markers [1,3], strong constitutive promoters [4], and sexual crossings [5,6]. Yet, even with these advances, the genetic system of *H. jecorina* presents significant technical challenges when compared to other model microbial organisms.

A significant body of research has been focused on expression of *H. jecorina* enzymes in heterologous hosts. One enzyme that has received particular focus is cellobiohydrolase I (the gene is referred to as '*cbh1*' and the protein as '*Cel7A*'), due to its enzymatic proficiency in cellulose depolymerization. However, production of Cel7A with native-like properties from heterologous expression systems has proven difficult. For example, Cel7A expression in *Pichia pastoris* yielded hyperglycosylated and misfolded protein with reduced activity [7,8], expression in *Ashbya gossypii* yielded catalytically inactive enzyme [9], and expression in *Aspergillus niger* var. *awamori* produced overglycosylated isoforms with reduced activities and altered thermal stability [10]. Numerous expression studies of *H. jecorina* Cel7A in *Saccharomyces cerevisiae* also show hyperglycosylation, low-level expression, and/or low-level secretion, although some other fungal cellobiohydrolases appear more amenable to yeast expression [11,12]. Dana *et al.* [13] have recently shown that this result is at least in part due to the failure of *S. cerevisiae* to correctly process the *N*-terminal glutamine of Cel7A. Whereas there have certainly been advances in the heterologous expression and secretion of cellobiohydrolases in yeast [14-19], the overall trend is clear - there remains a significant challenge in effectively expressing Cel7A enzymes in organisms other than the native species.

As Cel7A is the major enzymatic activity in the *H. jecorina* cellulase system, the wide variety of issues with heterologous expression of Cel7A is a significant concern for cellulase improvement. Without a simple, robust, and productive heterologous expression system capable of producing Cel7A with native characteristics, improvement of Cel7A for inclusion in new industrial cellulase formulations becomes very difficult. Because *H. jecorina* is a major commercial cellulase production host and because Cel7A produced by other heterologous hosts is not necessarily equivalent to *H. jecorina*-produced Cel7A, evaluating novel or engineered enzymes produced by *H. jecorina* itself promises to be a very valuable tool. However, using *H. jecorina* as an expression host for recombinant Cel7A presents additional problems. With the objective of engineering a single cellulase, it is imperative that the enzyme of choice be produced in an enzymatically 'clean' background. Many cellulase expression studies in *H. jecorina* use the powerful *cbh1* promoter, which is induced by the presence of many substrates, including lactose, cellulose, and sophorose (reviewed in [20]). However, the induction of Cel7A expression also results in the induction of the entire cellulase system, making the *cbh1* promoter less than ideal for expressing single enzymes. Moreover, in order to achieve very high titers of cellulases, research is frequently conducted on highly mutated strains, such as RUT-C30, which are extremely proficient enzyme producers. These de-repressed strains constitutively express large suites of enzymes, even when grown on glucose, making the detailed study of single enzymes difficult. Growing the wild-type strain, QM6a, on glucose results in complete repression of the cellulase system. The use of QM6a as an expression host has the distinct advantage of allowing high expression of the target heterologous protein while repressing expression of other cellulases.

Obviously, an *H. jecorina* strain in which the endogenous cellulases are deleted would be ideal for production and characterization of heterologous cellulases. However, given the slow nature of sequential gene deletion in *H. jecorina* and the sheer number of potentially 'contaminating' cellulases produced by this host, we instead worked to generate an expression system that would utilize catabolite repression of endogenous cellulases while providing robust expression of our single target cellulase, in this case, Cel7A. As glucose is a natural global repressor of the cellulolytic machinery in *H. jecorina* [21], with repression mediated through the Cre1 repressor protein [22], the use of promoters with strong activity in glucose-containing media provides a valuable tool for the simultaneous expression of singular enzymes with the global repression of endogenous cellulases. For example, *Tef1* was identified as a strong promoter in glucose-containing media [23] and was successfully used to drive expression of both Cel7A and EGI (endoglucanase I) [24]. Recently, it was shown that the promoters from two glycolytic pathway enzymes, namely, enolase and pyruvate decarboxylase, were constitutively active in glucose-containing media and were capable of expressing high levels of homologous xylanases in *H. jecorina* [4]. While the *pdc* promoter was reported to be slightly better than *eno* (83 *vs.* 82% of total protein), the level of precision of the densitometry used to measure relative protein levels is low enough that the two promoters may be considered equivalent in performance [4]. We chose to use *eno* initially and since it functions remarkably well for Cel7A expression, we have not pursued using the *pdc* promoter.

For the immediate purpose of expressing native and engineered Cel7A homologously, and with the long-term goal of developing an expression system generally capable of expressing and secreting important classes of single proteins, we utilized a QM6a strain deleted for the native *cbh1* gene as the host and used an integrating expression cassette driving the expression of Cel7A from the *eno* promoter. Using this system, we are capable of producing native Cel7A with very little, if any, background from native cellulases. This report describes, to the best of our knowledge, the first successful use of a glucose-active promoter to drive Cel7A expression in a *cbh1* deletion strain. Accordingly, this work represents a technical foundation for moving towards our ultimate goal of generating a robust cellulase expression and secretion host for detailed expression studies on various classes of glycolytic enzymes in *Trichoderma reesei*. While Li *et al.* expressed a xylanase in a similar system, they did not purify the enzyme to test its intrinsic kinetic properties or compared it to xylanases expressed in their native context [4].

Results and discussion

We set out to create a heterologous host capable of high levels of cellulase expression in the absence of contaminating endogenous cellulases. To achieve this goal, we replaced the *cbh1*-promoter sequence in our vector pTR50 [25] with the *eno* (enolase) promoter to generate a vector, called 'pTrEno' (Figure 1). Enolase is a glycolytic enzyme whose transcriptional level is constitutive in glucose-containing medium [4], a situation which simultaneously serves to repress endogenous cellulases. Furthermore, to avoid even the smallest amount of contaminating native Cel7A, we used strain AST1116, a QM6a derivative strain, deleted for the native *cbh1* gene, as our host strain. To summarize, this newly generated strain, JLT102A, has the native *cbh1* gene deleted and a chromosomally integrated *pEno-cbh1* cassette liberated from pTrEno.

The pTrEno vector was designed such that the *eno* promoter can be readily substituted and the coding sequence can be easily interchanged using either traditional restriction cloning or recombination-based cloning techniques, including Gibson Assembly [26]. As the *cbh1* 5′ homology region was deleted during the construction of pTrEno, the expression cassette liberated by restriction digest with *SbfI* and *XhoI* does not target via homology and instead serves as a random integration cassette. While specific chromosomal integration sites are more difficult to identify, non-homologous integration allows multiple cassettes to be incorporated,

a common phenomenon in *H. jecorina* [27,28]. Furthermore, random insertion can provide the mechanism for integrating into chromosomal transcriptional 'hotspots,' such as euchromatic regions potentially enabling heightened expression, while avoiding transcriptionally repressed heterochromatic regions [29]. Chromatin heterogeneity is found in virtually all eukaryotes from *S. cerevisae* to humans and can have dramatic effects on gene expression. Such site-specific integration effects on gene expression will likely be observed with both the cellulase genes and the antibiotic resistance gene (*hph* in this case) contained on the expression cassette. Accordingly, expression can be variable from clone to clone, and random integration can lead to strains with heightened expression. We have thus designed a versatile plasmid for high-level expression of homologous or heterologous enzymes in *H. jecorina* in the absence of endogenous cellulase expression.

After designing the plasmid, we next wanted to determine the transformation and expression efficiency of the pTrEno expression construct. *H. jecorina* strain AST1116 was transformed via electroporation and plated onto potato-dextrose agar plates with hygromycin for selection and the non-ionic, non-denaturing detergent, IGEPAL CA-630, as a colony size restrictor. Two distinct colony morphologies are observed during transformation: 'fast growers' and 'slow growers' (Figure 2C), where only fast growers appear to have the potential to be enzyme-expressing transformants. Yet, even within the fast-growing subset, screening the extracellular glucose-containing growth medium of fast-growing transformants by dot-blot protein immunoblot showed varied expression between transformants (Figure 2A,B). As suggested above, this result could be due to multiple integrations of the expression cassette or to chromosomal position effects. Sodium dodecyl sulfate polyacrylamide gel electrophoresis (SDS-PAGE) coupled with Western blot analysis confirms the results of the initial dot-blot screen, where four-of-four fast growers show a protein band consistent with Cel7A as compared with an immune-reactive band from wild-type QM6a grown in lactose-containing medium for cellulase induction. Slow growers produce no immuno-reactive proteins (Figure 2D).

Given the constitutive nature of the *eno* promoter, we next wanted to determine *eno*-driven Cel7A expression in media containing various carbon sources. The changes in composition of total protein with changes in carbon source are shown in Figure 3A. Using Western blot, it was found that Cel7A is expressed in all media tested (Figure 3B). Interestingly, the relative amounts of Cel7 vary by carbon source, in that it appears the use of xylose or glycerol specifically produces higher levels of Cel7A in a cleaner background. However, this preliminary observation should be confirmed in a more quantitative manner. Additionally, the

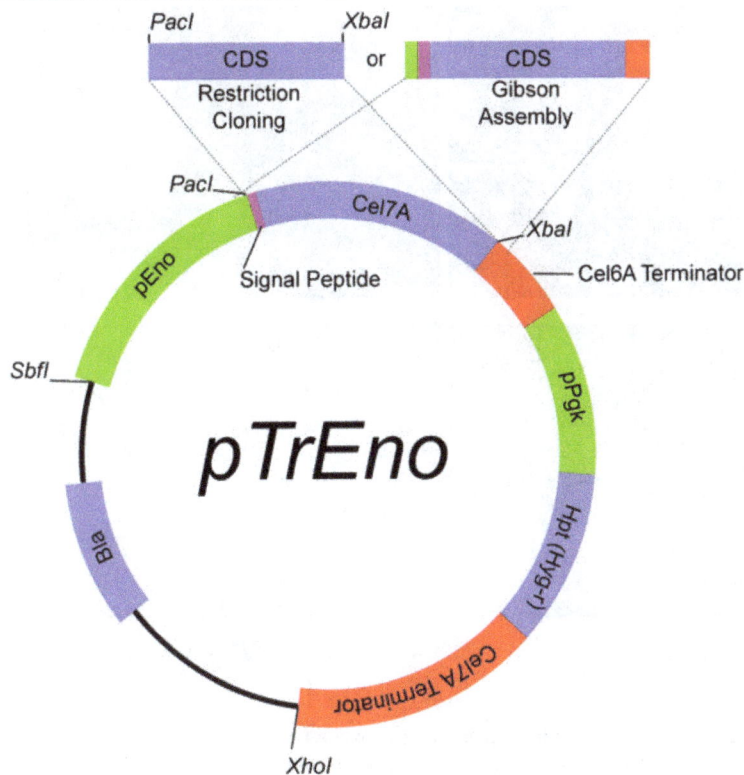

Figure 1 Schematic and features of the pTrEno expression plasmid.

apparent molecular weight distribution of Cel7A is slightly shifted between the various carbon sources, perhaps a consequence of differential glycosylation characteristics of the protein [30]. Further investigation of the fundamental differences in Cel7A purified from strains grown on various carbon-containing media will be of particular interest for optimizing this process.

Of primary importance to the design of this expression system was the notion that the strain should produce very little to no endogenous cellulases during the production of the *eno*-driven Cel7A. To examine this concern specifically, we performed a time-course growth of JLT102A coupled with multiplex Western blot using antibodies directed towards both Cel7A and Cel6A (*cbhII*) to analyze protein contents of extracellular growth medium containing either glucose or lactose (Figure 3C,D). Cel6A was specifically examined because it is second only to Cel7A in abundance in the *H. jecorina* secretome [31]. As expected, Cel7A was expressed in either glucose- or lactose-containing medium, whereas Cel6A could only be detected in lactose-containing medium. This lack of endogenous cellulases makes the purification of Cel7A much simpler and reduces the risks of cross-cellulase contamination during the measurement of activities. This outcome is of the utmost importance, as we move towards exploring and rapidly assessing phylogenetically diverse and mutant enzymes expressed in this *H. jecorina* system.

To validate that Cel7A enzyme produced from the *eno* promoter was functionally active, we concentrated the secreted enzyme from the growth medium and purified the enzyme using multiple fast protein liquid chromatography protocols. Enzymatic activity assays were carried out using dilute acid pretreated corn stover as the substrate. As can be seen in Figure 4, the performance of the *eno*-driven Cel7A (Figure 4B) is consistent with that of Cel7A purified from the wild-type QM6a secretome (Figure 4A).

Previous work in our lab evaluating wild-type rCel7A from a *cbh1*-delete, RUT-C30-based expression system [25] as well as native QM6a and RUT-C30 Cel7A protein from lactose fermentations, resulted in inconsistent specific activities on dilute acid-pretreated corn stover (Figure 4A) when assayed in a ternary enzyme system [32]. These results suggest either an inconsistency in protein processing, that is, glycosylation or trimming, or that purification from the high-cellulase background was itself variable, with low-levels of background endocellulase activity leading to variable observed activity. As the main goal of our work is to measure changes in Cel7A activity as a result of genetic manipulations and to screen new Cel7 exocellulases for enhanced properties, inconsistent measured activity from independent growth and purification steps was of great concern. To demonstrate that use of our new *eno*-driven expression system

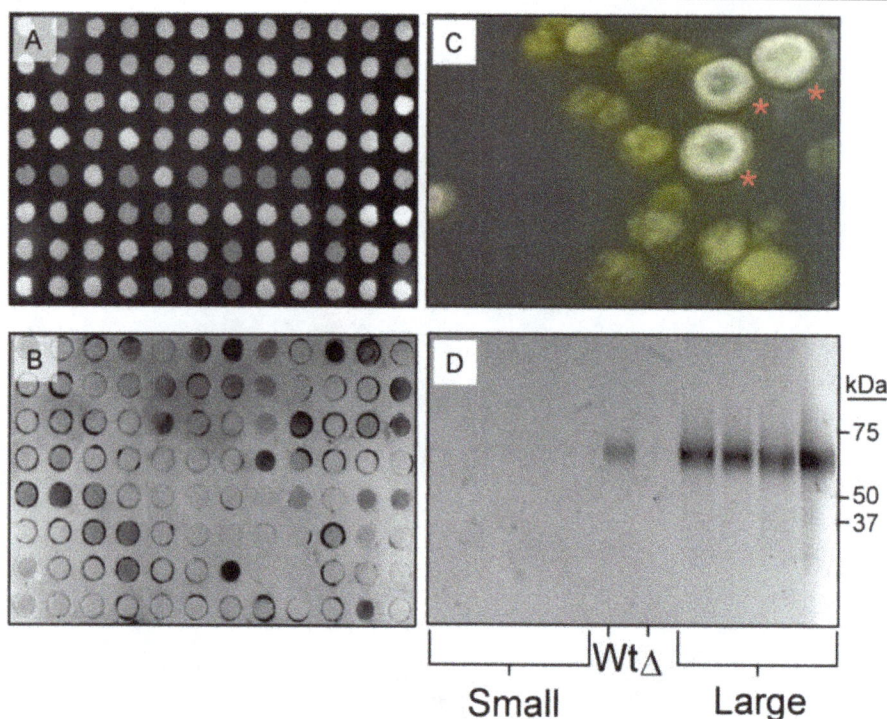

Figure 2 Rapid screening of the secretomes of potential transformants. Colonies from transformed plates were allowed to grow in liquid medium for 3 days prior to being screened by Western blot. **(A)** PVDF membrane illuminated with UV light to indicate successful transfer of broth and proteins to membrane. **(B)** Anti-Cel7A Western blot on the membrane shown in (A), showing numerous immunoreactive transformants. **(C)** We identified both 'small' and 'large' colonies after allowing transformed plates to incubate for beyond 3 days. **(D)** SDS-PAGE coupled Western blot highlights our observations that 'large' colonies are overwhelmingly more likely to be true Cel7A transformants expressing protein.

avoids this problem, we purified Cel7A from five independent *eno*-driven Cel7A fermentations, each grown under exacting conditions to minimize the impact of growth/stress parameters on enzyme activity. High stringency hydrolysis assays of these five purified proteins on pretreated corn stover clearly demonstrate nearly identical activities (Figure 4B). While the underlying mechanism(s) of activity inconsistency for the RUT-C30-strain-expressed Cel7A is still not entirely clear, the pTrEno system clearly demonstrates a much more consistent and stable system for evaluating differences in cellobiohydrolase activity.

Biophysical characterization of rCel7A was carried out for comparison with the native enzyme. Thermal stability was evaluated by differential scanning microcalorimetry (DSC). The eno-expressed protein showed no significant difference in thermal stability compared with the wild-type (Figure 5B), unlike Cel7A expressed from *Saccharomyces* or *Aspergillus*, which have shown significant differences compared to the wild-type Cel7A [33,34]. Similarly, the molecular weights of these enzymes as determined by SDS-PAGE (Figure 5A) show very similar masses for QM6a- and *eno*-expressed Cel7A, whereas molecular weights of *Aspergillus*- and *Saccharomyces*-expressed Cel7A are significantly higher, presumably due to increases in glycosylation.

Accuracy and consistency are important attributes of any expression system meant to compare enzyme activities. However, achieving this capability has proven very challenging in typical *cbh1*-promoter-driven systems using lactose as the inducer. Specifically, in our hands, growth and enzyme expression have proven quite variable using this system from trial to trial. For example, Figure 4A shows drastically different enzyme activities from independently expressed and purified enzymes from lactose-containing medium, even though they are identical in amino acid sequence. In contrast, when we performed five independent expressions and purifications of Cel7A derived from the *eno*-driven system, we see remarkable consistency in enzyme activity (Figure 4B). We suspect that much of the inconsistency with the lactose-induced wild-type Cel7A activity profiles arises from inconsistent purification despite the use of a rigorous purification scheme (Figure 5A). Miniscule amounts of other endogenous cellulases can drastically swing these PCS digestion curves making it very difficult to compare enzyme activities using this system. However, using the *eno*-driven system, much of this inconsistency falls away. For example, the average time it takes to achieve 80% digestion of PCS using four independently purified Cel7A preps from QM6a in lactose-containing medium is 55.1 h with a standard deviation of 12.0 h. In

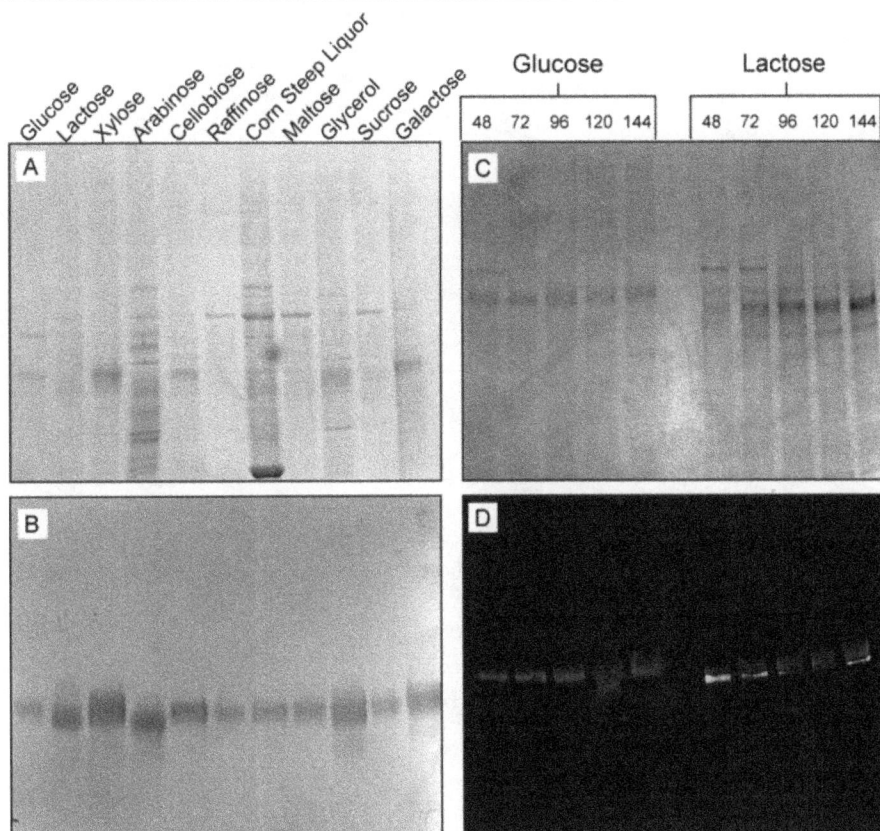

Figure 3 *eno*-driven Cel7A is constitutively expressed using numerous sole carbon sources and Cel6A is not expressed in glucose medium. **(A-B)** The same stock of JLT102A was used to inoculate media with varied carbon sources. Following 3-day growth, the secretome was analyzed via SDS-PAGE coupled Western blots. **(A)** Amido black staining highlights the total extracellular protein in each described media. **(B)** Anti-Cel7A Western blots show Cel7A expression in each described medium. **(C-D)** JLT102A was grown in MAG or MAL, and a time course of medium was taken for Western blot analyis using both Cel7A and Cel6A coupled with differently colored fluorescent secondary antibodies. **(C)** Amido black-stained PVDF membrane showing total protein as a loading control. **(D)** Multiplex Western blot using anti-Cel7A (red) and anti-Cel6A (green) shows constitutive *eno*-Cel7A expression and glucose repression of endogenous Cel6A.

contrast, five independently generated preparations of *eno*-driven Cel7A in glucose-containing medium take an average of 65.3 h. However, the standard deviation of this latter set is only 3.1 h. Given the variable and unpredictable nature of our Cel7A preparations from QM6a in lactose-containing medium, an exact comparison of the *eno*-driven Cel7A with the wild type form is exceedingly difficult to perform with any confidence. However, for future assessments of enzyme activities, the consistency provided by the *eno*-driven system will be of paramount importance in studies aimed at comparison of heterologous or mutant enzymes.

The work presented here springboards off of the work of Li *et al.* [4] and provides more functional characterization of the enzymes expressed. To summarize, Li *et al.* quite clearly showed that the *eno* promoter (among others) was capable of expressing a single xylanase to high levels and represents the initial identification and application of a powerful tool for *T. reesei* molecular biology-based pursuits.

However, this enzyme was not purified, and there were no biophysical or enzymatic activity comparisons to the native enzymes. This data is critical to understanding the effect of altering the gene expression environment (that is, promoter, carbon source) on the activity and biophysical state of the expressed enzyme. As such, the work presented here adds much-needed validation of the use of glycolytic promoters for the expression of cellulolytic enzymes in *H. jecorina* and provides a valuable tool to the field of fungal enzyme expression.

Conclusions

The portfolio of biomass-derived fuels and chemicals continues to expand, and accordingly, the ability to efficiently depolymerize cellulose remains a critical industrial challenge. Consequently, the need for identifying superior enzymes continues to be a priority, and perhaps no enzyme class is as valuable to this end goal as are the cellobiohydrolases. However, there are very few heterologous

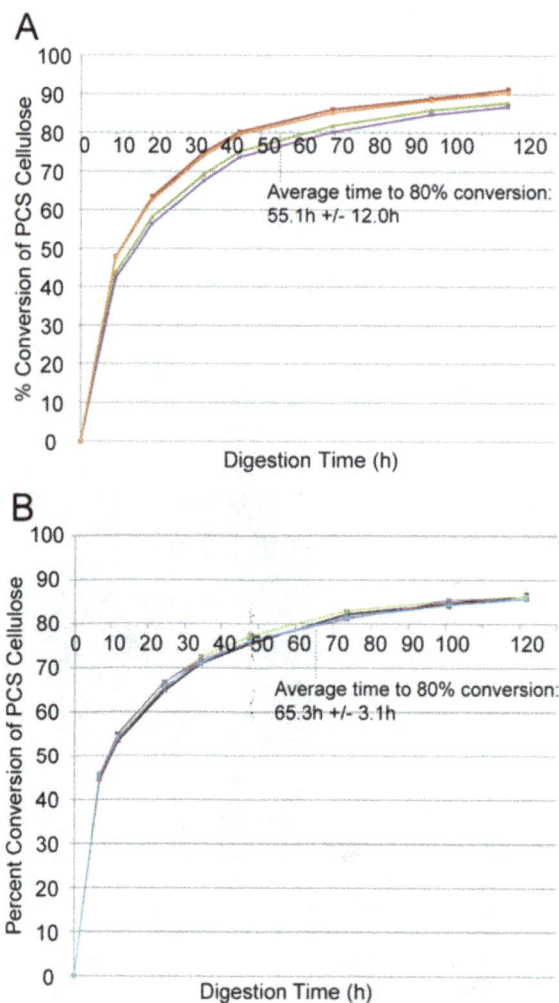

Figure 4 *eno*-expressed Cel7A provides consistent reproducibility in enzyme activity using pretreated corn stover as a substrate. **(A)** Four independent preparations of Cel7A from QM6a show very high levels of variability from batch to batch making enzyme activity assessments difficult. **(B)** Five independent preparations of *eno*-expressed Cel7A show remarkable consistency in enzyme activity assays. For both curves, error bars from triplicate assay digestions are included but are very difficult to see owing to the highly reproducible nature of these digestions.

expression systems that produce enzymes functionally equivalent to wild-type enzymes and are free from contaminating endogenous cellulases. The system we report here, using a glycolytic (*eno*) promoter-driven processing construct using the *cbh1* signal sequence in a *cbh1* deletion strain, provides a platform for detailed analysis of single heterologous cellulases produced in a native, industrially relevant host. Standard activity assays of heterologously expressed Cel7A show that these enzymes are comparable to wild-type Cel7A on pretreated corn stover and have similar thermal stability and glycosylation. More importantly, the eno-expression system permits heterologous Cel7A expression while repressing native cellulase production, making protein purification easier and, critically, eliminating variability in measured activity possibly caused by synergy with trace amounts of endocellulases. Future work will focus on increasing expression protein levels, exploiting

the expression of targeted and random mutations, and exploring the field of incorporating phylogenetically diverse enzymes into the *H. jecorina* secretome.

Methods

Media and growth conditions

Growth medium for Cel7A expression was a modified version of Mandels and Andreotti (MA) medium [35]. To make 1.0 L of MA, add 20.0 mL 50× MA salts, 5.0 g tryptone up to 737 mL with H_2O. Autoclave and then add 2.7 mL of separately sterilized 1.0 M $CaCl_2$ to minimize precipitation. Add 10 mL of filter sterilized micronutrient solution. Add 250 mL sterile 20% glucose or lactose to make 1.0 L of Mandels Andreotti minimal medium with 5% glucose (MAG) or Mandels Andreotti minimal medium with 5% lactose (MAL), respectively. Add hygromycin to a final concentration of 100 μg/mL as needed.

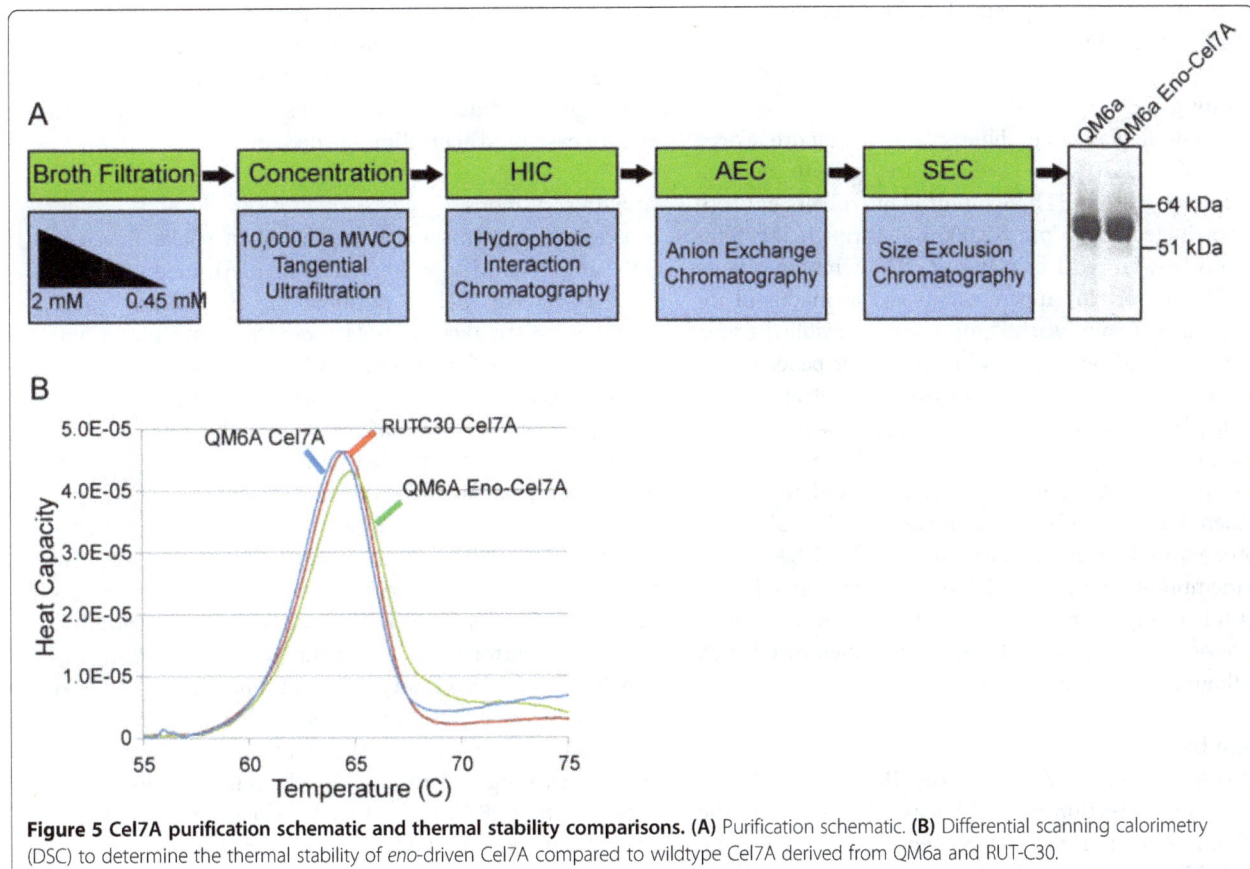

Figure 5 Cel7A purification schematic and thermal stability comparisons. (A) Purification schematic. **(B)** Differential scanning calorimetry (DSC) to determine the thermal stability of *eno*-driven Cel7A compared to wildtype Cel7A derived from QM6a and RUT-C30.

To prepare 1.0 L of 50× MA salts: combine 100.0 g KH_2PO_4, 70.0 g $(NH_4)_2SO_4$, 15.0 g urea, and 15.0 g $MgSO_4.7H_2O$, titrate to pH 5.5 with KOH. To prepare 1.0 L of micronutrient solution, add 500 mg $FeSO_4.7H_2O$, 160 mg $MnSO_4.H_2O$, 140 mg $ZnSO_4$, and 200 mg $CoCl_2$. Dissolve each component completely in order listed and then filter sterilize.

Growth medium for transformation outgrowth was complete medium lactose (CML), which consisted of 5.0 g/L yeast extract, 5.0 g/L tryptone, and 10.0 g/L lactose in a volume of 950 mL. The pH was adjusted to pH 7.5 with KOH and autoclaved. Following cooling, 50 mL of Clutterbuck's salt solution (per L: 120.0 g $NaNO_3$, 10.4 g KCl, 10.4 g $MgSO_4$, 30.4 g KH_2PO_4) was added. For spore production, potato dextrose (PD) plates were used and made according to the manufacturer's (Sigma Aldrich, St. Louis, MO, USA) recommendations. Hygromycin was added to the medium (after autoclaving) at a concentration of 100 µg/mL to make 'PDH' plates 'as required' following transformation, and IGEPAL CA-630 (similar to TritonX-100) was added at 0.1% as a colony restrictor to make 'PDHX' plates.

pTrEno construction

Vector pTR50 was PCR-amplified excluding the *cbh1* promoter and upstream homology region using primers (fwd:

ATGTATCGGAAGTTGGCCGTC, rev: TCTCGACG-CATTCGCGAA). The *eno* promoter was amplified directly from gDNA extracted from QM6a using primers (fwd: TTCGCGAATGCGTCGAGA*CCTGCAGG*-tgattccgtcctg-gattgc, rev: GACGGCCAACTTCCGATACAT*TTAAT-TAA*-tttgaagctatttcaggtggctgg).

These primers have 5′ 'tails' (capitalized) that are homologous to the ends of the PCR-linearized pTR50 described above and have the *SbfI* and *PacI* restriction sites incorporated, respectively (italicized). *In vitro* recombination was achieved using Gibson Assembly (New England Biolabs, Ipswich, MA, USA) according to the manufacturer's protocol.

PAGE

Culture broths were clarified via centrifugation and transferred to microcentrifuge tubes. Broths were diluted 3:1 in 4× LDS sample buffer (Life Technologies Corp., Carlsbad, CA, USA) with 50 µL/mL β-mercaptoethanol as a reducing agent. Samples were incubated at 95°C for 5 min prior to loading onto NuPAGE SDS gels with MOPS buffer, electrophoresed at 200 V constant for approximately 50 min and then transferred to a polyvinylidene difluoride (PVDF) membrane for Western blotting using the iBlot transfer system (Life Technologies Corp., Carlsbad, CA, USA).

Dot-blotting

Ninety-six-well glass fiber filter plates (Millipore Corp., Billerica, MA, USA) were used to clarify growth medium of *H. jecorina* cultures. Three hundred microliters of broth was centrifuged for 1 min ($2,000 \times g$) through the filter plate into 96-well receiver plates. PVDF membranes were cut to fit the dot blot apparatus, soaked in methanol for approximately 1 min, washed for 1 min in distilled water, and then overlaid on wetted Whatman filter paper cut to the same size and assembled on the dot blot apparatus. Typically, 100 µL of broth was loaded into each well, and a vacuum was applied until each well had the entirety of the broth pulled through. Blots were allowed to air dry and then were visualized and imaged under UV light (Fluorchem Q, Protein Simple, San Jose, CA, USA) to assure membrane transfer. Total protein detection is not achievable using this method due to the background fluorescence of media components. Blots were then reactivated in methanol and analyzed by Western blot.

Western blots

For single antibody Western blots (Figures 2B,D and 3B), immuno-detection of Cel7A was achieved using the SNAP i.d. Protein Detection System (Millipore Corp.). The PVDF membrane was blocked using SuperBlock PBS (Thermo Fisher Scientific Inc., Rockford, IL, USA) for 20 min. Rabbit anti-Cel7A polyclonal IgG was used as the primary antibody (1:20,000 dilution of crude serum), with alkaline phosphatase-conjugated goat anti-rabbit IgG (Thermo Fisher Scientific Inc.) as secondary. The alkaline phosphatase localization was visualized using BCIP/NBT (Life Technologies Corp., Carlsbad, CA, USA).

For the multiplex Western blot shown in Figure 3D, all solutions were sterile filtered to minimize background fluorescence. SDS-PAGE gels were transferred via standard wet tank transfer to PVDF membranes. All post-transfer solutions were from Protein Simple. Membranes were blocked in blocking solution for 1 h at room temperature and then washed four times for 2 min each in wash buffer. Polyclonal rabbit anti-Cel7A and monoclonal mouse anti-Cel6A antibodies were diluted 1:5,000 in blocking buffer, and the blot was incubated for 1 h. The blots were washed four times for 5 min in wash buffer. Goat anti-rabbit (red) and goat anti-mouse (green) Alexa Fluor conjugated antibodies were diluted 1:1000 in blocking buffer, added to the PVDF membrane and incubated at room temperature and covered to protect from ambient light with orbital shaking for 1 h. Membranes were washed three times for 10 s, followed by four washes for 5 min in wash buffer and finally washed two times for 5 min in final wash buffer. The blots were allowed to dry and then visualized using a FluorchemQ imaging system (Protein Simple). The total blot contrast was digitally adjusted evenly across the image planes to ensure ease of visualization of the bands.

Growth conditions

Small-scale growths were conducted in shake flasks at 30°C at 225 RPM in either MAG or MAL medium for 2 to 3 days. For Cel7A purification, *H. jecorina* spore stocks were streaked on potato dextrose agar plates and allowed to grow 2 to 3 days until a well 'lawned' plate of spores was achieved. Using the wide end of a sterile 1.0-mL pipette tip, an approximately 0.5-cm plug was extracted from the plate and deposited into 1.0 L of MAG medium in a 2.8-L shake flask. The culture was grown at 28°C with at 225 RPM for 24 h, after which the entire 1.0 L was transferred to 7.0 L of the same medium in a bioreactor. The bioreactors were 10-L working volume vessels manufactured by New Brunswick (Eppendorf Inc., Enfield, CT, USA) and controlled via New Brunswick's BioFlo3000 system. The total of 8.0 L was grown with mixing at 200 RPM with a combined Rushton (upper) and marine downflow (lower) style impellers (Rushton and Company, Gainesville, GA, USA), purged with 1.0 vol*vol^{-1}*min^{-1} of filtered air, kept at a strict 28°C, and pH controlled at 4.8 for 48 h. pH control was accomplished using 2.0 M HCl and KOH.

Transformation procedure

Transformations were achieved by way of spore electroporation, as modified from [6]. Spores from a frozen stock were spread evenly onto PD agar plates and allowed to grow at 30°C in the light (to enhance sporulation) for 2 to 3 days. The spores were harvested by gently spreading 2.0 to 3.0 mL ice-cold sterile distilled water on the plates to suspend the spores. The spores were moved to microcentrifuge tubes and centrifuged for 5 min at $2,000 \times g$ at 4°C. The spores harvested from up to five plates were pooled and suspended in 1.0 mL of ice-cold 10% glycerol. The spores were either used immediately for transformation or frozen at −80°C for future use. For transformation, 5 µg of pTrEno was digested with *SbfI* and *XhoI* and gel purified to isolate the Cel7A expression cassette. Purified DNA in 10 µL of water was mixed with 100 µL of spore suspension and placed in an ice-chilled 0.1-cm gap electrocuvette. The spores were electroporated using a BioRad Gene Pulser (Bio-Rad, Hercules, CA, USA) and the following conditions (1.8 kV, 25 µF, 800 Ω). Immediately following pulse, 1.0 mL of CML was added to the transformation. This cell suspension was then transferred to six-well tissue culture plates and incubated statically on a benchtop overnight (approximately 18 h). Microscopic visualization of the spores following this incubation shows spores just beginning to elongate and enter an active cellular growth stage. At this point, the cells were suspended by repeatedly pipetting up and down, and then 100 μL of cell suspension was plated onto potato dextrose with hygromycin and Triton X-100 (PDHX) plates. Transformants were allowed to grow for 2 to 3 days at 30°C in lighted incubators to enhance sporulation. Fast-growing colonies were selectively picked by hand for Cel7A expression testing. The selected colonies grew more rapidly and began to sporulate and turn green at a much faster rate compared to the slow-growing background colonies (Figure 2).

Initial screening of transformants and generation of clonal stocks

To screen initial single-colony transformants, 'plugs' were cored from agar plates using sterile disposable Pasteur pipettes and transferred to 2.0 mL of MAG medium supplemented with hygromycin (100 μg/mL) in sterile 24-well tissue culture plates. Cultures were grown statically at 30°C in lighted incubators for 3 days. Surface-lying mycelial mats were moved aside with sterile pipette tips, and broth was extracted for use in pre-screening for cellulase expression by dot blot analysis. Tissue culture plates were stored at 4°C until positive expressing transformants were identified. Following identification of positive expressing clones, the mycelial mats were transferred using sterile tweezers to the edge of PDH plates. These plates were incubated for 3 days at 30°C in lighted incubators to generate a lawn of spores. These spores were then struck out for single colonies on PDHX plates to ensure clonal populations. These colonies were again screened for Cel7A production, and positive expressing clones were again allowed to generate spore lawns on PD plates, and spore stocks were made using 20% glycerol. Stocks were frozen at −80°C.

Cel7A purification

Fermentation broths (approximately 8 to 10 L) were harvested and sequentially vacuum-filtered through the following series: (1) Miracloth (EMD Biosciences, St. Charles, MO, USA), (2) approximately 2-μm glass fiber filter, (3) 1.1-μm glass fiber, and (4) a 0.45-μM PES membrane. This filtered broth was then concentrated by tangential ultrafiltration with a 10,000-Da MWCO. The broths were roughly concentrated from 8.0 L to 150 to 200 mL. The final concentrated volume was exchanged with at least 1.0 L of 20 mM Bis-Tris pH 6.5 to remove residual peptides and other low molecular weight debris. This concentrate was then re-filtered to 0.2 μM. This filtrate was adjusted to 1.5 M $(NH_4)_2SO_4$ for hydrophobic interaction chromatography (HIC) and vacuum filtered through 0.2-μm PES, then loaded onto a 26/10 Phenyl Sepharose Fast Flow column. Buffer (A) was 20 mM Bis-Tris pH 6.5 and buffer (B) was 20 mM Bis-Tris pH 6.5, 2 M $(NH_4)_2SO_4$. After washing out the unbound

sample at 80% B, elution was via a descending buffer B gradient from 80% (1.6 M) to 0% over eight column volumes. Active fractions were identified by a pNP-lactose (pNPL) activity assay (pNPL at 2 mM in 50 mM acetate pH 5.0) where 100 μL of pNPL added to each well of a 96-well plate, followed by 25.0 μL of each fraction. The plate was then incubated 30 min at 45°C. Reactions were quenched with 25 μL 1.0 M $NaCO_3$ and the absorbance at 405 nm (A_{405}) was measured. Standard curve concentrations range from 0 to 250 μM pNP.

pNPL-active fractions were pooled and concentrated and then desalted and exchanged into 20 mM Bis-Tris pH 6.5 buffer using two sequential Superdex 25 Hi-Prep desalting columns. The desalted protein was loaded onto a Source 15Q 10/100 Tricorn anion exchange column and run at 0% to 50% B over 30 column volumes. Buffers were 20 mM Bis-Tris pH 6.5 (A) and the same buffer plus 1.0 M NaCl (B). pNP-L activity was followed again to identify active fractions. SDS-PAGE and αCel7A immunoblotting (described elsewhere) was performed to assess purity. The final stage of purification consisted of size exclusion chromatography (SEC) using a 26/60 Superdex 75 column and a 20 mM acetate pH 5.0 buffer with 100 mM NaCl in the mobile phase.

Differential scanning calorimetry

Thermal stability was evaluated by DSC using a Microcal model VP-DSC calorimeter (Microcal, Inc., Northampton, MA, USA). Data analysis was completed by Origin for DSC software (Microcal). Samples were prepared containing 50 μg/mL protein in a 20 mM acetate pH 5.0 buffer with 100 mM NaCl. Calorimeter scan rate was 60°C/h over a range of 30°C to 110°C.

Cel7A enzyme activity assay

Cellobiohydrolase activity is measured as the conversion of the cellulose fraction of a sample of a standard dilute acid-pretreated corn stover by the cellobiohydrolase used in conjunction with two other enzymes at standard loadings: (1) the endoglucanase *Acidothermus cellulolyticus* E1 (Cel5A, catalytic domain, Y245G mutant) loaded at 1.894 mg/g of biomass cellulose and (2) the chromatographically-purified *beta*-glucosidase from *Aspergillus niger*, loaded at 2.0 mg/g biomass cellulose.

The standard biomass substrate used in the activity assays is NREL dilute acid-pretreated corn stover P050921, washed first with water and then with 20 mM acetic acid/sodium acetate buffer, pH 5.0, until the pH of the (buffer) decantate is within 0.03 units of pH 5.00. From a slurry of this washed biomass material (approximately 9.0 mg biomass/mL of pH 5.0, 20 mM acetate buffer containing 0.02% sodium azide to retard microbial growth), a series of biomass substrate aliquots are prepared in 2.0-mL high-performance liquid chromatography (HPLC) vials, in such a way that each vial contains 8.5 mg biomass cellulose (which, given that the 'glucan' content of this batch of pre-treated stover is 59.1%, requires 14.38 mg of biomass per digestion vial). Biomass dry weights for each batch of assay vials were verified by dry weight determinations on a group of five samples co-pipetted into pre-tared vials. The acceptable relative standard deviation for a batch of biomass assay aliquots is 1% or less, with a preferred value of 0.8% or less. Adjustment of these biomass assay aliquots to a 1.7-mL final volume results in a cellulose concentration of 5.0 mg/mL.

Cellobiohydrolase assays were carried out in triplicate vials at 40°C, pH 5.0 in 20 mM azide-containing acetate buffer, with continuous mixing by inversion at 10 rpm while immersed in a water bath. At various times during the digestion, vials are removed from the rotator and representative 100-μL samples containing both solids and liquid are removed from the well-stirred contents and diluted 18-fold into glass HPLC vials. The primary digestion vials are immediately resealed and returned to the rotator in the assay 40°C water bath so that the assay digestions may continue. The vials containing the withdrawn and diluted samples of digestion mixture are then crimp-sealed and immersed in a boiling water bath for 10 min to denature the enzymes and terminate the reaction. The contents of the boiled time sample vials are then syringe-filtered (0.2-micron Acrodisc) into a third set of vials for sugar analysis by HPLC on a BioRad HPX-87H column operated at 65°C with 0.01 N H_2SO_4 as eluent at 0.6 mL/min and refractive index detection. Values for individual sugar concentrations in the digestion vials are back-calculated from the values measured by HPLC and then used to construct saccharification progress curves in terms of percent of conversion of biomass cellulose.

Abbreviations

AEC: anion exchange chromatography; CBH1: cellobiohydrolase I; CML: complete media with lactose; DSC: differential scanning microcalorimetry; E1: endoglucanase I (*Acidothermus cellulolyticus*); EG1: endogluconase I (*Hypocrea jecorina*); HIC: hydrophobic interaction chromatography; HPLC: high-performance liquid chromatography; MA: Mandels Andreotti minimal medium; MAG: Mandels Andreotti minimal medium with 5% glucose; MAL: Mandels Andreotti minimal medium with 5% lactose; MOPS: 3-(n-morpholino)propanesulfonic acid; NHEJ: non-homologous end joining; PD: potato dextrose; PDH: potato dextrose with hygromycin; PDHX: potato dextrose with Hygromycin and Triton X-100; pNP: p-Nitrophenol; pNP-L: 4-nitrophenyl β-D-lactopyranoside; PVDF: polyvinylidene difluoride; SEC: size exclusion chromatography.

Competing interests

The authors declare that they have no competing interests.

Authors' contributions

JGL was involved in initial conception, experimental design, strain development, enzyme expression, and authoring the manuscript. LET was involved in initial conception and all aspects of protein purification. JOB was responsible for assaying enzyme activities. TVW was primarily responsible for *H. jecorina* fermentations. SEH performed the DSC analysis. KP provided technical assistance on experimental optimizations. MEH was involved in the

initial conception and project leadership. SRD was involved with initial conceptualization, design of experimental objectives and protocols, technical direction, oversight, management of the work, and writing and editing of the manuscript. All authors read and approved the final manuscript.

Acknowledgements

This work was supported by the US Department of Energy under Contract No. DE-AC36-08GO28308. Funding for the work was provided by the DOE Office of Energy Efficiency and Renewable Energy, Bioenergy Technologies Office.

Author details

[1]Biosciences Center, National Renewable Energy Laboratory, 16253 Denver West Parkway, Golden, CO 80401, USA. [2]National Bioenergy Center, National Renewable Energy Laboratory, 16253 Denver West Parkway, Golden, CO 80401, USA.

References

1. Steiger MG, Vitikainen M, Uskonen P, Brunner K, Adam G, Pakula T, et al. Transformation system for *Hypocrea jecorina* (*Trichoderma reesei*) that favors homologous integration and employs reusable bidirectionally selectable markers. Appl Environ Microbiol. 2011;77:114–21.
2. Guangtao Z, Hartl L, Schuster A, Polak S, Schmoll M, Wang TH, et al. Gene targeting in a non-homologous end joining deficient *Hypocrea jecorina*. J Biotechnol. 2009;139:146–51.
3. Hartl L, Seiboth B. Sequential gene deletions in *Hypocrea jecorina* using a single blaster cassette. Curr Genet. 2005;48:204–11.
4. Li JX, Wang J, Wang SW, Xing M, Yu SW, Liu G. Achieving efficient protein expression in Trichoderma reesei by using strong constitutive promoters. Microb Cell Factories. 2012;11–6:84–93.
5. Seidl V, Seibel C, Kubicek CP, Schmoll M. Sexual development in the industrial workhorse *Trichoderma reesei*. Proc Natl Acad Sci U S A. 2009;106:13909–14.
6. Schuster A, Bruno KS, Collett JR, Baker SE, Seiboth B, Kubicek CP, et al. A versatile toolkit for high throughput functional genomics with *Trichoderma reesei*. Biotechnol Biofuels. 2012;5:1.
7. Godbole S, Decker SR, Nieves RA, Adney WS, Vinzant TB, Baker JO, et al. Cloning and expression of *Trichoderma reesei* cellobiohydrolase I in *Pichia pastoris*. Biotechnol Prog. 1999;15:828–33.
8. Boer H, Teeri TT, Koivula A. Characterization of *Trichoderma reesei* cellobiohydrolase Cel7A secreted from *Pichia pastoris* using two different promoters. Biotechnol Bioeng. 2000;69:486–94.
9. Ribeiro O, Wiebe M, Ilmen M, Domingues L, Penttilä M. Expression of *Trichoderma reesei* cellulases CBH1 and EGI in *Ashbya gossypii*. Appl Microbiol Biotechnol. 2010;87:1437–46.
10. Jeoh T, Michener W, Himmel ME, Decker SR, Adney WS. Implications of cellobiohydrolase glycosylation for use in biomass conversion. Biotechnol Biofuels. 2008;1:10.
11. Penttilä ME, Andre L, Lehtovaara P, Bailey M, Teeri TT. Efficient secretion of two fungal cellobiohydrolases by *Saccharomyces cerevisiae*. Gene. 1988;63:103–12.
12. Den Haan R, Mcbride JE, La Grange DC, Lynd LR, Van Zyl WH. Functional expression of cellobiohydrolases in *Saccharomyces cerevisiae* towards one-step conversion of cellulose to ethanol. Enzyme Microb Technol. 2007;40:1291–9.
13. Dana CM, Dotson-Fagerstrom A, Roche CM, Kal SM, Chokhawala HA, Blanch HW, et al. The importance of pyroglutamate in cellulase Cel7A. Biotechnol Bioeng. 2014;111(4):842–7.
14. Ilmén M, den Haan R, Brevnova E, McBride J, Wiswall E, Froehlich A, et al. High level secretion of cellobiohydrolases by *Saccharomyces cerevisiae*. Biotechnol Biofuels. 2011;4:30.
15. den Haan R, Kroukamp H, van Zyl JHD, van Zyl WH. Cellobiohydrolase secretion by yeast: current state and prospects for improvement. Process Biochem. 2013;48:1–12.
16. Voutilainen SP, Nurmi-Rantala S, Penttila M, Koivula A. Engineering chimeric thermostable GH7 cellobiohydrolases in *Saccharomyces cerevisiae*. Appl Microbiol Biotechnol. 2014;98:2991–3001.

17. Song JZ, Liu BD, Liu ZH, Yang Q. Cloning of two cellobiohydrolase genes from *Trichoderma viride* and heterogenous expression in yeast *Saccharomyces cerevisiae*. Mol Biol Rep. 2010;37:2135–40.

18. Xu LL, Shen Y, Hou J, Peng BY, Tang HT, Bao XM. Secretory pathway engineering enhances secretion of cellobiohydrolase I from *Trichoderma reesei* in *Saccharomyces cerevisiae*. J Biosci Bioeng. 2014;117:45–52.

19. Voutilainen SP, Murray PG, Tuohy MG, Koivula A. Expression of *Talaromyces emersonii* cellobiohydrolase Cel7A in *Saccharomyces cerevisiae* and rational mutagenesis to improve its thermostability and activity. Protein Eng Des Sel. 2010;23:69–79.

20. Mach RL, Zeilinger S. Regulation of gene expression in industrial fungi: *Trichoderma*. Appl Microbiol Biotechnol. 2003;60:515–22.

21. Elgogary S, Leite A, Crivellaro O, Eveleigh DE, Eldorry H. Mechanism by which cellulose triggers cellobiohydrolase I gene expression in *Trichoderma reesei*. Proc Natl Acad Sci U S A. 1989;86:6138–41.

22. Strauss J, Mach RL, Zeilinger S, Hartler G, Stoffler G, Wolschek M, et al. CreI, the carbon catabolite repressor protein from *Trichoderma reesei*. Febs Lett. 1995;376:103–7.

23. Nakari T, Alatalo E, Penttila ME. Isolation of *Trichoderma-reesei* genes highly expressed on glucose-containing media - characterization of the *tef1* gene encoding translation elongation-factor 1-alpha. Gene. 1993;136:313–8.

24. Nakari-Setala T, Penttilä M. Production of *Trichoderma reesei* cellulases on glucose-containing media. Appl Environ Microbiol. 1995;61:3650–5.

25. Singh A, Taylor LE, Vander Wall TA, Linger JG, Himmel ME, Podkaminer K, et al. Heterologous protein expression in *Hypocrea jecorina*: a historical perspective and new developments. Biotechnol Adv. 2014;33–1:142–54.

26. Gibson DG, Young L, Chuang RY, Venter JC, Hutchison CA, Smith HO. Enzymatic assembly of DNA molecules up to several hundred kilobases. Nat Methods. 2009;6:343–5.

27. Hazell BW, Te'o VSJ, Bradner JR, Bergquist PL, Nevalainen KMH. Rapid transformation of high cellulase-producing mutant strains of *Trichoderma reesei* by microprojectile bombardment. Lett Appl Microbiol. 2000;30:282–6.

28. Penttilä M, Nevalainen H, Ratto M, Salminen E, Knowles J. A versatile transformation system for the cellulolytic filamentous fungus *Trichoderma reesei*. Gene. 1987;61:155–64.

29. Seiboth B, Ivanova C, Seidl-Seiboth V. Trichoderma reesei: a fungal enzyme producer for cellulosic biofuels. In: Aurelio Dos Santos Bernardes M, editor. Biofuel production-recent developments and prospects. 1st ed. Rijeka: InTech; 2011. p. 309–40.

30. Stals I, Sandra K, Geysens S, Contreras R, Van Beeumen J, Claeyssens M. Factors influencing glycosylation of *Trichoderma reesei* cellulases. I: postsecretorial changes of the *O*- and *N*-glycosylation pattern of Cel7A. Glycobiology. 2004;14:713–24.

31. Rosgaard L, Pedersen S, Langston J, Akerhielm D, Cherry JR, Meyer AS. Evaluation of minimal *Trichoderma reesei* cellulase mixtures on differently pretreated barley straw substrates. Biotechnol Prog. 2007;23:1270–6.

32. Baker JO, Ehrman CI, Adney WS, Thomas SR, Himmel ME. Hydrolysis of cellulose using ternary mixtures of purified celluloses. Appl Biochem Biotechnol. 1998;70–2:395–403.

33. Qin YQ, Wei XM, Liu XM, Wang TH, Qu YB. Purification and characterization of recombinant endoglucanase of *Trichoderma reesei* expressed in *Saccharomyces cerevisiae* with higher glycosylation and stability. Protein Expr Purif. 2008;58:162–7.

34. Takashima S, Iikura H, Nakamura A, Hidaka M, Masaki H, Uozumi T. Overproduction of recombinant *Trichoderma reesei* cellulases by *Aspergillus oryzae* and their enzymatic properties. J Biotechnol. 1998;65:163–71.

35. Mandels M, Andreotti RE. Problems and challenges in cellulose to cellulase fermentation. Process Biochem. 1978;13:6.

8

Genetic basis of the highly efficient yeast *Kluyveromyces marxianus*: complete genome sequence and transcriptome analyses

Noppon Lertwattanasakul[1,5†], Tomoyuki Kosaka[2†], Akira Hosoyama[3†], Yutaka Suzuki[4†], Nadchanok Rodrussamee[1,6], Minenosuke Matsutani[2], Masayuki Murata[1], Naoko Fujimoto[1], Suprayogi[1], Keiko Tsuchikane[3], Savitree Limtong[5], Nobuyuki Fujita[3] and Mamoru Yamada[1,2*]

Abstract

Background: High-temperature fermentation technology with thermotolerant microbes has been expected to reduce the cost of bioconversion of cellulosic biomass to fuels or chemicals. Thermotolerant *Kluyveromyces marxianus* possesses intrinsic abilities to ferment and assimilate a wide variety of substrates including xylose and to efficiently produce proteins. These capabilities have been found to exceed those of the traditional ethanol producer *Saccharomyces cerevisiae* or lignocellulose-bioconvertible ethanologenic *Scheffersomyces stipitis*.

Results: The complete genome sequence of *K. marxianus* DMKU 3-1042 as one of the most thermotolerant strains in the same species has been determined. A comparison of its genomic information with those of other yeasts and transcriptome analysis revealed that the yeast bears beneficial properties of temperature resistance, wide-range bioconversion ability, and production of recombinant proteins. The transcriptome analysis clarified distinctive metabolic pathways under three different growth conditions, static culture, high temperature, and xylose medium, in comparison to the control condition of glucose medium under a shaking condition at 30°C. Interestingly, the yeast appears to overcome the issue of reactive oxygen species, which tend to accumulate under all three conditions.

Conclusions: This study reveals many gene resources for the ability to assimilate various sugars in addition to species-specific genes in *K. marxianus*, and the molecular basis of its attractive traits for industrial applications including high-temperature fermentation. Especially, the thermotolerance trait may be achieved by an integrated mechanism consisting of various strategies. Gene resources and transcriptome data of the yeast are particularly useful for fundamental and applied researches for innovative applications.

Keywords: *Kluyveromyces marxianus*, Thermotolerant yeast, Complete genome sequence, Transcriptome analysis, Xylose fermentation

Background

Along with rising concern about global warming and the rapid increase in fuel consumption, there is worldwide interest in the production of bioethanol from renewable resources [1]. For economically sustainable production of bioethanol, it is necessary to increase the types of biomass such as lignocellulosic materials that can be used without competing with food supplies. Accordingly, microbes that can efficiently convert various sugars in these kinds of biomass to ethanol must be developed.

A high-temperature fermentation (HTF) technology is expected to help reduce cooling cost, efficiently achieve simultaneous saccharification and fermentation, reduce the risk of contamination, and offer stable fermentation even in tropical countries [2,3]. Thermotolerant yeast *Kluyveromyces marxianus*, which is able to ferment various sugars, may be a suitable microbe for HTF with lignocellulosic hydrolysates [4,5].

* Correspondence: m-yamada@yamaguchi-u.ac.jp
†Equal contributors
[1]Applied Molecular Bioscience, Graduate School of Medicine, Yamaguchi University, Ube 755-8505, Japan
[2]Department of Biological Chemistry, Faculty of Agriculture, Yamaguchi University, Yamaguchi 753-8515, Japan
Full list of author information is available at the end of the article

K. marxianus is a haploid, homothallic, thermotolerant, hemiascomycetous yeast [6,7] and a close relative of *Kluyveromyces lactis*, a model Crabtree-negative yeast [8-11]. Both yeasts share the assimilating capability of lactose, which is absent from *Saccharomyces cerevisiae*. *K. marxianus* has a number of advantages over *K. lactis* or *S. cerevisiae*, including the intrinsic fermentation capability of various sugars at high temperatures [4,12,13], weak glucose repression that is preferable for mixed sugars such as hemicellulose hydrolysate, and fermentability of inulin [13,14]. However, its fermentation activity from xylose is extremely low compared to that of glucose. Recently, we developed a procedure that improves this disadvantageous trait (unpublished data) and increases the fermentation activity to slightly less than that of *Scheffersomyces stipitis* at around 30°C and a much higher activity at higher temperatures. Many biotechnological applications of *K. marxianus* have so far been achieved: production of various enzymes including heterologous proteins, aroma compounds or bioingredients, reduction of lactose content in food products, production of ethanol or single-cell protein, and bioremediation [13]. In addition, novel methods and genetic tools for genetic engineering have been developed on the basis of its high nonhomologous end-joining activity [15,16]. *K. marxianus* is thus a highly competent yeast for future developments. In order to facilitate such developments, its genomic information is essential. Draft genome sequences of three *K. marxianus* strains have been published [7,17,18], but no detailed analysis is available.

This study provides core information on *K. marxianus* DMKU 3-1042, which is one of the most thermotolerant strains in the same species isolated (3, unpublished data), including its ability to assimilate various sugars and the molecular basis of its thermotolerance and efficient protein productivity, in addition to the complete genome sequence.

Results

Genomic information and comparative genomics

The genome sequence of *K. marxianus* DMKU 3-1042 was precisely determined (less than one estimated error per chromosome) by nucleotide sequencing with three different sizes of shotgun libraries. Telomeric regions were further analyzed by transposon-insertion sequencing of corresponding fosmid clones. This strategy allowed us to determine the complete genome sequence of 11.0 Mb including all centromeric regions and boundary regions containing up to one to several sequence repeats (GGTGTACGGATTTGATTAGTTATGT) of telomeres. Optical mapping confirmed the genome organization except for three inverted regions, which were fixed in the final complete sequences (Additional file 1: Figure S1). There are eight chromosomes ranging in size from 0.9 to 1.7 Mb and a mitochondrial genome of 46 kb. The annotation process predicted 4,952 genes (Table 1), of which 98.0% were predicted to consist of a single exon (Additional file 2). The average gene density is 68.0% (Table 2). The average gene and protein lengths are 1.5 kb and 501 amino acids, respectively (Table 1).

Eukaryotic orthologous groups (KOG) database analysis led to the assignment of protein functions of about 72.4% of predicted genes (Additional file 1: Table S1) and protein domains were predicted in 3,584 gene models. UniProt and KAAS assignments led to the assignment of homologous genes of about 86.4% of predicted genes and KEGG Orthology number of 50.5%, respectively. The yeast shares 1,552 genes with *K. lactis*, *Ashbya gossypii*,

Table 1 General information of nuclear and mitochondrial genomes of *K. marxianus* DMKU 3-1042

	Length	CDS	Intron-containing CDS	tRNA	rDNA	na_length average	na_length maximum	aa_length average	aa_length maximum
Total	10,966,467	4,952	172	202	8[b]				
Average						1,505		501	
Chromosome									
1	1,745,387	803	25	31	0	1,473	12,174	491	4,058
2	1,711,476	808	29	28	0	1,440	6,168	480	2,056
3	1,588,169	706	21	24	0	1,553	8,004	517	2,668
4	1,421,472	624	26	25	0	1,593	14,742	531	4,914
5	1,353,011[a]	611	20	30	6[b]	1,529	9,018	509	3,006
6	1,197,921	537	18	17	0	1,500	8,709	500	2,903
7	963,005	438	19	10	0	1,454	8,310	484	2,770
8	939,718	414	12	15	0	1,523	9,381	507	3,127
Mitochondrion	46,308	11	2	22	2	978	1,491	326	497

[a]The length does not include that of most of rDNA. [b]Six rDNA copies in the genome sequence in database, but 140 rDNA copies by optical mapping. CDS, coding DNA sequence.

Table 2 General characteristics of 11 hemiascomycetous yeast genomes

Species	Genome size (Mb)	Average G + C content (%)	Total CDS	Total tRNA genes	Average gene density (%)	Average G + C in CDS (%)	Source
Kluyveromyces marxianus	10.97	40.12	4,952	202	68.00	41.69	This study
Kluyveromyces lactis	10.60	38.70	5,329	162	71.60	40.10	[20]
Saccharomyces cerevisiae	12.10	38.30	5,807	274	70.30	39.60	[20]
Candida glabrata	12.30	38.80	5,283	207	65.00	41.00	[20]
Yarrowia lipolytica	20.50	49.00	6,703	510	46.30	52.90	[20]
Scheffersomyces stipitis	15.40	41.10	5,841	-	55.90	42.70	[21]
Ashbya gossypii	9.12	51.70	4,776	220	77.10	52.80	[19]
Ogataea parapolymorpha	8.87	47.83	5,325	80	84.58	49.13	This study[a]
Debaryomyces hansenii	12.18	36.34	6,290	225	74.31	37.45	[20]
Clavispora lusitaniae	12.11	44.50	5,936	217	68.07	46.80	[22]
Schizosaccharomyces pombe	12.59	36.04	5,133	195	57.17	39.63	[23]

[a]Values were summarized by us using data from Joint Genome Institute (JGI, http://gold.jgi-psf.org). G + C, guanine + cytosine; CDS, coding DNA sequence.

Candida glabrata, *S. cerevisiae*, *Ogataea parapolymorpha*, *Debaryomyces hansenii*, *S. stipitis*, *Clavispora lusitaniae*, *Yarrowia lipolytica*, and *Schizosaccharomyces pombe* as hemiascomycetous yeasts [19-23]. The phylogenetic tree exhibits the closest location of *K. marxianus* to *K. lactis* and closer to *A. gossypii*, *C. glabrata*, and *S. cerevisiae* in the 11 yeasts (Figure 1). Consistent with this, *K. marxianus* shares 4,676; 3,826; 3,672; and 3,853 genes with *K. lactis*, *A. gossypii*, *C. glabrata*, and *S. cerevisiae*, respectively. On

the other hand, there are 193 genes specific for *K. marxianus* (Additional file 1: Table S2), which may be responsible for its species-specific characteristics, of which two thirds of the genes could not be assigned by the KOG database (Additional file 1: Table S3). There are 422 genes shared only between *K. marxianus* and *K. lactis* (Additional file 1: Table S4), which may be related to their genus-specific characteristics, such as production of β-galactosidase [24], assimilation of a wide variety of inexpensive

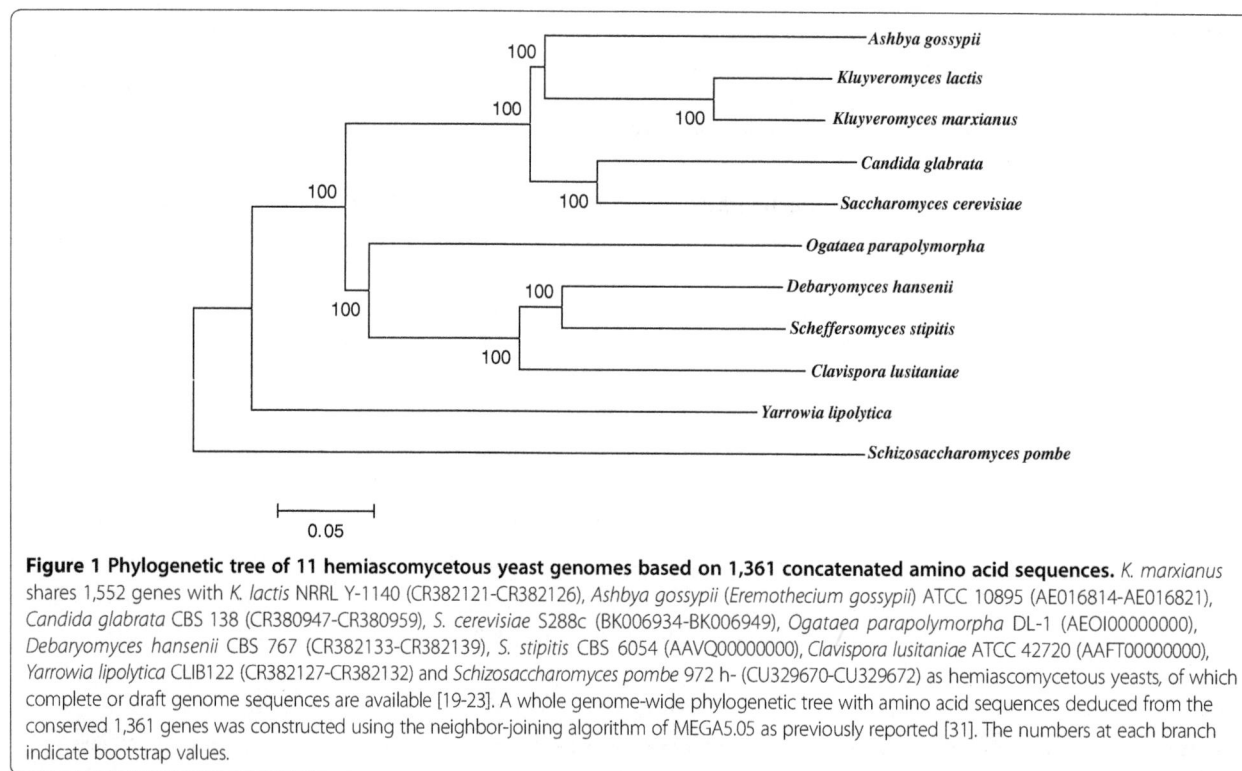

Figure 1 Phylogenetic tree of 11 hemiascomycetous yeast genomes based on 1,361 concatenated amino acid sequences. *K. marxianus* shares 1,552 genes with *K. lactis* NRRL Y-1140 (CR382121-CR382126), *Ashbya gossypii* (*Eremothecium gossypii*) ATCC 10895 (AE016814-AE016821), *Candida glabrata* CBS 138 (CR380947-CR380959), *S. cerevisiae* S288c (BK006934-BK006949), *Ogataea parapolymorpha* DL-1 (AEOI00000000), *Debaryomyces hansenii* CBS 767 (CR382133-CR382139), *S. stipitis* CBS 6054 (AAVQ00000000), *Clavispora lusitaniae* ATCC 42720 (AAFT00000000), *Yarrowia lipolytica* CLIB122 (CR382127-CR382132) and *Schizosaccharomyces pombe* 972 h- (CU329670-CU329672) as hemiascomycetous yeasts, of which complete or draft genome sequences are available [19-23]. A whole genome-wide phylogenetic tree with amino acid sequences deduced from the conserved 1,361 genes was constructed using the neighbor-joining algorithm of MEGA5.05 as previously reported [31]. The numbers at each branch indicate bootstrap values.

substrates [25], efficient productivity of heterologous proteins [26-28], and synthesis of a killer toxin against certain ascomycetous yeasts [29,30].

The two most attractive traits of *K. marxianus* for fermentation applications are thermotolerance and pentose assimilation capability, which are also found in *O. parapolymorpha* and *S. stipitis*, respectively. The number of genes that are shared between the two thermotolerant yeasts but absent from *K. lactis* is 30, including genes for three siderophore-iron transporters and three vacuolar proteins (Additional file 1: Table S5). Notably, there are 27 putative sugar transporters in the *K. marxianus* genome (Additional file 1: Table S6). Like *S. stipitis*, the initial xylose catabolism after its uptake in *K. marxianus* is accomplished by three genes, *XYL1*, *XYL2*, and *XKS1*, which are involved in the conversion of xylose to xylulose-5-phosphate as an intermediate in the pentose phosphate pathway (PPP). Genes for utilization of various other sugars and alcohol dehydrogenases are listed (Table 3).

Chromosomal segments including 4,277 genes, which retain the ancestral gene groupings, were found between *K. marxianus* and *K. lactis* (Additional file 1: Figure S2). The average of mapped segments to the chromosome is 57.9%, and the *K. marxianus* chromosome best covered by *K. lactis* chromosomal segments is chromosome 6, with 60.8% coverage.

Ribosomal DNA (rDNA) copy number and thermotolerance

Optical mapping allowed us to estimate at least 140 copies of the rDNA gene as a cluster on chromosome 5 (Additional file 1: Figure S1), which occupies 67.5% (0.9 Mb) of the chromosome. To examine the relationship between the rDNA copy number and its thermotolerance among *K. marxianus* strains, strains exhibiting different growth at different temperatures (Additional file 1: Figures S3, S4) were subjected to a test to determine the copy number of rDNA (Additional file 1: Table S7). As a result, the rDNA copy number is not correlated to the thermotolerance of the yeast, and at least 31 copies of rDNA are sufficient to support its thermotolerance.

Genes regulated under a static condition

The alteration of genome-wide gene expression was analyzed by transcription start site sequencing (TSS Seq) under four different conditions (Additional file 3): shaking condition in yeast extract peptone dextrose (YPD) medium at 30°C (30D) or 45°C (45D), static condition in YPD medium at 30°C (30DS), and shaking condition in yeast extract peptone xylose (YPX) medium at 30°C (30X) and was expressed as the ratio of 30DS/30D, 45D/30D, and 30X/30D using 30D as a control condition, which was evaluated by a statistical test (FDR < 0.05) (Additional file 1: Table S8) and summarized (Additional file 1: Figure S5).

The growth of *K. marxianus* under the static condition is much slower than that under the shaking condition [4,12]. Under the 30DS condition, there were 159 significantly upregulated genes (Additional file 1: Table S8), and the top-five significantly enriched GO terms were ribosome biogenesis, ribonucleoprotein complex biogenesis, rRNA processing, rRNA metabolic process, and noncoding RNA processing (Additional file 1: Table S9); their individual gene details are shown in Additional file 1: Table S10. Interestingly, 55% of upregulated gene products are located in the nucleus (Additional file 1: Figure S6), some of which are factors for ribosome biogenesis, ATP-dependent RNA helicases, RNA polymerases subunits, nucleolar complex proteins, components of exosome complex, DNA polymerase subunits, and chromatin assembly factor 1 subunits. Conversely, there were 154 significantly downregulated genes (Additional file 1: Table S8), and their most significantly enriched GO terms were ascospore formation, sexual sporulation, sexual sporulation resulting in the formation of a cellular spore, cell development, and reproductive process in single-celled organisms (Additional file 1: Tables S11, S12). The largest population consists of 43 genes for membrane proteins including 15 transporters for amino acids and other metabolites. Taken together, under the static condition, *K. marxianus* may increase the turnover of RNAs and proteins in addition to suppression of transporters and spore formation that depends on mitochondrial respiration activity [32].

Under the static condition, the oxygen level in cells may become low as cells proliferate so that the condition may affect oxygen-requiring biosynthetic pathways, such as those for heme, sterols, unsaturated fatty acids, pyrimidine, and deoxyribonucleotides [33]. As expected, almost all genes related to ergosterol biosynthesis, sterol biosynthesis, unsaturated fatty acids production, pyrimidine synthesis, and ribonucleotide reductase were largely upregulated under the 30DS condition (Additional file 1: Figure S7). However, unlike *S. cerevisiae*, the expression of *MDL1* for a putative mitochondrial heme carrier was not significantly altered [34]. In addition, *NPT1* for nicotinate phosphoribosyl transferase (Npt1) involved in the nicotinamide adenine dinucleotide phosphate (NAD) salvage pathway, *CYC7* for cytochrome *c* and *AAC* for ADP/ATP carrier in mitochondrial inner membrane were upregulated. The ADP/ATP carrier functions to exchange cytoplasmic adenosine diphosphate (ADP) for mitochondrial adenosine triphosphate (ATP) under aerobic conditions and vice versa under anaerobic conditions [33]. Taken together, these results suggest that the enhanced expression of most genes for several oxygen-dependent biosynthetic pathways or some genes related to the production and management of energy is crucial for the cellular metabolism of *K. marxianus* under the static condition. Notably, almost all genes described above

Table 3 Genes for utilization of sugars at their initial catabolism and genes for alcohol dehydrogenases in *K. marxianus*

	Product	UniProt gene	Sugar
KMLA_60412	Hexokinase	HXK2 (RAG5)	Glucose, fructose, mannose
KLMA_10763	Glucose-6-phosphate dehydrogenase	RAG2	
KLMA_50384	Mannose 6-phosphate isomerase	PMI40	Mannose
KLMA_20333	Galactokinase	GAL1	Galactose
KLMA_20331	Galactose-1-phosphate uridylyltransferase	GAL7	
KLMA_30099	Phosphoglucomutase	GAL5 (PGM2)	
KLMA_10683	Xylose reductase	XYL1	Xylose
KLMA_70044	Xylitol dehydrogenase	XYL2	
KLMA_80066	Xylulokinase	XKS1	
KLMA_30577	Arabinose dehydrogenase [NADP$^+$ dependent]	ARA1	Arabinose
KLMA_40310	Arabinose dehydrogenase [NAD dependent]	ARA2	
KLMA_80176	Xylose/arabinose reductase	YJR096W	
KLMA_10558	D-Arabitol-2-dehydrogenase	ARD2	
KLMA_10157	Probable ribokinase	RBK1	D-Ribose
KLMA_10176	Ribose-phosphate pyrophosphokinase 5	PRS5	
KLMA_10783	Sorbose reductase	SOU1	Mannitol, glucitol, L-sorbose
KLMA_10649	D-Lactate dehydrogenase [cytochrome], mitochondrial	DLD1	Lactate
KLMA_40583	D-Lactate dehydrogenase [cytochrome] 1, mitochondrial	DLD1	
KLMA_50301	D-Lactate dehydrogenase [cytochrome], mitochondrial	DLD1	
KLMA_60482	D-Lactate dehydrogenase [cytochrome] 2, mitochondrial	DLD2	
KLMA_10179	Glycerol-3-phosphate dehydrogenase [NAD(+)] 1	GPD1	Glycerol
KLMA_30722	Glycerol-3-phosphate dehydrogenase, mitochondrial	GUT2	
KLMA_60361	Glycerol uptake protein 1	GUP1	
KLMA_80411	Glycerol uptake/efflux facilitator protein	FPS1	
KLMA_80412	Glycerol kinase	GUT1	
KLMA_10427	Galactose/lactose metabolism regulatory protein GAL80	GAL80	Lactose
KLMA_20830	Lactose permease	LAC12	
KLMA_30010	Lactose permease	LAC12	
KLMA_30728	Lactose permease	LAC12	
KLMA_30011	Beta-glucosidase	-	Cellobiose
KLMA_20184	Endo-1,3(4)-beta-glucanase 1	DSE4	1,3-β-D-Glucan
KLMA_50517	Endo-1,3(4)-beta-glucanase 2	ACF2	
KLMA_10518	Inulinase	INU1	Sucrose, raffinose, inulin
KLMA_40102	Alcohol dehydrogenase 1	ADH1	
KLMA_40220	Alcohol dehydrogenase 2	ADH2	
KLMA_80306	Alcohol dehydrogenase 3	ADH3	
KLMA_20005	Alcohol dehydrogenase 4a	ADH4a	
KLMA_20158	Alcohol dehydrogenase 4b	ADH4b	
KLMA_40624	Alcohol dehydrogenase	ADH	
KLMA_80339	Alcohol dehydrogenase 6	ADH6	

NAD, nicotinamide adenine dinucleotide; NADP, nicotinamide adenine dinucleotide phosphate.

were upregulated not only under the 30DS condition but also under the 45D condition, suggesting that cells suffer from oxygen deficiency under the two conditions.

Metabolic changes were further analyzed by KEGG assignment. A number of genes for glycolysis after 1,3-bisphosphoglycerate, PPP, and tricarboxylic acid (TCA) cycle were relatively upregulated (Figure 2A,B). Enhanced PPP may provide nicotinamide adenine dinucleotide phosphate (NADPH) to cope with reactive oxygen species (ROS) generated under the condition, consistent with upregulation of ROS-scavenging genes (see Figure 3).

Enhancement of expression of genes related to the pathway from 2-phosphoglycerate to acetyl-CoA via acetaldehyde may indicate the possibility that cells under a static condition tend to increase in acetyl-CoA and NADPH production in the process of oxidation of acetaldehyde.

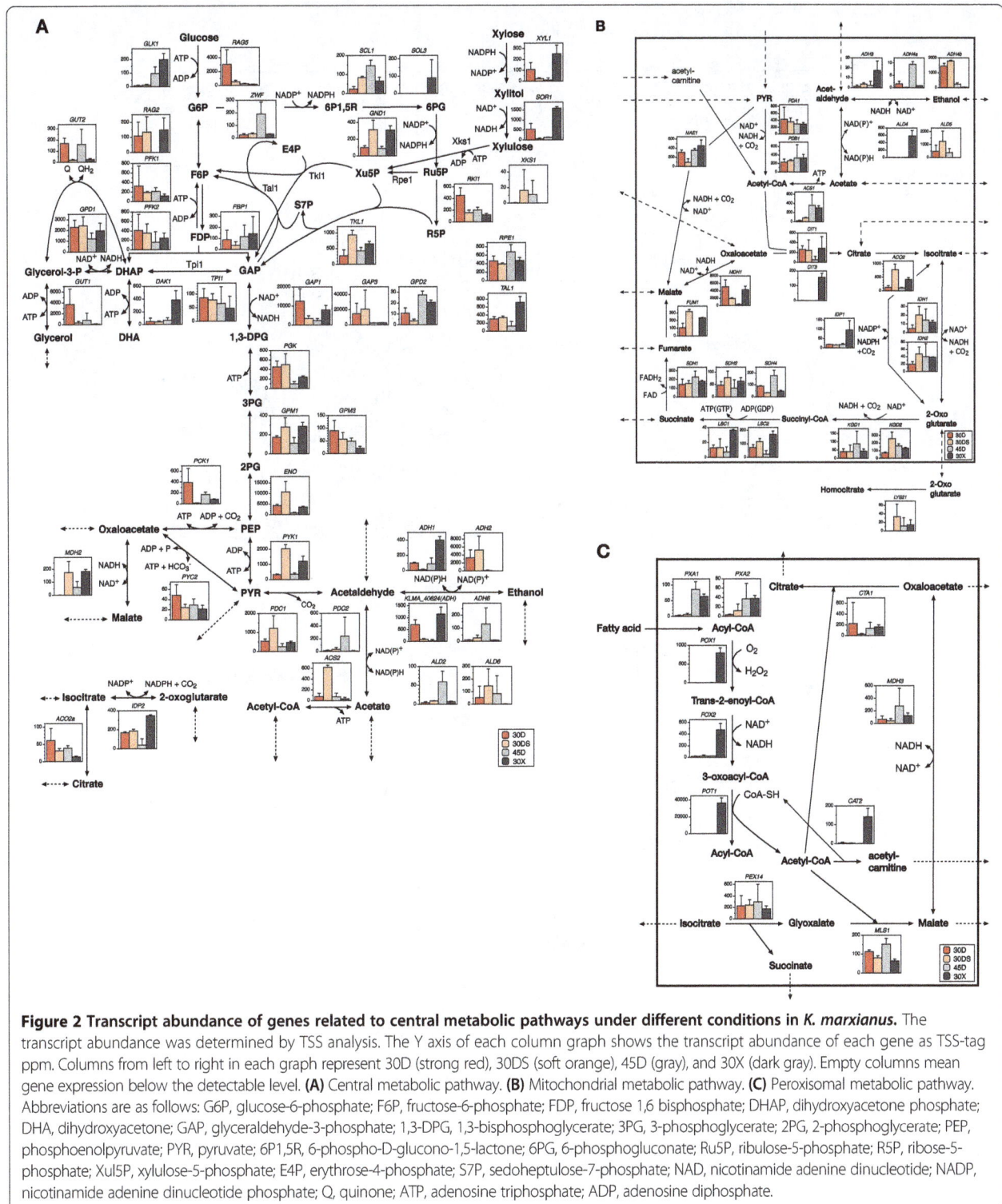

Figure 2 Transcript abundance of genes related to central metabolic pathways under different conditions in *K. marxianus.* The transcript abundance was determined by TSS analysis. The Y axis of each column graph shows the transcript abundance of each gene as TSS-tag ppm. Columns from left to right in each graph represent 30D (strong red), 30DS (soft orange), 45D (gray), and 30X (dark gray). Empty columns mean gene expression below the detectable level. **(A)** Central metabolic pathway. **(B)** Mitochondrial metabolic pathway. **(C)** Peroxisomal metabolic pathway. Abbreviations are as follows: G6P, glucose-6-phosphate; F6P, fructose-6-phosphate; FDP, fructose 1,6 bisphosphate; DHAP, dihydroxyacetone phosphate; DHA, dihydroxyacetone; GAP, glyceraldehyde-3-phosphate; 1,3-DPG, 1,3-bisphosphoglycerate; 3PG, 3-phosphoglycerate; 2PG, 2-phosphoglycerate; PEP, phosphoenolpyruvate; PYR, pyruvate; 6P1,5R, 6-phospho-D-glucono-1,5-lactone; 6PG, 6-phosphogluconate; Ru5P, ribulose-5-phosphate; R5P, ribose-5-phosphate; Xul5P, xylulose-5-phosphate; E4P, erythrose-4-phosphate; S7P, sedoheptulose-7-phosphate; NAD, nicotinamide adenine dinucleotide; NADP, nicotinamide adenine dinucleotide phosphate; Q, quinone; ATP, adenosine triphosphate; ADP, adenosine diphosphate.

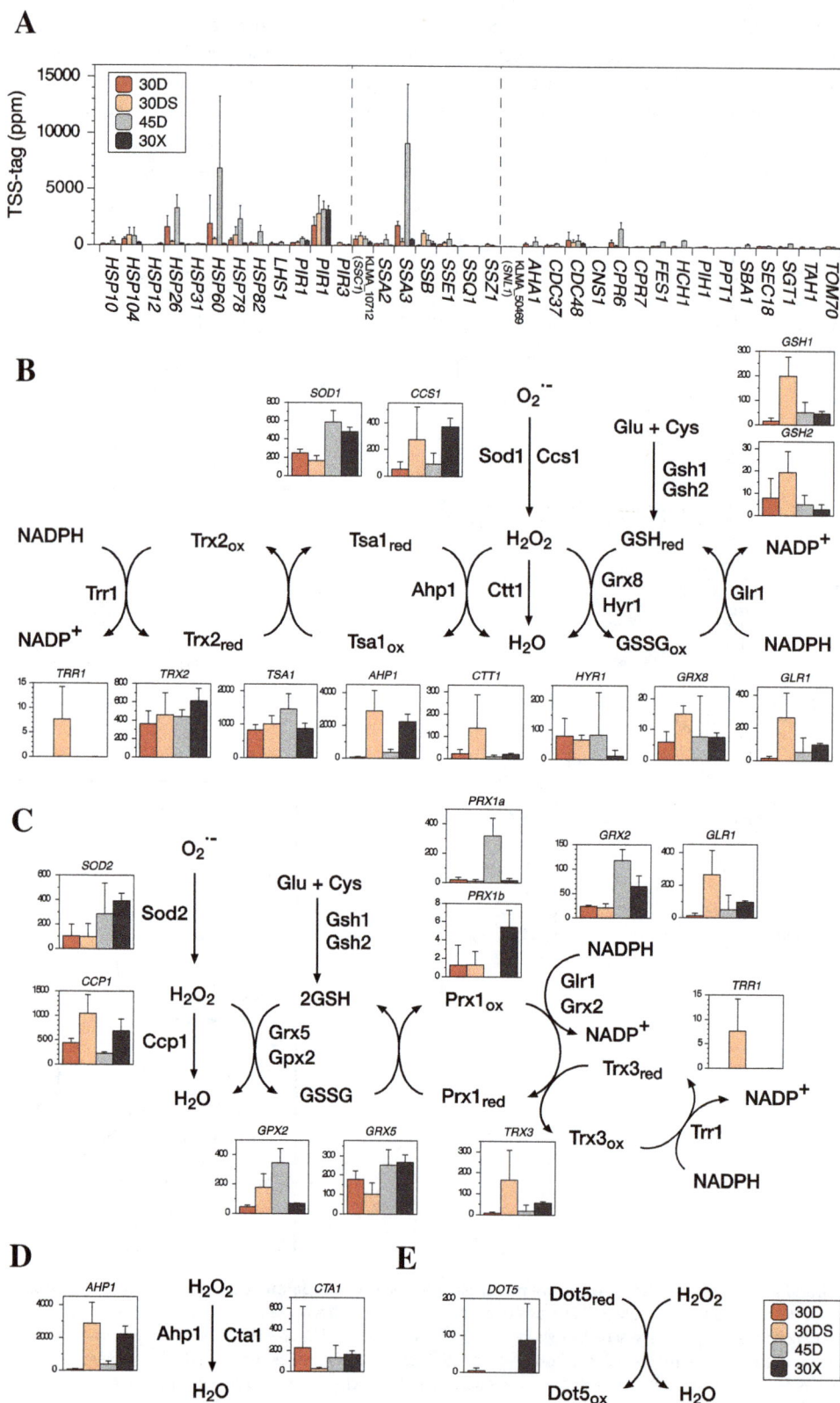

Figure 3 (See legend on next page.)

(See figure on previous page.)
Figure 3 Transcript abundance of genes related to oxidative stress response in *K. marxianus*. (A) Transcript abundance of heat shock genes and related genes listed in Additional file 1: Table S18 is represented as TSS-tag ppm. Enzyme reactions to scavenge ROS in **(B)** cytoplasm, **(C)** mitochondria, **(D)** peroxisome, and **(E)** nucleus are shown. Each column in the graph shows the transcript abundance of each gene as described in Figure 2.

The oxaloacetate-malate shuttle may contribute to the oxidation of NADH in cytoplasm. Many genes for the respiratory chain were downregulated, though compensatorily most of the chaperone-coding genes for cytochrome *c* oxidase were upregulated (Additional file 1: Figure S8).

Genes regulated under a high-temperature condition

To clarify the thermotolerant mechanism of *K. marxianus*, it is necessary to consider the expressional alteration of whole genomic genes at high temperature. Under the 45D condition, 508 genes were significantly downregulated (Additional file 1: Table S8) and the top-five significant GO terms were carbohydrate metabolic process, isoleucine biosynthetic process, small molecule metabolic process, monosaccharide metabolic process, and branched chain family amino acid biosynthetic process (Additional file 1: Tables S13, S14). Conversely, in 199 upregulated genes (Additional file 1: Table S8), the most significant GO terms were noncoding RNA processing, ribosome biogenesis, rRNA processing, ribonucleoprotein complex biogenesis, and rRNA metabolic process (Additional file 1: Tables S15, S16), and 45 genes were related to translation, transcription, DNA replication and repair, and protein and RNA degradations. Interestingly, *LYS21* for homocitrate synthase, which is linked to the key process of DNA damage repair in a nucleus [35] in addition to its involvement in lysine biosynthesis in the cytoplasm, was upregulated. Several genes for homologous recombination and nonhomologous end joining, which function in the repair of DNA double-stranded breaks, were also upregulated (Additional file 1: Figure S9). Therefore, it is assumed that *K. marxianus* copes with high temperatures by reducing central metabolic activities and reinforcing the synthesis and degradation of proteins and DNA repair.

In further analysis of the subcellular localization of products (Additional file 1: Figure S6), 21% of all upregulated genes are located in the nucleolus or nucleus (Additional file 1: Table S17), and are related to 18S rRNA preprocessing, 60S ribosomal subunit biogenesis, transcription process, and pre-rRNA processing (Additional file 1: Tables S15, S16). Conversely, products of several downregulated genes exist in the nucleolus, which are involved in 40S ribosomal subunit biogenesis and pre-18S rRNA processing (Additional file 1: Tables S13, S14). Notably, seven genes for mitochondrial ribosome subunits were downregulated. In addition, genes for DNA repair in nuclei and mitochondria, including a DNA damage sensor or chromosome transmission fidelity protein, were significantly upregulated (Additional file 1: Table S16).

Genes for glycolysis were remarkably downregulated, except for *GLK1* and *FBP1*, under the 45D condition (Figure 2A), which is consistent with relatively slow growth speed and low ethanol productivity at high temperatures [4]. Conversely, *ZWF* and *SOL1* in PPP in addition to *GLK1* were upregulated, indicating an increase in NADPH amount. In TCA cycle, downregulation of *CIT1*, *LSC2*, *FUM1*, and *MDH1* and upregulation of *IDH1*, *IDH2*, *KGD1*, and *KGD2* were found, which might lead to the accumulation of intermediates (Figure 2B and Additional file 1: Table S16). Most mitochondrial genomic genes were selectively expressed (Additional file 1: Figure S8). Additionally, some genes for succinate dehydrogenase (Sdh) and *bc*1 complex were upregulated, whereas some genes for NADH dehydrogenase (Ndh), Coenzyme Q biosynthesis, and cytochrome *c* oxidase including heme *a* synthesis were downregulated (Additional file 1: Figure S8). These findings allow us to speculate that *K. marxianus* scavenges H_2O_2 by the pathway of Sdh-*bc*1 complex-cytochrome *c* peroxidase and prevents the production of ROS by reduction of gene expression for Ndh and Coenzyme Q biosynthesis at high temperatures.

Heat shock proteins (Hsps) and chaperones are expected to be crucial for survival at high temperatures. The transcription of *HSP26*, *HSP60*, *HSP78*, *HSP82*, *SSA3*, and *CPR6* was enhanced under the 45D condition (Figure 3A and Additional file 1: Table S18), suggesting that both mitochondrial and cytoplasmic compartments need such Hsps at that temperature.

Genes regulated under a xylose-utilizing condition

Under the 30X condition, the top-five significant GO terms of significantly upregulated genes were fatty acid catabolic process, monocarboxylic acid catabolic process, cellular lipid catabolic process, fatty acid β-oxidation, and fatty acid oxidation (Additional file 1: Tables S19, S20). Conversely, the most significant GO terms of significantly downregulated genes were α-amino acid metabolic process, carboxylic acid metabolic process, lysine biosynthetic process, small molecule biosynthetic process, and oxoacid metabolic process (Additional file 1: Tables S21, S22). Therefore, the 30X condition may stimulate the degradation of lipid in peroxisome and keep a low level of amino acid synthesis, which is consistent with the slow growth of the yeast in xylose [4].

The phylogenetic tree of sugar transporters revealed that KLMA_50360, KLMA_50361, KLMA_50362, KLMA_50363, and KLMA_50364 share high similarity with xylose transporters predicted in *S. stipitis* (Figure 4) [21]. Unlike *S. stipitis*, however, these genes were not induced specifically under the 30X condition. Genes of several transporters exhibited specific induction under the 30X or 30D condition, suggesting that KLMA_60073, KLMA_80101, and KLMA_70145 are involved in xylose uptake.

Genes for the initial catabolism of xylose, PPP, the conversion of PEP to ethanol, the mitochondrial conversion of acetaldehyde to acetyl-CoA and TCA cycle were relatively upregulated (Figure 2A,B). A part of the ethanol produced in the cytoplasm may thus be consumed by conversion to acetyl-CoA in mitochondria through the ethanol-acetaldehyde shuttle followed by TCA cycle, which is consistent with low ethanol productivity in xylose medium [4]. In contrast, *ADH2* and *ADH4b* for alcohol dehydrogenases, *GUT1* for glycerol kinase and *GUT2* for the glycerol-3-phosphate shuttle were downregulated. TSS results also indicate the generation of acetyl-CoA from the fatty acids through the peroxisomal β-oxidation pathway (Figure 2C), indicating the possibility that fatty acids could be a subsidiary intracellular carbon source in xylose medium. Such supply of acetyl-CoA in xylose medium might result in more NADH production and generate more ATP, which is required for phosphorylation of xylulose and dihydroxyacetone. Additional ATP may also be supplied as a result of the *DAK1* upregulation in the cytoplasm (Figure 2A).

Discussion

As the first step to understanding the genetic basis of the highly efficient yeast *K. marxianus*, complete genome sequence and transcriptome analyses of its most thermotolerant strain were performed. The former analysis revealed many gene resources for the ability to assimilate various sugars in addition to species-specific genes. The latter clarified the molecular basis of attractive traits of the yeast. All information obtained here about the yeast will be useful for fundamental and applied research for innovative applications.

The thermotolerance as an attractive trait was investigated under the 45D condition, which revealed that *K. marxianus* seems to drastically change metabolic pathways from those under the 30D condition, that is, the enhancement of PPP and the attenuation of TCA cycle after the fumarate-producing step. The changes lead to the speculation that the former provides NADPH for scavenging ROS and that the latter deals with H_2O_2 via the electron transfer from Sdh to cytochrome *c* peroxidase (Figures 2 and 3). Consistent with these conjectures, a higher temperature generates more ROS, which causes

DNA damage [36]. Notably, the findings of the upregulation of genes for DNA double-stranded break repair and removal of uracil in DNA molecules suggest the occurrence of enhancement of double-stranded break or deamination of cytosines in DNA at high temperatures. In addition, ATP synthesis via oxidative phosphorylation may be greatly reduced due to the repression of *ATP3* for the gamma subunit of ATP synthase. The existence of additional strategies is guessed from the TSS analysis data for survival at high temperature: alteration of ribosome biogenesis including pre-rRNA processing presumably for stable and efficient protein synthesis, reduction of mitochondrial ribosome biogenesis probably for saving energy, reinforcement of checkpoints of DNA replication and spindle assembly, minimization of electron leakage in respiratory chain by reduction of NADH dehydrogenase, and Coenzyme Q or enhanced expression of Hsps and chaperones. Taken together, the thermotolerance of *K. marxianus* is likely achieved by systematic mechanisms consisting of various strategies. Especially, the yeast would mainly acquire ATP from glycolysis rather than TCA cycle at high temperatures, which could prevent the generation of ROS by minimization of mitochondrial activity.

Under a static condition, the growth and ethanol production of *K. marxianus* were low compared to those under a shaking condition [4], probably due to low ATP yield in mitochondria, which may be related to the enhanced expression of *AAC*. *K. lactis* bearing a null mutation of *AAC2* exhibits growth defect on glycerol, galactose, maltose, and raffinose [37]. In addition, the expression of *RAG5* for hexokinase was relatively low. In *K. lactis*, *RAG5* mutations [38] abolish the expression of *RAG1* for a low-affinity glucose transporter [39] and decrease the level of the 2.0-kb mRNA species of *HGT1* for a high-affinity glucose transporter [40]. Furthermore, NADH would be accumulated in the cytoplasm because of the downregulation of *GAP1* for glyceraldehyde-3-phosphate dehydrogenase. Contrarily, respiratory genes kept their transcriptional levels similar to those under the 30D condition (Additional file 1: Figure S8). Such situations may raise reactive oxygen species from the respiratory chain to cause oxidative stress. Consistently, cytoplasmic oxidative stress response genes were relatively strongly expressed (Figure 3B), especially glutathione-related genes depicted at the cytoplasm side were highly induced. Almost *HSPs*, however, were not upregulated under the 30DS condition as under the 30D condition (Figure 3A). These metabolic activities may lead the low level of cell proliferation under a static condition.

Regulation of gene expression in response to the level of oxygen is achieved via several transcription factors. Under a static condition, *K. marxianus* seems to increase glucose metabolism and shift to fermentation, implying a

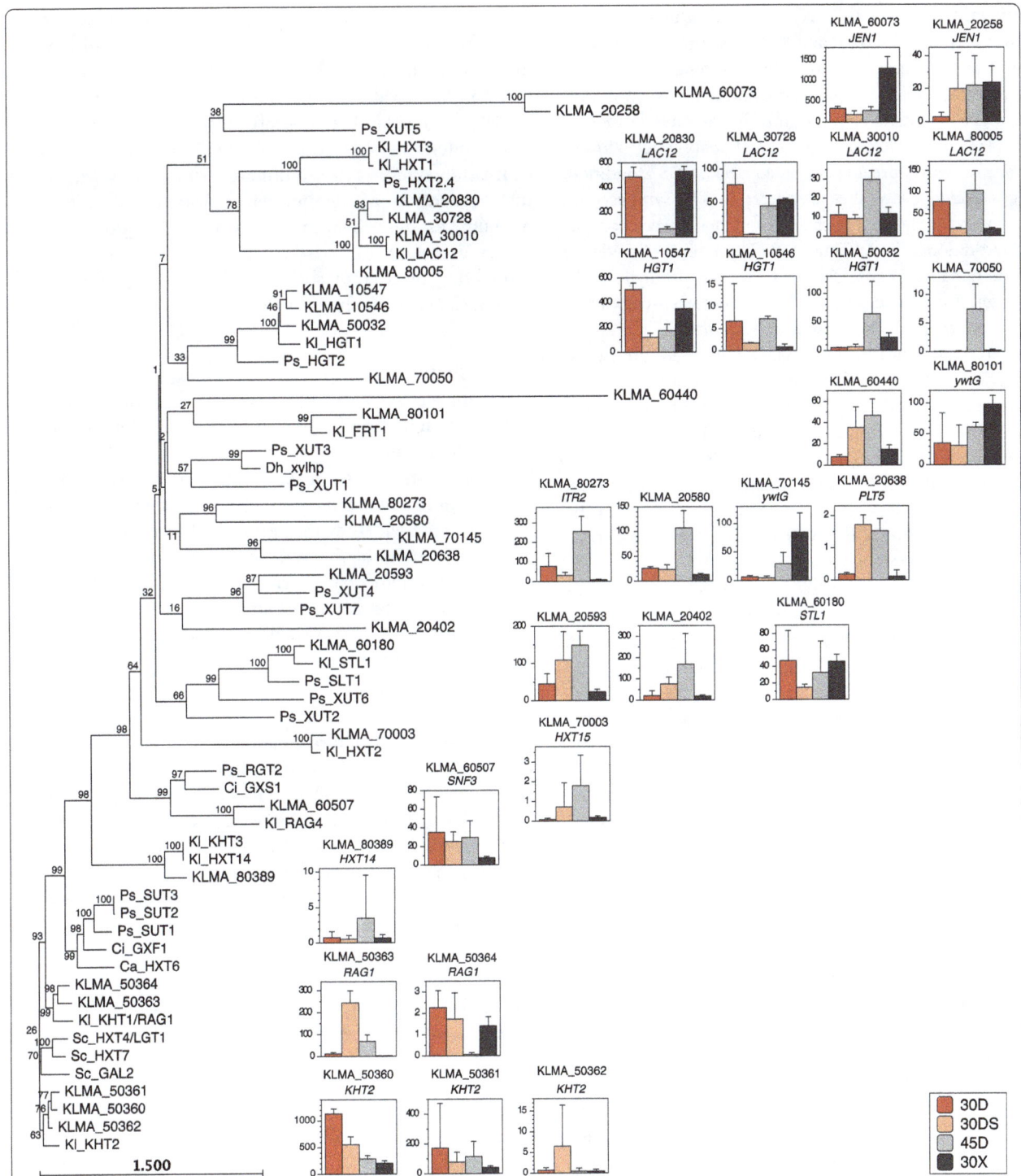

Figure 4 Phylogenetic tree of predicted sugar transporters with transcript abundance in _K. marxianus_. The phylogenetic tree was constructed with amino acid sequences of predicted sugar transporters by the neighbor-joining algorithm using CLC Sequence Viewer. Each column in the graph shows the transcript abundance of each gene as described in Figure 2. The accession numbers of amino acid sequences of sugar transporters are as follows: Ps_HGT2, XP_001382755; Sc_HXT7, NP_010629; Sc_GAL2, NP_013182; Dh_xylhp, AY347871; Ci_GXS1, GN107181; Ci_GXF1, GN107179; Ps_SUT1, XP_001387898; Ps_SUT2, XP_001384295; Ps_SUT3, XP_001386019; Ps_HXT2.4, XP_001387757; Ps_XUT1, XP_001385583; Ps_XUT2, XP_001387242; Ps_XUT3, XP_001387138; Ps_XUT4, XP_001386715; Ps_XUT5, XP_001385962; Ps_XUT6, XP_001386589; Ps_XUT7, XP_001387067; Ps_RGT2, XP_001386588; Ps_SLT1, XP_001383774; Kl_KHT1/RAG1, XP_453656; Kl_KHT2, GN107317; Kl_KHT3, XP_454897; Kl_FRT1, XP_454356; Kl_HGT1, XP_451484; Kl_HXT1, XP_455078; Kl_HXT14, XP_454897; Kl_HXT2, XP_453960; Kl_HXT3, XP_453088; Sc_HXT4/LGT1, NP_011960; Kl_STL1, XP_456249; Kl_RAG4, XP_455315; Kl_LAC12, XP_452193; Ca_HXT6, XP_719472.

connection between the oxygen- and glucose-sensing pathways. In *S. cerevisiae* and *K. lactis*, the transcription factor Hap1 mediates the induction of genes involved in their respiration, lipid metabolism, and oxidative stress response [11]. *K. lactis* Hap1 negatively regulates fermentation [41]. In contrast, the *HAP1* expression in *K. marxianus* was upregulated under the 30DS condition, and consistently, several genes related to ATP synthase and chaperones for respiratory chain components were upregulated (Additional file 1: Figure S8). These lines of evidence and the enhanced expression of genes for glycolytic pathway suggest differences in regulation of oxygen-responsive genes from those in *S. cerevisiae* and *K. lactis*.

The oxidative stress-response genes were found to be highly induced under the three conditions tested (Figures 2C and 4B,C,D,E), indicating that ROS is accumulated in the cytoplasm, mitochondria, and peroxisome under the 30DS and 30X conditions and in the cytoplasm and mitochondria under the 45D condition. Notably, Ahp1 in addition to Cta1 and Dot5 may be responsible for H_2O_2 detoxification in the peroxisome and nucleus, respectively.

Xylose assimilation capability as the second trait was examined under the 30X condition. It is known that *K. marxianus* tends to suffer from cofactor imbalance in xylose medium [42,43], and thus, its growth strongly depends on mitochondrial respiratory activity [44]. Interconvertibility of NAD^+ species for maintaining the redox balance is essential for growth efficiency and metabolite excretion [45], but the yeast is incapable of directly converting NAD^+ and NADPH into $NADP^+$ and NADH owing to lack of transhydrogenases [46]. Instead, redox-balancing mechanisms between the cytoplasm and mitochondria are probably used to resolve the NADH/NADPH imbalance. Reoxidation of cytosolic NADPH and its strong connection to oxidative stress in *K. lactis* have been reported [47,48]. In *S. cerevisiae*, five cytosolic-mitochondrial redox shuttles have been proposed [49]. Of these, genes for enzymes related to ethanol-acetaldehyde, citrate-oxoglutarate, and oxaloacetate-malate shuttles were relatively upregulated under the 30X condition (Figures 2A, B). The GABA shunt from 2-oxoglutarate to succinate that has been proposed in *S. cerevisiae* and *S. stipitis* [21,50] may not be so important due to the low expression of related genes (Additional file 1: Figure S10), suggesting that *K. marxianus* uses different shuttles for resolving the cofactor imbalance from those of the two yeasts. In addition to TCA cycle intermediates by these shuttles, acetyl-CoA might be transferred from peroxisome on the basis of TSS data. Eventually, mitochondrial activity may be enhanced, which tends to increase the leakage of electrons to generate ROS, which is consistent with elevation of expression of oxidative stress-response genes.

The last trait is efficient protein productivity, which meets the demand for fast growth and high yield biomass [51]. The yeast has been exploited as a cell factory to obtain valuable enzymes, showing retention of activity over a large temperature range [52]. TSS results under the 30D and 30X conditions reveal high expression of *INU1* for inulinase, which is useful for the production of recombinant proteins in culture medium, as described in previous studies [26,27,53]. These useful characteristics may allow simultaneous production of ethanol and valuable proteins, thus, reducing the cost of ethanol production.

Conclusions

The complete sequences of *K. marxianus* DMKU 3-1042 nuclear and mitochondrial genomes have been determined, which reveal many genes for the cells to cope with a high temperature and to assimilate a wide variety of sugars including xylose and arabinose in addition to species-specific genes. The present study thus provides the molecular basis of attractive traits of the yeast for industrial applications including high-temperature fermentation, and information of its gene resources and transcriptome data, which are particularly useful for fundamental and applied researches for innovative applications.

Methods
Strains, media, and culture conditions

The yeast strain used in this work was *K. marxianus* DMKU 3-1042 strain, which has been deposited in the NITE Biological Resource Center (NBRC) under the deposit numbers NITE BP-283 and NBRC 104275. Media used were YP (1% *w/v* yeast extract and 2% *w/v* peptone) supplemented with one of two different carbon sources: YPD, with 2% *w/v* glucose, or YPX, with 2% *w/v* xylose.

Genome sequencing, assembly, and annotation

Two plasmid libraries with average insert sizes of 3 and 5 kb were generated in pTS1 (Nippon Gene, Tokyo, Japan) and pUC118 (Takara Bio. Inc., Otsu, Shiga, Japan) plasmid vectors, respectively, while a fosmid library with an average insert size of 40 kb was constructed in pCC1FOS (EPICENTRE, Illumina Inc., San Diego, CA, USA). Shotgun sequencing was performed on an ABI 3730*xl* DNA Analyzer (Applied Biosystems Co., Thermo Fisher Scientific, Inc., Foster City, CA, USA). Gaps were closed by the sequencing of gap-spanning PCR products. Telomeric regions were further analyzed by transposon-insertion sequencing of corresponding fosmid clones with Template Generation System II (Finnzymes, Thermo Fisher Scientific, Inc., Foster City, CA, USA). Genome assemblies were validated by optical mapping (OpGen, Gaithersburg, MD, USA). The genome was finally assembled into nine ungapped contigs corresponding to eight chromosomes and a mitochondrion with and average coverage of 11.1x. The mean error rate was estimated to be less than 6×10^{-8}.

The rRNA-encoding regions were identified by a BLASTN program using the ribosomal sequences of *K. marxianus*, while the tRNA-coding regions were predicted by the ARAGORN program [54]. Proteins-coding gene prediction was performed by combining the Glimmer 3.02 program with a self-training dataset and six-frame prediction by using *in silico* Molecular Cloning (in silico biology, Inc., Yokohama, Kanagawa, Japan) and manual identification [55,56]. Intron prediction was performed by the AUGUSTUS program using the coding DNA sequences (CDSs) of *Ashbya gossypii* as the reference for pattern learning and manual correction of each CDS [57,58]. Functional annotation of the predicted CDSs was performed by BLASTP searching against the nonredundant (nr) database [59] with an E-value threshold of 10^{-10}. Protein domains were predicted using the InterProScan program against various domain libraries (Prints, Prosite, PFAM, ProDom, SMART). Protein functions were assigned by KOG database [60]. Assignment of UniProt number was performed by a BLASTP program against the UniProt database [61] with an E-value threshold of 10^{-10}. UniProt gene name, cellular localization, and Gene Ontology were assigned based on UniProt database information. The assignment of KO number of KEGG Orthology was performed by the KAAS program [62,63]. Individual annotations were then summarized according to KEGG Orthology and KEGG metabolic pathways.

Transcriptome analysis in *K. marxianus*

Cells grown in 100-ml Erlenmeyer flasks in 30 ml YPD medium at 30°C, 160 rpm for 18 h were inoculated at the initial OD_{660} of about one into sequential batch culture, which was conducted in 300-ml Erlenmeyer flasks with 100 ml YPD at 30°C or 45°C under the shaking condition (30D and 45D) or the static condition at 30°C (30DS) or with 100 ml YPX at 30°C under the shaking condition (30X). The cells were further cultivated under each condition for 6 h and immediately subjected to RNA isolation. Each culture condition for TSS Seq experiments is under the same condition as performed previously [4]. Total RNA from cells was isolated by the hot phenol method [64] and was purified using an RNeasy Midi Kit (QIAGEN, Hilden, Germany) with RNase free DNase I (QIAGEN) according to the manufacturer's instructions. Transcriptional starting site (TSS) Seq analysis was performed using the extracted total RNA, providing precise information on TSSs and their expression levels in a high-throughput manner [65]. TSS-tag counts were divided by the total number of uniquely

and perfectly (with no mismatch) mapped TSS-tags to calculate TSS-tag ppm (parts per million). Experiments on each culture condition and TSS Seq analysis were performed in triplicate. Statistical testing was performed by edgeR of R package using the TSS-tag counts data of TSS analysis with a cutoff value as a false discovery rate smaller than 0.05 indicating significantly changed genes [66]. The GO enrichment test was performed by topGO of R package [67]. The TSS data was submitted to the Gene Expression Omnibus database (GEO, http://www.ncbi.nlm.nih.gov/geo/) under the accession number GSE66600.

Nucleotide sequence accession numbers

The complete genome sequence of *K. marxianus* DMKU 3-1042 has been deposited in DDBJ/EMBL/GenBank under accession no. AP012213-AP012221.

Quantitative real-time PCR (qPCR) analysis

Primers for qPCR were designed by using Primer Express software version 3.0 (Applied Biosystems Co.). A pair of primers, Km-rDNA-F1: 5'-GATCGGGTGGTGT TTTTCTTATG-3' and Km-rDNA-R1: 5'-TCCCCCCA-GAACCCAAAG-3', was designed to amplify the 18S rDNA gene. The reaction produced a 71-bp PCR product. Probe for 18S rDNA, Km-rDNA-probe1: 5'-CCC ACTCGGCACCTTACGAGAAATCA-3' was labeled at the 5'-end with 6-carboxyfluorescein (FAM) as a reporter and at the 3'-end with dihydrocyclopyrroloindole tripeptide minor groove binder (MGB) as a quencher. Real-time PCR was performed using a TaqMan® Universal Master Mix II (Applied Biosystems Co.) and Applied Biosystems 7300 Real-Time PCR system (Applied Biosystems Co.). The 18S rDNA was amplified using genomic DNA isolated from 10 strains of *K. marxianus*. The C_T value was determined by the instrument's software and adjusted manually as necessary. Concentration and DNA quality were measured by using Qubit™ dsDNA HS Assay Kits (Invitrogen Ltd., Paisley, UK) with Qubit® Fluorometer and by gel electrophoresis and converted to the number of copies by using the molecular weight of the DNA. The equation $C_T = m$ (log quantity) + b from the equation for a line (y = mx + b) was constructed by plotting the standard curve of log quantity versus its corresponding C_T value. The 18S rDNA copy numbers were determined by the absolute quantitation method, by which total copies were first calculated using the following equation: total 18S rDNA copies = $10^{((C_T - b)/m)}$. The number of 18S rDNA copies per genome was then determined by the following equation: 18S rDNA copies per genome = (Total copies of 18S rDNA)/(Total copy of genomic DNA). The genome size of 11.0 Mb of *K. marxianus* DMKU 3-1042 was used for all calculations.

Additional files

> **Additional file 1: Supplementary information.** A file containing 10 supplementary figures and 22 supplementary tables.
>
> **Additional file 2: Gene annotation.** A file that contains locus tags and their corresponding products.
>
> **Additional file 3: Transcriptome analysis results.** A file that contains locus tags, their corresponding products and TSS-tag count in part per million (ppm) as used under the different conditions described in the study.

Abbreviations

1,3-DPG: 1,3-bisphosphoglycerate; 2PG: 2-phosphoglycerate; 3PG: 3-phosphoglycerate; 6P1,5R: 6-phospho-D-glucono-1,5-lactone; 6PG: 6-phosphogluconate; ADP: Adenosine diphosphate; ATP: Adenosine triphosphate; DHA: Dihydroxyacetone; DHAP: Dihydroxyacetone phosphate; E4P: Erythrose-4-phosphate; F6P: Fructose-6-phosphate; FDP: Fructose 1,6 bisphosphate; G6P: Glucose-6-phosphate; GAP: Glyceraldehyde-3-phosphate; Hsp: Heat shock protein; HTF: High-temperature fermentation; KOG: Eukaryotic orthologous groups; NAD: Nicotinamide adenine dinucleotide; NADP: Nicotinamide adenine dinucleotide phosphate; NBRC: NITE Biological Resource Center; Ndh: NADH dehydrogenase; NR: Nonredundant; PEP: Phosphoenolpyruvate; PPP: Pentose phosphate pathway; PYR: Pyruvate; qPCR: quantitative real-time PCR; R5P: Ribose-5-phosphate; rDNA: ribosomal DNA; ROS: Reactive oxygen species; Ru5P: Ribulose-5-phosphate; S7P: Sedoheptulose-7-phosphate; Sdh: Succinate dehydrogenase; TSS Seq: Transcription start site sequencing; Xul5P: Xylulose-5-phosphate.

Competing interests

The authors declare that they have no competing interests.

Authors' contributions

AH, N Fujit, and KT performed a whole-genome sequencing and optical mapping. NL, TK, NR, M Mu, N Fujim, and S contributed to the gene annotation. YS performed the transcription start site (TSS) analysis. M Ma and TK contributed to the analysis of TSS data and most of other bioinformatics analysis. NL performed most of other wet lab experiments. SL provided the *K. marxianus* DMKU 3-1042 strain and its general information. NL, TK, and MY prepared the manuscript. MY organized this research project. All authors have read and approved the final version of the manuscript.

Acknowledgements

We thank Drs. Shimoi, Matsushita, and Akada for their helpful discussions. This work was supported by the Program for Promotion of Basic Research Activities for Innovative Biosciences, New Energy and Industrial Technology Development Organization (NEDO), the Special Coordination Funds for Promoting Science & Technology, Ministry of Education, Culture, Sports, Science & Technology (MEXT), and the Advanced Low Carbon Technology Research and Development Program, Japan Science and Technology Agency (JST). This work was partially performed as a collaborative research in the Asian Core Program and in the Core to Core Program, which was supported by the Scientific Cooperation Program agreed by the Japan Society for the Promotion of Science (JSPS), the National Research Council of Thailand (NRCT), and universities involved in the program.

Author details

[1]Applied Molecular Bioscience, Graduate School of Medicine, Yamaguchi University, Ube 755-8505, Japan. [2]Department of Biological Chemistry, Faculty of Agriculture, Yamaguchi University, Yamaguchi 753-8515, Japan. [3]National Institute of Technology and Evaluation, Shibuya-ku, Tokyo 151-0066, Japan. [4]Department of Medical Genome Sciences, The University of Tokyo, Chiba 277-8562, Japan. [5]Department of Microbiology, Faculty of Science, Kasetsart University, Bangkok 10900, Thailand. [6]Department of Biology, Faculty of Science, Chiang Mai University, Chiang Mai 50200, Thailand.

References

1. Hahn-Hägerdal B, Galbe M, Gorwa-Grauslund MF, Lidén G, Zacchi G. Bio-ethanol – the fuel of tomorrow from the residues of today. Trends Biotechnol. 2006;24:549–56.
2. Banat IM, Nigam P, Singh D, Marchant R, McHale AP. Review: ethanol production at elevated temperatures and alcohol concentrations: Part I – Yeasts in general. World J Microbiol Biotechnol. 1998;14:809–21.
3. Limtong S, Sringiew C, Yongmanitchai W. Production of fuel ethanol at high temperature from sugar cane juice by a newly isolated *Kluyveromyces marxianus*. Biores Technol. 2007;98:3367–74.
4. Rodrussamee N, Lertwattanasakul N, Hirata K, Suprayogi, Limtong S, Kosaka T, et al. Growth and ethanol fermentation ability on hexose and pentose sugars and glucose effect under various conditions in thermotolerant yeast *Kluyveromyces marxianus*. Appl Microbiol Biotechnol. 2011;90:1573–86.
5. Nonklang S, Abdel-Banat BM, Cha-aim K, Moonjai N, Hoshida H, Limtong S, et al. High-temperature ethanol fermentation and transformation with linear DNA in the thermotolerant yeast *Kluyveromyces marxianus* DMKU 3–1042. Appl Environ Microbiol. 2008;74:7514–21.
6. Lachance MA. *Kluyveromyces* van der Walt. In: Kurtzman CP, Fell JW, Boekhout T, editors. The Yeasts. A Taxonomic Study. 5th ed. Amsterdam: Elsevier; 2011. p. 471–81.
7. Llorente B, Malpertuy A, Blandin G, Artiguenave F, Wincker P, Dujon B. Genomic exploration of the hemiascomycetous yeasts: 12. *Kluyveromyces marxianus* var. *marxianus*. FEBS Lett. 2000;487:71–5.
8. Fukuhara H. *Kluyveromyces lactis* - a retrospective. FEMS Yeast Res. 2006;6:323–4.
9. Schaffrath R, Breunig KD. Genetics and molecular physiology of the yeast *Kluyveromyces lactis*. Fungal Genet Biol. 2000;30:173–90.
10. González-Siso MI, Freire-Picos MA, Ramil E, González-Domínguez M, Rodríguez Torres A, Cerdán ME. Respirofermentative metabolism in *Kluyveromyces lactis*: insights and perspectives. Enzyme Microb Technol. 2000;26:699–705.
11. Rodicio R, Heinisch JJ. Yeast on the milky way: genetics, physiology and biotechnology of *Kluyveromyces lactis*. Yeast. 2013;30:165–77.
12. Lertwattanasakul N, Rodrussamee N, Suprayogi, Limtong S, Thanonkeo P, Kosaka T, et al. Utilization capability of sucrose, raffinose and inulin and its less-sensitiveness to glucose repression in thermotolerant yeast *Kluyveromyces marxianus* DMKU 3–1042. AMB Express. 2011;1:20.
13. Fonseca GG, Heinzle E, Wittmann C, Gombert AK. The yeast *Kluyveromyces marxianus* and its biotechnological potential. Appl Microbiol Biotechnol. 2008;79:339–54.
14. dos Santos VC, Bragança CRS, Passos FJV, Passos FML. Kinetics of growth and ethanol formation from a mix of glucose/xylose substrate by *Kluyveromyces marxianus* UFV-3. Antonie Van Leeuwenhoek. 2013;103:153–61.
15. Abdel-Banat BM, Nonklang S, Hoshida H, Akada R. Random and targeted gene integrations through the control of non-homologous end joining in the yeast *Kluyveromyces marxianus*. Yeast. 2010;27:29–39.
16. Hoshida H, Murakami N, Suzuki A, Tamura R, Asakawa J, Abdel-Banat BM, et al. Non-homologous end joining-mediated functional marker selection for DNA cloning in the yeast *Kluyveromyces marxianus*. Yeast. 2014;31:29–46.
17. Jeong H, Lee DH, Kim SH, Kim HJ, Lee K, Song JY, et al. Genome sequence of the thermotolerant yeast *Kluyveromyces marxianus* var. *marxianus* KCTC 17555. Eukaryot Cell. 2012;11:1584–5.
18. Suzuki T, Hoshino T, Matsushika A. Draft genome sequence of *Kluyveromyces marxianus* strain DMB1, isolated from sugarcane bagasse hydrolysate. Genome Announcements. 2014;2:e00733–14.
19. Dietrich FS, Voegeli S, Brachat S, Lerch A, Gates K, Steiner S, et al. The *Ashbya gossypii* genome as a tool for mapping the ancient *Saccharomyces cerevisiae* genome. Science. 2004;304:304–7.
20. Dujon B, Sherman D, Fischer G, Durrens P, Casaregola S, Lafontaine I, et al. Genome evolution in yeasts. Nature. 2004;430:35–44.
21. Jeffries TW, Grigoriev IV, Grimwood J, Laplaza JM, Aerts A, Salamov A, et al. Genome sequence of the lignocellulose-bioconverting and xylose-fermenting yeast *Pichia stipitis*. Nat Biotechnol. 2007;25:319–26.
22. Butler G, Rasmussen MD, Lin MF, Santos MA, Sakthikumar S, Munro CA, et al. Evolution of pathogenicity and sexual reproduction in eight *Candida* genomes. Nature. 2009;459:657–62.
23. Wood V, Gwilliam R, Rajandream MA, Lyne M, Lyne R, Stewart A, et al. The genome sequence of *Schizosaccharomyces pombe*. Nature. 2002;415:871–80.
24. Rubio-Texeira M. Endless versatility in the biotechnological applications of *Kluyveromyces LAC* genes. Biotechnol Adv. 2006;24:212–25.

25. Lane MM, Morrissey JP. *Kluyveromyces marxianus*: a yeast emerging from its sister's shadow. Fungal Biol Rev. 2010;24:17–26.

26. Rocha SN, Abrahão-Neto J, Cerdán ME, Gombert AK, González-Siso MI. Heterologous expression of a thermophilic esterase in *Kluyveromyces* yeasts. Appl Microbiol Biotechnol. 2011;89:375–85.

27. Rocha SN, Abrahão-Neto J, Cerdán ME, González-Siso MI, Gombert AK. Heterologous expression of glucose oxidase in the yeast *Kluyveromyces marxianus*. Microb Cell Fact. 2010;9:4.

28. van Ooyen AJ, Dekker P, Huang M, Olsthoorn MM, Jacobs DI, Colussi PA, et al. Heterologous protein production in the yeast *Kluyveromyces lactis*. FEMS Yeast Res. 2006;6:381–92.

29. Abranches J, Mendonça-Hagler LC, Hagler AN, Morais PB, Rosa CA. The incidence of killer activity and extracellular proteases in tropical yeast communities. Can J Microbiol. 1997;43:328–36.

30. Jablonowski D, Schaffrath R. Zymocin, a composite chitinase and tRNase killer toxin from yeast. Biochem Soc Trans. 2007;35:1533–7.

31. Matsutani M, Hirakawa H, Yakushi T, Matsushita K. Genome-wide phylogenetic analysis of *Gluconobacter*, *Acetobacter*, and *Gluconacetobacter*. FEMS Microbiol Lett. 2010;315:122–8.

32. Codón AC, Gasent-Ramírez JM, Benítez T. Factors which affect the frequency of sporulation and tetrad formation in *Saccharomyces cerevisiae* baker's yeasts. Appl Environ Microbiol. 1995;61:630–8.

33. Snoek ISI, Steensma HY. Why does *Kluyveromyces lactis* not grow under anaerobic conditions? Comparison of essential anaerobic genes of *Saccharomyces cerevisiae* with the *Kluyveromyces lactis* genome. FEMS Yeast Res. 2006;6:393–403.

34. Kwast KE, Lai LC, Menda N, James III DT, Aref S, Burke PV. Genomic analyses of anaerobically induced genes in *Saccharomyces cerevisiae*: functional roles of Rox1 and other factors in mediating the anoxic response. J Bacteriol. 2002;184:250–65.

35. Scott EM, Pillus L. Homocitrate synthase connects amino acid metabolism to chromatin functions through Esa1 and DNA damage. Genes Dev. 2010;24:1903–13.

36. Hori A, Yoshida M, Shibata T, Ling F. Reactive oxygen species regulate DNA copy number in isolated yeast mitochondria by triggering recombination-mediated replication. Nucleic Acids Res. 2009;37:749–61.

37. Flores CL, Rodríguez C, Petit T, Gancedo C. Carbohydrate and energy-yielding metabolism in non-conventional yeasts. FEMS Microbiol Rev. 2000;24:507–29.

38. Prior C, Mamessier P, Fukuhara H, Chen XJ, Wésolowski-Louvel M. The hexokinase gene is required for transcriptional regulation of the glucose transporter gene *RAG1* in *Kluyveromyces lactis*. Mol Cell Biol. 1993;13:3882–9.

39. Chen XJ, Wésolowski-Louvel M, Fukuhara H. Glucose transport in the yeast *Kluyveromyces lactis*. II. Transcriptional regulation of the glucose transporter gene *RAG1*. Mol Gen Genet. 1992;233:97–105.

40. Billard P, Menart S, Blaisonneau J, Bolotin-Fukuhara M, Fukuhara H, Wésolowski-Louvel M. Glucose uptake in *Kluyveromyces lactis*: role of the *HGT1* gene in glucose transport. J Bacteriol. 1996;178:5860–6.

41. Bao WG, Guiard B, Fang ZA, Donnini C, Gervais M, Passos FML, et al. Oxygen-dependent transcriptional regulator Hap1p limits glucose uptake by repressing the expression of the major glucose transporter gene *RAG1* in *Kluyveromyces lactis*. Eukaryot Cell. 2008;7:1895–905.

42. Zhang B, Zhang L, Wang D, Gao X, Hong J. Identification of a xylose reductase gene in the xylose metabolic pathway of *Kluyveromyces marxianus* NBRC1777. J Ind Microbiol Biotechnol. 2011;38:2001–10.

43. Lulu L, Ling Z, Dongmei W, Xiaolian G, Hisanori T, Hidehiko K, et al. Identification of a xylitol dehydrogenase gene from *Kluyveromyces marxianus* NBRC1777. Mol Biotechnol. 2003;53:159–69.

44. Lertwattanasakul N, Suprayogi, Murata M, Rodrussamee N, Limtong S, Kosaka T, et al. Essentiality of respiratory activity for pentose utilization in thermotolerant yeast *Kluyveromyces marxianus* DMKU 3-1042. Antonie Van Leeuwenhoek. 2013;103:933–45.

45. Marres CA, de Vries S, Grivell LA. Isolation and inactivation of the nuclear gene encoding the rotenone-insensitive internal NADH: ubiquinone oxidoreductase of mitochondria from *Saccharomyces cerevisiae*. Eur J Biochem. 1991;195:857–62.

46. van Dijken JP, Scheffers WA. Redox balances in the metabolism of sugars by yeasts. FEMS Microbiol Rev. 1986;32:199–224.

47. Tarrio N, Becerra M, Cerdan ME, Gonzalez-Siso MI. Reoxidation of cytosolic NADPH in *Kluyveromyces lactis*. FEMS Yeast Res. 2006;6:371–80.

48. Gonzalez-Siso MI, Garcia-Leiro A, Tarrio N, Cerdan ME. Sugar metabolism, redox balance and oxidative stress response in the respiratory yeast *Kluyveromyces lactis*. Microb Cell Fact. 2009;8:46.

49. Bakker BM, Overkamp KM, van Maris AJ, Kötter P, Luttik MA, van Dijken JP, et al. Stoichiometry and compartmentation of NADH metabolism in *Saccharomyces cerevisiae*. FEMS Microbiol Rev. 2001;25:15–37.

50. Cao J, Barbosa JM, Singh NK, Locy RD. GABA shunt mediates thermotolerance in *Saccharomyces cerevisiae* by reducing reactive oxygen production. Yeast. 2013;30:129–44.

51. Groeneveld P, Stouthamer AH, Westerhoff HV. Super life – how and why 'cell selection' leads to the fastest-growing eukaryote. FEBS J. 2009;276:254–70.

52. Foukis A, Stergioua P, Theodoroua LG, Papagianni M, Papamichael EM. Purification, kinetic characterization and properties of a novel thermo-tolerant extracellular protease from *Kluyveromyces marxianus* IFO 0288 with potential biotechnological interest. Biores Technol. 2012;123:214–20.

53. Raimondi S, Uccelletti D, Amaretti A, Leonardi A, Palleschi C, Rossi M. Secretion of *Kluyveromyces lactis* Cu/Zn SOD: strategies for enhanced production. Appl Microbiol Biotechnol. 2010;86:871–8.

54. Laslett D, Canback B. ARAGORN, a program to detect tRNA genes and tmRNA genes in nucleotide sequences. Nucleic Acids Res. 2004;32:11–6.

55. Salzberg SL, Delcher AL, Kasif S, White O. Microbial gene identification using interpolated Markov models. Nucleic Acids Res. 1998;26:544–8.

56. Delcher AL, Bratke KA, Powers EC, Salzberg SL. Identifying bacterial genes and endosymbiont DNA with Glimmer. Bioinformatics. 2007;23:673–9.

57. Stanke M, Waack S. Gene prediction with a hidden Markov model and a new intron submodel. Bioinformatics. 2003;19:ii215–25.

58. Stanke M, Schöffmann O, Morgenstern B, Waack S. Gene prediction in eukaryotes with a generalized hidden Markov model that uses hints from external sources. BMC Bioinformatics. 2006;7:62.

59. Altschul SF, Madden TL, Schäffer AA, Zhang J, Zhang Z, et al. Gapped BLAST and PSI-BLAST: a new generation of protein database search programs. Nucleic Acids Res. 1997;25:3389–402.

60. Tatusov RL, Fedorova ND, Jackson JD, Jacobs AR, Kiryutin B, Koonin EV, et al. The COG database: an updated version includes eukaryotes. BMC Bioinformatics. 2003;4:41.

61. Apweiler R, O'onovan C, Magrane M, Alam-Faruque Y, Antunes R, Bely B, et al. Reorganizing the protein space at the Universal Protein Resource (UniProt). Nucleic Acids Res. 2012;40:D71–5.

62. Moriya Y, Itoh M, Okuda S, Yoshizawa AC, Kanehisa M. KAAS: an automatic genome annotation and pathway reconstruction server. Nucleic Acids Res. 2007;35 Suppl 2:W182–5.

63. Aoki-Kinoshita KF, Kanehisa M. Gene annotation and pathway mapping in KEGG. Methods Mol Biol. 2007;396:71–91.

64. Aiba H, Adhya S, de Crombrugghe B. Evidence for two functional gal promoters in intact *Escherichia coli* cells. J Biol Chem. 1981;256:11905–10.

65. Tsuchihara K, Suzuki Y, Wakaguri H, Irie T, Tanimoto K, Hashimoto S, et al. Massive transcriptional start site analysis of human genes in hypoxia cells. Nucleic Acids Res. 2009;37:2249–63.

66. Robinson MD, McCarthy DJ, Smyth GK. edgeR: a Bioconductor package for differential expression analysis of digital gene expression data. Bioinformatics. 2010;26:139–40.

67. Alexa A, Rahnenfuhrer J. topGO: Enrichment Analysis for Gene Ontology. In: R Package Version 2. 2000.

9

A new generation of versatile chromogenic substrates for high-throughput analysis of biomass-degrading enzymes

Stjepan Krešimir Kračun[1†], Julia Schückel[1†], Bjørge Westereng[1,2,3], Lisbeth Garbrecht Thygesen[3], Rune Nygaard Monrad[4], Vincent G H Eijsink[2] and William George Tycho Willats[1*]

Abstract

Background: Enzymes that degrade or modify polysaccharides are widespread in pro- and eukaryotes and have multiple biological roles and biotechnological applications. Recent advances in genome and secretome sequencing, together with associated bioinformatic tools, have enabled large numbers of carbohydrate-acting enzymes to be putatively identified. However, there is a paucity of methods for rapidly screening the biochemical activities of these enzymes, and this is a serious bottleneck in the development of enzyme-reliant bio-refining processes.

Results: We have developed a new generation of multi-coloured chromogenic polysaccharide and protein substrates that can be used in cheap, convenient and high-throughput multiplexed assays. In addition, we have produced substrates of biomass materials in which the complexity of plant cell walls is partially maintained.

Conclusions: We show that these substrates can be used to screen the activities of glycosyl hydrolases, lytic polysaccharide monooxygenases and proteases and provide insight into substrate availability within biomass. We envisage that the assays we have developed will be used primarily for first-level screening of large numbers of putative carbohydrate-acting enzymes, and the assays have the potential to be incorporated into fully or semi-automated robotic enzyme screening systems.

Keywords: Chromogenic substrates, Carbohydrate-active enzymes, High-throughput screening, Biomass degradation, Plant cell walls, Lytic polysaccharide monooxygenases

Background

Polysaccharide-degrading enzymes including glycosyl hydrolases (GHs) and lytic polysaccharide monooxygenases (LPMOs) are abundant in nature and of major biotechnological importance [1-4]. In particular, the effective utilization of lignocellulosic materials for second-generation biofuels and chemical production is heavily reliant on enzymes for plant cell wall deconstruction [5-7]. Advances in genomics, proteomics, and associated bioinformatics have enabled the identification of vast numbers of putative enzymes currently assigned to over 130 families in the carbohydrate-active enzyme (CAZy) database [8]. However, it is estimated that the activity of

no more than 20% of the enzymes in CAZy can be reliably predicted with confidence [9-13], and the gulf between identification and characterisation is rapidly widening. There is therefore a pressing need for fast, cheap and facile techniques to empirically screen enzyme activities and preferably methods that can be integrated with semi- or fully automated colony picking and protein expression systems.

Several techniques already exist for assaying GH and LPMO activities, although all have some limitations. Oligosaccharide products can be analysed using high-performance chromatography, and information on product identities obtained by mass spectrometry [14]. This is a powerful and quantitative approach but low throughput and requires highly specialised equipment and personnel. GH activities can also be monitored by measuring the generation of reducing ends, for example, using the

* Correspondence: willats@plen.ku.dk

†Equal contributors

[1]Department of Plant and Environmental Sciences, Thorvaldsensvej 40, Frederiksberg C 1871, Denmark

Full list of author information is available at the end of the article

Nelson-Somogyi [15], the 3,5-dinitrosalicylic acid [16], p-hydroxybenzoic acid hydrazide (PAHBAH) [17] and 3-methyl-2-benzothiazolinone hydrazone (MBTH) [18] methods. However, these reducing end assays also have limited throughput and can be prone to side reactions [19]. Chromogenic substrates are available for screening some enzyme activities. For example, para-nitrophenyl (pNP) glycosides can be useful for rapidly assaying GH activities, although these small artificial compounds are of limited use for the study of high molecular weight and sometimes crystalline substrates such as chitin, cellulose and some arabinoxylans [6,20]. However, an elegant new method for the colorimetric detection of chitinase and cellulase activities was recently described [21]. This technique is based on the fact that the oligomeric products of chitinases and cellulases can be modified by chito-oligosaccharide oxidase (ChitO) and the mutant ChitO ChitO-Q268R, respectively, producing hydrogen peroxide (H_2O_2). The amount of H_2O_2 released is then monitored using a second enzyme, horseradish peroxidase, together with a peroxidase substrate [21]. Azo-dyed and azurine cross-linked (AZCL) polysaccharides are also very widely used for GH screening in a variety of assay types [22,23]. AZCL substrates are produced by cross-linking polysaccharides to render them insoluble and then dyeing them. When exposed to an enzyme with appropriate activity, small dyed oligosaccharide products are released [24]. Assays can be performed in multi-well (usually 96) plates, and activity is detected by the change in colour of enzyme reaction supernatants which can be measured spectrophotometrically. Alternatively, AZCL substrates can be incorporated into agar plates, and activity can be monitored by the formation of coloured halos as the dyed oligosaccharides diffuse into the gel. AZCL-based assays have found widespread usage because of their speed, ease of use and relatively low cost, but they do have some disadvantages, especially in microplate assays [25]. The substrates are only available in one colour (dark blue), limiting throughput since only one substrate can be tested per reaction. Also, they are powders which must be individually weighed into the wells of multi-well plates and undigested substrate particles can interfere with subsequent measurement of coloured supernatants.

We have developed a new generation of chromogenic polysaccharide hydrogel (CPH) substrates based on chlorotriazine dyes that, when used in conjunction with a 96-well filter plate, form a high-throughput assay system. Each substrate can be produced in one of the four colours, and different coloured substrates can be combined in a single well. The use of dyed insoluble polysaccharide substrates has been known for some time, such as in agar plates [26-28] and use of dyed cellulose and starch [29] derivatives. This type of substrates has been used in agar plates [30,31], and this is the first time that

a high-throughput assay such as the one we describe has been developed and established. We show here that this methodology can be applied to a wide variety of polysaccharides and proteins and demonstrate its potential for screening GHs, LPMOs and proteases. Importantly, we have also produced chlorotriazine-dyed biomass samples which provide information about substrate availability within the complex polymer mixtures typically encountered by enzymes in industrial contexts.

Results and discussion
Development of novel chromogenic hydrogel substrates
CPH substrates were produced by first dyeing polysaccharides with one of the four chlorotriazine dyes (red, blue, green or yellow) via nucleophilic aromatic substitution (S_NAr) [32]. The polysaccharides were then cross-linked with 1,4-butanediol diglycidyl ether via base-catalysed epoxide opening. The resulting materials are hydrogels which can be easily dispensed using syringes into 96-well filter plates (Figure 1A,B) and be stored for at least 24 months at 4°C without loss of function. A list of the CPH substrates made to date is shown in Table 1, and the enzymes used to test them are detailed in Table 2.

We developed a rapid assay using the CPH substrates based on 96-well filter plates. Solutions containing enzymes or appropriate buffer controls are added to the wells containing CPH substrates to form the 'reaction plate' (Figure 1C). The plate is then sealed and incubated under the desired conditions (Figure 1D). If an enzyme is active against a particular substrate, then soluble, dyed oligosaccharide products are released (some undigested insoluble substrate may remain). The reaction plate is then placed on top of an ordinary 96-well plate (the 'product plate'), and the two plates are either centrifuged together or placed in a 96-well plate vacuum manifold so that the liquid phase in the reaction plate is transferred through the filter in the bottom to the product plate below (Figure 1E). The absorbance of the products is then measured using a multi-well plate spectrophotometer. The fact that each CPH substrate can be made in four colours may be exploited to increase the throughput of the technique since multiple substrates can be present in a single well. In a theoretical example shown in Figure 1F, a blue CPH-galactomannan and yellow CPH-xylan substrate are combined in each well, so that mannanase and xylanase activities are detected by the release of blue and yellow products, respectively. If both activities are present then a green product is produced. This general approach can be extended so that up to four substrates are combined.

The soluble products (examples of which are shown in Figure 2A) have distinct spectral properties, and examples of spectral scans of products from red, blue, green and yellow CPH-2-hydroxyethyl-cellulose (CPH-HE cellulose)

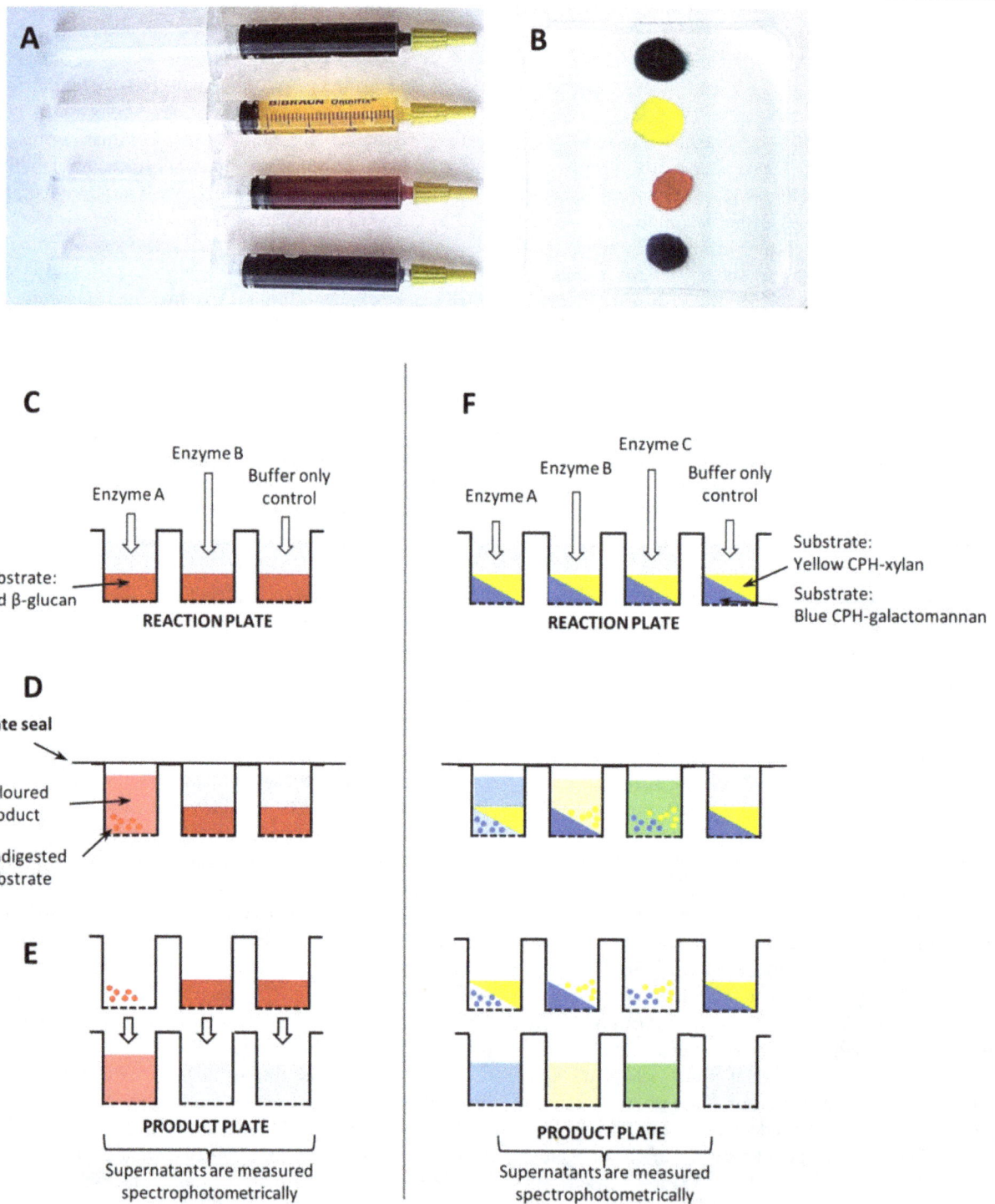

Figure 1 Chromogenic polysaccharide hydrogel (CPH) substrates. **(A)** Examples of CPH substrates loaded into syringe applicators for easy dispensing into microplate wells. **(B)** CPH substrates in four colours, from top green, yellow, red and blue. **(C)** Diagram of part of a reaction plate containing a CPH substrate (red β-glucan) and different enzymes and a control buffer. **(D)** Diagram showing part of a reaction plate during the incubation period. **(E)** Transfer of the supernatant into the product plate prior to spectrophotometric measurement. Result: enzyme A has glucanase activity, and enzyme B does not. **(F)** Diagram showing the simultaneous use of two different coloured CPH substrates (yellow xylan and blue galactomannan). Result: enzyme A has mannanase activity but no xylanase activity, enzyme B has xylanase but no mannanase activity, and enzyme C has both xylanase and mannanase activity.

Table 1 List of CPH substrates made and used in the study along with substrate sources

Substrate	Source
CPH-2-hydroxyethylcellulose (CPH-HE cellulose)	N/A
CPH-amylopectin	Potato
CPH-amylose	Potato
CPH-arabinan	Sugar beet
CPH-arabinoxylan	Wheat
CPH-casein	Bovine milk
CPH-curdlan	*Alcaligenes faecalis*
CPH-dextran	*Leuconostoc* spp.
CPH-galactomannan	Carob
CPH-laminarin	*Laminaria digitata*
CPH-lichenan	Icelandic moss
CPH-pachyman	*Poria cocos*
CPH-pectic galactan	Potato
CPH-pullulan	*Aureobasidium pullulans*
CPH-rhamnogalacturonan I	Potato
CPH-rhamnogalacturonan I (-Gal)[a]	Potato
CPH-rhamnogalacturonan	Soy bean
CPH-xylan	Beechwood
CPH-xyloglucan	Tamarind
CPH-β-glucan from barley	Barley
CPH-β-glucan from oat	Oat
CPH-β-glucan from yeast	Yeast
ICB-Arabidopsis	Rosette leaves from *Arabidopsis thaliana* Col-0 (adult plant)
ICB-bagasse	*Saccharum officinarum* (dried adult plant, stem and leaves)
ICB-fenugreek seeds	*Trigonella* spp. seeds
ICB-hemp	*Cannabis* spp. (dried adult plant, stem and leaves)
ICB-lupin seeds	*Lupinus angustifolius* seeds
ICB-lupin seeds	*Lupinus angustifolius* seeds
ICB-tobacco	Leaves from *Nicotiana benthamiana* (young plant)
ICB-wheat straw	*Triticum* spp. (dried adult plant, stem and leaves)
ICB-willow	*Salix* spp. (dried adult plant, milled)

[a](β-1,4-D-galactan side chains removed with endo-galactanase *gal*). N/A, not applicable.

digested with an endo-cellulase (*cel1*) are shown in Figure 2B. For comparison, the spectrum of products released by digestion of AZCL-HE cellulose with the same enzyme is also shown in Figure 2B. When different coloured substrates are combined in a single reaction with more than one enzyme, then the colour of the products reflects the relative activities of the enzyme mixture

against each substrate. Examples of the products of mixed substrates are shown in Figure 2C,D and demonstrate that increments of 10% in colour contribution can be clearly resolved by visual inspection of the products and by differences in the spectra. Linear regression analysis showed that when two substrates with different dyes were present in known ratios in the range from 9:1 to 1:9, this ratio could be estimated within ±5% of the true value based on visible light absorbance spectra of the unmixed coloured supernatants and the spectrum recorded for the mixture. This was done for all two-colour combinations (Additional file 1: Figure S1).

Stability, reproducibility, detection limits and dose response

The reproducibility of the CPH substrate assays was tested by setting up a series of replicate reactions and measuring the variability of the products (Additional file 2: Figure S2). A series of substrates was made in various colours and added in sextuplet to wells in a 96-well reaction plate. Enzymes were added, and after 30-min incubation, the products were collected and quantified at the appropriate wavelength for each colour (Additional file 2: Figure S2A). This procedure was repeated three more times, so that variability both with single plates and across plates was assessed. The quantified products from these experiments (Additional file 2: Figure S2B) are shown in Additional file 2: Figure S2C and demonstrate a high degree of reproducibility (with a standard error of mean (SEM) no greater than 7%). We also tested dose response by varying both enzyme concentration and reaction time (Figure 3). The effect of enzyme concentration is shown for four differently coloured versions of the CPH-galactomannan substrate treated for 1 h with mannanase (*man*) used between 0 and 2 U/mL (Figure 3A,B). The increased colour production can be clearly seen in the product plate (Figure 3B) and quantification of the products revealed that although the absolute absorbance values were different, each substrate yielded an essentially linear response with increasing enzyme concentration (Figure 3C). The release of products over time was assessed by making four different coloured versions of CPH-xyloglucan, digesting with xyloglucanase *xg* at 0.25 U/mL and measuring the absorbances of products formed in 20-min intervals (Figure 3D). In parallel, we also set up the same reaction with undyed xyloglucan and measured the release of reducing ends using the MBTH referred to previously (black line in Figure 3D). Absorbance values were measured at the appropriate wavelengths for each substrate, and again, although the absolute absorbance values varied between the differently coloured versions of the substrate, the trend was the same for each, and this trend closely followed the production of reducing ends from the undyed xyloglucan. The detection

Table 2 Enzymes used in the study

Code name	Description	CAZy family	Source
ara	Endo-arabinase (*Aspergillus niger*)	GH43	Megazyme
cel1	Endo-cellulase (endo-β-1,4-glucanase) (*Trichoderma longibrachiatum*)	GH7	Megazyme
cel2	Cellulase (endo-β-1,4-glucanase) (*Bacillus amyloliquefaciens*)	GH5	Megazyme
Gal	Endo-β-1,4-D-galactanase (*Aspergillus niger*)	GH53	Megazyme
glc	Endo-β-1,3-glucanase (*Trichoderma* sp.)	GH16	Megazyme
lic	Lichenase (endo-β-1,3(4)-glucanase) (*Bacillius* sp.)	GH16	Megazyme
man	Endo-β-1,4-mannanase (*Cellvibrio japonicus*)	GH26	Megazyme
nz1	Endo-β-1,4-xylanase (*Aspergillus aculeatus*)	GH10	Novozymes
nz2	Endo-β-1,4-xylanase (*Thermomyces lanuginosus*)	GH11	Novozymes
nz3	Endo-β-1,4-glucanase (*Aspergillus aculeatus*)	GH5	Novozymes
nz4	Proprietary fungal endo-β-1,4-mannanase	GH5	Novozymes
ply1	Macerase™ Pectinase (*Rhizopus* sp.)	N/A	Calbiochem
ply2	Pectolyase Y-23 (*Aspergillus japonicus*)	N/A	Duchefa Biochemie
ply3	Pectolyase (*Aspergillus japonicus*)	N/A	Sigma
pec1	Pectate lyase (*Cellvibrio japonicus*)	PL10	Megazyme
pec2	Pectate lyase (*Aspergillus* sp.)	N/A	Megazyme
pol1	Endo-polygalacturonase M2 (*Aspergillus niger*)	GH28	Megazyme
pol2	Endo-polygalacturonase M1 (*Aspergillus niger*)	GH28	Megazyme
rgh	Rhamnogalacturonan hydrolase (*Aspergillus aculeatus*)	GH28	Novozymes
xg	Xyloglucanase (*Paenibacillus* sp.)	GH5	Megazyme
xyl1	β-xylanase, M4 (*Aspergillus niger*)	GH11	Megazyme
xyl2	Endo-β-1,4-xylanase M1 (*Trichoderma viride*)	GH11	Megazyme
NcLPMO9C	Lytic polysaccharide monoxygenase (*Neurospora crassa*)	AA9	N/A
Broth	Culture broth from *Phanerochaete chrysosporium* (3d after inoculation)	N/A	N/A
Proteinase K	Proteinase K (*Tritirachium album*)	N/A	Sigma
Trypsin	Trypsin from bovine pancreas	N/A	Sigma
Elastase	Elastase from porcine pancreas	N/A	Sigma

Code, source and carbohydrate-active enzyme (CAZy) database family. N/A, not applicable.

limit of the substrates is between 100 and 500 (mU/mL)/h depending on the substrate and the specific activity of the enzymes.

Most of the enzymes used in the work described here are well-characterized and commercially available (described in Table 2), and data from CPH substrates was in good agreement with previous findings. However, we also validated the technique by making a direct comparison between the performance of the CPH substrates and activity data determined by the PAHBAH method to measure the release of reducing ends [17] (Figure 4). Four purified enzymes were tested: two xylanases, a glucanase and a mannanase (designated *nz1*, *nz2*, *nz3* and *nz4*, respectively) on CPH substrates of blue CPH-arabinoxylan, yellow CPH-xylan, red CPH-β-glucan and green CPH-galactomannan as shown in Figure 4A. Product plate results (Figure 4B) showed that *nz1* and *nz2* had activity against both CPH-arabinoxylan and CPH-xylan

with visibly greater colour production for *nz2* against both substrates, and neither enzyme appeared to have activity against CPH-β-glucan or CPH-galactomannan. *nz3* and *nz4* had activity against CPH-β-glucan and CPH-galactomannan, respectively, with no or very little side activity observable. The information about relative activity obtained on the release of reducing ends using the same enzymes with pure (undyed, non-cross-linked) polysaccharides (Figure 4C) was in close agreement with the findings from the CPH substrates with the same major activities detected and very low side activities (Figure 4C).

To verify that the substrates can perform within a wide range of pH values, CPH-HE cellulose and CPH-rhamnogalacturonan were treated with a cellulase (*cel2*) and a pectinase (*pec1*) with activity optima at acidic and basic pH values, respectively. From this experiment, it was apparent that at least for these substrates, the CPH polysaccharides are compatible with a wide range of pH

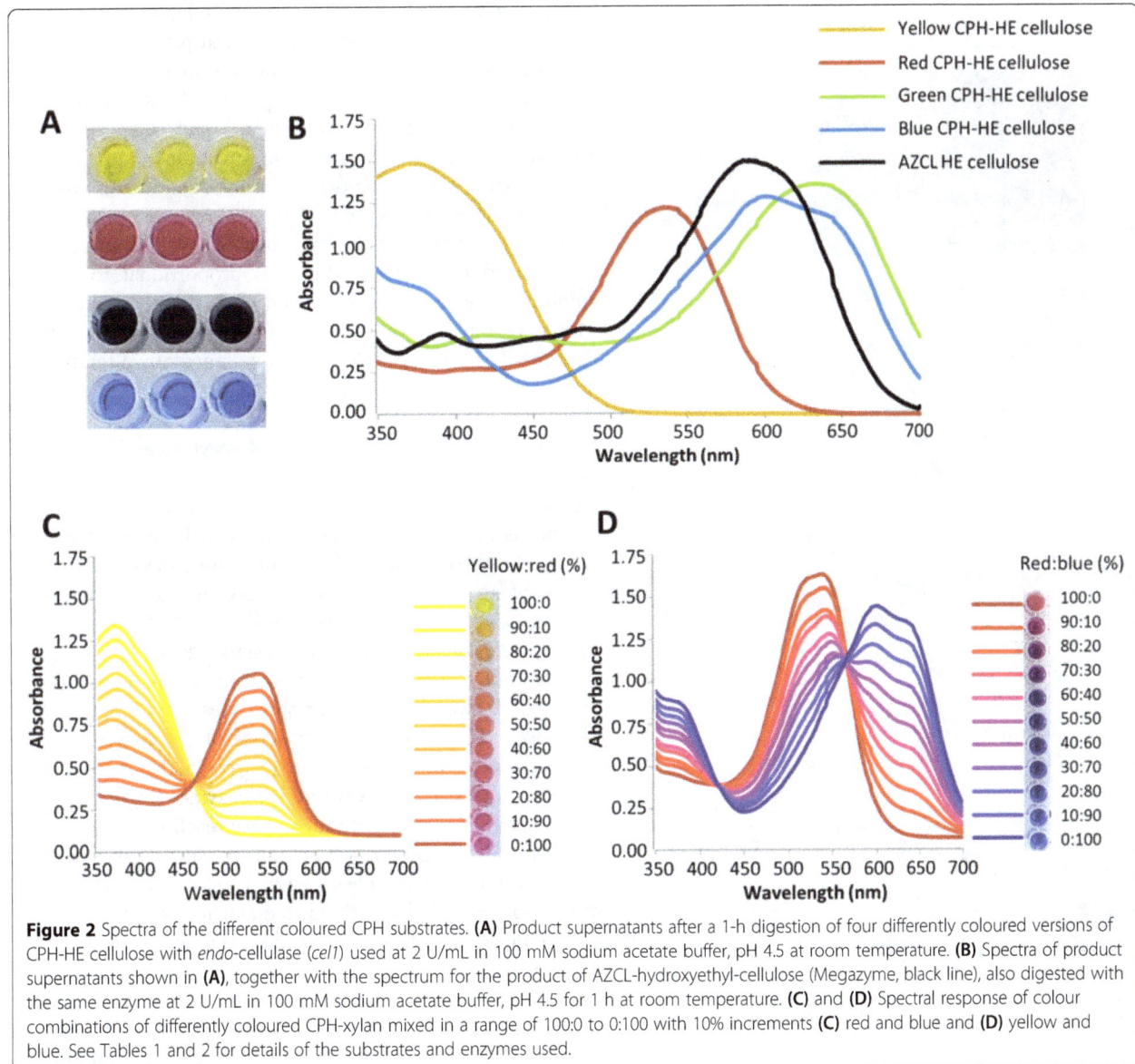

Figure 2 Spectra of the different coloured CPH substrates. **(A)** Product supernatants after a 1-h digestion of four differently coloured versions of CPH-HE cellulose with *endo*-cellulase (*cel1*) used at 2 U/mL in 100 mM sodium acetate buffer, pH 4.5 at room temperature. **(B)** Spectra of product supernatants shown in **(A)**, together with the spectrum for the product of AZCL-hydroxyethyl-cellulose (Megazyme, black line), also digested with the same enzyme at 2 U/mL in 100 mM sodium acetate buffer, pH 4.5 for 1 h at room temperature. **(C)** and **(D)** Spectral response of colour combinations of differently coloured CPH-xylan mixed in a range of 100:0 to 0:100 with 10% increments **(C)** red and blue and **(D)** yellow and blue. See Tables 1 and 2 for details of the substrates and enzymes used.

conditions (Additional file 3: Figure S3A). Thermostability of the substrates was tested with CPH-xylan and a thermophilic xylanase (*xyl3*) showing the stability of CPH substrates at high temperatures (Additional file 3: Figure S3B).

We were interested to assess in detail the nature of the products released from CPH substrates. To do this, we used matrix-assisted laser desorption/ionization-time-of-flight-mass spectrometry (MALDI-ToF-MS) to analyse products resulting from the digestion of CPH-xylan (Additional file 4: Figure S4A) and undyed cross-linked xylan (Additional file 4: Figure S4B) with xylanase *xyl1*. The spectrum from the CPH-xylan substrate displayed more near-baseline noise than that of the undyed substrate, and this was presumably due to the presence of the dye molecules. However, in both cases, a range of

pentose-containing oligosaccharide products were released that were consistent with xylan fragments. From the mass profiles, we determined that in many cases, xylan oligomers were attached to one or more hydrolysed linker molecules, implying that a significant proportion of the linker was not in fact spanning between polysaccharide chains in either the dyed or undyed version of these substrates (shown schematically in Additional file 4: Figure S4C). Despite this apparent incomplete cross-linking, the substrates were still insoluble and still effective substrates for *xyl1*.

We also tested the storage stability of CPH substrates. A set of reaction plates was made, sealed and stored at 4°C for 1 year. A series of enzyme digestions were made on these plates and compared to identical reactions made in freshly produced reaction plates. As shown in

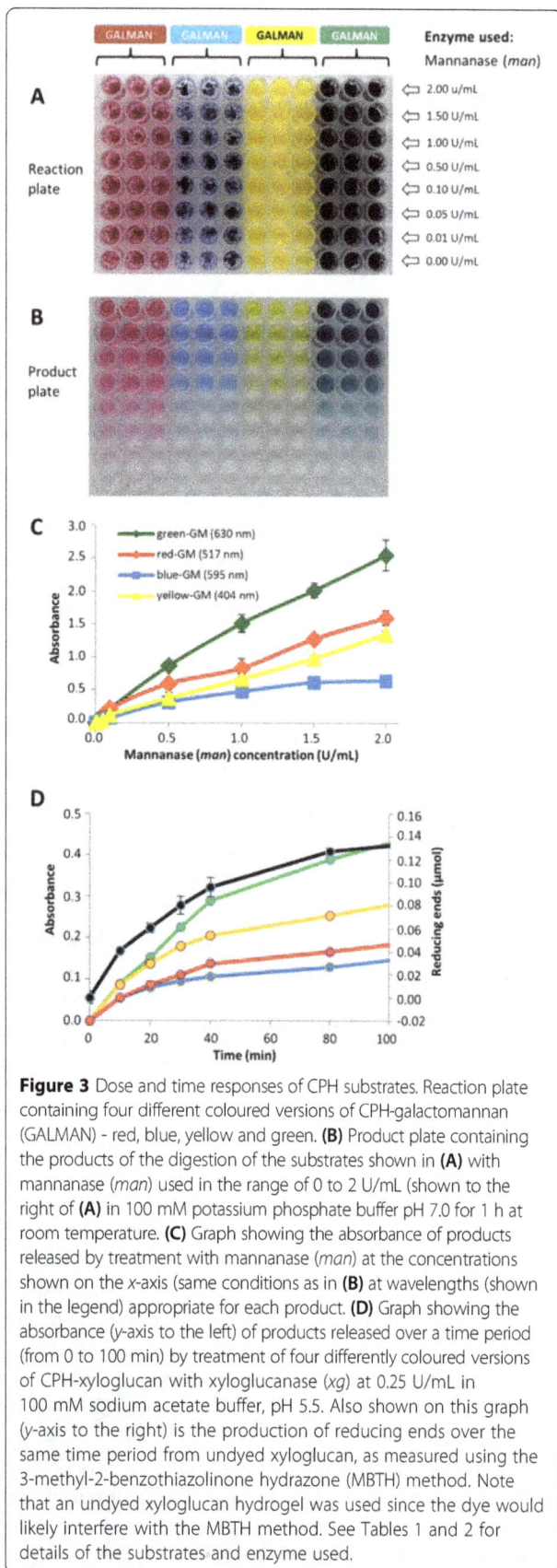

Figure 3 Dose and time responses of CPH substrates. Reaction plate containing four different coloured versions of CPH-galactomannan (GALMAN) - red, blue, yellow and green. **(B)** Product plate containing the products of the digestion of the substrates shown in **(A)** with mannanase (*man*) used in the range of 0 to 2 U/mL (shown to the right of **(A)** in 100 mM potassium phosphate buffer pH 7.0 for 1 h at room temperature. **(C)** Graph showing the absorbance of products released by treatment with mannanase (*man*) at the concentrations shown on the *x*-axis (same conditions as in **(B)** at wavelengths (shown in the legend) appropriate for each product. **(D)** Graph showing the absorbance (*y*-axis to the left) of products released over a time period (from 0 to 100 min) by treatment of four differently coloured versions of CPH-xyloglucan with xyloglucanase (*xg*) at 0.25 U/mL in 100 mM sodium acetate buffer, pH 5.5. Also shown on this graph (*y*-axis to the right) is the production of reducing ends over the same time period from undyed xyloglucan, as measured using the 3-methyl-2-benzothiazolinone hydrazone (MBTH) method. Note that an undyed xyloglucan hydrogel was used since the dye would likely interfere with the MBTH method. See Tables 1 and 2 for details of the substrates and enzyme used.

Additional file 5: Figure S5, there was high degree of consistency between the fresh and stored plates.

Taken together, these data indicate that the assays based on CPH substrates are reproducible, stable over time and can tolerate a range of pH and temperature conditions. Like all chromogenic substrates, the CPH polysaccharides are intended for semi-quantitative analysis, but nevertheless, the CPH dose-response profiles show that, for at least the examples tested, the amount of coloured products produced is proportional to the amount of enzyme present and/or incubation times. Moreover, our results from MALDI-ToF analysis suggest that the dying process does not fundamentally alter the nature of the products released.

Multiplexed assays using mixtures of enzymes and substrates

The ability to produce CPH substrates in four different colours is an important feature that can be used to increase the throughput of CPH substrate assays. An example of an assay involving multiple mixed substrates and enzymes is shown in Figure 5. Red CPH-β-glucan, yellow CPH-xylan, green CPH-galactomannan and blue CPH-arabinoxylan were added to wells either alone or mixtures of two, three or four substrates together in the same well (Figure 5A). The *nz* series of enzymes previously characterized (see Figure 4) were added to wells alone or as mixtures of two or three enzymes (Figure 5A). The resulting product plate from these reactions is shown in Figure 5B. Single enzymes acting on single substrates produced the expected corresponding colour of product - but importantly, individual enzyme activities could be resolved by their coloured products even when multiple substrates were present. For example, the product of well A5 had the qualitatively same spectral response as the products in wells B5 and D5, despite the fact that whereas in A5, only red CPH-β-glucan was present, in B5 and D5, the red CPH-β-glucan was mixed with one and three other substrates, respectively (Figure 5C). A similar effect was observed for wells A11, B11 and D11 which produced essentially qualitatively the same spectral responses of a green product, although only well A11 of the substrate plate contained green galactomannan alone (Figure 5D). Note that there are quantitative differences in the absorbance values, and these reflect the fact that when substrates mixed in a single well, then the amount of each individual substrate is proportionally lower since the total volume of substrate material is the same in each well. Thus, the absorbance value for products from well D11 is lower than for well A11 because in D11, the green galactomannan was combined with three other substrates (Figure 5D).

Substrate mixtures also enable multiple enzyme activities to be detected simultaneously. For example, wells G4, G6 and G12 of the substrate plate all contained the

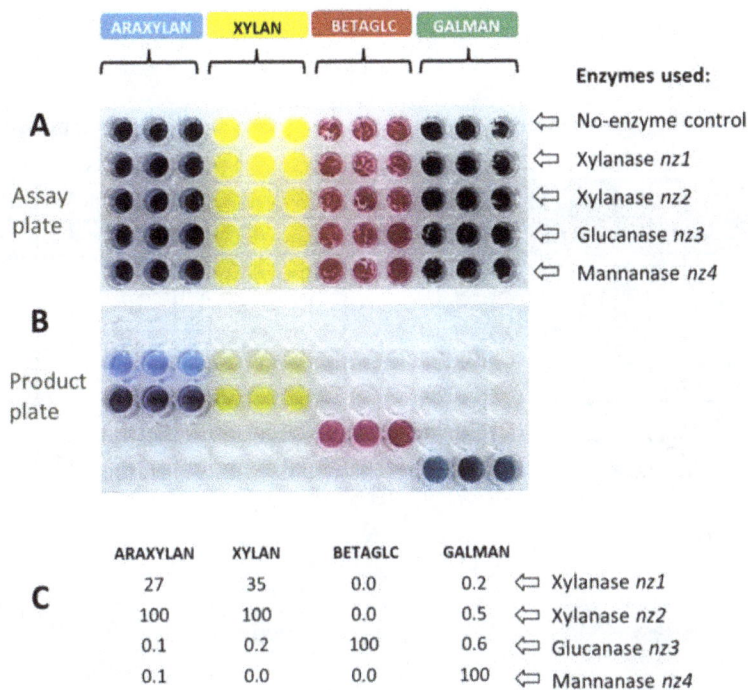

Figure 4 Direct comparison of the output from CPH substrates with the measurement of reducing ends. **(A)** Each well in the reaction plate contained one of the four different CPH substrates: blue CPH-arabinoxylan (ARAXYLAN); yellow CPH-xylan, (XYLAN); red CPH-β-glucan (BETAGLC) and green CPH-galactomannan (GALMAN). The substrates were treated with the enzymes: xylanase *nz1*; xylanase *nz2*; glucanase *nz3* or mannanase (*nz4*) as shown to the right (note that all wells in one row contained the enzymes or control buffer indicted). All enzymes were used at 5 µg/mL in 50 mM sodium acetate buffer, pH 5.5 and incubated at 50°C for 10 min. **(B)** Product plate containing the product of the reaction shown in **(A)**. **(C)** Table showing the relative activities of the enzymes *nz1* to *nz4* as determined by the production of reducing ends when native versions of the polysaccharides shown in **(A)** were treated with enzymes *nz1* to *nz4* used at 5 µg/mL in 50 mM sodium acetate buffer, pH 5.5 at 50°C for 10 min. The highest value was set to 100, and all other values adjusted accordingly. In this case, the PAHBAH reducing end assay was used. See Tables 1 and 2 for details of the substrates and enzyme used.

same mixture of yellow xylan, green galactomannan and red β-glucan. When xylanase *nz1* alone was added (well G4), then the expected yellow product was produced, but the contribution of both glucanase *nz3* and xylanase *nz1* (well G6) are apparent by the orange product with its distinct spectral response. When mannanase *nz4* was also added (well G12), the contribution of the green product can also be detected by a distinct spectrum (Figure 5E).

Detecting enzymatic activities in microbial broths
We also tested the utility of the CPH substrates for assessing enzymatic activity in broths of *Penicillium expansum* and *Colletotrichum acutatum* cultured on apple pomace liquid media (Additional file 6: Figure S6). Samples of the crude broths were taken once daily from 1 to 10 days after inoculation. The broths were applied directly to 96-well assay plates containing a variety of CPH substrates including CPH-galactomannan (Additional file 6: Figure S6A), CPH-lichenan, CPH-β-glucan from barley (Additional file 6: Figure S6B and S6C, respectively) and arabinoxylan (Additional file 6: Figure S6D). As expected, broths from

both fungi showed a general increase over time in their ability to degrade the substrates, but there were striking differences in the relative activities produced by the two species. After 3 days, there was a sharp increase in mannanase activity in the *C. acutatum* broth (Additional file 6: Figure S6A), whereas the *P. expansum* broth had notably increased activity against lichenan and β-(1,3),(1-4)-glucan from barley (Additional file 6: Figure S6B and S6C). The degradation results for arabinoxylan were similar for both fungi (Additional file 6: Figure S6D).

CPH substrates can also be used in a simple agar gel system
We considered that for some applications, it might be preferable to use a very simple gel-based assay in which enzyme activity is detected by the formation of a coloured halo (Figure 6). This approach is conceptually similar to assays in which particulate AZCL substrates are embedded within agar. Our method differs in that instead of being distributed through the agar, the CPH substrates are contained within a well, either made at the casting stage or subsequently once the agar has set

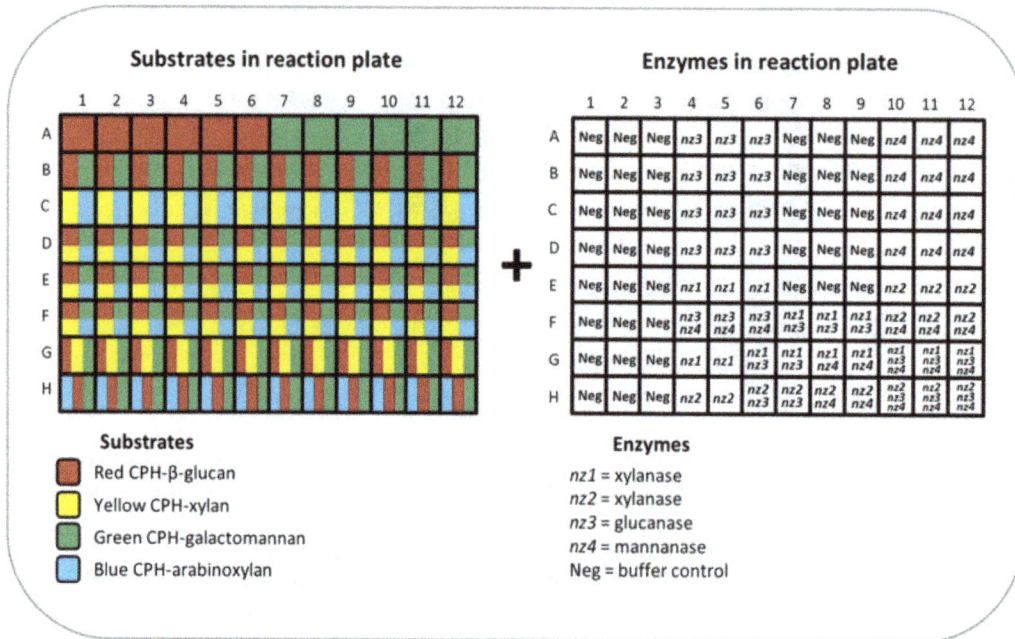

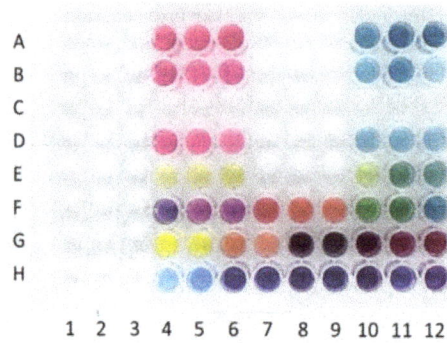

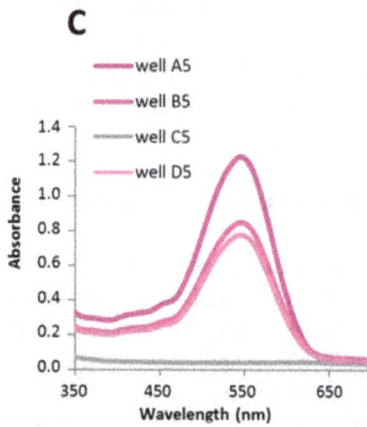

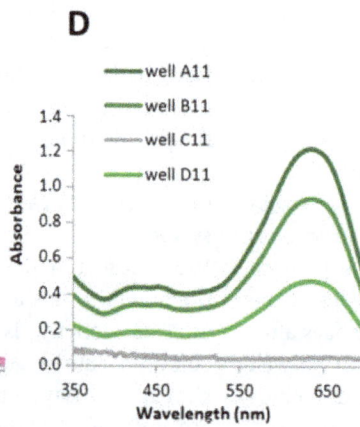

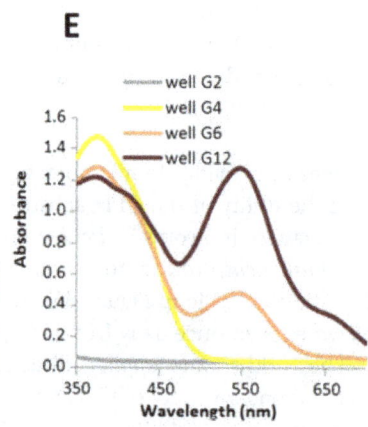

Figure 5 (See legend on next page.)

(See figure on previous page.)
Figure 5 Multiplexed assays using the CPH substrates in different substrate and enzyme concentrations. **(A)** Setup of the reaction plate: CPH substrates used in the experiment were red CPH-β-glucan, green CPH-galactomannan, yellow CPH-xylan and blue CPH-arabinoxylan, and the distribution of these substrates within the reaction plate is shown to the left. Each well contained a single substrate or two, three or four substrates mixed together. Note that the total volume of substrate was the same in each well, and when substrates were mixed in wells, they were present in equal amounts. The distribution of enzymes in the reaction plate is shown to the right, and the enzymes used were xylanase *nz1*, xylanase *nz2*, glucanase *nz3* and mannanase *nz4*. Note that the total amount of enzyme was the same in each well, and when enzymes were mixed in wells, they were present in equal amounts. Some wells contain buffer only ('Neg'). The reaction buffer used was 50 mM sodium acetate buffer, pH 5.5. **(B)** Image of the product plate containing the products of the reactions shown in **(A)** after a 10-min reaction at 50°C. **(C)** to **(E)** spectra of selected wells as shown in the legends. See Tables 1 and 2 for details of the substrates and enzyme used.

(Figure 6A). If an enzyme has activity against the CPH substrate, then small, soluble products are released that diffuse through the agar and are seen as halos around the wells (Figure 6B). Typically, overnight (18 h) incubations are used, and representative experiments using an *endo*-glucanase (*glc*) and a xyloglucanase (*xg*) are shown in Figure 6C,D. In these experiments, four differently coloured versions of pachyman (Figure 6C) and xyloglucan (Figure 6D) were made and treated with *glc* (see Table 2) and *xg* (see Table 2), respectively. Halos were generated from each substrate version by the respective enzymes, and not by buffer controls.

Using a similar rationale to that described for multi-well plate-based assays, substrates can also be mixed. In the example shown in Figure 6E, all of the wells were filled with the same mixture of three substrates (red CPH-pachyman, blue CPH-galactomannan and yellow CPH-xylan). However, the colours of the halo produced after 18 h depend on the enzyme used, and when more than one enzyme was present, intermediate coloured halos were formed. These experiments showed that although simple, the agar plate-based assays using CPH substrates have the capacity to be highly information-rich.

Analysis of LPMO and protease activities

We further extended the scope of the CPH substrates by assessing if they were compatible with classes of degradative enzyme other than glycosyl hydrolases, specifically an LPMO and different proteases (Figure 7). We used *Nc*LPMO9C, an enzyme that is the first LPMO to be described with activity against both cellulosic and hemicellulosic substrates [4]. Consistent with previous findings, *Nc*LPMO9C released products from CPH substrate versions of CPH-HE cellulose and CPH-lichenan, but not CPH-xylan (Figure 7A). The release of products from CPH-HE cellulose and CPH-lichenan was dependent on the presence of the reducing agent ascorbic acid, since assays lacking ascorbic acid yielded no or negligible products. These data indicate that the activity we observed from *Nc*LPMO9C was indeed mediated by an oxidative reaction. The proteases trypsin, elastase and proteinase K were tested using CPH-casein, and all three enzymes yielded degradation products (Figure 7B).

Insoluble chromogenic biomass substrates

Enzymes that degrade or modify biomass feedstocks very rarely, if ever, encounter single polysaccharides in isolation. Rather, enzymes usually act upon highly complex mixtures of polysaccharides that are physically intermeshed and sometimes covalently linked. Ideally, assays for screening enzymes intended for biomass deconstruction should reflect this complexity and heterogeneity, and we therefore further extended the concept of chromogenic substrates by producing a series of dyed versions of biomass materials typically used as biofuel/biorefinery feedstocks. Such insoluble chromogenic biomass (ICB) substrates cannot be used to resolve individual enzyme activities since multiple polymers are dyed with the same colour within a single ICB substrate type. However, they can provide information about the ability of an enzyme, cocktail or microbial broth to release oligomeric products from a complex biomass matrix *per se* and as such are a useful addition to the screening toolbox.

ICB substrates are based on raw biomass samples or alcohol-insoluble residue (AIR) preparations which are standard crude preparations of polysaccharides widely used as the starting point for biomass analysis, for example, saccharification assays [33,34]. As a proof of concept, we made six ICB substrates of vegetative material from *Arabidopsis*, tobacco, hemp, willow, wheat straw and bagasse. We also made a further two ICB substrates from fenugreek and lupin seeds (the sources of biomass samples are listed in Table 1). This set of samples was chosen because they represent distinct cell wall types with different polymer compositions and architectures. The *Arabidopsis* and tobacco samples were made from leaf and stem material that is rich in the 'type I' primary cell walls found in the younger green parts of the eudicot species. The wheat and bagasse samples predominantly contain the 'type II' cell walls typical of commelinid monocots in which β-(1,3),(1,4)-D-glucan and glucuronoarabinoxylans (GAXs) are the prominent hemicelluloses with generally lower levels of pectin. The hemp and willow samples were prepared from older mature organs, which are characterized by secondary cell walls which contain proportionally higher levels of cellulose and xylan than primary cell walls. The lupin and fenugreek seeds are dominated by storage polysaccharides,

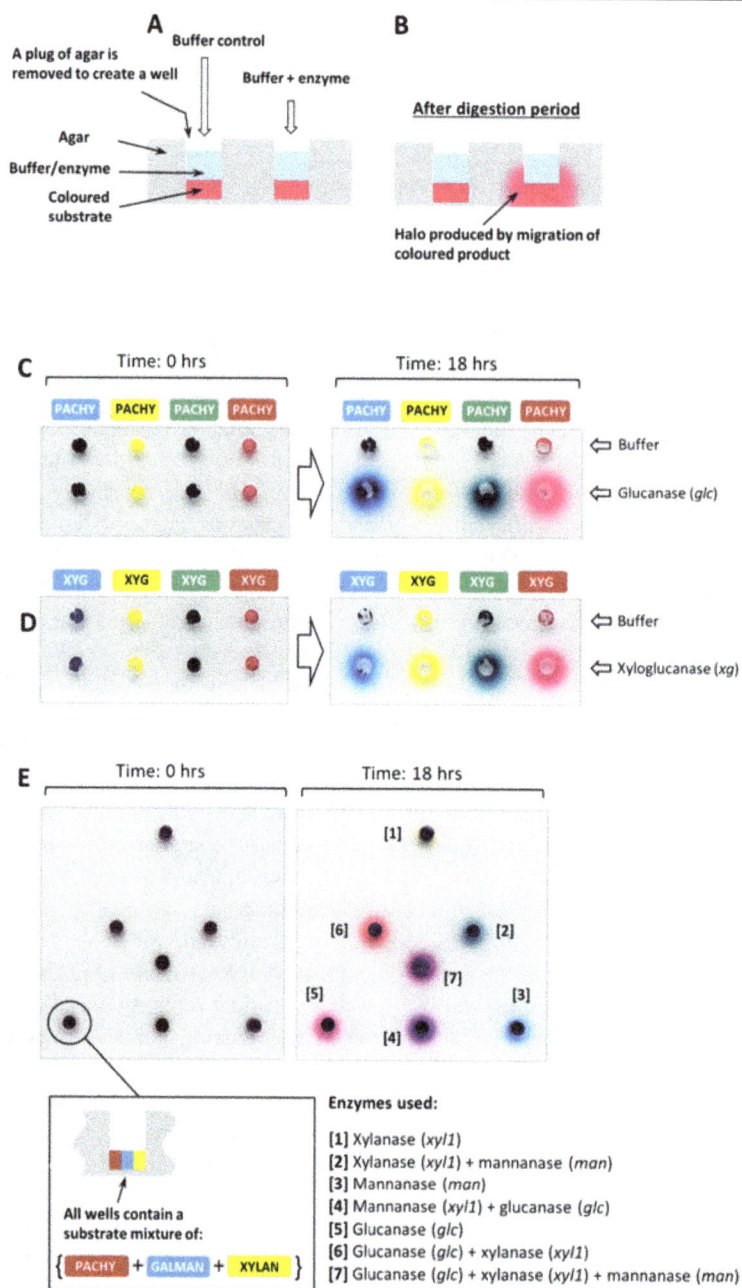

Figure 6 Using CPH substrates in simple agarose plate assays. **(A)** Schematic illustration of the assay using agarose plates. A well is formed in the agarose (either during or after casting), and CPH substrate is added together with an enzyme-containing solution. **(B)** Diagram showing the situation after the digestion period. If the enzyme has activity against the CPH substrate, then a halo is formed by the migration of soluble products. **(C)** and **(D)** examples of agarose plate assays with a glucanase (*glc*) used to digest four differently coloured version of CPH-pachyman (PACHY), and a xyloglucanase (*xg*) used to digest four differently coloured versions of CPH-xyloglucan (XYG), respectively. **(E)** An example of an agar plate assay containing mixed CPH substrates. All the wells contained red CPH-pachyman (PACHY), blue CPH-galactomannan (GALMAN) and yellow CPH-xylan (XYLAN) mixed in approximately equal proportions. Note that different coloured halos are generated depending on what enzyme or combinations of enzymes were present in wells and as indicated in the legend. See Tables 1 and 2 for details of the substrates and enzyme used.

and lupin seeds are distinctive because β-(1,4)-D-galactan is highly abundant.

The eight ICB substrates were treated separately with a range of enzymes, and the resulting products quantified (Additional file 7: Figure S7). The overall product profiles for *Arabidopsis* and tobacco material were similar, with relatively high signals obtained for pectinolytic enzymes (*ply2* and *ply3*) (Additional file 7: Figures S7A and S7B). Some activity was also observed for xyloglucanase (*xg*) and cellulases (*cel1* and *cel2*), and these data

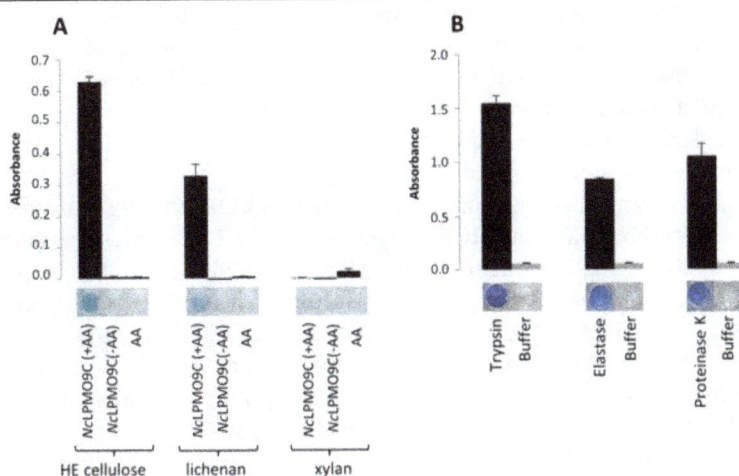

Figure 7 Using CPH substrates to analyse LPMO and protease activities. **(A)** The LPMO NcLPMO9C was tested using three different CPH substrates, from left to right; blue CPH-HE cellulose; blue CPH-lichenan and blue CPH-xylan. NcLPMO9C was tested with and without ascorbic acid (AA, used at 3.2 mM), and ascorbic acid alone was used as a control. Note that activity was observed for the two CPH substrate that contained β-linked glucan (blue CPH-HE cellulose; blue CPH-lichenan) but not blue CPH-xylan. All incubations were in 100 mM potassium phosphate buffer, pH 7.0 incubated for 1 h at 50°C. **(B)** Blue CPH-casein was treated with 0.1 U/mL trypsin, elastase and proteinase K (containing 10 mM Ca^{2+}) in 100 mM sodium phosphate pH 7.0 for 20 min at room temperature. See Tables 1 and 2 for details of the substrates and enzyme used.

broadly reflect, at least in qualitative terms, the composition of type I primary cell walls. The product profiles from the two commelinid monocot species, wheat and bagasse, were also similar to each other (Additional file 7: Figures S7C and S7D). Compared to *Arabidopsis* and tobacco, much higher signals were obtained by xylanase (*xyl1* and *xyl2*) treatment and this is consistent with the high levels of GAXs in type II cell walls. Interestingly, although the type II cell walls of grasses contain low levels of pectin, higher signals were obtained with the pectolyases *ply2* and *ply3* than for *Arabidopsis* and tobacco. One possible explanation for this is that the signals obtained from ICB substrates are a function of both substrate abundance and accessibility and certain pectins may be less abundant but more accessible in type II cell walls compared to their type I counterparts. Relatively high signals were obtained for hemp and willow (Additional file 7: Figures S7E and S7F respectively) from the xylanases *xyl1* and *xyl2*, and this is consistent with the fact that these species contain type II secondary walls. The unusually high level of β-(1,4)-D-galactan in lupin seeds was readily apparent by comparison of the products produced from the fenugreek and lupin ICB samples (Additional file 7: Figures S7G and S7H, respectively), with abundant product formation by endo-β-(1-4)-galactanase (*gal*) digestion of lupin but not fenugreek ICB samples.

We envisage that, when used in combination, the CPH substrates of defined polysaccharides and ICB substrates can provide insight into what substrate is degraded by an enzyme, to what extent that substrate is available for degradation within a given biomass type. Some caution is required in interpreting data from ICB substrates

because we cannot determine to what extent each individual polysaccharide within a biomass sample is dyed, and they are most likely not dyed to an equal extent. Nevertheless, we noted that all the enzymes we used did yield products to some degree, suggesting that most if not all, the polymers were in fact dyed.

We were also interested to analyse what oligomeric products were released from ICB substrates and performed MALDI-ToF-MS analysis of products from hemp and bagasse ICB substrates (Additional file 8: Figure S8A and S8B, respectively) treated with xylanases *xyl1* and *nz2*, respectively. The spectra showed that the enzymes released products similar to those that would be expected for the corresponding natural feedstocks. As expected, xylanases *xyl1* release pentose sugars with or without glucuronic acids depending on the feedstock. Xylans in monocotyledons are typified by arabinosyl- and (4-methyl)-glucuronic acid substituents [35], whereas dicot hardwood xylan contains mainly 4-methyl-glucuronic acid substitutions [36]. Both these types of substitutions limit xylanase activity [37], and consistent with this, we observed longer oligosaccharides which would not be expected in linear xylan devoid of substitutions. Another observation from Additional file 8: Figure S8 is that there are no acetyl-ester modifications present which are common on xylan in these materials [38] and that is due to the high pH used during substrate synthesis which causes hydrolysis of all esters.

Conclusions

We show here that a new generation of chromogenic substrates have considerable potential for screening single enzymes, enzyme cocktails and microbial broths relevant for biomass deconstruction. Several other chromogenic substrate-based assays have been described, but the CPH and ICB substrates have some distinct features that make them a valuable addition to existing screening technology. Because they are hydrogels rather than powders, the CPH substrates are convenient to handle and quick and easy to distribute into assay plates. The ability to produce both CPH and ICB substrates in several colours supports high-throughput multiplexed assays whereby multiple enzymes or side activities of a single enzyme can be detected in a single reaction well. This means that CPH-based assays have the capacity to be highly information-rich compared to other simple assay systems. Furthermore, to our knowledge, the ICB substrates represent the first attempt to produce a rapid assay system in which the natural multi-polymer complexity of biofuel feedstocks is represented. Our data also showed that both CPH and ICB substrates have potential for LPMO and protease screening.

However, in common with other insoluble chromogenic substrates, the CPH and ICB reagents also have limitations. For example, ester bonds are not preserved, so esterase activities cannot be tested. Also, as with AZCL polysaccharides, the CPH and ICB substrates are not reactive with *exo*-acting enzymes, presumably because of steric hindrance of cleavage sites and alternative methods may need to be employed for assaying that class of GHs. It is also important to recognise that although spectrophotometric analysis of the CPH and ICB products provides quantitative information about relative activity levels, these assays are not intended to replace fully quantitative biochemical techniques to measure absolute activity values. We envisage that the 96-well assays we have developed will be used primarily for first-level screening of large numbers of putative CAZymes, and the assays have the potential to be incorporated into fully or semi-automated robotic enzyme screening systems.

Methods
Reagents, enzymes and microorganisms
Amylose, 2-hydroxyethyl-cellulose (Sigma 434965, molar 2-hydroxyethyl substitution 2.5 mol per mol cellulose), curdlan, laminarin, amylopectin, pectolyase from *A. japonicus* and α-amylase from bovine pancreas were obtained from Sigma (Brøndby, Denmark). All other polysaccharides were obtained from Megazyme (Bray, Ireland). The enzymes used are listed with the supplier in Table 2. The dyes reactive red 4, reactive blue 4, reactive green 19 and reactive yellow 2, cross-linker 1,4-butanediol diglycidyl ether, NaOH and all salts for buffers were obtained from Sigma (Brøndby, Denmark). Two pathogenic fungi *C. acutatum* (isolate SA 0-1) and *P. expansum* (isolate IK2020) and the apple pomace media

were kindly provided by Birgit Jensen and Daniel Buchvaldt Amby (Department of Plant and Environmental Sciences, Faculty of Science, University of Copenhagen, 1871 Frederiksberg, Denmark).

Production of CPH and ICB substrates
Synthesis of CPH substrates was performed according to a protocol modified from Ten *et al.* [30]. The polymer was mixed with 0.5 M NaOH (concentrations from 3% to 20% w/V), and the sample was dissolved by shaking (110 rpm) at 60°C (or room temperature). Samples were then dyed by adding 0.5 g of chlorotriazine dye followed by incubation at 60°C (or at room temperature for substrates soluble at room temperature) for 0.5 to 4 h with shaking. Cross-linking of the dyed polysaccharide solutions was achieved, after being cooled down to room temperature if needed, by adding 1,4-butanediol-diglycidyl ether. The cross-linker concentration ranged from 1.2% to 16% (V/V) and was optimized to provide maximum hydrogel responsiveness for enzyme digestion whilst preserving optimal physical consistency for handling and the optimal amount it varied greatly from polysaccharide to polysaccharide. The reaction mixture was vortexed vigorously for 2 min and then left to stand for 48 to 96 h at room temperature without agitation for hydrogel formation. To enable more efficient purification and handling, the resulting hydrogels were homogenized within the tube using a spatula to a paste-like material and the material was transferred onto a nylon membrane mesh (31-µm mesh) placed over a Büchner funnel where it was washed with boiling sterile water until no more free dye was released, left to stand until there was no more water draining from the funnel, collected into a fresh tube, and stored at 4°C. In the case of the casein protein substrate, the last wash was with 0.01% NaN$_3$ in sterile water. Production of ICB substrates was based on the same dyeing chemistry as for the CPH substrates, but the cross-linking step was omitted. Briefly, 2 g of plant material (freeze-dried and crushed raw and not pre-treated) and 500 mg of dye were mixed and suspended in 10 to 20 mL of 0.5 M NaOH (the volume was adjusted so that the suspension was free-flowing). The reaction mixture was shaken for 4 h at room temperature. After cooling to room temperature, the substrates were cleaned similarly as for the CPH substrates above. After an isopropanol wash, samples were stored suspended in isopropanol at 4°C.

Effects of dye concentration have not been studied in detail, especially because dyes used were of technical grade, but the general observation was that there is no observable change in substrate properties depending upon dye concentration within the concentration range tested (5% to 20%, w/V).

Enzyme assay in 96-well filter plates

Enzyme activity assays in 96-well plate format were performed by transferring 100 μL CPH or 1 mg ICB substrates into a 96-well 'reaction' plate (filter-plate MSHVN4510, Millipore, VWR, Herlev, Denmark) where they were washed once with water. Buffer (100 mM, sodium acetate or potassium phosphate buffer, pH dependent on the individual enzyme) containing enzyme (0.5 to 2 U/mL) was added to each well to a final volume of 150 μL. The plate was sealed using adhesive PCR plate seals (Thermo Scientific, VWR, Herlev, Denmark) and incubated for 10 min to 1 h at room temperature with shaking. The reaction plate was then place on top of another 96-well plate (the 'product plate', MSCPNUV40, Millipore, Herlev, Denmark), and supernatants containing products were transferred by centrifugation at 2,700 g for 10 min or by using a vacuum manifold. Absorbances were measured at the λ_{max} for each substrate colour (404 nm for yellow, 517 nm for red, 595 nm for blue and 630 nm for green) using a plate reader (SpectraMax M5, Molecular Devices, Sunnyvale, USA).

CPH assays in agar plates

Agar plates were cast containing 23 mM Britton-Robinson buffer (pH 6.0), 1% agar and 0.01% NaN$_3$. Holes were made in the agar with a metal cylinder borer (diameter 6 mm). CPH substrates (prepared as described above) were diluted (300 mg CPH substrate with 500 μL water) and homogenized using a sample disruptor (TissueLyser II, Qiagen AB, Sollentuna, Sweden) at 30 Hz for 15 min to form a fluid-like gel suspension. Sixty microliters of each substrate was transferred to each hole. The wells were topped with 0.75% agar. To perform the assay, 20 μL 10 U/mL of enzyme in appropriate buffer was added and the plate was incubated overnight at 30°C, after which halos were examined.

Analysis of culture broth from fungi

Two pathogenic fungi *C. acutatum* (isolate SA 0-1) and *P. expansum* (isolate IK2020) were cultivated following the protocol described in Vidal-Melgosa *et al.* 2015 [39]. Supernatants were collected every day, frozen in liquid nitrogen and stored at –80°C until use. To assay enzyme activities in the broths, 5 μL of broth was added to 100 mM sodium acetate buffer pH 4.5 to a final volume of 155 μL. This solution was then used to perform CPH substrate assays as describe above with an incubation time of 1 h and at room temperature. Four independent replicates of each sample were performed.

Mass spectrometry

For mass spectrometry analysis of products, 2 μL of a 9 mg/mL mixture of 2,5-dihydroxybenzoic acid (DHB) in 30% acetonitrile was applied to a MTP 384 ground steel target plate TF (Bruker Daltonics GmbH, Bremen, Germany). One-microliter sample was then mixed into the DHB droplet and dried under a stream of air. The samples were analysed with an Ultraflex2 MALDI-ToF/ToF instrument (Bruker Daltonics GmbH, Bremen, Germany) equipped with a Nitrogen 337-nm laser beam. The instrument was operated in positive acquisition mode and controlled by the FlexControl 3.3 software package. All spectra were obtained using the reflectron mode with an acceleration voltage of 25 kV, a reflector voltage of 26, and pulsed ion extraction of 40 ns in the positive ion mode. The acquisition range used was from *m/z* 300 to 3,000. The data was collected from averaging 400 laser shots, with the lowest laser energy necessary to obtain sufficient signal-to-noise ratios. Peak lists were generated from the MS spectra using Bruker FlexAnalysis software (Bruker Daltonics GmbH, Bremen, Germany) (Version 3.3).

Additional files

Additional file 1: Figure S1. Linear regression analysis of 2-colour combinations of CPH substrates. The graphs show linear regression analysis from spectra of two-colour combinations of degradation products (the supernatant) from CPH-xylan mixed in ratios from 0% to 100% for each. The deviation of the data points from the line shows deviation from linearity. The results of this analysis show that the true ratio can be determined using linear regression within a ±5% error margin.

Additional file 2: Figure S2. Reproducibility of the high-throughput assay using CPH substrates. (A) Four identical reaction plates were set up as shown. The substrates in the reaction plate are shown to the left, and the enzymes in the reaction plate are shown to the right. The substrates used were red CPH-arabinan, green CPH-xylan, yellow CPH-HE cellulose, blue CPH-β-glucan, red CPH-pectic galactan, green CPH-lichenan, red CPH-xyloglucan and yellow CPH-galactomannan. The enzymes used were arabinanase (*ara*), xylanase (*xyl1*), glucanase (*glc*), galactanase (*gal*), lichenanase (*lic*) xyloglucanase (*xg*) and mannanase (*man*). (B) The four product plates containing the products from the four separate reaction plates. (C) Graph showing the mean absorbances from the product plates (measured at λ_{max} for each substrate colour). See Tables 1 and 2 for details of the substrates and enzyme used. The reaction was performed for 30 min at room temperature in 100 mM sodium acetate buffer pH 4.5 for *ara*, *xyl1*, *glc* and *gal*, in 100 mM sodium phosphate buffer pH 7.0 for *lic* and *man* and in 100 mM sodium acetate buffer pH 5.5 for *xg*.

Additional file 3: Figure S3. pH and temperature stability of CPH substrates. (A) Graph showing pH responses for green CPH-HE cellulose and green CPH-rhamnogalacturonan treated with a cellulase (*cel2*) and a pectinase (*pec1*), respectively, to show the usage of the CPH substrates at a pH range between pH 3.0 to 10.0. The assay was performed with 1 U/mL enzyme concentration in 100 mM Britton-Robinson buffer at room temperature for 30 min. (B) Graph showing a temperature stability test. Green CPH-xylan was treated with a thermophilic xylanase (*xyl3*, 0.1 U/mL) in sodium acetate buffer pH 6.0 for 1 h from 20°C to 80°C. See Tables 1 and 2 for details of the substrates and enzyme used.

Additional file 4: Figure S4. Mass spectra of products from CPH-xylan digestion. (A) Mass spectrum of dyed CPH-xylan digested by xylanase *xyl1* where the oligomers are composed of *x* pentose units and *y* hydrolysed linker units (Pent$_x$linker$_y$). (B) Mass spectrum of undyed cross-linked xylan digested by xylanase *xyl1* where the oligomers are composed of *x* pentose units and *y* hydrolysed linker units (Pent$_x$linker$_y$). (C) Theoretical scheme showing the possible origins of the products observed in (A) and (B).

Additional file 5: Figure S5. Storage stability of CPH substrates. The storage stability of CPH substrates was tested by comparing the output from fresh and year-old reaction plates. (A) Images showing products from fresh and year-old reaction plates in which each well contained a mixture of the substrates shown to the right. The enzymes used are indicated at the top. (B) Another second example of an experiment comparing the output from fresh and year-old plates. Each well contained a mixture of the substrates shown to the right. The enzymes used are indicated at the top. See Tables 1 and 2 for details of the substrates and enzyme used.

Additional file 6: Figure S6. Using CPH substrates to analyse enzyme activity in crude fungal broths. The two pathogenic fungi *Penicillium expansum* and *Colletotrichum acutatum* were cultivated over 10 days, and enzyme activities in the culture broth were analysed using four different CPH substrates: (A) yellow CPH-galactomannan, (B) yellow CPH-lichenan, (C) yellow CPH-β-glucan from barley, (D) yellow CPH-arabinoxylan. As a negative control, the substrates were also treated with broth alone (grey lines). The reaction was incubated in 100 mM sodium acetate buffer pH 4.5 for 1 h at room temperature. See Table 2 for details of the substrates used.

Additional file 7: Figure S7. Insoluble chromogenic biomass substrates. A range of insoluble chromogenic biomass substrates were produced (all using a red-coloured dye) and treated with a range of enzymes (shown on x-axes). The absorbance values are means from three wells measured at 517 nm. Enzymes were used at 10 U/mL for 24 h at room temperature. The assay was performed using 100 mM sodium acetate buffer pH 4.5, except for the enzymes *ply1*, *ply2*, *ply3*, *pol1* and *xg* used at pH 5.5, for *cel2* pH 6.0, for *pec3* 100 mM sodium phosphate pH 7.0 and for *pec1* sodium carbonate pH 10.0. The substrates used were (A) tobacco, (B) *Arabidopsis*, (C) wheat straw, (D) bagasse, (E) hemp, (F) willow, (G) fenugreek seeds and (H) lupin seeds. See Table 2 for details of the enzyme used.

Additional file 8: Figure S8. MALDI-ToF MS spectra of released products during enzyme treatments of insoluble chromogenic biomass substrates. (A) *MALDI-ToF MS spectra* of the products released from hemp insoluble chromogenic biomass substrates treated with xylanase *xyl1*. (B) *MALDI-ToF MS spectra* of the products released from bagasse treated with xylanase *nz2*. MeGlcA$_x$Pent$_y$ represent oligosaccharides with x number of 4-methyl-glucuronic acid and y number of xylose residues; Pent$_x$ = pentose oligosaccharides (most likely (arabino-) xylooligosaccharides, arabinosylations limiting complete degradation by the endoxylanases *xyl1* and *nz2*) containing x number of pentoses; asterisk (*) represents sodium adducts, which commonly occur for uronic acids.

Abbreviations

AZCL: azurine cross-linked; CPH: chromogenic polysaccharide hydrogel; GAX: glucuronoarabinoxylan; GH: glycosyl hydrolase; ICB: insoluble chromogenic biomass; LPMO: lytic polysaccharide monooxygenase; MALDI-ToF-MS: matrix-assisted laser desorption/ionization-time-of-flight-mass spectrometry; SEM: standard error of mean.

Competing interests

The authors declare that they have no competing interests.

Authors' contributions

SKK synthesised the CPH and ICB substrates, performed synthetic optimisations required for their optimal physicochemical performance, purified them and performed some of the analytical experiments. JS designed the assay development for using CPH and ICB substrates in a 96-well plate format and developed the performance in agar plates, did the optimisation and validation of the assays and studied the stability of the substrates as well as the reproducibility of the assay performance. BW performed the mass spectrometry analysis and provided some of the raw biomass material and LGT performed the linear regression analysis. RNM provided enzymes *nz1* to *nz4* and provided reducing end assay activity information, VGHE participated in designing the study and assisted with writing the manuscript and WGTW wrote the manuscript and designed the study. All authors read and approved the final manuscript.

Acknowledgements

This work was funded by the Danish Council for Strategic Research (DSF, SET4Future project), by the Norwegian Research Council, grant numbers 214613 and 221568 and by the Seventh Framework Programme of the European Union (FP7 2007-2013) under Grant Agreements N263916 (WallTraC project) and N308363 (project Waste2Go). This paper reflects the authors' views only. The European Union is not liable for any use that may be made of the information contained herein. We thank Birgit Jensen and Daniel Buchvaldt Amby for providing the two fungi *Colletotrichum acutatum* (isolate SA 0-1) and *Penicillium expansum* (isolate IK2020) and their assistance in cultivation. A patent application related to the assays described in this study (patent application number PA 2013 70507) has been filed.

Author details

[1]Department of Plant and Environmental Sciences, Thorvaldsensvej 40, Frederiksberg C 1871, Denmark. [2]Department of Chemistry, Biotechnology and Food Science, Norwegian University of Life Sciences, Chr. M. Falsens vei 1., Aas 1432, Norway. [3]University of Copenhagen, Faculty of Science, Rolighedsvej 23, Frederiksberg C 1958, Denmark. [4]Novozymes A/S, Krogshoejvej 36, Bagsværd 2880, Denmark.

References

1. Tamayo-Ramos JA, Orejas M. Enhanced glycosyl hydrolase production in Aspergillus nidulans using transcription factor engineering approaches. Biotechnol Biofuels. 2014;7. doi:Artn 103
2. Li LL, McCorkle SR, Monchy S, Taghavi S, van der Lelie D. Bioprospecting metagenomes: glycosyl hydrolases for converting biomass. Biotechnol Biofuels. 2009;2. doi:Artn 10
3. Dimarogona M, Topakas E, Christakopoulos P. Recalcitrant polysaccharide degradation by novel oxidative biocatalysts. Appl Microbiol Biot. 2013;97(19):8455–65. doi:10.1007/s00253-013-5197-y.
4. Agger JW, Isaksen T, Varnai A, Vidal-Melgosa S, Willats WGT, Ludwig R, et al. Discovery of LPMO activity on hemicelluloses shows the importance of oxidative processes in plant cell wall degradation. Proc Natl Acad Sci U S A. 2014;111(17):6287–92. doi:10.1073/pnas.1323629111.
5. Sanchez OJ, Cardona CA. Trends in biotechnological production of fuel ethanol from different feedstocks. Bioresource Technol. 2008;99(13):5270–95. doi:10.1016/j.biortech.2007.11.013.
6. Sweeney MD, Xu F. Biomass converting enzymes as industrial biocatalysts for fuels and chemicals: recent developments. Catalysts. 2012;2(2):244–63. doi:10.3390/Catal2020244.
7. Sims REH, Mabee W, Saddler JN, Taylor M. An overview of second generation biofuel technologies. Bioresource Technol. 2010;101(6):1570–80. doi:10.1016/j.biortech.2009.11.046.
8. Cantarel BL, Coutinho PM, Rancurel C, Bernard T, Lombard V, Henrissat B. The carbohydrate-active enZymes database (CAZy): an expert resource for glycogenomics. Nucleic Acids Res. 2009;37(Database issue):D233–8. doi:10.1093/nar/gkn663.
9. Davies G, Henrissat B. Structures and mechanisms of glycosyl hydrolases. Structure. 1995;3(9):853–9. doi:10.1016/S0969-2126(01)00220-9.
10. Henrissat B. A classification of glycosyl hydrolases based on amino-acid-sequence similarities. Biochem J. 1991;280:309–16.
11. Henrissat B, Bairoch A. New families in the classification of glycosyl hydrolases based on amino-acid-sequence similarities. Biochem J. 1993;293:781–8.
12. Henrissat B, Bairoch A. Updating the sequence-based classification of glycosyl hydrolases. Biochem J. 1996;316:695–6.
13. Henrissat B, Davies G. Structural and sequence-based classification of glycoside hydrolases. Curr Opin Struc Biol. 1997;7(5):637–44. doi:10.1016/S0959-440x(97)80072-3.
14. Pena MJ, Tuomivaara ST, Urbanowicz BR, O'Neill MA, York WS. Methods for structural characterization of the products of cellulose-and xyloglucan-hydrolyzing enzymes. Cellulases. 2012;510:121–39. doi:10.1016/B978-0-12-415931-0.00007-0.
15. Nelson N. A photometric adaptation of the Somogyi method for the determination of glucose. J Biol Chem. 1944;153(2):375–80.
16. Miller GL. Use of dinitrosalicylic acid reagent for determination of reducing sugar. Anal Chem. 1959;31(3):426–8. doi:10.1021/Ac60147a030.

17. Lever M. Colorimetric and fluorometric carbohydrate determination with para-hydroxybenzoic acid hydrazide. Biochem Med Metab B. 1973;7(2):274–81. doi:10.1016/0006-2944(73)90083-5.

18. Anthon GE, Barrett DM. Determination of reducing sugars with 3-methyl-2-benzothiazolinonehydrazone. Anal Biochem. 2002;305(2):287–9.

19. Gusakov AV, Kondratyeva EG, Sinitsyn AP. Comparison of two methods for assaying reducing sugars in the determination of carbohydrase activities. Int J Anal Chem. 2011. doi:Artn 283658

20. Biely P, Mastihubova M, Tenkanen M, Eyzaguirre J, Li XL, Vrsanska M. Action of xylan deacetylating enzymes on monoacetyl derivatives of 4-nitrophenyl glycosides of beta-D-xylopyranose and alpha-L-arabinofuranose. J Biotechnol. 2011;151(1):137–42. doi:10.1016/j.jbiotec.2010.10.074.

21. Ferrari AR, Gaber Y, Fraaije MW. A fast, sensitive and easy colorimetric assay for chitinase and cellulase activity detection. Biotechnol Biofuels. 2014;7. doi:Artn 37 doi:10.1186/1754-6834-7-37.

22. Li LL, Taghavi S, McCorkle SM, Zhang YB, Blewitt MG, Brunecky R et al. Bioprospecting metagenomics of decaying wood: mining for new glycoside hydrolases. Biotechnol Biofuels. 2011;4. doi:Artn 23 doi:10.1186/1754-6834-4-23

23. Nyyssonen M, Tran HM, Karaoz U, Weihe C, Hadi MZ, Martiny JBH et al. Coupled high-throughput functional screening and next generation sequencing for identification of plant polymer decomposing enzymes in metagenomic libraries. Front Microbiol. 2013;4. doi:Artn 282 doi:10.3389/Fmicb.2013.00282

24. Leemhuis H, Kragh KM, Dijkstra BW, Dijkhuizen L. Engineering cyclodextrin glycosyltransferase into a starch hydrolase with a high exo-specificity. J Biotechnol. 2003;103(3):203–12. doi:10.1016/S0168-1656(03)00126-3.

25. Zantinge JL, Huang HC, Cheng KJ. Microplate diffusion assay for screening of beta-glucanase-producing microorganisms. Biotechniques. 2002;33(4):798.

26. Ceska M. A new type of reagent for the detection of molecular varieties of some hydrolytic enzymes: detection of -amylase isoenzymes. Biochem J. 1971;121(3):575–6.

27. Huang JS, Tang J. Sensitive assay for cellulase and dextranase. Anal Biochem. 1976;73(2):369–77.

28. Leisola M, Linko M. Determination of the solubilizing activity of a cellulase complex with dyed substrates. Anal Biochem. 1976;70(2):592–9.

29. Kamaryt J, Zemek J, Kuniak L. Determination of the molecular weight of alpha-amylase, as the enzyme-inhibitor complex, using thin layer gel filtration on Sephadex. J Clin Chem Clin Biochem Zeitschrift fur klinische Chemie und klinische Biochemie. 1982;20(6):451–5.

30. Ten LN, Im WT, Kim MK, Kang MS, Lee ST. Development of a plate technique for screening of polysaccharide-degrading microorganisms by using a mixture of insoluble chromogenic substrates. J Microbiol Meth. 2004;56(3):375–82. doi:10.1016/j.mimet.2003.11.008.

31. Ten LN, Im WT, Kim MK, Lee ST. A plate assay for simultaneous screening of polysaccharide- and protein-degrading micro-organisms. Lett Appl Microbiol. 2005;40(2):92–8. doi:10.1111/j.1472-765X.2004.01637.x.

32. Lippa KA, Roberts AL. Nucleophilic aromatic substitution reactions of chloroazines with bisulfide (HS-) and polysulfides (S(n)2(-)). Environ Sci Technol. 2002;36(9):2008–18. doi:10.1021/Es011255v.

33. Biswal AK, Soeno K, Gandla ML, Immerzeel P, Pattathil S, Lucenius J et al. Aspen pectate lyase PtxtPL1-27 mobilizes matrix polysaccharides from woody tissues and improves saccharification yield. Biotechnol Biofuels. 2014;7. doi:Artn 11 doi:10.1186/1754-6834-7-11

34. Van Acker R, Vanholme R, Storme V, Mortimer JC, Dupree P, Boerjan W. Lignin biosynthesis perturbations affect secondary cell wall composition and saccharification yield in Arabidopsis thaliana. Biotechnol Biofuels. 2013;6. doi:Artn 46 doi:10.1186/1754-6834-6-46

35. Darvill JE, Mcneil M, Darvill AG, Albersheim P. Structure of plant-cell walls.11. Glucuronoarabinoxylan, a 2nd hemicellulose in the primary-cell walls of suspension-cultured sycamore cells. Plant Physiol. 1980;66(6):1135–9. doi:10.1104/Pp.66.6.1135.

36. Degroot B, Vandam JEG, Vantriet K. Alkaline pulping of hemp woody core - kinetic modeling of lignin, xylan and cellulose extraction and degradation. Holzforschung. 1995;49(4):332–42. doi:10.1515/hfsg.1995.49.4.332.

37. Smith DC, Forsberg CW. Alpha-glucuronidase and other hemicellulase activities of fibrobacter-succinogenes-S85 grown on crystalline cellulose or ball-milled barley straw. Appl Environ Microb. 1991;57(12):3552–7.

38. Biely P, Cziszarova M, Uhliarikova I, Agger JW, Li XL, Eijsink VGH, et al. Mode of action of acetylxylan esterases on acetyl glucuronoxylan and acetylated oligosaccharides generated by a GH10 endoxylanase. Bba-Gen Subjects. 2013;1830(11):5075–86. doi:10.1016/j.bbagen.2013.07.018.

39. Vidal-Melgosa S, Pedersen HL, Schückel J, Arnal G, Dumon C, Amby DB et al. A new versatile microarray-based method for high-throughput screening of carbohydrate-active enzymes. The Journal of biological chemistry. 2015. doi:10.1074/jbc.M114.630673.

10

Deconstructing the genetic basis of spent sulphite liquor tolerance using deep sequencing of genome-shuffled yeast

Dominic Pinel[1,3], David Colatriano[1], Heng Jiang[1,4], Hung Lee[2] and Vincent JJ Martin[1]*

Abstract

Background: Identifying the genetic basis of complex microbial phenotypes is currently a major barrier to our understanding of multigenic traits and our ability to rationally design biocatalysts with highly specific attributes for the biotechnology industry. Here, we demonstrate that strain evolution by meiotic recombination-based genome shuffling coupled with deep sequencing can be used to deconstruct complex phenotypes and explore the nature of multigenic traits, while providing concrete targets for strain development.

Results: We determined genomic variations found within *Saccharomyces cerevisiae* previously evolved in our laboratory by genome shuffling for tolerance to spent sulphite liquor. The representation of these variations was backtracked through parental mutant pools and cross-referenced with RNA-seq gene expression analysis to elucidate the importance of single mutations and key biological processes that play a role in our trait of interest. Our findings pinpoint novel genes and biological determinants of lignocellulosic hydrolysate inhibitor tolerance in yeast. These include the following: protein homeostasis constituents, including Ubp7p and Art5p, related to ubiquitin-mediated proteolysis; stress response transcriptional repressor, Nrg1p; and NADPH-dependent glutamate dehydrogenase, Gdh1p. Reverse engineering a prominent mutation in ubiquitin-specific protease gene *UBP7* in a laboratory *S. cerevisiae* strain effectively increased spent sulphite liquor tolerance.

Conclusions: This study advances understanding of yeast tolerance mechanisms to inhibitory substrates and biocatalyst design for a biomass-to-biofuel/biochemical industry, while providing insights into the process of mutation accumulation that occurs during genome shuffling.

Keywords: Evolutionary engineering, Genome shuffling, Reverse engineering, Complex trait, Tolerance, Yeast

Background

Mapping genotype to phenotype for complex traits and using these data for the rational design of biocatalysts is a natural progression in an increasingly sophisticated biotechnology industry. Unfortunately, current technologies do not allow for the rapid creation of industrially relevant microorganisms or the ability to access and understand multigenic phenotypic traits. Traditionally, strain improvement has been based on a repetitive cycle of random mutagenesis and selection to improve the phenotypic traits of industrial microbes [1]. Advanced

DNA sequencing technology now allows for rapid sequencing of the genomes of these industrial strains to identify the mutations that confer improved phenotypes. However, in resequencing the genomes of randomly evolved strains, a small number of potentially productive mutations are often accompanied by a background of non-productive mutations [2-4]. Extensive functional characterizations of individual genotypic variations are therefore needed to unravel which mutations are associated with the phenotype of interest. Furthermore, our ability to deconstruct complex, multigenic traits is still limited. Possible solutions to these problems include sequencing pools of independent mutants [5], backcrossing non-productive mutations prior to genome resequencing, or combining intercrossing with pool sequencing to assign quantitative trait loci [6] in order to hone

* Correspondence: vincent.martin@concordia.ca
[1]Department of Biology, Centre for Structural and Functional Genomics, Concordia University, 7141 Sherbrooke Street West, Montréal, Québec H4B 1R6, Canada
Full list of author information is available at the end of the article

in on productive mutations. Nonetheless, resolving such data into manageable and testable hypotheses can be insurmountably challenging.

In this study, we aimed to mitigate these shortcomings by sequencing a strain created by genome shuffling from a known background strain and tracking mutations throughout the evolving population. This allows for important mutations to be ranked and novel gene targets to be acquired from background mutations. Moreover, genome shuffling (GS) is an alternative to classical strain improvement that is a means to accelerate the evolution of industrial strains in the laboratory and minimize the accumulation of non-productive mutations. The rationale behind GS is to rapidly combine beneficial mutations and cross out deleterious ones, which can be achieved in *Saccharomyces cerevisiae* by recursive pool-wise mating of mutant populations (Figure 1A) [7-10]. This strain engineering technique is particularly powerful to address multigenic, complex phenotypes such as resistance to ethanol, lactic acid, heat and low pH or production of compounds like tylosin or taxol (reviewed in [11]). Theoretically, the background of non-productive or deleterious mutations can be minimized by attenuating mutagen dosage, screening for parental strains that contain productive mutations, followed by trait-enhanced mutant strain recombination to combine mainly productive mutations into a single strain. Furthermore, by its very nature, GS brings interacting mutations together into single strains. Although the utility of GS has been demonstrated repeatedly through phenotypic observation, the nature of the mutations accumulated during the strain evolution has not been tracked through genome resequencing. Sequencing GS isolates, therefore, should yield access to determinants of multigenic traits at single nucleotide resolution, while minimizing non-productive variation discovery. Tracking mutations throughout the population of genome-shuffled strains can then be used to further increase the possibility of finding productive mutations.

Microbial tolerance to lignocellulosic hydrolysates is a complex, multigenic trait that is of significant importance to a biomass-to-fuel/chemical industry. The pretreatment of lignocellulose to fermentable sugars yields many by-products that are inhibitory to fermenting yeasts. The main sources of inhibition come from osmotic pressure, reactive oxygen species (ROS) damage or compounds that include furan aldehydes, primarily furfural and 5-(hydroxymethyl)-2-furaldehyde (HMF), phenolics and organic acids, especially acetic, formic and levulinic acids [12-16]. The biological factors implicated in the tolerance of yeast to lignocellulose fermentation inhibitors have been reviewed [12,13,17]. Ultimately, engineering productive industrial biocatalysts with tolerance traits will be a pervasive biotechnological problem, and rationally engineering these traits will require an understanding of interacting genes and biological processes that affect tolerance. Currently, a lack of knowledge on the multiple cellular processes and genes involved in microbial tolerance to lignocellulosic hydrolysates makes rational engineering of strains resistant to these substrates implausible [8,18,19].

In a previous study [8], we evolved a strain of *S. cerevisiae*, R57, through genome shuffling (Figure 1A,B) that is capable of survival, growth and ethanol productivity in hardwood spent sulphite liquor (HWSSL), a highly inhibitory lignocellulosic substrate generated by the acid bisulphite pulping process [20,21], to levels of tolerance previously unreported. HWSSL contains sugars, lignosulphonates, inhibitory compounds, residual pulping chemicals, ammonia and sulphite [22] and can contain heavy metal ions (iron, chromium, nickel and copper) that originate from the corrosion of pulping and bleaching equipment [23,24]. Some of the major constituents are approximated at the following levels (% w/v): 0.83% to 1.45% hexose sugars, 1.7% to 2.1% pentose sugars, 0.18% to 0.5% furfural, 0.9% to 1.0% acetic acid, 0.5% to 0.7% sulphate, 1% ammonia and 17% lignosulphonate [8,25]. In evolving R57 through GS, it was hypothesized that beneficial, tolerance-conferring mutations were combined through recursive population-wise meiotic recombination, which yielded a progression towards higher tolerance of the hydrolysate displayed with each subsequent round of GS (Figure 1) [8]. Strain R57 showed improved cross-tolerance to several known inhibitors of lignocellulosic hydrolysates [8], supporting the theory that R57 harbours multiple tolerance-conferring mutations. In this study, we resequenced the genome of R57 in order to discover the mutations that were accumulated to confer hydrolysate tolerance, combined with profiling the relative abundance of R57 mutations in the heterogeneous parental GS populations to probe for the relative phenotypic effect of each discovered mutation and explore mutation recombination in GS-evolved pools.

This study describes a relatively cost-effective way to explore combinatorial space of productive mutations within a single genome. To our knowledge, this is the first study to resequence the genome of a strain evolved through GS and the first resequencing project for a yeast strain specifically evolved for its tolerance to a complex mixture of inhibitors in lignocellulosic substrates.

Results and discussion
Genome sequencing of GS-evolved strain R57
The genome of *S. cerevisiae* strain R57 was resequenced in an effort to pinpoint genetic changes associated with its tolerance to HWSSL. Both the parental haploid CEN.PK113-7D and mutant diploid R57 were sequenced and compared at approximately 100-fold and approximately 350-fold coverage per nucleotide (Additional file 1: Data S1),

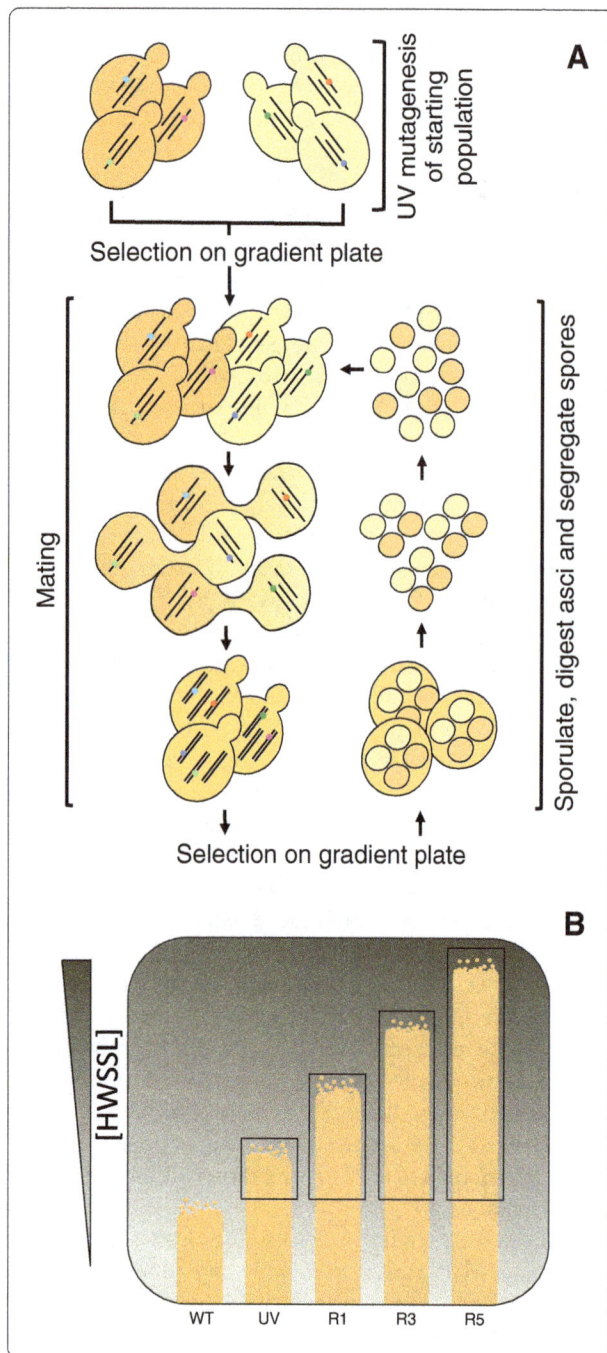

Figure 1 Meiotic recombination-mediated genome shuffling by recursive breeding for HWSSL tolerance. (A) A recursive mating methodology was used to create the HWSSL strains and populations used in this study. Large pools of UV mutants and genome-shuffled populations were screened on HWSSL gradient agar plates prior to each round of shuffling. **(B)** Portions of each population that showed more tolerance than the reference (WT) (black boxes) were scraped from gradient plates and used for genome shuffling (different rounds of genome shuffling are depicted - round 1 (R1), round 3 (R3) and round 5 (R5)). Initial UV mutant populations (UV) of each haploid mating type showing enhanced HWSSL tolerance were scraped and used to begin the recursive breeding methodology. Selection on HWSSL gradient plates was carried out between each round of GS in order to enrich the mating pool for strains showing the tolerance phenotype. A portion of each mutant pool (UV through five rounds of GS) was frozen for population sequencing (see 'Results and discussion'). Individual colonies showing the highest tolerance to HWSSL were isolated from the frontier of growth. HWSSL, hardwood spent sulphite liquor.

respectively, which allows for meaningful mutation prediction [2]. The relative level of sequence read coverage per chromosome between the strains is similar and suggests the absence of aneuploidy (Additional file 1: Data S1). Insertion/deletion (indel) and copy number variation (CNV) analysis returned no detectable differences between the wild type (WT) and R57 after visual inspection. All of the mutations discovered from the mutation analysis are single nucleotide polymorphisms (SNPs) and were confirmed by Sanger sequencing.

Twenty-one point mutations were found that could affect at least 17 genes, based on location within open reading frames (ORFs) or untranslated regions (UTRs). These include 16 SNPs affecting 12 ORFs, 14 of which lead to missense mutations, with the remaining 2 leading to silent mutations (Table 1). The five mutations not found within ORFs are all located in 5′ or 3′ UTRs. A heterogeneous SNP lies 43 bps 3′ of *BCS1* and is predicted to be part of the 5′ UTR of *YDR374C* [26] and in the 3′ UTR of *WIP1* [27]. This mutation is not included in subsequent analyses due to the ambiguity of the affected gene. Gene ontology categories and interaction maps for the affected genes were generated (Additional file 2: Methods and Additional file 3: Figures S1 and S2), and the mutation analysis results are summarized in Table 1.

Predicting protein function and phenotypic effect for missense mutations in R57

Functional prediction of altered primary protein structure was carried out using the SIFT (Sorting Intolerant From Tolerant) algorithm [28-30] (Table 1). Ten of the 14 missense mutations are expected to affect protein function, with 4 predicted as tolerated by the protein. All but two of the mutations are heterozygous and therefore are expected to have a dominant effect if they contribute to the R57 phenotype. The two homozygous mutations are

Table 1 Point mutations discovered in GS-evolved strain R57

Gene	Chr	Mutation	Gene function	Genotype	SIFT score
NRG1	IV	137C > A (P46Q)	Transcriptional repressor, stress tolerance	Homo	0
UBP7[a]	IX	2466 T > A (N822K)	Ubiquitin-specific protease	Homo	0.33
ART5[a]	VII	454C > A (L152I)	Regulates endocytosis and turnover of cell-surface proteins by targeted ubiquitination	Hetero	0.17
SSA1[a]	I	91C > A (Q31K)	ATP-ase, protein folding, heat shock, HSP70	Hetero	0
GDH1[b]	XV	47C > T (S16F)	Glutamate synthesis from ammonia	Hetero	0
GDH1[b]		68 T > G (F23C)		Hetero	0
ARO1[b]	IV	1283C > T (S428F)	Catalyzes biosynthesis of chorismate leading to aromatic amino acids	Hetero	0
ARO1[b]		1284C > T (Silent)		Hetero	-
STE5[b]	IV	512C > T (S171F)	Pheromone-response scaffold protein, forms MAPK cascade complex	Hetero	0
STE5[b]		2649 T > C (Silent)		Hetero	-
MAL11[b]	VII	310C > T (P104S)	Alpha-glucoside symporter, with high affinity for trehalose	Hetero	0
MAL11[b]		482 T > A (M161K)		Hetero	0.02
GSH1[c]	X	T > A (73 bp 5′ UTR)	Glutamylcysteine synthetase, glutathione biosynthesis	Hetero	-
PBP1[c]	VII	T > C (191 bp 5′ UTR)	Controls mRNA poly(A), stress granule formation and translation control	Hetero	-
FIT3[c]	XV	C > T (42 bp 3′ UTR)	Iron transport	Hetero	-
NOP58[c]	XV	A > T (25 bp 3′ UTR)	Pre-rRNA processing and rRNA synthesis	Hetero	-
YNL058C[d]	XIV	7A > G (K3E)	Unknown function	Hetero	0.42
DOP1[d]	IV	40A > T (N14Y)	Endosome to Golgi transport, ER organization, cell polarity and morphogenesis	Hetero	0.05
TOF2[d]	XI	2141 C > T (S714L)	rDNA silencing, stimulates Cdc14p for mitotic rDNA separation	Hetero	0.27
SGO1[d]	XV	575C > A (S192Y)	Chromosomal segregation and stability	Hetero	0.03

Bold font represents alleles that gain in frequency over GS evolution. [a]Gene group containing genes that are related to protein homeostasis. [b]Gene group containing genes bearing more than one mutation in R57. [c]Gene group containing UTR mutations. [d]Gene group containing alleles with limited evidence for a phenotypic linkage to HWSSL tolerance.

located within *NRG1* and *UBP7*. Homozygous mutations at these loci suggest that they have been enriched at a high enough population density during GS evolution that they were able to mate with the opposite mating type and may therefore be important to the HWSSL tolerance trait or that spores were insufficiently segregated during GS (Figure 1A), promoting homozygosity.

Several genes contain multiple mutations in R57. Genes bearing more than one mutation in R57 suggest that the affected gene is important to the phenotype and the mutations have accumulated due to GS evolution and screening. Two missense mutations affect both *GDH1* and *MAL11*. Gdh1p bears two mutations in R57, S16F and F23C. Visual inspection of the mapping alignment shows that the two mutations are never located on the same sequence read, and therefore, both versions of *GDH1* are mutated in R57. The close proximity of these two mutations suggests that a mutation in this region of *GDH1* may yield a phenotypic trait that has been selected for through GS evolution. The R57 *MAL11* gene contains two missense mutations (leading to P104S and M161K). Cloning and sequencing of *MAL11* from R57 shows that the two mutations are not located on the same allele. *STE5* and *ARO1* each bear a second silent mutation that

may be present due to close genetic linkage to the productive mutation or lead to a non-obvious phenotypic modification, such as altered mRNA stability, and thereby arise in R57 through selection.

Analysis of mutation loci in GS-evolved heterogeneous populations

We deeply sequenced the R57 mutation loci within the pooled mutant populations generated during the genome shuffling experiments [8]. The mutation loci were PCR amplified from DNA extracted from the heterogeneous populations and sequenced. There were a total of 3.76×10^6 reads with an average read length of 104 bp and an average fold coverage of 1.35×10^5 reads per nucleotide sequenced. Eighteen SNPs were called (Figure 2). We assessed the frequency of each sequence read that contains a mutation within the total number of sequence reads spanning each locus (Figure 2). These data were used to predict mutations that may have arisen in single parental strains, due to similar prevalence within a population, or may be epistatic, and to probe for changes in SNP frequency between populations that could indicate the relative influence of those alleles on phenotype. Several mutant alleles increase in frequency over GS evolution, and we

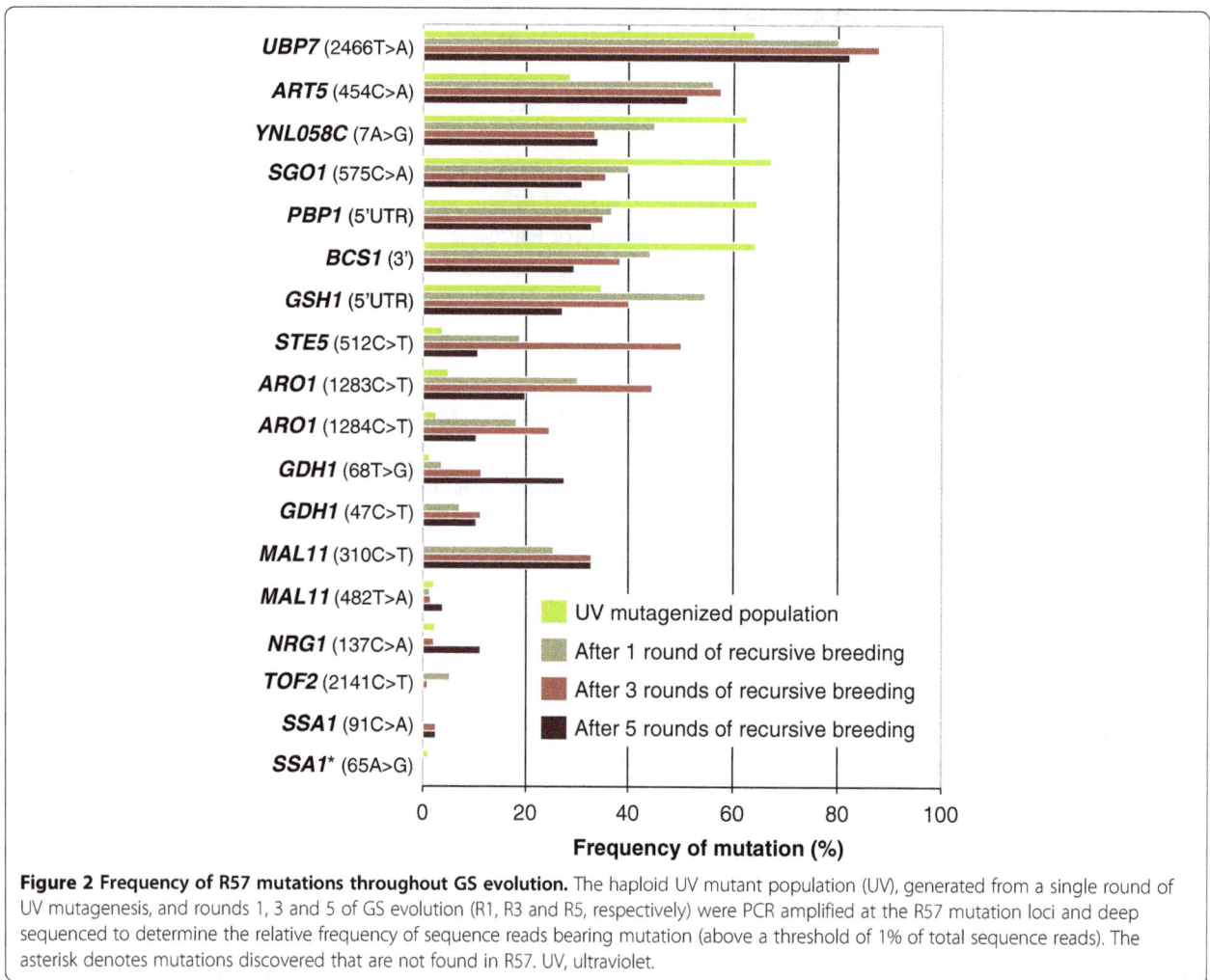

Figure 2 Frequency of R57 mutations throughout GS evolution. The haploid UV mutant population (UV), generated from a single round of UV mutagenesis, and rounds 1, 3 and 5 of GS evolution (R1, R3 and R5, respectively) were PCR amplified at the R57 mutation loci and deep sequenced to determine the relative frequency of sequence reads bearing mutation (above a threshold of 1% of total sequence reads). The asterisk denotes mutations discovered that are not found in R57. UV, ultraviolet.

hypothesize that this increase is due to a beneficial phenotypic effect that is generated through GS evolution and repeated selection. These include the *ART5*, *UBP7*, *SSA1*, *STE5*, *NRG1*, *MAL11* (leading to Mal11p - P104S) and both *GDH1* mutations (Figure 2). This analysis also yields an additional SNP that falls within the area of PCR amplification, affecting *SSA1* 26 bp upstream of the R57 mutation and leading to a D22G missense mutation, which further supports a determinant role for *SSA1* in the observed phenotype. The *NOP58*, *STE5* (silent), *DOP1*, *FIT3* 3′ UTR mutations were not located at a high enough density within the sequenced mutant populations to surpass our detection threshold.

Five mutations (located upstream of *BCS1* and *PBP1* and in the ORFs of *YNL058C*, *SGO1* and *UBP7*) are represented at approximately 60% frequency in the UV mutant population. Due to the virtually identical representation of these five mutations within the UV mutant population, we hypothesized that they may arise from a single, highly tolerant haploid strain. The mutated loci identified in R57 were sequenced from seven random, discernible colonies

selected from the frontier of growth for the UV mutant populations on HWSSL gradient agar plates (Figure 1B). Each of the single colonies contained the *BCS1*, *PBP1*, *YNL058C*, *SGO1* and *UBP7* mutations, corroborating our hypothesis that a single mutant strain present after UV mutagenesis was likely enriched to represent a large portion of that population. Assuming this mutant strain harboured at least one particularly productive mutation, propagation of these alleles to strain R57 may be a likely outcome. Heterogeneous population sequencing shows that of these five mutations, only the *UBP7* mutation increased in frequency through GS evolution while the other four decreased in frequency (Figure 2). We hypothesize that this finding is indicative of a set of four non-productive or less important mutations found with the productive *UBP7* mutation within a single genome, and when meiotic recombination occurs, linkage to the productive mutation diminishes until it reaches a steady state within the evolving population.

The mutation in the 5′ UTR of *GSH1* was identified in approximately 30% of sequence reads generated for this locus (Figure 2) in the UV population and was also identified in each of the seven UV mutant isolates that contain BCS1, PBP1, YNL058C, SGO1 and UBP7 mutations. It is therefore likely that all six mutations arose in a single mutant strain. The reason for the discrepancy in allele frequency between GSH1 and the other five mutations from UV mutant population sequencing is unknown but may be a population sequencing artefact. Unlike BCS1, PBP1, YNL058C and SGO1, the 5′ UTR GSH1 mutation increases in frequency in the first three rounds of evolution (Figure 2) and therefore, as with UBP7, more likely contributes to the tolerance phenotype than the other four mutations.

Several mutations that comprise a large part of the first three rounds of GS, and are therefore likely playing a determinant role in HWSSL tolerance, decrease in population frequency in the fifth round of GS (UBP7, ART5, ARO1). We hypothesize this occurs due to competition from strains bearing mutations that were rare in the initial mutant pools (that is, NRG1 and GDH1) or strains harbouring rare recombination events of multiple mutations that have resulted in augmented fitness.

Analysis of R57 SNPs in isolates of GS round 5 heterogeneous population

To identify possible combinations or permutations of mutations enriched through evolution, 20 strains isolated from the growth frontier of the fifth round of recursive GS (Figure 1B lane R5) were sequenced via the Sanger method at each of the mutation loci. All of the strains show heterozygosity in at least one of the mutation loci. The results show a heterogeneous population of mutations that are found together in one strain (Figure 3). Of these isolates, ≥70% contain at least one mutation in UBP7, ART5 and either of the GDH1 mutation loci, which further supports determinant roles for these genes on the phenotype. Several of the strains contain very few of the R57 mutations, which may indicate that not all R57 mutations are needed for HWSSL tolerance and likely that other unidentified mutations present within the round 5 population contribute to the tolerance trait. The percentages of mutated alleles within mutant populations as enumerated by population sequencing and single colony sequencing are similar (Figure 4); discrepancies in these frequencies may be due to the relatively large difference in sample size. General trends as to allele frequency are easily apparent and support our ability to generalize allele frequency within GS mutant populations by sequencing PCR-amplified mutation loci. The GS population sequencing data support determinant roles in HWSSL tolerance for a large portion of the mutated R57 genes based on their pervasiveness throughout the evolving populations. However, the most highly enriched mutations that increase over the strain evolution are of particular interest for reverse engineering studies. These include UBP7, GDH1, ART5, ARO1, STE5 and MAL11.

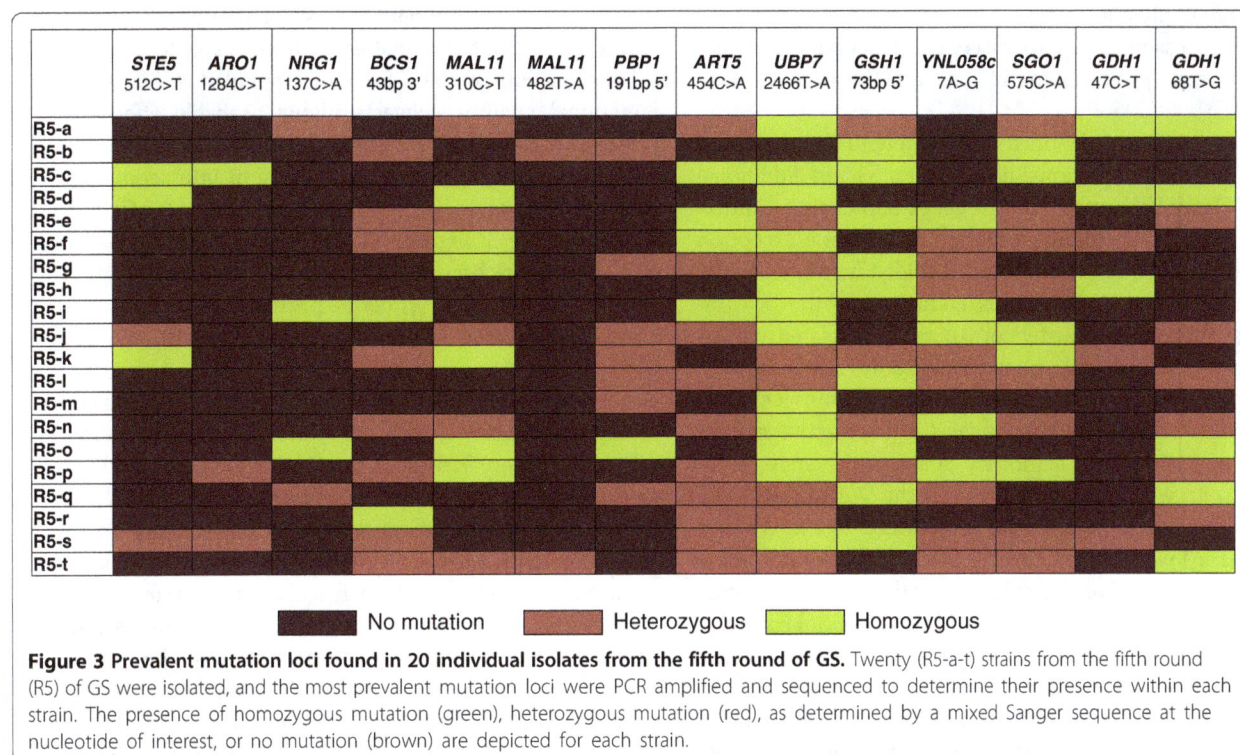

	STE5 512C>T	ARO1 1284C>T	NRG1 137C>A	BCS1 43bp 3′	MAL11 310C>T	MAL11 482T>A	PBP1 191bp 5′	ART5 454C>A	UBP7 2466T>A	GSH1 73bp 5′	YNL058c 7A>G	SGO1 575C>A	GDH1 47C>T	GDH1 68T>G
R5-a														
R5-b														
R5-c														
R5-d														
R5-e														
R5-f														
R5-g														
R5-h														
R5-i														
R5-j														
R5-k														
R5-l														
R5-m														
R5-n														
R5-o														
R5-p														
R5-q														
R5-r														
R5-s														
R5-t														

■ No mutation   ■ Heterozygous   ■ Homozygous

Figure 3 Prevalent mutation loci found in 20 individual isolates from the fifth round of GS. Twenty (R5-a-t) strains from the fifth round (R5) of GS were isolated, and the most prevalent mutation loci were PCR amplified and sequenced to determine their presence within each strain. The presence of homozygous mutation (green), heterozygous mutation (red), as determined by a mixed Sanger sequence at the nucleotide of interest, or no mutation (brown) are depicted for each strain.

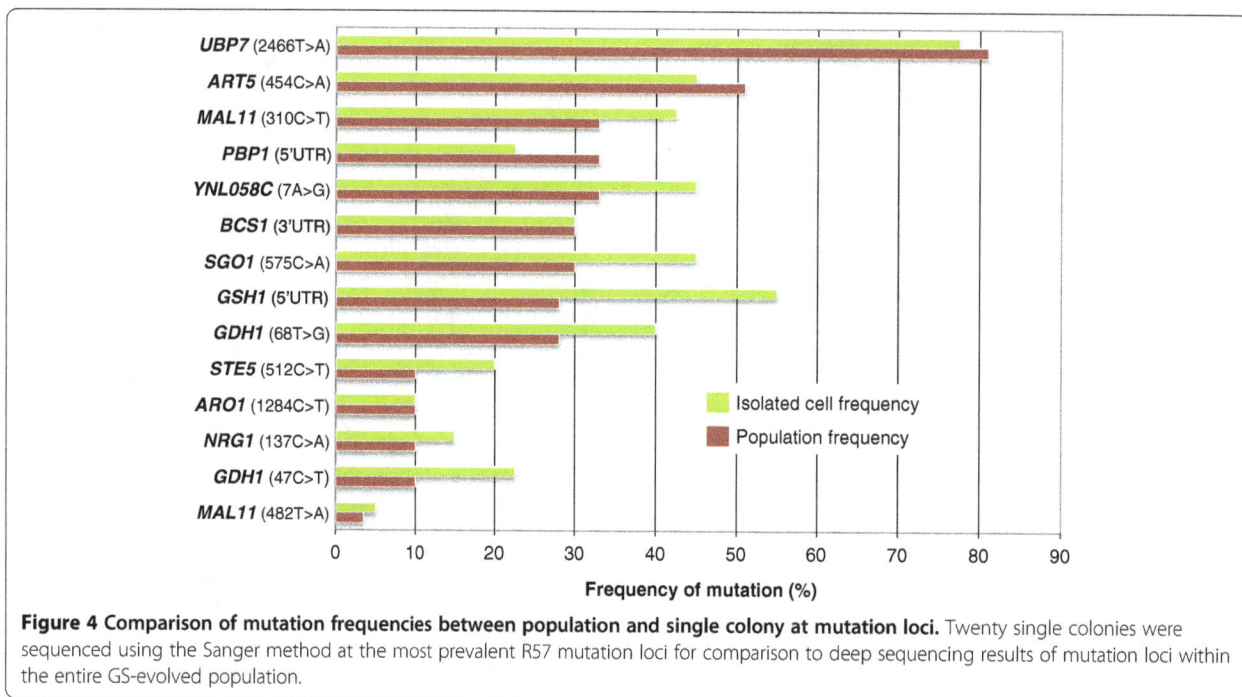

Figure 4 Comparison of mutation frequencies between population and single colony at mutation loci. Twenty single colonies were sequenced using the Sanger method at the most prevalent R57 mutation loci for comparison to deep sequencing results of mutation loci within the entire GS-evolved population.

RNA-seq gene expression analysis

To measure the impact of the mutations on gene expression in strain R57 and to probe for biological processes related to HWSSL tolerance, the gene transcription profile of strain R57 was compared to the WT diploid under control conditions (growth in defined medium, see 'Materials and methods') (Additional file 4: Data S2). Functional clustering was performed on the differentially expressed genes to discover enriched functional roles of gene products and biological pathways of interest (Figure 5).

These analyses identified 149 differentially expressed genes (>2-fold) (Additional file 3: Figure S3 and Additional file 4: Data S2). None of the 16 genes harbouring a mutation (Table 1) are found in this group with the exception of NRG1, which is upregulated 3-fold. Clustering of the 131 upregulated R57 genes as compared to the WT includes the major cluster of translation-related genes, mainly associated with ribosome biogenesis and translation regulation and 15 genes related to monosaccharide metabolism. These findings suggest a more active metabolism of R57 in early stationary phase, which may be related to growth differences between the WT and R57 [8]. Indeed, R57 displays a similar growth rate to the WT but reaches a lower optical density (OD) at stationary phase under non-inhibitory conditions with residual glucose remaining in the R57 medium (Additional file 3: Figure S4). The remaining upregulated enrichment clusters include genes related to cell wall organization, the cell membrane, ubiquitin-like (UBL) conjugation and organic acid synthesis pathways. Only 18 genes were downregulated under

non-inhibitory conditions, resulting in 3 clusters of genes that are highly enriched (Figure 5). These include genes related to NADH/alcohol metabolism and metal-ion metabolism or are associated with cellular membrane transport or lipid metabolism.

UBP7 and the ubiquitin-mediated protein homeostasis machinery are determinants of HWSSL tolerance

UBP7 bears a mutation that gains in frequency to represent a large portion of the GS-evolved population (Figure 2). Functional analysis suggests highly probable effects on protein structure and function due to this mutation, and UBP7 is known to have a high degree of interaction with mutated R57 genes (Table 1, Additional file 2: Methods, Additional file 3: Figures S1 and S2). The RNA-seq data also shows enrichment clustering for increased expression of genes that encode proteins related to the ubiquitin-mediated proteolytic machinery (UBL conjugation) (Figure 5). However, the genes within these UBL-conjugated gene clusters are associated with diverse biological processes, many of which do not play a direct role in the ubiquitination of proteins or ubiquitin-mediated proteolysis. Two notable exceptions, UBI2 and UBI3, which show increased expression in R57 relative to the WT (2.7- and 2.8-fold, respectively; Additional file 4: Data S2), encode ubiquitin fused to ribosomal proteins [31] and are responsible for generating ubiquitin as a fusion protein that is then cleaved to yield free ubiquitin by deubiquitinases. Most of the UBL conjugation cluster genes encode proteins that are regulated by this mechanism. Many of these genes are stress-tolerance related

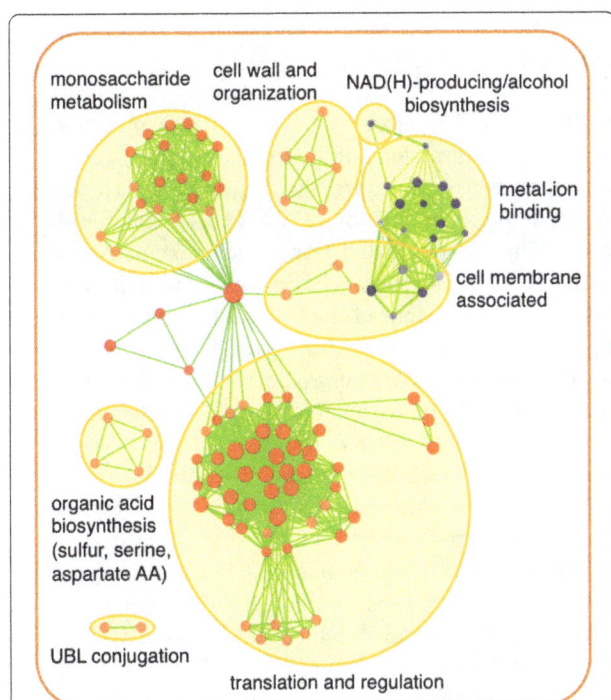

Figure 5 Enrichment clustering of differentially expressed genes from RNA-seq analysis. WT and R57 were compared for constitutive differential expression of genes when grown in SD medium. Gene lists were compiled for significantly ($P < 0.05$) upregulated >2-fold or significantly downregulated <2-fold differentially expressed genes. Red colours represent upregulation, while blue colours represent downregulation. Darker shades represent a relatively higher confidence of enrichment score. Larger node sizes represent relatively larger numbers of differentially regulated genes associated with the given ontology category as compared to the full gene. Smaller distance between nodes denotes a higher degree of relationship between ontology categories, while thicker edge lines (green) denote a relatively higher degree of similarity between category nodes in terms of the degree of overlap between the specific gene sets they are associated with. AA, amino acid; UBL, ubiquitin-like.

(*RHR2*, *PUN1*, *ENA5*, *PDR5*, *PDR12*). This suggests that stress tolerance genes showing increased expression due to HWSSL exposure may be also differentially controlled at the protein level by a modified ubiquitination machinery. Altogether, a significant portion of the HWSSL tolerance trait shown by R57 seems to be a direct result of changes in ubiquitin-mediated proteolytic pathways.

Protein damage and aggregation are likely a source of toxicity in cells exposed to lignocellulosic hydrolysates and have at least been partially shown to arise due to ROS damage from furan aldehyde exposure [32]. Cells regulate protein quality through destruction of misfolded or damaged polypeptides largely through selective, energy-dependent labelling with ubiquitin leading to digestion by the 26S proteasome complex [33]. *UBP7* encodes a ubiquitin-specific protease that cleaves ubiquitin-protein fusions [34], and as such, it is part of this ubiquitin-induced

signalling machinery of the cell [35-37]. The cell's requirement for available ubiquitin increases during stress exposure [38]. Deubiquitinating enzymes act to recover ubiquitin from ubiquitin-protein conjugates and may therefore have a direct bearing on cellular protein and ubiquitin homeostasis [37]. It has already been shown that mutations within a deubiquitinase enzyme, *UBP6*, can dramatically change steady-state ubiquitin levels within a cell [39], which is known to affect tolerance to a variety of stressors [40-42] and yeast prion toxicity [42]. Furthermore, upregulation of *UBP13*, another yeast deubiquitinating enzyme, is beneficial to cells under cold stress and suggests that altering ubiquitin-induced signalling may be a viable path towards other forms of stress tolerance [43].

In order to test the role of the *UBP7* 2466 T > A mutation in hydrolysate tolerance, we replaced both WT copies of *UBP7* with this gene variant in a diploid WT CEN.PK background. The homozygous *UBP7* 2466 T > A strain was able to colonize a higher concentration of HWSSL on gradient agar plate screening (Figure 6) compared to the WT, but its tolerance to HWSSL is still below that of R57. The phenotype conferred by the *UBP7* mutant does not reconstitute the full HWSSL tolerance displayed by R57 and supports the hypothesis that the high level of tolerance shown by R57 is a result of several mutations incorporated through GS.

One role of ubiquitination is the internalization of cell surface proteins [44-46]. This function relates to Art5p, which belongs to the ART (arrestin-related trafficking) family of proteins that are believed to function as adaptors for Rsp5p, a ubiquitin ligase that promotes endocytosis of plasma membrane proteins, including transporters, targeting damaged or unneeded plasma membrane proteins for vacuolar degradation [47]. Mutation of *ART5* may represent a way for R57 to regulate destruction of proteins damaged by HWSSL stress or direct changes to the plasma membrane in order to respond more efficiently to the toxic HWSSL environment. As might be expected of a leucine to isoleucine mutation, like that found in the R57 Art5p L152I protein, this change is expected to be tolerated by Art5p (Table 1). Nevertheless, the high and increasing frequency of the *ART5* 454C > A mutation shown in the mutant pool sequencing experiment (Figure 2), the differential regulation of cell surface remodelling genes between the WT and R57 (Additional file 4: Data S2) and the proven role of ubiquitin-mediated degradation-machinery gene *UBP7* in HWSSL tolerance suggest that this mutation might also play a determinant role in HWSSL tolerance.

Nrg1p as a determinant in the inhibitor tolerance trait

As the sole transcription factor-encoding gene located amongst R57 mutations, the *NRG1* 137C > A may result in the most pervasive phenotypic consequences. The homozygous P46Q mutation of Nrg1p was predicted to

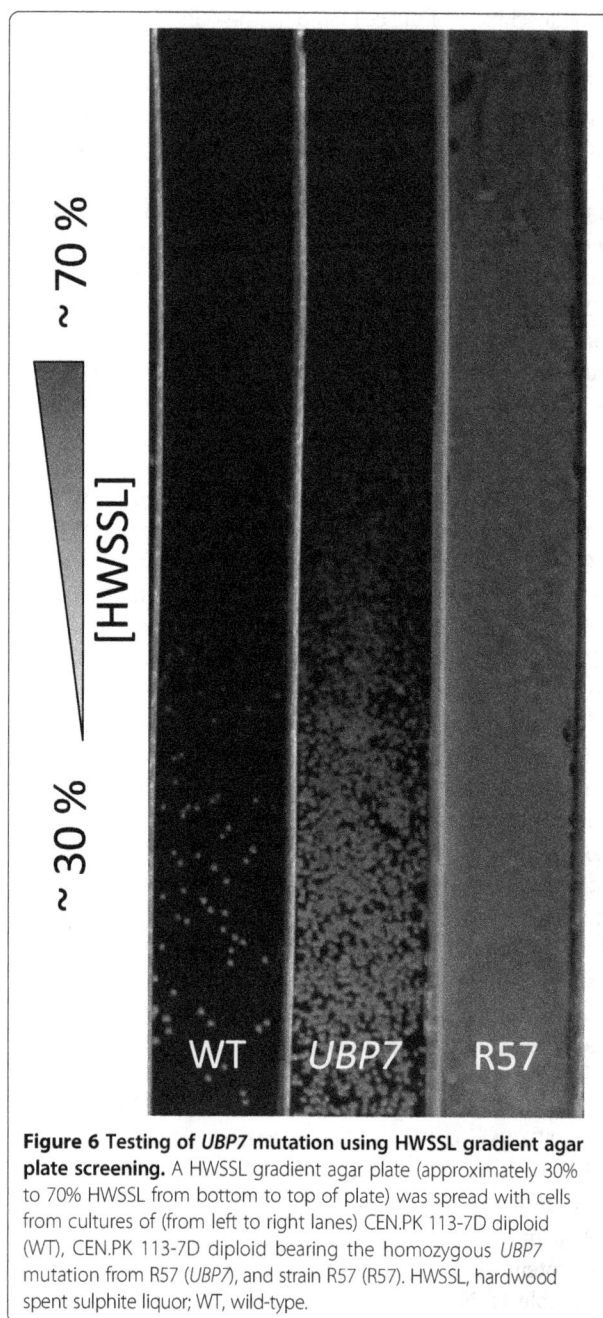

Figure 6 Testing of *UBP7* mutation using HWSSL gradient agar plate screening. A HWSSL gradient agar plate (approximately 30% to 70% HWSSL from bottom to top of plate) was spread with cells from cultures of (from left to right lanes) CEN.PK 113-7D diploid (WT), CEN.PK 113-7D diploid bearing the homozygous *UBP7* mutation from R57 (*UBP7*), and strain R57 (R57). HWSSL, hardwood spent sulphite liquor; WT, wild-type.

pool-wise recursive breeding, when two copies of the allele are more likely present in single strains.

Nrg1p recruits the Tup1p-Cyc8p complex to repress gene expression. Therefore, the Nrg1p P46Q mutation may decrease repression of Nrg1p-controlled genes. *NRG1*, which self-regulates its transcription [48-51], was upregulated 3-fold in R57 over WT (Additional file 4: Data S2). The closely related transcription factor *NRG2* was similarly upregulated approximately 5-fold in R57 over WT. In addition, four of the five most highly upregulated genes in R57 are known to be regulated by Nrg1p. These genes include *CWP1* (approximately 13-fold upregulated), YLR015C (approximately 44-fold upregulated), *GAT3* (approximately 23-fold upregulated) and *TDA6* (approximately 14-fold upregulated) (Additional file 4: Data S2). In *S. cerevisiae*, the most significant transcriptional responses governed by Nrg1p are all related to a multitude of stress conditions [52]. One of the main functional gene categories controlled by Nrg1p is related to peroxide tolerance, which is a specific trait of R57 [8]. Additionally, Vyas *et al.* reported that Nrg1-2p regulates a set of stress response genes and Δnrg1Δnrg2 deletion mutants exhibit tolerance to oxidative stress and salt exposure [53], a trait that is also shared by R57 [8].

R57 also displays acetic acid tolerance after pre-exposure to HWSSL [8]. Nrg1p can directly repress genes that are activated by the downstream action of protein kinase Snf1p [54], which stimulates upregulation of stress responsive genes [55] and the transcription activator Haa1p, which imparts acetic acid tolerance [56]. Many of the genes highlighted as members of enriched clusters that show increased expression in R57 are known constituents of the Haa1p regulon including *AQR1*, *HSP26*, *MSN4*, *PDR12*, *PDR16*, *SPI1*, *SUR2*, *SSE2*, *TDA6*, *TPO1*, *TPO2* and *TPO3* (Additional file 4: Data S2). Genes like Haa1p-dependent *TPO2* and *TPO3* (Nrg1p-regulated [49]) show a prominent effect on acetic acid tolerance [57]. Overall, our data support a strong role for Nrg1p in the control of hydrolysate inhibitor tolerance.

A determinant role for *GDH1* in HWSSL tolerance

The presence of mutations predicted as non-tolerated in both copies of *GDH1* and their close proximity (Table 1) strongly suggest a determinant role for *GDH1* in HWSSL tolerance. Furthermore, population sequencing shows a steady increase in mutant allele frequency at both loci (Figure 2). *GDH1* encodes an NAD(P)H-dependent glutamate dehydrogenase that catalyses the reductive amination of α-ketoglutarate to yield glutamate, responsible for the majority of cellular nitrogen in *S. cerevisiae* via ammonium assimilation [58]. *GDH1* is recognized as a determinant of resistance to acetic acid [59,60]. Likewise, recent proteomics studies using a strain that is tolerant of furfural, phenol and acetic acid show a downregulated

be non-tolerated with a SIFT score of 0.00 (Table 1), as proline is strictly conserved at this position and its substitution would result in altered protein function. SNP analysis of the GS-evolved populations showed a diminished prevalence of this mutation after mating of the UV-treated haploid population (Figure 2), leading to the hypothesis that the *NRG1* mutation results in a recessive, loss-of-function mutation. The *NRG1* 137C > A allele gains prevalence in the GS populations after that point, present at a frequency of approximately 2% after three rounds and approximately 11% after five rounds of

nitrogen assimilation machinery, including Gdh1p [59]. It is believed this occurs in order to slow growth and allow stress tolerance mechanisms to protect the cell more effectively. A decrease in biomass yield exhibited by R57 relative to the WT under permissive conditions (Additional file 3: Figure S4) suggests that metabolism has shifted to a state of decreased resource utilization efficiency. As *GDH1* is a central hub of nitrogen metabolism, microbial substrate tolerance engineering studies that focus on this gene are warranted, especially on the HWSSL substrate that was used to evolve strain R57, which is generally high in ammonia content (approximately 1% *w/v* [22]) and could lead to ammonia toxicity [61]. Bayer *et al.* recently showed that by increasing expression variability of *GDH1* alone, one can tune the metabolism of a cell so that it responds more efficiently to limiting or toxic levels of ammonia [62].

The NADPH cofactor requirement of *GDH1* is also a major consideration in attempting to explain the consequences of the *GDH1* mutations. Yeast detoxifies the furan aldehyde inhibitors found in lignocellulosic hydrolysates by way of NADH/NADPH requiring enzymes [13]. Differential gene expression analysis between the WT and R57 show increased expression of *GRE3* (Additional file 4: Data S2) in R57, encoding for a methylglyoxal reductase that can reduce furan aldehyde inhibitors via NADH [14]. The NAD(P)H cofactor usage of R57 may be modulated by a modified GDH1p, providing the reducing equivalents needed to detoxify the HWSSL furan aldehyde inhibitors.

Sequencing supports determinant HWSSL tolerance roles for mutated genes *SSA1*, *ARO1*, *MAL11* and *GSH1*

Given the GS methodology, it is likely that several of the accumulated mutations influence hydrolysate tolerance. This study was able to generate more restricted evidence that the *SSA1*, *ARO1*, *MAL11* and *GSH1* mutations found in R57 may be affecting HWSSL tolerance.

SSA1, also related to protein homeostasis, bears a mutation in R57. Ssa1p is a member of the heat shock 70 (Hsp70) family of proteins, which consists of highly conserved, broad specificity, essential protein chaperones (for reviews, see [63,64]). The tendency of harsh conditions to damage proteins and lead to aggregation [33] likely makes the role of Hsp-encoding genes important for HWSSL tolerance. The R57 Q31K mutation of Ssa1p is located at a highly conserved residue, as predicted by SIFT (Table 1). Although not highly represented in population sequencing, the discovery of an adjacent *SSA1* mutation (D22G) that is also present within the tolerant population (Figure 2) suggests that this region of Ssa1p may have bearing on the R57 phenotype. The structure of Ssa1p contains two distinct domains, the

nucleotide-binding domain (NBD) that is responsible for binding and hydrolyzing ATP and the substrate-binding domain which can bind short hydrophobic segments of incompletely folded or unfolded polypeptides, in order to prevent adverse aggregation [63]. The Q31K mutation of Ssa1p is located in the NBD, and when the *Escherichia coli* Hsp70 homolog DnaK is used as a structural reference, the mutation is shown to lie adjacent to the ATP-binding pocket [65]. Although the Q31K and D22G mutations in Ssa1p have not been studied thus far, residues in this area of the protein have been shown to influence its NBD function and folding activity [65,66].

Furthermore, Aro1p, which catalyses steps 2 to 6 of the chorismate pathway leading to synthesis of aromatic amino acids [67], harbours a predicted phenotype-conferring mutation that gains in frequency within sequenced mutant pools. After an acetic acid challenge, aromatic amino acid synthesis and tryptophan synthesis in particular are pathways that are found to be upregulated [60], while mutants auxotrophic for aromatic amino acid synthesis show acetic acid sensitivity [68] and deletion of *ARO1* leads to sensitivity to osmotic and ethanol stress [69].

Mal11p bears two mutations that pervade and increase in the GS-evolved HWSSL-tolerant population. Mal11p is a trehalose-H$^+$ symporter [70,71] and could be related to osmotic stress protection via trehalose transport or pH stasis due to its proton requirement. Although the effect of *MAL11* on tolerance to industrial processes has not been demonstrated, it is a common trend for the genome of industrial yeast strains to show a loss or reduction of *MAL11* genes [72-74].

Finally, the mutation in the 5′ UTR of *GSH1* may potentially be affecting redox homeostasis by influencing glutathione levels; *GSH1* can influence tolerance to lignocellulosic hydrolysate inhibitors in this way [75]. Glutathione is comprised of glycine, cysteine and glutamate and is a major redox buffer of the cell, cycling between its reduced and oxidized form, relying on NADP(H) for recycling. Hydrogen peroxide induces *GSH1* transcription but relies on the presence of intercellular amino acid pools, namely glutamate, glutamine and lysine to induce glutathione production [76]. Therefore, modified glutamate assimilation via *GDH1* mutation and the 5′ UTR mutation in *GSH1* both potentially affect intercellular glutamate pools and concomitant expression of *GSH1*. Between the WT and R57, genes that lead to cysteine biosynthesis (*HOM2*, *MET16*, *MET17* and *MET3*), along with genes leading to glutamate and lysine biosynthesis (*HOM3*, *LYS9* and *LYS12*), are upregulated (Additional file 4: Data S2). This finding suggests that R57 has upregulated pathways towards glutathione precursor generation as part of its physiology and constitutes a possible link to the *GSH1* 5′ UTR mutation.

GS population sequencing provides evidence against prominent phenotypic roles of isolated mutations

Although some of the remaining R57 mutations may be of interest based on the known functions of the affected genes, sequencing of GS populations did not support a determinant effect on HWSSL tolerance for every mutation. Namely, the mutations located in the 5′ UTR of *PBP1*, 3′ of *BCS1* and in *SGO1* and YNL058C seem to be linked in a single mutant that comprised a large proportion of the UV mutant population, but decreased in frequency throughout GS evolution. The verified tolerance-conferring effect of the *UBP7* mutation to which they were initially linked, and which gained in frequency over population evolution, suggests that these linked mutations were merely carried through to strain R57 from UV mutagenesis. The *TOF2* mutation gained in frequency within the GS-evolved strains at round 1 but diminished in later rounds, while the *DOP1* and *FIT3* 3′ UTR mutations were not highly represented in the GS populations and may be relatively specific to R57, which devalues these genes as potential tolerance-associated determinants. One exception is the mutation in Ste5p, which increased in frequency throughout GS, suggesting that it may play a role in the evolution of R57. However, since Ste5p is involved in mating, which is essential to the GS method used to generate R57, a strain with a modified mating behaviour may play a role in GS evolution but is likely not linked to stress tolerance.

Conclusions

GS theory asserts that beneficial mutations accumulate within strains to rapidly evolve a trait of interest while simultaneously eliminating detrimental mutations. Our data suggest that during GS evolution, there is accumulation of complementary mutations in key cellular processes through recursive genetic recombination or by the accumulation of single mutations in crucial genes that confer a fitness advantage. Mutations that lead to a large fitness advantage, such as *UBP7*, may become highly represented in the initial mutant pool and lead to over-representation of non-productive mutations found in the same strain through genetic linkage. The decrease in the frequency of these non-productive mutations during GS evolution suggests that mutations of lesser or no impact on the trait of interest can be crossed out of final strains.

As a workflow, meiotic recombination-mediated GS of *S. cerevisiae*, combined with genome resequencing, population sequencing of mutation loci and RNA-seq transcriptional profiling, generated complementary results that provided novel insights into tolerance to a specific lignocellulosic hydrolysate, along with gene targets that can be used for strain engineering. The assortment of processes and genes involved in inhibitor tolerance could not have been rationally determined prior to this study.

This study provides insights into the multiple biological processes that act in concert to establish tolerance to a multi-inhibitory substrate. Our strongest evidence supports determinant roles for Ubp7p and Art5p, related to ubiquitin-mediated proteolysis; stress response transcriptional repressor, Nrg1p; and NADPH-dependent glutamate dehydrogenase, Gdh1p. However, important roles in hydrolysate tolerance are supported for several of the mutations discovered in the GS-evolved strains and populations (*SSA1*, *ARO1*, *MAL11* and *GSH1*), and the potential phenotypic impact of each mutation has not been ruled out. Therefore, a subsequent study is ongoing in our lab to examine the effect of each mutation discovered in R57 and explore potential epistatic effects between the mutations reported here. As whole genome sequencing becomes ever more accessible, and as the biotechnology industry requires biocatalysts with increasingly complex traits, GS followed by analyses like those carried out in this study stands to have a rapid and profound effect on our understanding of complex multigenic traits.

Materials and methods

Strains and materials

The *S. cerevisiae* CEN.PK strains, supplied by EUROSCARF (Institute for Molecular Biosciences, Frankfurt, Germany), were used as the WT reference and progenitor strains for genome-shuffled mutant populations, including prototrophic diploid strain CEN.PK 122 and haploid strains CEN.PK 113-1A (*MAT* α) and 113-7D (*MAT* a). The haploid CEN.PK strains were used in a previous study to generate HWSSL-tolerant strain R57.

The HWSSL used for all experiments was provided by Tembec Inc. (Temiscaming, Quebec, Canada). HWSSL was adjusted to pH 5.5 with 10 M NaOH and contained, on average (*w/v*), 0.076% arabinose, 2% xylose, 0.16% galactose, 0.24% glucose, 0.43% mannose, 1% acetic acid, 0.18% furfural and 0.11% HMF.

Sequencing of WT and R57 strains

Genomic DNA of WT haploid CEN.PK113-7D and the CEN.PK-derived R57 diploid yeast strains were isolated from overnight cultures grown in YPD using the DNeasy Blood & Tissue Kit (Qiagen, Toronto, Ontario, Canada). Library construction and sequencing were done at the Michael Smith Genome Sciences Centre using an Illumina 1G Genome Analyzer (Illumina, Inc., San Diego, CA, USA). Two lanes were sequenced for the WT strain and four for strain R57. All genome sequence data and RNA-seq read data from this publication have been submitted to the National Center for Biotechnology Information (NCBI) sequence read archive under BioProject # PRJNA231093.

Sequencing alignment and mutation calling

Sequencing reads were aligned to the WT CEN.PK113-7D reference genome obtained from the NCBI (PRJNA52955) [77] and cross-referenced with read alignments obtained in our laboratory using our WT consensus sequence that was created using the S288c genome sequence as an alignment backbone. Alignments were performed both with Bowtie using standard parameters [78] and CLC Genomics Workbench version 5.1 with default parameters and ignoring non-specific matches. SNP and indel calling were both performed with Maq version 0.7.1 [79] and CLC Genomics Workbench for verification and visualization. To eliminate false positives in mutation calling, the DNA sequencing reads obtained from the WT were subjected to the same variation calling protocol as strain R57. Variations that were called when CEN.PK113-7D Illumina reads were aligned onto the CEN.PK113-7D consensus genome sequence and those that corresponded to variations called for R57 reads aligned onto the CEN. PK113-7D consensus sequence were discarded as false positives. False positive SNP calls were likely derived from sequence-specific miscall errors [80] or due to alignments in non-specific regions of the genome or to areas of low complexity [81]. The coverage requirement for SNP calling for WT reads was also lowered to $\leq$5-fold in order to ensure that SNPs that may result due to misalignment could be easily identified. These miscalls were edited out of the final mutation list by manual inspection and visualization with CLC Genomics Workbench. Maq SNP analysis was performed using the cns2snp command, and SNPs were called if the region upon which they were mapped returned a genome copy score = 1 and carried a Phred-like quality score $\geq$40. SNP analysis was corroborated with CLC Genomics Workbench SNP Detection function with a quality score of $\geq$40 for the central base and $\geq$30 for the surrounding bases. The threshold for variation at a specific base in the genome needed for SNP calling was lowered to $\geq$10% of reads, yielding an aberrant base call for the CEN.PK113-7D read alignment from $\geq$35% for R57 SNP calling in order to maintain stringency on positive variation calling. Therefore, if a SNP was called for more than 35% of the reads in R57 but was also called for 10% or more of the reads for the WT control, it was discarded as a false SNP call. Indel analysis was also performed with Maq using the Indelpe command and verified with CLC Genomics Workbench, both under standard parameters and compared to WT reads for control, as described above. Copy number variation was assessed using CNV-seq [82] to compare WT to R57 reads with a log2 threshold of 0.75, below the level used to detect reliable CNVs in CEN.PK [77]. The mutations identified were verified by Sanger sequencing and compared to the other Mat parental WT, CEN.PK113-1A, to ensure the mutations were not present in either of the WT parental

haploid strains. Locations of mutations were assessed using the CEN.PK consensus sequence as a guide.

Protein impact assessment of ORF-located mutations

Mutations in ORFs were examined by translating DNA sequences using ExPASy translate [83] and performing a BLAST comparison using the BLAST2Seq software program hosted by NCBI. Mutational impact assessment was done with the SIFT program [28,29] using recommended best practices [28]. Homologous proteins with <90% identity were chosen for comparison of the degree of conservation of the amino acid position in question for up to 100 homologous proteins. Amino acid substitutions were predicted as leading to a phenotype if the SIFT score was $\leq$0.05 and tolerated if the score is >0.05.

Sequencing of R57 SNPs in GS heterogeneous populations

The following heterogeneous populations of cells were used to track R57 SNP frequency through GS evolution: WT diploid CEN.PK122, a pooled UV mutagenized population (three pools of CEN.PK113-1A and two pools CEN. PK113-7D UV mutants), along with cells from the population of the first, third and fifth rounds of recursive GS obtained from our previous study that generated R57 [8]. Cells from each population were selected from above the frontier of WT growth as observed by screening on HWSSL gradient agar plates (Figure 1B), yielding populations that were enriched for tolerance to HWSSL, as described [8]. A sample of 10^9 cells, suspended in phosphate-buffered saline (PBS), from each population was spread onto a single lane of the gradient plate and incubated for 6 days at 30°C. Each plate contained two lanes spread with CEN.PK 122 cells for comparison to each individual GS population, in duplicate. Each plate was screened in biological duplicates.

Cells that grew to higher HWSSL concentrations than the WT on the gradient plates were scraped, suspended in PBS and adjusted to approximately 4×10^8 cells/mL using a haemocytometer. For each population, DNA from approximately 4×10^7 cells was extracted using a DNeasy Blood & Tissue Kit (Qiagen, Toronto, Ontario, Canada). Five microlitres of each genomic DNA preparation was used as template for PCR with primers specific for each of the 20 SNP regions, located at approximately 50 bp from either end of the SNP. The primers were designed with sequencing adapter attachment for use with the Ion Torrent Personal Genome Machine (Life Technologies, Carlsbad, CA, USA) according to the manufacturer's instructions (Additional file 2: Table S1). All PCR products were gel purified and quantified in triplicate using the Promega Quantifluor dsDNA system (Promega Corporation, Madison, WI, USA). The PCR products were diluted to 16 pmol, and 20-μL samples from each reaction were pooled to make four pools (one UV mutagenized and three GS pools). Pools of PCR products were sequenced using a 316 chip with the 200-bp kit (following the Ion Torrent protocols).

Sequencing SNPs from isolates of the round 5 GS heterogeneous population

Twenty isolated colonies picked from the growth frontier of a HWSSL gradient plate of the round 5 GS population were streak purified on 50% (*v/v*) HWSSL and 2% agar (*w/v*) Petri plates. Colonies isolated from the plates were grown overnight in 5 mL YPD broth at 30°C, and DNA was extracted with a DNeasy Blood & Tissue Kit for use as a template in the PCR amplification of the mutated gene region. Primers used for amplification are described in Additional file 2: Table S2, and PCR products were sequenced using the Sanger method.

Transcriptome analysis

WT diploid strain CEN.PK122 and HWSSL-tolerant R57 were used in RNA-seq experiments. The WT and R57 strains were grown in 50 mL synthetic defined (SD) medium (yeast nitrogen base without amino acids 0.17% *w/v*, ammonium sulphate 0.5% *w/v*, glucose 2% *w/v*) overnight at 30°C under semi-fermentative conditions (sealed 125-mL flasks shaken at 100 rpm) to early stationary phase (Additional file 3: Figure S4). Cultures were normalized to an $OD_{600\ nm}$ of 3, and two independently grown cell samples of these cultures were used for RNA extraction. For each sample, cells from 5 mL of culture were harvested by centrifugation at $1,800 \times g$ and 4°C and frozen in liquid N_2 until RNA isolation. RNA extracts were prepared using the RNeasy Plant Mini Kit (Qiagen) according to manufacturer's specifications for use with yeast, in which frozen cells were suspended in lysis buffer and disrupted with a mini bead beater (Precellys 24, Bertin Technologies, Montigny-le-Bretonneux, France) at 4°C. Prior to sequencing, RNA quality was confirmed using an Agilent 2100 Bioanalyzer (Agilent Technologies, Santa Clara, CA, USA).

RNA sequencing was performed at the McGill/Genome Quebec Innovation Centre in duplicate on the Illumina Genome Analyzer *IIx* and Illumina HiSeq 2000 for the WT *vs.* R57. DNA libraries were subjected to 36 or 50 cycles of sequencing on the Illumina Genome Analyzer *IIx* and Illumina HiSeq 2000, respectively.

RNA-seq differential transcription analysis and statistical comparisons were performed with CLC Genomics Workbench version 5.1. The cDNA sequence reads from RNA-seq were trimmed to remove Illumina sequencing adaptors as well as unreliable read ends, and alignments were performed using the CEN.PK113-7D genome sequence and associated GTF file [77] as the backbone for alignment mapping and quantitation. Significance values

for differential expression were computed using Baggerly's test [84]. The samples were then FDR-corrected in order to eliminate non-productive leads from the expression results. Gene transcripts showing differential expression with a corrected *P* value of <0.05 and a >2-fold increase were used for functional clustering and enrichment mapping of differentially expressed genes.

Functional annotation clustering was executed with DAVID Bioinformatics Resources 6.7 [85]. Clusters of up or down expressed genes with gene ontology (GO) term enrichment scores of ≥1.3 (equivalent to a non-log scale value of 0.05) are reported, unless stated otherwise. Enrichment maps of ontology categories from clustering were generated with the Enrichment Map 1.2 software plug-in for Cytoscape 2.8 [86,87]. All functional annotations presented were derived from SGD [88] or the DAVID server unless otherwise referenced. Transcription factor binding analysis was done through the YEASTRACT database [89-91].

Reconstitution of the *UBP7* mutation in WT

To determine if the *UBP7* mutation was contributing to the HWSSL tolerance phenotype of R57 as predicted, the mutation was introduced into WT and the resulting strain was tested for growth on HWSSL. The WT allele was replaced in CEN.PK113-7D via homologous recombination of a DNA cassette containing the mutated *UBP7* sequence flanked by a kanamycin resistance marker. Sanger sequencing of the PCR-amplified region was used to confirm that the transformants harboured the mutation. Homozygous diploid strains of the WT and *UBP7* mutant were created by mating type switching using the YCp50::HO plasmid [92] and mating haploid strains of opposite mating type. The *UBP7* homozygous diploid mutant and WT diploid strains along with R57 were tested for their tolerance to HWSSL in parallel, as previously described [8].

Additional files

Additional file 1: Data S1. A summary mapping report for CEN.PK 113-7D.

Additional file 2: Supplemental methods. A document on assessing functional interaction of mutations. (DOCX 31 kb)

Additional file 3: Supplemental figures. Figure S1. Interaction map of genes affected by mutation in R57. **Figure S2.** Ontology categories associated with R57 genes affected by mutation. **Figure S3.** Differentially expressed genes comprising enrichment clusters based on biological function between the WT and R57. **Figure S4.** Cell growth and glucose consumption of WT and R57 mutant strains.

Additional file 4: Data S2. RNA-seq results for differentially expressed genes between WT *vs* R57 without HWSSL exposure.

Abbreviations

CNV: copy number variation; GS: genome shuffling; HMF: 5-(hydroxymethyl)-2-furaldehyde; HWSSL: hardwood spent sulphite liquor; indel: insertion/deletion; NBD: nucleotide-binding domain; UBL: ubiquitin-like.

Competing interests
The authors declare that they have no competing interests.

Authors' contributions
DP and VM conceived and designed the study. DP, DC and HJ acquired the data. DP, DC and VM analysed and interpreted the data. Drafting of the manuscript was done by DP and DC. Critical revision of the manuscript for intellectual content was performed by DP, HL and VM. VM and HL obtained funding. All authors have read and approved of the final version of this manuscript.

Acknowledgements
An NSERC Strategic Project (GHGPJ322381), the NSERC Bioconversion Network (NETGP350246-07), the AAFC Agricultural Bioproducts Innovation Program (ABTP_000159), BioFuelNet and a Canada Research Chair to V.J.J.M supported this research. Dominic Pinel was supported by a graduate scholarship from Le Fonds Québécois de la Récherche sur la Nature et les Technologies.

Author details
[1]Department of Biology, Centre for Structural and Functional Genomics, Concordia University, 7141 Sherbrooke Street West, Montréal, Québec H4B 1R6, Canada. [2]School of Environmental Sciences, University of Guelph, Guelph, Ontario N1G 2 W1, Canada. [3]Current address: Energy Biosciences Institute, University of California, Berkeley, Berkeley, CA 94704, USA. [4]Current address: Crabtree Nutrition Laboratories, McGill University Health Center, Montreal, Quebec H3A 1A1, Canada.

References

1. Oud B, van Maris AJ, Daran JM, Pronk JT. Genome-wide analytical approaches for reverse metabolic engineering of industrially relevant phenotypes in yeast. FEMS Yeast Res. 2012;12:183–96.
2. Smith DR, Quinlan AR, Peckham HE, Makowsky K, Tao W, Woolf B, et al. Rapid whole-genome mutational profiling using next-generation sequencing technologies. Genome Res. 2008;18:1638–42.
3. Le Crom S, Schackwitz W, Pennacchio L, Magnuson JK, Culley DE, Collett JR, et al. Tracking the roots of cellulase hyperproduction by the fungus Trichoderma reesei using massively parallel DNA sequencing. Proc Natl Acad Sci U S A. 2009;106:16151–6.
4. Sarin S, Bertrand V, Bigelow H, Boyanov A, Doitsidou M, Poole RJ, et al. Analysis of multiple ethyl methanesulfonate-mutagenized Caenorhabditis elegans strains by whole-genome sequencing. Genetics. 2010;185:417–30.
5. Harper MA, Chen Z, Toy T, Machado IM, Nelson SF, Liao JC, et al. Phenotype sequencing: identifying the genes that cause a phenotype directly from pooled sequencing of independent mutants. PLoS One. 2011;6:e16517.
6. Parts L, Cubillos FA, Warringer J, Jain K, Salinas F, Bumpstead SJ, et al. Revealing the genetic structure of a trait by sequencing a population under selection. Genome Res. 2011;21:1131–8.
7. Zhang YX, Perry K, Vinci VA, Powell K, Stemmer WPC, del Cardayre SB. Genome shuffling leads to rapid phenotypic improvement in bacteria. Nature. 2002;415:644–6.
8. Pinel D, D'Aoust F, del Cardayre SB, Bajwa PK, Lee H, Martin VJJ. Saccharomyces cerevisiae genome shuffling through recursive population mating leads to improved tolerance to spent sulfite liquor. Appl Environ Microbiol. 2011;77:4736–43.
9. Patnaik R, Louie S, Gavrilovic V, Perry K, Stemmer WPC, Ryan CM, et al. Genome shuffling of Lactobacillus for improved acid tolerance. Nat Biotechnol. 2002;20:707–12.
10. Dai MH, Copley SD. Genome shuffling improves degradation of the anthropogenic pesticide pentachlorophenol by Sphingobium chlorophenolicum ATCC 39723. Appl Environ Microbiol. 2004;70:2391–7.
11. Biot-Pelletier D, Martin VJJ. Evolutionary engineering by genome shuffling. Appl Microbiol Biotechnol. 2014;98(9):3877–87.
12. Almeida JRM, Modig T, Petersson A, Hahn-Hägerdal B, Lidén G, Gorwa-Grauslund MF. Increased tolerance and conversion of inhibitors in lignocellulosic hydrolysates by Saccharomyces cerevisiae. J Chem Technol Biotechnol. 2007;82:340–9.
13. Liu ZL. Molecular mechanisms of yeast tolerance and in situ detoxification of lignocellulose hydrolysates. Appl Microbiol Biotechnol. 2011;90:809–25.
14. Liu ZL, Blaschek HP. Lignocellulosic biomass conversion to ethanol by Saccharomyces. In: Vertes A, Qureshi N, Yukawa H, Blaschek H, editors. Biomass to biofuels: strategies for global industries. West Sussex, U. K: John Wiley & Sons, Ltd; 2010. p. 17–36.
15. Palmqvist E, Hahn-Hägerdal B. Fermentation of lignocellulosic hydrolysates. I: Inhibition and detoxification. Biores Technol. 2000;74:17–24.
16. Richardson TL, Harner NK, Bajwa PK, Trevors JT, Lee H. Approaches to deal with toxic inhibitors during fermentation of lignocellulosic substrates. Acs Sym Ser. 2011;1067:171–202.
17. Pinel D, Gawand P, Mahadevan R, Martin VJJ. 'Omics' technologies and systems biology for engineering Saccharomyces cerevisiae strains for lignocellulosic bioethanol production. Biofuels. 2011;2:659–75.
18. Gorsich SW, Slininger PJ, Liu ZL. Physiological responses to furfural and HMF and the link to other stress pathways. J Biotechnol. 2005;118:S91–1.
19. Petersson A, Almeida JRM, Modig T, Karhumaa K, Hahn-Hägerdal B, Gorwa-Grauslund MF, et al. A 5-hydroxymethyl furfural reducing enzyme encoded by the Saccharomyces cerevisiae ADH6 gene conveys HMF tolerance. Yeast. 2006;23:455–64.
20. Keating JD, Panganiban C, Mansfield SD. Tolerance and adaptation of ethanologenic yeasts to lignocellulosic inhibitory compounds. Biotechnol Bioeng. 2006;93:1196–206.
21. Pinel D, Martin VJJ. Meiotic recombination-based genome shuffling of Saccharomyces cerevisiae and Scheffersomyces stiptis for increased inhibitor tolerance to lignocellulosic substrate toxicity. In: Patnaik R, editor. Engineering complex phenotypes in industrial strains. 1st ed. Hoboken, New Jersey: John Wiley & Sons, Inc; 2012. p. 233–50.
22. Helle SS, Murray A, Lam J, Cameron DR, Duff SJ. Xylose fermentation by genetically modified Saccharomyces cerevisiae 259ST in spent sulfite liquor. Bioresour Technol. 2004;92:163–71.
23. Olsson L, HahnHagerdal B. Fermentation of lignocellulosic hydrolysates for ethanol production. Enzyme Microb Tech. 1996;18:312–31.
24. Parajó JC, Domínges H, Domínguez JM. Biotechnological production of xylitol. Part 3: operation in culture media made from lignocellulose hydrolysates. Bioresour Technol. 1998;66:25–40.
25. Helle S, Duff S. Supplementing spent sulfite liquor with a lignocellulosic hydrolysate to increase pentose/hexose co-fermentation efficiency and ethanol yield. Final report-Natural Resources Canada-Tembec Industries; 2004. http://www.lifesciencesbc.ca/files/dufffinal_report.pdf.
26. Yassour M, Kaplan T, Fraser HB, Levin JZ, Pfiffner J, Adiconis X, et al. Ab initio construction of a eukaryotic transcriptome by massively parallel mRNA sequencing. Proc Natl Acad Sci U S A. 2009;106:3264–9.
27. Nagalakshmi U, Wang Z, Waern K, Shou C, Raha D, Gerstein M, et al. The transcriptional landscape of the yeast genome defined by RNA sequencing. Science. 2008;320:1344–9.
28. Kumar P, Henikoff S, Ng PC. Predicting the effects of coding non-synonymous variants on protein function using the SIFT algorithm. Nat Protoc. 2009;4:1073–81.
29. Ng PC, Henikoff S. SIFT: predicting amino acid changes that affect protein function. Nucleic Acids Res. 2003;31:3812–4.
30. Sim NL, Kumar P, Hu J, Henikoff S, Schneider G, Ng PC. SIFT web server: predicting effects of amino acid substitutions on proteins. Nucleic Acids Res. 2012;40:W452–7.
31. Ozkaynak E, Finley D, Solomon MJ, Varshavsky A. The yeast ubiquitin genes: a family of natural gene fusions. EMBO J. 1987;6:1429–39.
32. Modig T, Lidén G, Taherzadeh MJ. Inhibition effects of furfural on alcohol dehydrogenase, aldehyde dehydrogenase and pyruvate dehydrogenase. Biochem J. 2002;363:769–76.
33. Goldberg AL. Protein degradation and protection against misfolded or damaged proteins. Nature. 2003;426:895–9.
34. Hochstrasser M. Ubiquitin-dependent protein degradation. Annu Rev Genet. 1996;30:405–39.
35. Kimura Y, Tanaka K. Regulatory mechanisms involved in the control of ubiquitin homeostasis. J Biochem. 2010;147:793–8.
36. Hershko A, Ciechanover A. The ubiquitin system. Annu Rev Biochem. 1998;67:425–79.
37. Mukhopadhyay D, Riezman H. Proteasome-independent functions of ubiquitin in endocytosis and signaling. Science. 2007;315:201–5.

38. Finley D, Ozkaynak E, Varshavsky A. The yeast polyubiquitin gene is essential for resistance to high temperatures, starvation, and other stresses. Cell. 1987;48:1035–46.

39. Hanna J, Meides A, Zhang DP, Finley D. A ubiquitin stress response induces altered proteasome composition. Cell. 2007;129:747–59.

40. Leggett DS, Hanna J, Borodovsky A, Crosas B, Schmidt M, Baker RT, et al. Multiple associated proteins regulate proteasome structure and function. Mol Cell. 2002;10:495–507.

41. Hanna J, Leggett DS, Finley D. Ubiquitin depletion as a key mediator of toxicity by translational inhibitors. Mol Cell Biol. 2003;23:9251–61.

42. Chernova TA, Allen KD, Wesoloski LM, Shanks JR, Chernoff YO, Wilkinson KD. Pleiotropic effects of Ubp6 loss on drug sensitivities and yeast prion are due to depletion of the free ubiquitin pool. J Biol Chem. 2003;278:52102–15.

43. Hernández-López MJ, Garcia-Marqués S, Randez-Gil F, Prieto JA. Multicopy suppression screening of Saccharomyces cerevisiae identifies the ubiquitination machinery as a main target for improving growth at low temperatures. Appl Environ Microbiol. 2011;77:7517–25.

44. Kolling R, Hollenberg CP. The ABC-transporter Ste6 accumulates in the plasma membrane in a ubiquitinated form in endocytosis mutants. EMBO J. 1994;13:3261–71.

45. Hein C, Springael JY, Volland C, Haguenauer-Tsapis R, Andre B. NPI1, an essential yeast gene involved in induced degradation of Gap1 and Fur4 permeases, encodes the Rsp5 ubiquitin-protein ligase. Mol Microbiol. 1995;18:77–87.

46. Hicke L, Riezman H. Ubiquitination of a yeast plasma membrane receptor signals its ligand-stimulated endocytosis. Cell. 1996;84:277–87.

47. Lin CH, MacGurn JA, Chu T, Stefan CJ, Emr SD. Arrestin-related ubiquitin-ligase adaptors regulate endocytosis and protein turnover at the cell surface. Cell. 2008;135:714–25.

48. Goh WS, Orlov Y, Li J, Clarke ND. Blurring of high-resolution data shows that the effect of intrinsic nucleosome occupancy on transcription factor binding is mostly regional, not local. PLoS Comput Biol. 2010;6:e1000649.

49. Harbison CT, Gordon DB, Lee TI, Rinaldi NJ, Macisaac KD, Danford TW, et al. Transcriptional regulatory code of a eukaryotic genome. Nature. 2004;431:99–104.

50. Lee TI, Rinaldi NJ, Robert F, Odom DT, Bar-Joseph Z, Gerber GK, et al. Transcriptional regulatory networks in Saccharomyces cerevisiae. Science. 2002;298:799–804.

51. Workman CT, Mak HC, McCuine S, Tagne JB, Agarwal M, Ozier O, et al. A systems approach to mapping DNA damage response pathways. Science. 2006;312:1054–9.

52. Zhu C, Byers KJ, McCord RP, Shi Z, Berger MF, Newburger DE, et al. High-resolution DNA-binding specificity analysis of yeast transcription factors. Genome Res. 2009;19:556–66.

53. Vyas VK, Berkey CD, Miyao T, Carlson M. Repressors Nrg1 and Nrg2 regulate a set of stress-responsive genes in Saccharomyces cerevisiae. Eukaryot Cell. 2005;4:1882–91.

54. Kuchin S, Vyas VK, Carlson M. Snf1 protein kinase and the repressors Nrg1 and Nrg2 regulate FLO11, haploid invasive growth, and diploid pseudohyphal differentiation. Mol Cell Biol. 2002;22:3994–4000.

55. Mayordomo I, Estruch F, Sanz P. Convergence of the target of rapamycin and the Snf1 protein kinase pathways in the regulation of the subcellular localization of Msn2, a transcriptional activator of STRE (Stress Response Element)-regulated genes. J Biol Chem. 2002;277:35650–6.

56. Mira NP, Becker JD, Sá-Correia I. Genomic expression program involving the Haa1p-regulon in Saccharomyces cerevisiae response to acetic acid. OMICS. 2010;14:587–601.

57. Fernandes AR, Mira NP, Vargas RC, Canelhas I, Sá-Correia I. Saccharomyces cerevisiae adaptation to weak acids involves the transcription factor Haa1p and Haa1p-regulated genes. Biochem Biophys Res Commun. 2005;337:95–103.

58. Magasanik B. Ammonia assimilation by Saccharomyces cerevisiae. Eukaryot Cell. 2003;2:827–9.

59. Ding MZ, Wang X, Liu W, Cheng JS, Yang Y, Yuan YJ. Proteomic research reveals the stress response and detoxification of yeast to combined inhibitors. PLoS One. 2012;7:e43474.

60. Mira NP, Palma M, Guerreiro JF, Sá-Correia I. Genome-wide identification of Saccharomyces cerevisiae genes required for tolerance to acetic acid. Microb Cell Fact. 2010;9:79.

61. Hess DC, Lu W, Rabinowitz JD, Botstein D. Ammonium toxicity and potassium limitation in yeast. PLoS Biol. 2006;4:e351.

62. Bayer TS, Hoff KG, Beisel CL, Lee JJ, Smolke CD. Synthetic control of a fitness tradeoff in yeast nitrogen metabolism. J Biol Eng. 2009;3:1.

63. Wegele H, Müller L, Buchner J. Hsp70 and Hsp90–a relay team for protein folding. Rev Physiol Biochem Pharmacol. 2004;151:1–44.

64. Verghese J, Abrams J, Wang Y, Morano KA. Biology of the heat shock response and protein chaperones: budding yeast (Saccharomyces cerevisiae) as a model system. Microbiol Mol Biol Rev. 2012;76:115–58.

65. Jones GW, Masison DC. Saccharomyces cerevisiae Hsp70 mutations affect [PSI+] prion propagation and cell growth differently and implicate Hsp40 and tetratricopeptide repeat cochaperones in impairment of [PSI+]. Genetics. 2003;163:495–506.

66. Loovers HM, Guinan E, Jones GW. Importance of the Hsp70 ATPase domain in yeast prion propagation. Genetics. 2007;175:621–30.

67. Duncan K, Edwards RM, Coggins JR. The pentafunctional arom enzyme of Saccharomyces cerevisiae is a mosaic of monofunctional domains. Biochem J. 1987;246:375–86.

68. Bauer BE, Rossington D, Mollapour M, Mamnun Y, Kuchler K, Piper PW. Weak organic acid stress inhibits aromatic amino acid uptake by yeast, causing a strong influence of amino acid auxotrophies on the phenotypes of membrane transporter mutants. Eur J Biochem. 2003;270:3189–95.

69. Yoshikawa K, Tanaka T, Furusawa C, Nagahisa K, Hirasawa T, Shimizu H. Comprehensive phenotypic analysis for identification of genes affecting growth under ethanol stress in Saccharomyces cerevisiae. Fems Yeast Research. 2009;9:32–44.

70. Stambuk BU, Panek AD, Crowe JH, Crowe LM, de Araujo PS. Expression of high-affinity trehalose-H+ symport in Saccharomyces cerevisiae. Biochim Biophys Acta. 1998;1379:118–28.

71. Jules M, Guillou V, Francois J, Parrou JL. Two distinct pathways for trehalose assimilation in the yeast Saccharomyces cerevisiae. Appl Environ Microbiol. 2004;70:2771–8.

72. Babrzadeh F, Jalili R, Wang C, Shokralla S, Pierce S, Robinson-Mosher A, et al. Whole-genome sequencing of the efficient industrial fuel-ethanol fermentative Saccharomyces cerevisiae strain CAT-1. Mol Genet Genomics. 2012;287:485–94.

73. Carreto L, Eiriz MF, Gomes AC, Pereira PM, Schuller D, Santos MA. Comparative genomics of wild type yeast strains unveils important genome diversity. BMC Genomics. 2008;9:524.

74. Dunn B, Levine RP, Sherlock G. Microarray karyotyping of commercial wine yeast strains reveals shared, as well as unique, genomic signatures. BMC Genomics. 2005;6:53.

75. Ask M, Mapelli V, Hock H, Olsson L, Bettiga M. Engineering glutathione biosynthesis of Saccharomyces cerevisiae increases robustness to inhibitors in pretreated lignocellulosic materials. Microb Cell Fact. 2013;12:87.

76. Stephen DW, Jamieson DJ. Amino acid-dependent regulation of the Saccharomyces cerevisiae GSH1 gene by hydrogen peroxide. Mol Microbiol. 1997;23:203–10.

77. Nijkamp JF, van den Broek M, Datema E, de Kok S, Bosman L, Luttik MA, et al. De novo sequencing, assembly and analysis of the genome of the laboratory strain Saccharomyces cerevisiae CEN.PK113-7D, a model for modern industrial biotechnology. Microb Cell Fact. 2012;11:36.

78. Langmead B, Trapnell C, Pop M, Salzberg SL. Ultrafast and memory-efficient alignment of short DNA sequences to the human genome. Genome Biol. 2009;10:R25.

79. Li H, Ruan J, Durbin R. Mapping short DNA sequencing reads and calling variants using mapping quality scores. Genome Res. 2008;18:1851–8.

80. Nakamura K, Oshima T, Morimoto T, Ikeda S, Yoshikawa H, Shiwa Y, et al. Sequence-specific error profile of Illumina sequencers. Nucleic Acids Res. 2011;39:e90.

81. Oyola SO, Otto TD, Gu Y, Maslen G, Manske M, Campino S, et al. Optimizing Illumina next-generation sequencing library preparation for extremely AT-biased genomes. BMC Genomics. 2012;13:1.

82. Xie C, Tammi MT. CNV-seq, a new method to detect copy number variation using high-throughput sequencing. Bmc Bioinformatics. 2009;10:80.

83. Artimo P, Jonnalagedda M, Arnold K, Baratin D, Csardi G, de Castro E, et al. ExPASy: SIB bioinformatics resource portal. Nucleic Acids Res. 2012;40:W597–603.

84. Baggerly KA, Deng L, Morris JS, Aldaz CM. Differential expression in SAGE: accounting for normal between-library variation. Bioinformatics. 2003;19:1477–83.

85. da Huang W, Sherman BT, Lempicki RA. Systematic and integrative analysis of large gene lists using DAVID bioinformatics resources. Nat Protoc. 2009;4:44–57.

86. Merico D, Isserlin R, Bader GD. Visualizing gene-set enrichment results using the Cytoscape plug-in enrichment map. Methods Mol Biol. 2011;781:257–77.

87. Merico D, Isserlin R, Stueker O, Emili A, Bader GD. Enrichment map: a network-based method for gene-set enrichment visualization and interpretation. PLoS One. 2010;5:e13984.

88. Cherry JM, Hong EL, Amundsen C, Balakrishnan R, Binkley G, Chan ET, et al. *Saccharomyces* Genome Database: the genomics resource of budding yeast. Nucleic Acids Res. 2012;40:D700–5.

89. Abdulrehman D, Monteiro PT, Teixeira MC, Mira NP, Lourenco AB, dos Santos SC, et al. YEASTRACT: providing a programmatic access to curated transcriptional regulatory associations in *Saccharomyces cerevisiae* through a web services interface. Nucleic Acids Res. 2011;39:D136–40.

90. Monteiro PT, Mendes ND, Teixeira MC, d'Orey S, Tenreiro S, Mira NP, et al. YEASTRACT-DISCOVERER: new tools to improve the analysis of transcriptional regulatory associations in *Saccharomyces cerevisiae*. Nucleic Acids Res. 2008;36:D132–6.

91. Teixeira MC, Monteiro P, Jain P, Tenreiro S, Fernandes AR, Mira NP, et al. The YEASTRACT database: a tool for the analysis of transcription regulatory associations in *Saccharomyces cerevisiae*. Nucleic Acids Res. 2006;34:D446–51.

92. Russell DW, Jensen R, Zoller MJ, Burke J, Errede B, Smith M, et al. Structure of the *Saccharomyces cerevisiae* HO gene and analysis of its upstream regulatory region. Mol Cell Biol. 1986;6:4281–94.

11

Untreated *Chlorella homosphaera* biomass allows for high rates of cell wall glucan enzymatic hydrolysis when using exoglucanase-free cellulases

Marcoaurélio Almenara Rodrigues[1], Ricardo Sposina Sobral Teixeira[3], Viridiana Santana Ferreira-Leitão[2,3] and Elba Pinto da Silva Bon[3*]

Abstract

Background: Chlorophyte microalgae have a cell wall containing a large quantity of cellulose I_α with a triclinic unit cell hydrogen-bonding pattern that is more susceptible to hydrolysis than that of the cellulose I_β polymorphic form that is predominant in higher plants. This study addressed the enzymatic hydrolysis of untreated *Chlorella homosphaera* biomass using selected enzyme preparations, aiming to identify the relevant activity profile for the microalgae cellulose hydrolysis. Enzymes from *Acremonium cellulolyticus*, which secretes a complete pool of cellulases plus β-glucosidase; *Trichoderma reesei*, which secretes a complete pool of cellulases with low β-glucosidase; *Aspergillus awamori*, which secretes endoglucanases and β-glucosidase; blends of *T. reesei-A. awamori* or *A. awamori-A. cellulolyticus* enzymes; and a purified *A. awamori* β-glucosidase were evaluated.

Results: The highest initial glucan hydrolysis rate of 140.3 mg/g/h was observed for *A. awamori* enzymes with high β-glucosidase, low endoglucanase, and negligible cellobiohydrolase activities. The initial hydrolysis rates when using *A. cellulolyticus* or *T. reesei* enzymes were significantly lower, whereas the results for the *T. reesei-A. awamori* and *A. awamori-A. cellulolyticus* blends were similar to that for the *A. awamori* enzymes. Thus, the hydrolysis of *C. homosphaera* cellulose was performed exclusively by the endoglucanase and β-glucosidase activities. X-ray diffraction data showing negligible cellulose crystallinity for untreated *C. homosphaera* biomass corroborate these findings. The *A. awamori-A. cellulolyticus* blend showed the highest initial polysaccharide hydrolysis rate of 185.6 mg/g/h, as measured by glucose equivalent, in addition to the highest predicted maximum glucan hydrolysis yield of 47% of total glucose (w/w). *T. reesei* enzymes showed the lowest predicted maximum glucan hydrolysis yield of 25% (w/w), whereas the maximum yields of approximately 31% were observed for the other enzyme preparations. The hydrolysis yields were proportional to the enzyme β-glucosidase load, indicating that the endoglucanase load was not rate-limiting.

Conclusions: High rates of enzymatic hydrolysis were achieved for untreated *C. homosphaera* biomass with enzymes containing endoglucanase and β-glucosidase activities and devoid of cellobiohydrolase activity. These findings simplify the complexity of the enzyme pools required for the enzymatic hydrolysis of microalgal biomass decreasing the enzyme cost for the production of microalgae-derived glucose syrups.

Keywords: *Chlorella homosphaera*, Chlorophyte cell wall crystallinity, *Acremonium cellulolyticus*, *Trichoderma reesei* and *Aspergillus awamori* cellulases, Microalgae biomass enzymatic hydrolysis, Microalgae glucose sugar syrups

* Correspondence: elba1996@gmail.com
[3]Federal University of Rio de Janeiro, Institute of Chemistry, Department of Biochemistry, Enzyme Technology Laboratory, 21941-909 Rio de Janeiro, RJ, Brazil
Full list of author information is available at the end of the article

Background

In the pursuit of a new, renewable, and environmentally friendly feedstock supply, the use of seed plants, energy crops, grasses, and agricultural wastes has raised serious concerns regarding the use of land and water for food versus feedstock for chemicals and fuel production. Alternatively, attention has been turned to the use of algae, in particular microalgae, as feedstock for the production of renewable chemicals and biofuels, as they do not compete for arable land [1] despite some skeptical views [2,3]. Microalgae are versatile photosynthetic organisms that can adapt to and tolerate a variety of environmental conditions and can be cultivated in non-freshwater sources including salt and wastewater [4,5] besides being able to use flue gases as a source of carbon, sulfur, and nitrogen [6]. The calorific values of microalgae can be increased by growing them in tubular bioreactors [7] or in cheaper, low nitrogen media [8] because nitrogen starvation triggers lipid and carbohydrate accumulation [9]. Several products can be obtained from the same algal biomass using an integrated biorefinery concept whereby different biomass processing methods can be sequentially employed, including lipid extraction, algae biomass polysaccharide hydrolysis for sugar syrup production and subsequent fermentation [10], followed by biodigestion or pyrolysis for the production of biodiesel, bioethanol, methane, and syngas [11-13]. Furthermore, milder conditions for microalgae, with the aim of biomass polysaccharide hydrolysis, are necessary for the production of sugar syrups in comparison with lignocellulosic biomass processing [14,15].

The carbohydrates in green algae, primarily cellulose, come chiefly from the cell wall components although depending on the growth conditions in which starch accumulates in the chloroplasts [16]. Early studies on the algal cell wall [17] classified the algae species into three groups according to their cell wall constituents. Group 1 includes chlorophyte algae, in which native cellulose is the major component of the cell walls and is usually highly crystalline. These algae belong to the Cladophorales order. Group 2 includes the chlorophyte algae, which have cell walls that contain a large quantity of mercerized-like cellulose, which is presumably a derivative of native cellulose. This cellulose has a low degree of crystallinity, and the chains are randomly oriented. Most algae fall into this category. Group 3 contains algae for which the walls contain a well-oriented and highly crystalline skeletal substance, which is neither native nor mercerized cellulose, as the major constituent [17]. Three types of cellulose were identified in the algal system as follows: the I_α-rich, broad microfibril, 0.6 nm-oriented type; the I_β-dominant flat-ribbon, 0.53 nm-oriented type; and the I_β-dominant, small, random-oriented type. The first type appears to occur in more primitive organisms

than the other types. The three types of algal cellulose correlate well with the arrangements of cellulose synthesizing complexes, i.e., a multiple-row linear type, a consolidated rosette type, and an isolated rosette type, respectively [18,19]. Linear-type terminal complexes are found in algae belonging to the Chlorophyta division whereas the rosette terminal complex is found in algae belonging to the Charophyta division and land plants [18-21]. Cellulose I_α differs from I_β not only in the hydrogen-bonding pattern, and consequently in the crystalline unit cell [22,23], but also in its stability and susceptibility to hydrolysis. Cellulose I_α, which possesses a triclinic unit cell, is metastable and converts into the monoclinic form (I_β) after annealing in dilute alkali at 260°C [24]; the theoretical density of the monoclinic unit cell is slightly greater than that of the triclinic unit cell [25]. These differences in the stability, hydrogen-bonding pattern, and distribution of the two crystalline phases in the microfibril may render cellulose I_α more prone to enzymatic hydrolysis, as indicated by FTIR and electron diffraction data [25,26]. The X-ray diffraction patterns of cellulose in plants and algae differ in the presence of well-resolved and narrow peaks, especially at 2θ's of 14° and 16°, which are not common for native cellulose obtained from higher plants [27].

Chlorella homosphaera is a microalga that belongs to the Chlorophyta division. The microalgae from this genus divide by autosporulation and therefore possess lytic activity [28-31] that leads to the enzymatic dissolution of their inner cell wall polysaccharides. The *Chlorella* cell wall is formed by an inner and outer layer and possesses unusual polysaccharides other than cellulose and hemicellulose [32,33]. The inner layer is composed of a matrix and a rigid fibrillar structure, shows high cellulose content, and is susceptible to degradation by cellulolytic enzymes [34]. The outer cell wall, which is resistant to treatment with a number of enzymes, presents two types of ultrastructure, one of which is a trilaminar structure with several resistant components including the biopolymer algaenan, a non-hydrolyzable aliphatic biopolymer composed of long-chain, even-carbon-numbered, ω9-unsaturated ω-hydroxy fatty acid monomers that vary in their chain lengths from 30 to 34 carbon atoms. These monomers are intermolecularly ester-linked to form linear chains in which the unsaturated carbons act as the starting position of ether cross-linking [35].

This study evaluates the enzymatic hydrolysis of polysaccharides from untreated *C. homosphaera* biomass with enzyme preparations that were produced by the fungi *Trichoderma reesei*, *Aspergillus awamori*, and *Acremonium cellulolyticus*, in addition to the mixtures of enzymes produced by *T. reesei* and *A. awamori* or *A. cellulolyticus* and *A. awamori*. The focus was to identify the enzyme activities necessary for the efficient hydrolysis of

C. homosphaera cell wall polysaccharides because each enzyme preparation showed a different enzyme activity profile and load concerning its endoglucanase, cellobiohydrolase, and β-glucosidase content, aside from the differences regarding their xylanolytic enzyme profiles. This work also reports on the initial hydrolysis rates, the effect of the β-glucosidase load on the hydrolysis rates, and the final hydrolysis yields. The XRD crystallinity profile of the *C. homosphaera* cell wall from untreated algal biomass is also presented.

Results and discussion

C. homosphaera biomass sugar composition

Data for the total sugar content and sugar composition of *C. homosphaera* biomass dry weight for the biomass total hydrolysis by 1 mol/L H_2SO_4 followed by high-performance anion exchange chromatography and pulse amperometric detection (HPAEC) analysis is presented in Table 1. *C. homosphaera* biomass possessed a carbohydrate content of 59.7% with a glucose contribution of 53.9%. Considering that amyloglucosidase hydrolysable starch (AHS) contributed to 7% (*w/w*) of the total glucose, non-AHS polysaccharides corresponded to 46.1% of the *C. homosphaera* biomass. As such, glucose constituted 90.4% of all carbohydrates present in *C. homosphaera* (85.5% non-AHS polysaccharides) besides the minor contributions of galactose (5.4%), arabinose (2.6%), mannose (1.3%), and xylose (0.5%), which makes this microalga a glucose-rich biomass. This high carbohydrate content is caused by the growth conditions used in this study; after nitrate depletion, within the fifth day of cultivation, the microalga was kept under nitrogen starvation for 15 days before harvest. Nitrogen starvation reportedly triggers the accumulation of carbohydrates, primarily starch, and lipids in both *Chlorella* [9,16,36] and *Chlamydomonas reinhardtii* [37,38], increasing their calorific value [8]. The low AHS content of *C. homosphaera* late stationary grown phase cells found in this study suggests that reserve carbohydrates, which may have accumulated during the exponential

phase of growth as suggested by the Lugol's solution coloration used for cell counting (data not shown), were converted into structural carbohydrates and incorporated into the algal cell wall [37-39] primarily as a glucan type, most likely cellulose. Accordingly, cellulose was the primary polysaccharide present in *Chlorella fusca* [30,31] and *Coelastrum sphaericum* [35]; however, a similar low starch content was found in the chlorophyte *Chlorococcum humicola* [40].

These results resemble those previously reported for the cell wall sugar composition of some Chlorophytes [41-43], for which glucose, galactose, mannose, arabinose, xylose, rhamnose, and sometimes fucose were detected. A similar composition was also found in the TFA-hydrolyzed microalgal biomass of *C. reinhardtii* [37] and mucilage sheaths of *C. sorokiniana*, although sucrose was among the saccharides that were found [44]. Interestingly, glucose and mannose are the monosaccharides that are reportedly found in the rigid fibrillar structure of the inner layer of the cell wall, and all the others were found only in the matrix structure of the inner layer, with glucose as the primary saccharide [42,43]. Because we estimated the sugar composition of the *C. homosphaera* total biomass and not that of the purified cell walls, we expected that non-cell wall polysaccharides such as starch would contribute to the total glucose composition, as already noted; nevertheless, the chlorophyte microalgae *Scenesdesmus obliquus* possesses a cell wall that is enriched with glucose [45].

Biomass enzymatic hydrolysis

Hydrolysis experiments were performed using enzymes that were produced by selected fungal strains according to their different profiles in relation to cellulase, β-glucosidase, and xylanase activities, as follows: *Acremonium* cellulase (Meiji Seika Co., Japan), a preparation possessing a complex set of biomass-hydrolyzing enzyme activities; exoglucanase (4.3 filter paper unit (FPU)/mL) and endoglucanase (67.1 IU/mL); xylanase (208.0 IU/mL) and β-glucosidase (29.2 IU/mL); a T. reesei RUT C-30 preparation containing exoglucanase (1.2 FPU/mL), endoglucanase (37.3 IU/mL), xylanase (228.0 IU/mL), and low β-glucosidase content (1.4 IU/mL); the *A. awamori* 2B.361 U2/1 enzyme was devoid of exoglucanase but exhibited xylanase (117.0 IU/mL), β-glucosidase (8.4 IU/mL), and low endoglucanase (1.9 IU endo-β-1,4 glucanase (CMCase)/mL).

The *A. awamori*-*T. reesei* blend contained exoglucanase (5.3 IU FPU/mL), endoglucanase (20.5 IU/mL), xylanase (202.0 IU/mL), and β-glucosidase (69.5 IU/mL) activities; the supernatants of individual fungal cultures were concentrated by ultrafiltration using a 30-kDa membrane before blending. The *A. awamori*-*A. cellulolyticus* blend possessed exoglucanase (2.0 FPU/mL), endoglucanase

Table 1 The sugar composition determined by the HPAEC-PAD of *C. homosphaera* after acidic hydrolysis with 1 mol/L H_2SO_4

Sugar	% Biomass	% Total sugar
Glucose	53.9 (5.4)	90.4 (9.0)
Galactose	3.2 (0.2)	5.4 (0.3)
Arabinose	1.6 (0.1)	2.6 (0.2)
Mannose	0.8 (0.1)	1.3 (0.1)
Xylose	0.3 (0.0)	0.5 (0.0)
Total	59.7 (5.6)	100.2

Hydrolysis and HPAEC-PAD analysis conditions are described in the 'Materials and methods' section. The numbers shown in the brackets are the standard deviations of the means.

(32.5 IU/mL), xylanase (159.7 IU/mL), and β-glucosidase (18.2 IU/mL) activities.

Sugar analyses

Table 2 shows the sugar composition and yield released from the *C. homosphaera* biomass after 48 h of enzymatic hydrolysis. The overall analysis of the effectiveness of the different enzyme preparations in hydrolyzing the algae cell wall polysaccharides indicates that the *A. awamori-A. cellulolyticus* enzyme blend released the highest amounts of glucose and galactose, besides releasing mannose, and was the only preparation that was able to release the pentoses xylose and arabinose and thus able to attack pentose-containing polysaccharides, besides glucan and galactose and/or mannose-containing polysaccharides. The *A. awamori-T. reesei* blend was able to hydrolyze glucan aside from the galactose and/or mannose-containing polysaccharides, and all the other enzyme preparations were solely able to hydrolyze glucan and galactose-containing polysaccharides, which were noteworthy in that the *A. cellulolyticus* enzymes released the largest amount of galactose. Partially purified β-glucosidase was solely able to release limited amounts of glucose in comparison with the non-enzymatic control experiments; the small detected galactose release can be considered non-enzymatic when taking into account the data for the non-enzymatic galactose release. No cellobiose was detected in the hydrolysate for all enzyme preparations. Because *A. awamori* can also produce amylases [46], one could argue that the high glucose content found in the hydrolysates of *A. awamori* and the *T. reesei-A. awamori* blend might result from starch hydrolysis. To test this hypothesis, 50 mg/mL starch solution was incubated under the same conditions used for algae biomass hydrolysis for each enzyme preparation; however, all preparations were unable to hydrolyze starch (data not shown). Because the

C. homosphaera starch content was not converted into glucose, all the measured glucose was solely released from the cell wall polysaccharides.

Glucose and mannose are reportedly found in the rigid fibrillar structure of the inner layer of the *C. homosphaera* cell wall, and galactose, arabinose, and xylose are present in the matrix of the cell wall inner layer [42,43]. Thus, the absence of mannose in all enzyme preparations but its presence in the *A. awamori-A. cellulolyticus* and *A. awamori-T. reesei* blends might indicate that these enzyme preparations are able to hydrolyze only the glucan and galactose-containing polysaccharides in the matrix of the inner cell wall. However, the presence of mannose in the hydrolysates resulting from the use of *A. awamori-A. cellulolyticus* and *A. awamori-T. reesei* blends indicate that these preparations were able to attack the rigid fibrillar structure of the inner cell wall layer. Moreover, because xylose and arabinose, in addition to glucose, galactose, and mannose, were detected in the hydrolysates resulting from the use of the *A. awamori-A. cellulolyticus* blend, this preparation was also able to attack different glycosidic bonds or a different structural polysaccharide. The use of enzyme blends showed the synergistic effect of the enzymes produced by these two fungi. The importance of hemicellulases in the polysaccharide hydrolysis in the algae inner layer cell wall was not assessed in this study. However, even when considering that substances other than glucose monosaccharides represent less than 10% of the total algae biomass sugar, key glucosidic bonds involving galactose, arabinose, mannose, and xylose may hinder high-yield glucan hydrolysis. In this context, it is noteworthy that the best performing enzyme preparation, i.e., the *A. awamori-A. cellulolyticus* blend that resulted in the highest glucose hydrolysis yield, also released the highest amount of galactose, besides releasing mannose, arabinose, and xylose, most likely because of the action of the hemicellulose-degrading enzyme complex.

Table 2 The sugar composition of hydrolyzed *C. homosphaera* biomass after 48 h of enzymatic treatment

Enzyme preparation	Glucose		Galactose		Mannose		Total yield (%)
	g/L	Yield (%)	g/L	Yield (%)	g/L	Yield (%)	
T. reesei	6.6 (0.0)	23.4 (0.1)	0.2 (0.0)	0.5 (0.0)	0.0	0.0	24.0 (0.1)
A. cellulolyticus	9.9 (0.4)	35.0 (1.3)	0.5 (0.0)	1.8 (0.0)	0.0	0.0	36.7 (1.3)
A. awamori	10.0 (0.2)	35.4 (0.6)	0.3 (0.0)	1.2 (0.0)	0.0	0.0	35.4 (0.6)
T. reesei-A. awamori	9.4 (0.2)	33.1 (0.7)	0.1 (0.0)	0.4 (0.0)	0.1 (0.0)	0.1 (0.0)	33.5 (0.7)
A. cellulolyticus-A. awamori[a]	11.4 (1.2)	40.2 (4.2)	0.6 (0.0)	2.0 (0.1)	0.0 (0.0)	0.1 (0.0)	42.4 (4.3)
β-Glucosidase	2.1 (0.1)	7.6 (0.3)	0.1 (0.0)	0.3 (0.0)	0.0	0.0	7.9 (0.3)
Control	1.2 (0.1)	4.2 (0.3)	0.1 (0.0)	0.4 (0.0)	0.0	0.0	4.6 (0.4)

The results are shown as the sugar concentration (g/L) and hydrolysis yield in terms of the total sugar concentration (%). The hydrolysis and HPAEC-PAD analysis conditions are described in the 'Materials and methods' section. As a control, the sugar released from the biomass that was incubated without any enzyme preparation is also shown. Numbers shown in brackets are standard deviations of the means.
[a]Xylose and arabinose at a low concentration (0.03 g/L; 0.12% yield) were released solely for the hydrolysis performed by the *A. cellulolyticus-A. awamori* enzyme blend.

Nevertheless, interesting results were obtained in this study with regard to the selective hydrolysis of structures in the algae inner cell wall, and conclusive data regarding the specificity of the selected enzyme pool towards the structural polysaccharides could only be gathered from hydrolysis experiments that were performed with purified components of the cell wall. Those results would be compared to sequentially TFA and H_2SO_4-hydrolyzed cell wall components because TFA hydrolyzes only the matrix of the inner cell wall and H_2SO_4 hydrolyzes both the matrix and the inner cell wall fibrillar structure [47]. This question will be a matter of further investigation.

These results are consistent with the time course of algae biomass hydrolysis shown in Figures 1 and 2. The glucose and total reducing sugar yields curves are very similar when the biomass was hydrolyzed with the enzymes of *T. reesei*, *A. awamori*, or a mixture of both. However, when the enzymes of *A. cellulolyticus* or the *A. awamori-A. cellulolyticus* blend were used, the amount of released glucose was noticeably lower than that of the total reduced sugar, suggesting that non-glucan-type polysaccharides were hydrolyzed.

C. homosphaera biomass enzymatic hydrolysis

Figures 1 and 2 present time course profiles for the release yield of total reducing sugars and of glucose, in terms of total sugar and total glucose, respectively, on the enzymatic hydrolysis of *C. homosphaera* biomass. In Figure 1, curves for the *A. cellulolyticus*, *A. awamori*, *A. cellulolyticus-A. awamori* blend, partially purified β-glucosidase, and the control were best fitted into a mono-exponential model, and the curves for *T. reesei* and the *T. reesei-A. awamori* blend were best fitted into

a double exponential model. In Figure 2, the curve for *T. reesei* was best fitted into a double exponential model, and the others were best fitted into a mono-exponential model. Figure 1 shows that the initial rates and final yields for the release of reducing sugars for the *A. cellulolyticus-A. awamori* blend were the highest followed by *A. awamori* and *T. reesei-A. awamori* blend enzyme preparations, which were comparable to one another, plateauing at a 10-h reaction time. The enzymes from *A. cellulolyticus* exhibited a lower initial rate, plateauing with comparable hydrolysis yields at an 18-h reaction time. Enzymes from *T. reesei* showed the smallest initial rate and did not reach the yield plateau within the time frame of the hydrolysis reaction. The partially purified β-glucosidase, which is devoid of cellobiohydrolase, endoglucanase, and xylanase activities, showed a discrete ability to hydrolyze the *C. homosphaera* biomass with a low initial rate and maximum yield, indicating that the β-glucosidase excreted by *A. awamori* was able to attack the algae biomass glucan in addition to its well-known action on cellobiose and short chain oligosaccharides [48]. However, the results showed that the synergistic action of endoglucanases and β-glucosidase is essential to attain higher hydrolysis yields. The *C. homosphaera* biomass also showed a small amount of sugar release unrelated to the enzymatic action (control) for up to 10 h of incubation, possibly because of the reaction conditions, including the pH and temperature, or because of the residual action of native autolytic enzymes. It is well established that chlorophytes reproduce by autospore formation, in which daughter cells are enclosed inside the mother cell wall; autolytic enzymes are used to degrade the mother cell wall and release the daughter cells, which then may use the hydrolyzed monosaccharides as

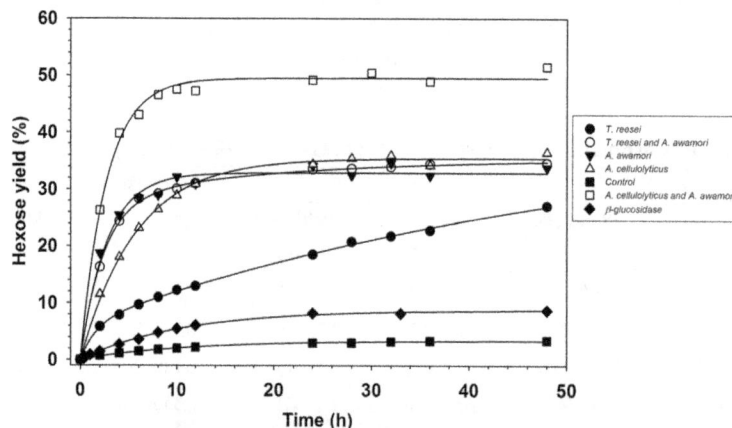

Figure 1 A time course for the release of total reducing sugars from the *C. homosphaera* biomass (50 mg d.w./mL) in 50 mM citrate buffer, pH 4.8, at 50°C. The biomass was hydrolyzed using the enzyme preparations from *T. reesei* (filled circles), *A. cellulolyticus* (open triangles), *A. awamori* (inverted filled triangles), the *T. reesei-A. awamori* blend (open circles), the *A. cellulolyticus-A. awamori* blend (open squares), and partially purified β-glucosidase (diamonds). Biomass suspended in buffer without enzyme preparation was used as a control (filled squares). The data were fitted into an exponential function as described in the 'Materials and methods' section. Standard deviation was less than 10% of the mean value, and bars were omitted for clarity.

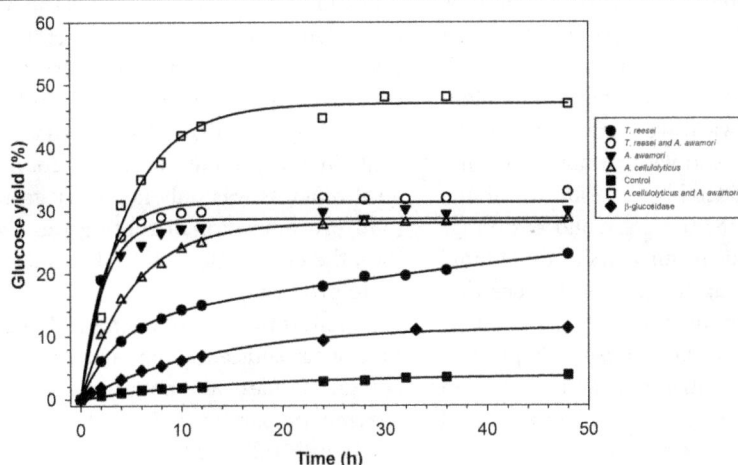

Figure 2 The time course for the release of glucose from *C. homosphaera* biomass (50 mg d.w./mL) in 50 mM citrate buffer, pH 4.8, at 50°C. The biomass was hydrolyzed using the enzyme preparations from *T. reesei* (filled circles), *A. cellulolyticus* (open triangles), *A. awamori* (inverted filled triangles), the *T. reesei-A. awamori* blend (open circles), the *A. cellulolyticus-A. awamori* blend (open squares), and partially purified β-glucosidase (diamonds). Biomass suspended in buffer without enzyme preparation was used as the control (filled squares). The data were fitted into an exponential function as described in the 'Materials and methods' section. Standard deviation was less than 10% of the mean value and bars were omitted for clarity.

carbon and energy sources [29-31,34]. Figure 2 shows the algae biomass hydrolysis pattern, including the initial rates and final yields for glucose release for all enzyme preparations. The general enzyme preparation performances were quite similar, as expected considering that glucan is the most abundant component of the *C. homosphaera* cell wall, except for the fact that the initial rate of glucose release for the *A. awamori-A. cellulolyticus* blend was considerably smaller than the release of total reducing sugar, which was consistent with the results shown in Table 2.

Table 3 summarizes the data for the predicted kinetic parameters regarding initial hydrolysis rates and maximum hydrolysis yields. The data are related to the total

sugar (hexose equivalents) and total glucose (glucose) hydrolysis yields. The initial hydrolysis rate of *A. awamori* preparation, when considering hexose equivalents, was 114.3 mg hexose/g total sugar/h, and the rate for the *T. reesei-A. awamori* blend was 115.8 mg hexose/g total sugar/h. Enzymes from *A. cellulolyticus* showed a lower initial velocity of 58.8 mg hexose/g total sugar/h); nevertheless, the maximum predicted hydrolysis yield of 336.2 mg/g was similar to that for *A. awamori* (311.7 mg/g) and for the *T. reesei-A. awamori* blends (335.4 mg/g). Enzymes from *T. reesei* showed the smallest initial rate (27.5 mg hexose/g sugar/h); because the hydrolysis plateau was not reached within the time frame

Table 3 Predicted kinetic parameters for *C. homosphaera* enzymatic hydrolysis yields

Enzyme preparation	Predicted parameters					
	Hexose equivalents (SE)			Glucose (SE)		
	V_0 (mg/g/h)	P_{max} (mg/g)	$t\frac{1}{2}$ (h)	V_0 (mg/g/h)	P_{max} (mg/g)	$t\frac{1}{2}$ (h)
T. reesei[a]	27.5 (6.3)	256.8 (38.9)	12.9	33.7 (4.3)	255.56 (40.2)	5.9
A. cellulolyticus	58.8 (2.1)	336.2 (3.2)	4.0	61.82 (2.7)	311.1 (3.5)	3.5
A. awamori	114.3 (9.0)	311.7 (9.0)	1.9	140.3 (16.9)	318.2 (6.3)	1.6
T. reesei and *A. awamori*	115.8 (16.1)	335.4 (27.3)	2.2	149.8 (29.2)	348.0 (11.3)	1.6
A. cellulolyticus and *A. awamori*	185.6 (16.7)	494.6 (8.6)	1.8	103.5 (7.9)	470.8 (9.3)	3.2
β-Glucosidase	8.4 (0.4)	88.0 (1.9)	7.3	8.9 (0.5)	116.0 (2.6)	9.0
Control	3.1 (0.1)	33.6 (0.5)	7.4	2.5 (0.3)	37.3 (1.5)	10.4

The data were fitted into an exponential model as described in the materials and methods, and the parameters were estimated as follows: the V_0 was estimated by the limit of the first derivative of the fitting function when time → 0, P_{max} was estimated by the limit of the fitting function when time → ∞, $t\frac{1}{2}$ was estimated using the curve fitting model by calculating the predicted $P_{max}/2$. The number in brackets accounts for the standard error of the estimation.
[a]Experimental results were used for the calculations because the parameters predicted by the fitting function showed very large standard errors. The V_0 was calculated using the first experimental result, i.e., the hydrolysis yield after two hours of incubation. The P_{max} was estimated using the 48 h hydrolysis yield. The $t\frac{1}{2}$ was estimated by calculating the $P_{max}/2$ and estimating the corresponding time using the fitting function.

of the experiment and the predicted maximum value by the model was unreliable, the mean experimental result (±standard deviation) at 48 h of incubation was used as the data for the hydrolysis yield of 256.8 mg/g. The highest initial rate and maximum yield were attained with the *A. awamori-A. cellulolyticus* blend with an initial rate and a final yield that were considerably superior to all the other enzymes preparations (185.6 mg/g/h and 494.6 mg/g, respectively), suggesting that the enzymes of these fungi act synergistically in hydrolyzing the polysaccharides that each enzyme preparation cannot hydrolyze by itself as already shown by the HPAEC-PAD. The partially purified β-glucosidase showed a low initial rate and maximum yield and the biomass incubated without enzymes released some sugar at a very low initial rate and yield.

Following the same trend, the enzyme pool of *A. awamori* and the blends of *A. awamori-T. reesei* and *A. awamori-A. cellulolyticus* took approximately 2 h to accomplish 50% of the maximum hydrolysis yield ($t\frac{1}{2}$). *A. cellulolyticus* enzymes took 4 h and *T. reesei* enzymes took 13 h to achieve 50% of the 48 h hydrolytic value (Table 3). The predicted kinetic parameters for the enzymatic hydrolysis results when using all enzyme preparations, and taking into account the glucose concentration measurements, were similar to those observed for the measurement of the total reducing sugars, as previously discussed, except for the *A. awamori-A. cellulolyticus* blend, which showed a similar initial rate for glucose release when compared with the enzyme preparations of *A. awamori* and the *A. awamori-T. reesei* blend. Because the glucose yield was considerably higher (470.8 mg glucose/g sugar), the $t\frac{1}{2}$ was larger for the *A. awamori-A. cellulolyticus* blend (Figures 1 and 2 and Table 3). The initial velocities found for *T. reesei* and *A. cellulolyticus* enzymes are close to those previously reported by Harun and Danquah [40] for *Chlorococum humicola* when the same enzyme load was applied. Nevertheless, those found for *A. awamori* and *A. awamori-T. reesei* were 1.5 times higher and those for the *A. awamori-A. cellulolyticus* blend were 2.4 higher. Moreover, with the exception of the *T. reesei* enzyme pool, we used the Marquardt non-linear regression method, which is a much more reliable procedure for estimating the initial velocity (Table 3). These results suggest that the *A. awamori* enzyme pools efficiently hydrolyze *C. homosphaera* biomass, and the results observed for the *A. awamori-T. reesei* blend resulted predominantly from the contribution of the *A. awamori* enzyme pool. The *A. awamori-A. cellulolyticus* blend hydrolyzes the *C. homosphaera* biomass synergistically to attain a high rate and yield for hydrolyses.

The maximum hydrolytic yield found for all enzyme pools were either similar [40] or lower, sometimes considerably lower (near 50%), than those previously reported [16]. However, in reports in which a higher hydrolytic yield was found, the algae biomass showed a

fourfold higher starch content [16], and the material was subjected to severe treatment beforehand, such as sonication, disc milling, or hydrothermal, acid, or alkali pretreatment before enzymatic hydrolysis [14,16,38,40,45], which may have disrupted the structure of the outer cell wall. In the present work, the room temperature dried algal biomass was only manually ground, which is a very mild procedure for disrupting the algae cell wall; therefore, the enzymatic hydrolysis results reported herein are quite promising.

The maximum yield was near 50 mg of reducing sugar/g total sugar, indicating that approximately 50% of the algae polysaccharides were not accessed by either of the studied enzyme preparations. Considering the structure of the whole cell wall and that the outer cell wall is resistant to treatments with a number of enzymes, we could speculate that only the inner cell wall layer polysaccharides from either the rigid fibrillar structure or the matrix were selectively or entirely hydrolyzed by the studied enzyme preparations. As such, the residual non-hydrolyzed polysaccharide fraction, from the higher yield hydrolysis experiments, would be a component of the outer cell wall. In reports in which the starch content was similar to ours, a comparable maximum yield was found [40]. Other studies about algae biomass pretreatments are under way, in our laboratory, to increase the maximum hydrolysis yields.

The hydrolytic efficiency of the *A. awamori*, *T. reesei*, and *A. cellulolyticus* enzyme pools and blends

Because the enzyme pool of *A. awamori* has mostly β-glucosidase activity in addition to endoglucanase activity and negligible cellobiohydrolase activity, a lower initial hydrolysis rate and a lower maximum hydrolysis yield were expected in comparison with the complete cellulases plus β-glucosidase preparations of *A. cellulolyticus* and the blend of *T. reesei* and *A. awamori* enzymes. In fact, we would expect a pattern of hydrolysis similar to the partially purified β-glucosidase, but with a higher initial rate and yield because it possesses endoglucanase activity and the partially purified β-glucosidase does not. However, a similar rate and hydrolytic efficiency were observed for the *A. awamori* enzyme preparation and the *A. awamori-T. reesei* blend in comparison with the *A. cellulolyticus* enzymes, indicating that exoglucanase (CBH) activity is not necessary for the hydrolysis of this microalgae biomass.

The use of the *A. awamori-A. cellulolyticus* enzyme blend resulted in a hydrolysis reaction medium with the highest β-glucosidase load (90.3 IU/g d.w. biomass) followed by the *A. cellulolyticus* enzymes (49.0 IU/g) and by both the *A. awamori* enzymes and the *T. reesei-A. awamori* enzyme blend (22.3 IU/g). The *T. reesei* enzymes showed the lowest β-glucosidase load (8.7 IU/g). With the exception of *A. cellulolyticus*, the initial hydrolysis rates and final yields correlated positively with the β-glucosidase. To understand further the importance of β-glucosidase activity, we conducted two sets of hydrolysis assays with *T. reesei*, *A. awamori*, *A. cellulolyticus*, and a blend of *A. awamori-T. reesei* enzyme preparations. In the first set, the assays were performed with the same β-glucosidase load for 3 h because more striking differences regarding the initial hydrolysis yields were observed within this time interval. A control experiment, with the same load of a partially purified β-glucosidase and devoid of endo- and exoglucanase activities, was conducted. Figure 3 shows that after 3 h of hydrolysis, no significant differences among the means of hydrolyzed hexose equivalents or glucose were observed for any enzyme preparation. However, the *A. cellulolyticus* enzymes showed lower mean values, mostly for glucose, which were significant when compared with the *A. awamori* enzyme, confirming the previous results shown in Figures 1 and 2, where despite having a higher β-glucosidase load, it showed a lower initial rate but a similar final hydrolysis yield.

The hydrolysis results from the control experiments were surprisingly high, accounting for over 40% and 50% of the total hexose and glucose released, respectively (Figure 3). The data were even higher than that presented in Figures 1 and 2, which could be related to differences in the uneven manual grinding of the algae biomass, affecting the particle sizes as we observed in preliminary assays in our laboratory (data not shown). The *C. homosphaera* biomass also showed a small amount of biomass hydrolysis not caused by enzymatic action as previously found (Figures 1 and 2). The different enzyme preparations contained different endoglucanase loads (Figure 3),

which might have played an important role in the hydrolysis results, although no correlation between the endoglucanase load and initial rates and final yields was found. For this reason, all enzyme preparations in the second set of assays were set to a fixed endoglucanase load (1.5 IU/g dry biomass) and increasing β-glucosidase loads as follows: 7.5, 15.0, and 22.5 IU/g dry biomass, which were achieved through purified β-glucosidase supplementation. Figure 4 shows that after 36 h of incubation, all enzymes showed an increase in the hydrolysis yield, especially when the β-glucosidase load increased from 7.5 to the 15 IU/g that was significant for all enzyme preparations. The *T. reesei* preparation showed no significant increase for the hydrolysis yield of reducing sugar when the β-glucosidase load increased from 15.0 to 22.5 IU/g; however, a significant increase in the hydrolysis yield of glucose was found. The same pattern was observed for the *A. cellulolyticus* preparation. However, no significant increase in the glucose hydrolysis yield was found for the *A. awamori* enzyme and the *A. awamori-T. reesei* blend; nevertheless, the reducing sugar yield increase was significant. An analysis among the groups showed that the hydrolysis yields for *A. awamori* enzyme preparations were the highest even when compared with experiments presenting a higher β-glucosidase load. Moreover, the *A. awamori* preparation was more responsive to the increase in the β-glucosidase load (Figure 4), indicating a unique catalytic feature for the *A. awamori* enzymes.

Data for the predicted kinetics parameters regarding the initial hydrolysis rates and maximum yields of each enzyme preparation with the three β-glucosidase load

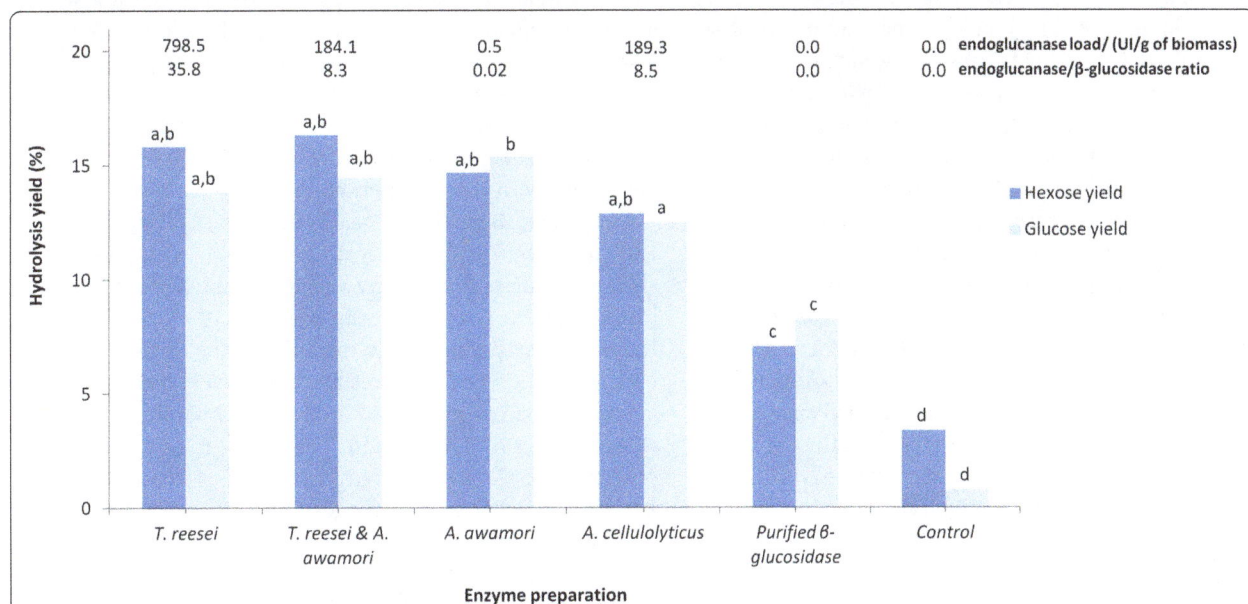

Figure 3 The hydrolysis yield of *C. homosphaera* biomass after 3 h of enzymatic treatment. *C. homosphaera* biomass (50 mg/mL) was incubated in 50 mM citrate buffer pH 4.8 at 50°C with the indicated enzyme preparation at a final β-glucosidase activity of 23.3 IU/g d.w. biomass load. Biomass without enzyme was incubated under the same conditions as the control. Letters a, b, c, and d indicate differences in the means at 0.05 level.

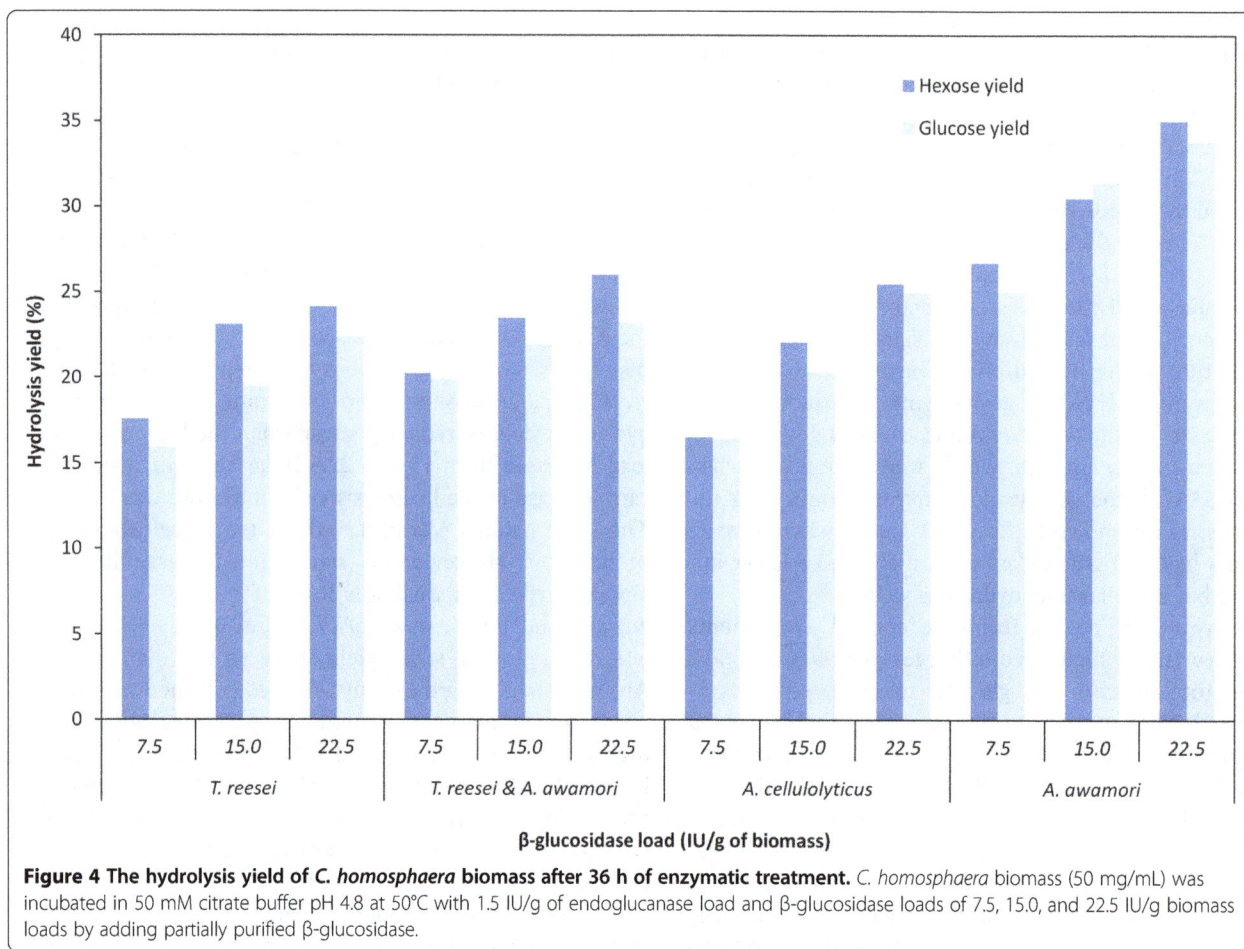

Figure 4 The hydrolysis yield of *C. homosphaera* biomass after 36 h of enzymatic treatment. *C. homosphaera* biomass (50 mg/mL) was incubated in 50 mM citrate buffer pH 4.8 at 50°C with 1.5 IU/g of endoglucanase load and β-glucosidase loads of 7.5, 15.0, and 22.5 IU/g biomass loads by adding partially purified β-glucosidase.

(Additional file 1: Figure S1 and Additional file 2: Figure S2) are shown in Table 4. The predicted maximum hydrolysis yield for all enzyme preparations followed the same pattern presented in Figure 4. The analysis within the group for the initial hydrolysis rates indicated that *T. reesei* and *A. cellulolyticus* preparations showed a significant increase when the β-glucosidase load was increased from 7.5 to 15 IU/g. A small increase was also found when the β-glucosidase load was increased from 15 to 22.5 IU/g, but the model failed to predict any significant difference in the initial rate values. *A. awamori* and *A. awamori-T. reesei* blend preparations showed no significant increase after the increment of the β-glucosidase load. An analysis among the groups showed that the *A. awamori* preparation showed the highest initial rates of reducing sugar release that were significant when compared with their counterparts, which showed no significant difference among them. The predicted initial rates for glucose release in the *A. awamori* preparation showed a large standard error and are not reliable, but the predicted means suggest higher initial rates when compared with the others.

HPAEC-PAD analyses showed that the released monosaccharides consisted of only glucose and galactose for all hydrolysis conditions, i.e., 1.5 IU endoglucanase load/ g (d.w.) biomass and 7.5, 15.0, and 22.5 IU β-glucosidase load/g (d.w.) biomass. The enzyme preparation of *A. awamori* showed the highest sugar hydrolysis yield, and the increment in the β-glucosidase load increased the amount of sugar that was hydrolyzed. These results are consistent with those presented in Figure 4.

Altogether, these results clearly show that for the same endoglucanase load, the hydrolysis yield of *C. homosphaera* biomass is positively affected by the β-glucosidase load. Moreover, even in the absence of *A. awamori* enzymes, the β-glucosidase load affects the initial rate of hydrolysis. This finding also shows that a very small endoglucanase load is needed to attain high hydrolysis yields within just 12 h of incubation. The higher rates and yields observed for *A. awamori* enzymes suggest that the enzyme prepared from this fungus have a higher ability to hydrolyze *C. homosphaera* biomass.

C. homosphaera cell wall studies

The fact that the *A. awamori* enzymes, which are devoid of exoglucanase activity, hydrolyzed *C. homosphaera* biomass at high rates and that the β-glucosidase activity

Table 4 The predicted kinetic parameters for *C. homosphaera* enzymatic hydrolysis yields at a 1.5 IU/g endoglucanase load and final β-glucosidase loads of 7.5, 15.0 and 22.5 IU/g

Enzyme preparation	β-Glucosidase load (IU/g dry biomass)	Predicted parameters					
		Hexose equivalents (SE)			Glucose (SE)		
		V_0 (mg/g/h)	P_{max} (mg/g)	$t^{1/2}$ (h)	V_0 (mg/g/h)	P_{max} (mg/g)	$t^{1/2}$ (h)
T. reesei	7.5	48.2 (7.5)	203.3 (10.1)	9.5	59.1 (17.5)	184.8 (9.2)	9.3
	15.0	61.0 (16.9)	286.6 (19.1)	9.6	78.0 (17.8)	228.0 (8.9)	9.0
	22.5	93.4 (13.8)	275.1 (19.4)	6.8	91.9 (8.5)	253.5 (3.1)	7.8
A. cellulolyticus	7.5	53.2 (5.9)	176.0 (5.7)	6.5	59.4 (7.9)	190.7 (12.8)	6.4
	15.0	95.5 (13.1)	254.8 (11.6)	7.8	96.5 (10.6)	241.8 (12.2)	8.2
	22.5	98.3 (12.7)	271.1 (14.4)	5.6	129.9 (32.7)	292.2 (4.6)	8.2
A. awamori	7.5	165.3 (24.7)	273.8 (20,7)	1.5	239.4 (100.3)	249.4 (12.8)	1.0
	15.0	176.8 (29.6)	325.8 (14.0)	1.9	254.6 (181.1)	314.2 (15.1)	2.2
	22.5	158.5 (6.8)	452.5 (24.2)	3.6	246.9 (477.9)	329.2 (39.8)	2.4
T. reesei and *A. awamori*	7.5	63.9 (9.2)	226.0 (9.0)	4.2	71.6 (20.8)	226.3 (8.1)	8.2
	15.0	74.1 (15,2)	250.5 (12.4)	4.4	69.3 (17.3)	251.14 (12.0)	7.8
	22.5	95.6 (25.1)	290.6 (13.7)	5.3	88.2 (26.4)	248.7 (9.2)	12.4

The data were fitted into an exponential model as described in the 'Materials and methods' section, and the parameters were estimated as described in Table 3. The numbers in brackets account for the standard error of the estimation.

load influences the hydrolysis yields and sometimes the initial rate, raised a question regarding the nature of the microalgal cell wall glucan. Chlorophytes are known to be enriched with cellulose type I_α, which is more susceptible to degradation than cellulose type I_β, which is predominant in Charophyta and land plants [22,25,26,49]. Although the XRD of microalgae cellulose is generally similar to that of land plants, and except for some peaks at 14° and 16° that are not found in land plants [27], our results showed otherwise. Indeed, Figure 5 shows the X-ray diffractometry of *C. homosphaera* biomass and, for comparison, that of pure crystalline cellulose. The figure shows that, contrary to the data for crystalline cellulose,

C. homosphaera biomass has a very low crystallinity, such that it was not possible to determine its degree of crystallinity. However, it was possible to identify a clear peak near 15°, which is typical for microalgae [27]. Remarkably, the XRD of *C. homosphaera* closely resembles that of the ball-milled XRD of microcrystalline cellulose after enzymatic hydrolysis, as found by Teixeira et al. [50], who also showed that cellobiohydrolase activity is not required to efficiently hydrolyze amorphous cellulose. These results suggest that the cell wall polysaccharides of *C. homosphaera* is mostly amorphous and, as such, *C. homosphaera* would be categorized as belonging to group 2, as established by Nicolai and Preston [17].

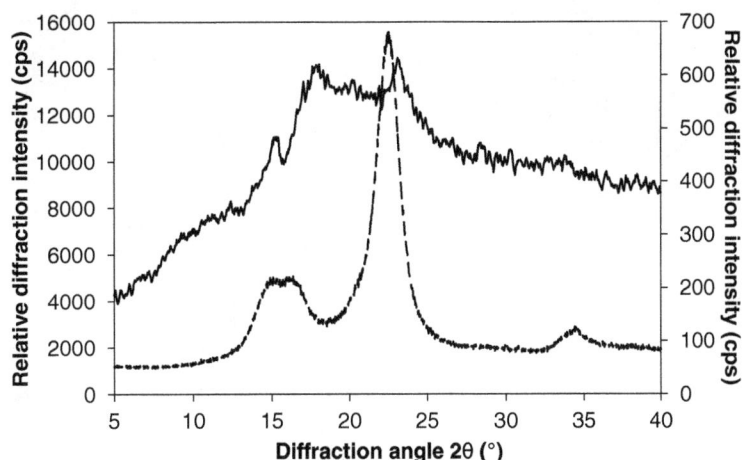

Figure 5 Diffractograms of *C. homosphaera* biomass (continuous line, right axis) and Avicel Fluka (split line, left axis). X-ray diffraction was performed using a Rigaku MiniFlex diffractometer and filtered copper Kα radiation.

However, the low crystallinity may also be an artifact of biomass preparation; under stress, chlorophytes secrete a mucilaginous substance that was observed during the centrifugation process. In addition, the presence of algaenan in the cell wall may create structural changes and amorphous zones during the drying process, as suggested by Mihranyan [27]. However, regardless of whether the centrifugation and drying processes altered the cell wall polysaccharide structure, the dried *C. homosphaera* biomass showed such a low crystallinity that the enzyme mixtures devoid of exoglucanase activity were able to efficiently hydrolyze 40.5% of the algal glucan. In terms of biomass processing and enzyme use, this feature poses a great advantage for the biorefinery concept because it simplifies raw material processing and enzyme blend complexity, thereby reducing processing costs.

Figure 6A shows the scanning electron microscopy of *C. homosphaera* cells and cell fragments that were dried and hand-milled, and Figure 6B shows the *C. homosphaera* cells after enzymatic hydrolysis when using the *A. awamori-T. reesei* enzyme blend. The images suggest that the enzyme action resulted in a more homogenous material and that an amorphous and sticky material that surrounded the undigested algae was removed by enzymatic hydrolysis.

Conclusions

Enzyme mixtures of *A. awamori*, which have β-endoglucanase and β-glucosidase activities and are devoid of cellobiohydrolase activity, efficiently hydrolyze untreated *C. homosphaera* biomass, reaching 42.3% of the cell wall glucan because of its very low crystallinity, and the *A. awamori-A. cellulolyticus* enzyme blend attain even higher hydrolysis rates and yields of approximately 50%. This finding is most likely explained by the synergistic action of both enzyme preparations that allowed the hydrolysis of different

polysaccharides that would be otherwise inaccessible. The β-glucosidase load was a determinant in the hydrolysis yield and in some cases for the initial hydrolysis rate of all the enzyme preparations.

Materials and methods

C. homosphaera cultivation

C. homosphaera cells were grown in inorganic W.C. medium [51] with continual aeration under white fluorescent light, with a 12-h photoperiod and a photosynthetic active radiation of 60 μmol/m^2/s in average irradiance. The cells were cultivated in five carboys of 5 L each for 21 days and collected at the stationary phase of growth by centrifugation, then dried cold, ground manually on a mortar, and stored in the freezer until use, as previously described [52]. Under this growth condition, an average of 300 mg/L of dry weight biomass was obtained at the end of cultivation. Cell growth was followed by cell counting and by light scattering at 750 nm, to avoid the absorption of any cell chromophore, using a previously calibrated dry weight curve. Each growth phase was determined by plotting the logarithm of the cell concentration against the cultivation time.

Fungal strains and enzyme preparations

Laboratory enzyme preparations were produced by *A. awamori* 2B.361 U2/1 and *T. reesei* Rut-C30 and deposited in the fungal culture collection of the National Institute of Quality Control in Health (INCQS 40259-40251, respectively) of the Oswaldo Cruz Foundation (http://www.incqs.fiocruz.br). Both fungi were propagated on potato dextrose agar (PDA) plates at 30°C for 7 days, until dense sporulation was observed. The spores were collected by adding 2 mL of sterilized distilled water to the plate, followed by gentle scraping. A standardized spore suspension

Figure 6 Scanning electron microscopy of ground *C. homosphaera* biomass (A) before enzymatic hydrolysis and (B) after 48 h of enzymatic hydrolysis using the *A. awamori-T. reesei* blend. Samples were adhered to carbon tape, sputter-coated with 28 nm gold using an Emitech/K550 model and observed via SEM as described in the Materials and methods section. Magnifications of ×1,200 (A) and ×6,000 (B) were chosen to show the action of the enzyme blend more clearly on *C. homosphaera* biomass. The scale bars represent 50 μm (A) and 10 μm (B).

presenting 10^7 spores/mL in 20% (v/v) glycerol was maintained at −20°C. For enzyme production, *T. reesei* Rut C30 was cultivated in a liquid medium containing the following (in g/L): 30.0 lactose, 6.0 yeast extract, 0.3 urea, 0.6% (v/v) corn steep liquor, plus the following salts (in g/L): 1.4 $(NH_4)_2SO_4$, 2.0 KH_2PO_4, 0.3 $CaCl_2$, and 0.3 $MgSO_4.7H_2O$, and the following trace elements (mg/L): 5.0 $FeSO_4.7H_2O$, 20 $CoCl_2$, 1.6 $MnSO_4$, and 1.4 $ZnSO_4$, with an initial pH of 6.0. *A. awamori* was cultivated in a liquid medium containing the following (in g/L): 12.0 yeast extract, 30.0 wheat bran, 1.2 $NaNO_3$, 3.0 KH_2PO_4, 6.0 K_2HPO_4, 0.2 $MgSO_4.7H_2O$, and 0.05 $CaCl_2$, with an initial pH of 7.0 [53]. Enzyme production by both fungi took place in 1,000-mL Erlenmeyer flasks containing 300 mL of growth medium inoculated with 1% (v/v) of the spore suspension. The cultures were incubated in a rotary shaker (Innova 4340, New Brunswick, Edison, NJ, USA) at 30°C and 200 rpm. The commercial enzyme *Acremonium* cellulase from *A. cellulolyticus* was kindly provided by Meiji Seika Pharma Co., Japan.

A. awamori β-glucosidase was partially purified as follows. Fungal culture supernatant was concentrated by ultrafiltration using a 100-kDa membrane (Amicon Filtration System-Stirred Cells). A 10 mL volume of the retentate, which contained 227 mg of protein and a total β-glucosidase activity of 5672 IU, was subsequently fractionated by gel filtration on a Sephadex G-75 Column (3.0 × 62.5 cm) and pre-equilibrated with 50 mM sodium acetate buffer (pH 5.0) containing 0.15 M NaCl. The sample was eluted using a flow rate of 20 mL/h, and 5.0 mL aliquots were collected and screened for β-glucosidase activity and protein concentration by measuring the absorbance at 280 nm. Fractions presenting β-glucosidase activity were pooled and concentrated by ultrafiltration with a 30-kDa membrane. This preparation was shown to be free of FPase, CMCase, and xylanase activities. The protein concentration was also measured according to Bradford [54]. β-glucosidase preparations from gel filtration and ion-exchange chromatography were analyzed by SDS-PAGE, Native PAGE, and zymogram with an 8% polyacrylamide gel [55].

The following enzyme preparations were used to hydrolyze the *C. homosphaera* biomass: enzymes from *T. reesei* RUT-C30, *A. awamori*, *A. cellulolyticus*, and blends of *T. reesei*-*A. awamori* and *A. awamori*-*A. cellulolyticus*-excreted enzymes. The *T. reesei*-*A. awamori* blend was made by taking into account the FPase and β-glucosidase of each individual enzyme preparation, which were mixed in the correct proportion to achieve the final FPase: β-glucosidase activities ratio of 1:2. Moreover, hydrolysis experiments were performed with a purified *A. awamori* β-glucosidase preparation to evaluate the effect of this enzyme alone on the initial biomass hydrolysis rate. The aforementioned fungal enzymes were chosen because of

their complementary activities as follows: (i) *A. awamori* 2B.361 U2/1, which produces high levels of glucoamylase and β-glucosidase, but lacks exoglucanase activity (8.4 IU/mL of β-glucosidase, 1.9 IU/mL of CMCase, and no FPase activity); (ii) *T. reesei* RUT C-30, which produces high levels of cellulases but low levels of β-glucosidase (1.4 IU/mL of β-glucosidase, 37.3 IU/mL of CMCase, and 1.2 IU/mL of FPase); (iii) *Acremonium* cellulase (Meiji Seika Co., Japan), which contains high CBH levels in addition to a complex set of biomass-hydrolyzing enzyme activities (29.2 IU/mL of β-glucosidase, 67.1 IU/mL of CMCase, and 4.3 IU/mL of FPase); and (iv) a blend of *A. awamori* and *T. reesei* after the concentration of each enzyme by ultrafiltration with a 30 kDa membrane (20.5 IU/mL of β-glucosidase, 69.5 IU/mL of CMCase, and 5.5 IU/mL of FPase).

FPAse, CMCase, and β-glucosidase activity determinations

The activities of *A. awamori* 2B.361 U2/1 and *T. reesei* Rut-C30 laboratory enzymes and the *A. cellulolyticus* commercial enzyme were determined according to standard IUPAC procedures and expressed as international units (IU), as described by Ghose (1987) [56]. Several dilutions of the enzyme pools were prepared by diluting them with 50 mM citrate buffer, pH 4.8. Enzyme activities were estimated by incubating the diluted enzyme preparations with the relevant substrate in a water bath at 50°C for the time indicated. FPAse was determined by incubating exactly 6 cm^2 of Whatman 1 filter paper for exactly 1 h. The reaction was stopped by adding 3,4-dinitrosalicylic acid (DNS) and centrifuging to remove paper debris; the reducing sugar content was then determined. The endo-β1,4-glucanase activity was determined by incubating a fresh 2% (w/v) carboxymethyl cellulose CMC 7 L2 solution in 50 mM citrate buffer, pH 4.8, for exactly 30 min, followed by the reducing sugar determination. The β-glucosidase activity was determined by incubating a fresh 15 mM cellobiose solution in 50 mM citrate buffer, pH 4.8, for exactly 30 min. The enzymatic reaction was stopped by incubating in a boiling water bath for 5 min; the glucose content was then determined. As controls, the reducing sugar contents of the paper and of the enzyme dilutions, in addition to the glucose contents of the enzyme dilutions, were determined. The enzyme activities were estimated using the calculations described in [56].

C. homosphaera biomass enzymatic hydrolysis

The experiments were performed in 50-mL glass Erlenmeyer flasks that were stoppered with glass to minimize evaporation, with a 30-mL reaction mixture containing 50 mg/mL cell mass powder suspended in 50 mM citrate buffer, pH 4.8, and an enzyme load of 10 FPU (filter paper activity) units/g dry biomass. The measured β-glucosidase loads for the *T. reesei* and *A. cellulolyticus* preparations and the *A. awamori*-*A. cellulolyticus* blend were 8.7, 49.0, and 90.3 IU/g biomass, respectively. Because the *A. awamori* enzymes exhibited no FPAse, the enzyme load was equalized with the β-glucosidase load of the *T. reesei*-*A. awamori* blend, for 22.3 IU/g dry biomass. The mixtures were incubated at 50°C in a rotatory shaker (Innova 4340, New Brunswick, Edison, NJ, USA), and sampling was performed at 0, 2, 4, 6, 8, 12, 24, 28, 32, 36, and 48 h, followed by incubation for 5 min in a boiling water bath to halt enzyme action. Sugar analyses were performed in the supernatants of centrifuged samples.

To investigate the influence of the β-glucosidase activity on the initial enzymatic biomass hydrolysis and final yield, two sets of assays were designed. In the first, the biomass powder was incubated for 3 h under the same conditions as described above, with an enzyme load of 22.3 IU β-glucosidase/g dry biomass for all the enzymes. This time frame was chosen because the kinetics for the hydrolysis experiments using the evaluated enzyme preparations (*A. cellulolyticus*, *T. reesei*, *A. awamori*, and the *A. awamori*-*T. reesei* blend) with the same FPAse load but different β-glucosidase loads showed significant rate differences up to 3 h of hydrolysis, i.e., where the most β-glucosidase hydrolysis dependency was observed. For the controls, a sample of biomass powder was submitted to hydrolysis with partially purified *A. awamori* β-glucosidase and another biomass sample was incubated without any enzymes under the same conditions. As in the first set, the reaction medium of each enzyme preparation had a different endoglucanase load, and in the second set of the assay, each reaction medium of the evaluated enzyme preparations was set to have an endoglucanase load of 1.5 IU endoglucanase/g dry biomass. The β-glucosidase load was then adjusted by adding partially purified β-glucosidase to final load values of 7.5, 15.0, and 22.5 I.U/g dry biomass. All enzymatic assays were performed in triplicate.

AHS determination

The AHS in *C. homosphaera* biomass was estimated on the basis of previously described procedures [57,58]. Soluble saccharides, lipids, and pigments were extracted by incubating 50 mg of biomass powder with 3 mL of 80% ethanol (v/v) for 30 min in a water bath at 70°C. The supernatant was discarded, and the procedure was repeated three times. The pellet was washed twice in 50 mM citrate buffer, pH 4.8, and resuspended in 0.9 mL of the same buffer; 0.1 mL of Novozyme amyloglucosidase AMG 300 L solution was then added to give a final enzyme activity load of 40 IU. The suspension was incubated for 24 h at 50°C, then centrifuged; the supernatant glucose content was then estimated. Sodium azide was added to prevent contamination.

Nitrate determination

The nitrate remaining in the culture medium was determined by UV spectrophotometry as described in [59]. A 10 mL aliquot of cell suspension was withdrawn from culture cell flasks under sterile conditions. After determining the cell suspension turbidity at 750 nm, the cells were centrifuged and the clear supernatant was used to determine the nitrate content. For each 1 mL of supernatant, 20 μL of 1 mol/L HCl solution was added to prevent any interference from hydroxide and carbonate ions, and the absorbances at 220 and 275 nm were measured, assuring that the absorbance at 275 nm was less than 10% of the absorbance at 220 nm. The reading at 275 nm was used to discount the interference of organic materials that might be present in the sample. The net absorbance was calculated using the following equation: Net Abs = Abs(220) – 2 × Abs (275). The nitrate concentration was determined by means of an analytical curve of the net absorbance against the nitrate concentration, using a 20 μg/mL NaNO₃ solution as a standard. All glassware was thoroughly rinsed with hot 1 mol/L HCl solution and distilled water before use, to eliminate any trace of carbonate and detergent.

Sugar determination

The *C. homosphaera* total biomass sugar content was determined on the basis of the procedure described in [41]. A total of 4 mg of dry cell powder was incubated with 2 mL of 1 mol/L H_2SO_4 at 100°C for 6 h. The samples were then neutralized with $BaCO_3$ and centrifuged, and the supernatants were used for total sugar determination by the phenol sulfuric method [60] with an analytical curve using a 0.2 mg/mL glucose solution as a standard. For the control, the same procedure was performed with pure glucose powder. The reducing sugar levels in hydrolysates were estimated by the 3,5-dinitrosalicylic method [61] by means of an analytical curve with a 2.0 mg/mL glucose solution as a standard. The glucose concentration was estimated using a YSI 2730 glucose analyzer (Yellow Springs Incorporated, Ohio, USA). The sugar composition was determined by HPAEC-PAD analysis.

HPAEC-PAD analysis

A sugar composition determination was performed in an Ion Chromatography System 5000 (ICS-5000, Dionex Ltd., Canada) equipped with Chromeleon 6.8 (Dionex Ltd., Canada) software. The column system consisted of a CarboPac PA1 (4 × 50 mm, Thermo Scientific Ltd., USA) pre-column and a CarboPac PA1 (4 × 250 mm, Thermo Scientific Ltd., USA) analytical column. The chromatographic conditions were as follows: the samples were automatically injected at 15°C with an injection volume of 10 μL, a furnace temperature of 30°C, and a pressure of 1,050 to 1,150 psi. The detection was performed through a pulse amperometric detector specific for monosaccharides at 30°C. The mobile phase was a step gradient (0% to 85% to 0%) of 300 μmol/mL NaOH solution (reagent grade type I) and ultra-pure 0.2 μm filtered degassed water (with a resistivity of at least 18 MΩ) at a flow rate of 1.0 mL/min. The postcolumn solvent was a 400 μmoL/mL NaOH solution at a flow rate of 0.3 mL/min. The total running time was 50 min, and all solvents were kept under a N_2 atmosphere.

Analytical curves (2.0 to 250.0 mg/L) were constructed for each monosaccharide to be determined using standards chosen based on previously reported *Chlorella* cell wall sugar compositions [42]. The sugar content was estimated by measuring the peak area.

Curve fit analysis

The hydrolysis time curves were fitted in a double exponential curve, which was rejected when it did not converge or when the predicted parameters had P values greater than 0.05, which resulted in a large standard error. Under these circumstances, a mono-exponential model was used. Fittings were performed using a non-linear regression (Levengerg-Marquardt Algorithm) with SigmaPlot 10.0 software (Systat Software Inc., San Jose, CA, USA) for Windows. A fitting analysis was used to estimate the initial hydrolytic activity rates, the maximum hydrolysis yield, and the time needed to reach 50% hydrolysis for each enzymatic preparation ($t½$).

Statistical analysis

The results obtained from the assays to test the dependency of biomass hydrolysis yields on β-glucosidase activity were analyzed by one-way ANOVA with MINITAB 15.0 software for Windows (Minitab Inc., PA, USA). The differences among the means were verified using the Fisher test. The means were considered to be significantly different when $P \leq 0.05$.

Cell wall X-ray diffractometry

The crystallinity of the *C. homosphaera* dried biomass was evaluated by X-ray diffraction with a Rigaku MiniFlex diffractometer and filtered copper Kα radiation ($\lambda = 0.1542$ nm) using a monochromator at a 30 KV voltage and 15 mA electric current, with a speed of approximately 2°/min and scanning at an angle (2θ) from 2° to 60°. For comparison, a diffractogram of cellulose microcrystalline (Avicel PH101, Sigma-Aldrich) was also performed. Each sample's crystallinity index (CrI) was calculated according to the Segal method [62].

SEM

Scanning electron microscopy (SEM-FEI/Inspect S50 model) was used to investigate the microalgae biomass morphology before and after 48 h of enzymatic treatment with the *A. awamori-T. reesei* enzyme blends. Samples were adhered to carbon tape, sputter-coated with 28 nm gold with an Emitech/K550 model and observed via SEM with an acceleration voltage of 20 KV and a working distance of approximately 19 mm. Several images were obtained from different areas of the samples (at least 20 images per sample) to guarantee the reproducibility of the results.

Additional files

Additional file 1: Figure S1. The time course for the release of total reducing sugars from *C. homosphaera* biomass (50 mg d.w./mL) in 50 mM citrate buffer, pH 4.8, at 50°C. The biomass was hydrolyzed using enzyme preparations with a 1.5 IU/g endoglucanase load from *T. reesei* **(A)**, *A. cellulolyticus* **(B)**, *A. awamori* **(C)**, and a *T. reesei-A. awamori* blend **(D)** at final β-glucosidase loads of 7.5 (filled circle), 15.0 (empty circle), and 22.5 IU/g (triangles). The data were fitted into an exponential function as described in the Materials and methods section. Standard deviation was less than 10% of the mean value and bars were omitted for clarity.

Additional file 2: Figure S2. A time course for the release of glucose from *C. homosphaera* biomass (50 mg d.w./mL) in 50 mM citrate buffer, pH 4.8, at 50°C. The biomass was hydrolyzed using the enzyme preparations with a 1.5 IU/g endoglucanase load from *T. reesei* **(A)**, *A. cellulolyticus* **(B)**, *A. awamori* **(C)**, and the *T. reesei-A. awamori* blend **(D)** at final β-glucosidase loads of 7.5 (filled circle), 15.0 (empty circle), and 22.5 IU/g (triangles). The data were fitted into an exponential function as described in the Materials and methods section. Standard deviation was less than 10% of the mean value and bars were omitted for clarity.

Abbreviations

SEM: scanning electron microscopy; CrI: crystallinity index; FTIR: Fourier transform infrared spectroscopy; AHS: amyloglucosidase hydrolyzable starch; XRD: X-ray diffraction; FPAse: filter paper activity; FPU: filter paper unit; CMCase: endo-β-1,4 glucanase; INCQS: Instituto Nacional de Controle de Qualidade em Saúde; IU: international unit; IUPAC: International union of pure and applied Chemistry; TFA: trifluoracetic acid; EG: endoglucanase; CBH: cellobiohydrolase; HPAEC-PAD: high-performance anion exchange chromatography and pulse amperometric detection.

Competing interests

The authors declare that they have no competing interests.

Authors' contributions

MAR performed the cultivation of *C. homosphaera* and the preparation of the algae biomass, study coordination, and data analysis, trained and supervised the experimental work performed by undergraduate students for tasks related to routine experimental work, and was involved in the manuscript preparation, revision, and discussion. VSFL was involved in the determination and analysis of X-ray diffraction data for *C. homosphaera* biomass and the scanning electron microscopy analysis of *C. homosphaera* biomass before and after enzymatic hydrolysis, manuscript preparation, revision, and discussion. RSST participated in the *A. awamori* and *T. reesei* RUT C-30 strain maintenance, propagation, and cultivation for enzyme production, enzyme activity determination, partial purification of β-glucosidase from *A. awamori*, trained and supervised the experimental work performed by undergraduate students on tasks related to routine experimental work, and was involved in the manuscript preparation, revision, and discussion. EPSB participated in the study outline and supervision, critical analysis of the results, manuscript revision, and final approval. All authors have read and approved the final manuscript.

Acknowledgements

The authors are grateful to FINEP from the Brazilian Ministry of Science, Technology and Innovation for financial support (Grant No. 01.09.0566.001421/08). The authors express gratitude to the undergraduate students Mr. Michel Quintal Nunes and Mr. João Carlos Lopes do Nascimento Junior for the cultivation of *C. homosphaera* and for their technical assistance throughout this work. The authors are also grateful to Daniel Santos Pereira, Maria Alice Santos Cerullo, and Marcello Martins Torquato de Carvalho for performing the chromatographic analyses.

Author details

[1]Federal University of Rio de Janeiro, Institute of Chemistry, Department of Biochemistry, Applied Photosynthesis Laboratory, Athos Avenida da Silveria Ramos, 149-Technology Centre, Block A, Room 532, University City, Rio de Janeiro, RJ 21941-909, Brazil. [2]National Institute of Technology - Ministry of Science, Technology and Innovation, Biocatalysis Laboratory, 20081-312 Rio de Janeiro, RJ, Brazil. [3]Federal University of Rio de Janeiro, Institute of Chemistry, Department of Biochemistry, Enzyme Technology Laboratory, 21941-909 Rio de Janeiro, RJ, Brazil.

References

1. John RP, Anisha GS, Nampoothiri KM, Pandey A. Micro and macroalgal biomass: a renewable source for bioethanol. Bioresour Technol. 2011;102:186–93.
2. Van Beilen JB. Why microalgal biofuels won't save the internal combustion machine. Biofuels, Bioprod Biorefining. 2010;4:41–52.
3. Walker DA. Biofuels - for better or worse? Ann Appl Biol. 2010;156:319–27.
4. Chinnasamy S, Bhatnagar A, Hunt RW, Das KC. Microalgae cultivation in a wastewater dominated by carpet mill effluents for biofuel applications. Bioresour Technol. 2010;101:3097–105.
5. Ellis JT, Hengge NN, Sims RC, Miller CD. Acetone, butanol, and ethanol production from wastewater algae. Bioresour Technol. 2012;111:491–5.
6. Pittman J. The potential of sustainable algal biofuel production using wastewater resources. Bioresour Technol. 2011;102:17–25.
7. Scragg A, Illman A, Carden A, Shales S. Growth of microalgae with increased calorific values in a tubular bioreactor. Biomass Bioenergy. 2002;23:67–73.
8. Illman A, Scragg A, Shales S. Increase in Chlorella strains calorific values when grown in low nitrogen medium. Enzyme Microb Technol. 2000;27:631–5.
9. Ho S-H, Chen C-Y, Chang J-S. Effect of light intensity and nitrogen starvation on CO2 fixation and lipid/carbohydrate production of an indigenous microalga Scenedesmus obliquus CNW-N. Bioresour Technol. 2012;113:244–52.
10. Harun R, Danquah MK, Forde GM. Microalgal biomass as a fermentation feedstock for bioethanol production. J Chem Technol Biotechnol. 2010;85:199–203.
11. Amin S. Review on biofuel oil and gas production processes from microalgae. Energy Convers Manag. 2009;50:1834–40.
12. Costa JAV, De Morais MG. The role of biochemical engineering in the production of biofuels from microalgae. Bioresour Technol. 2011;102:2–9.
13. Demirbas MF. Biofuels from algae for sustainable development. Appl Energy. 2011;88:3473–80.
14. Harun R, Danquah MK. Influence of acid pre-treatment on microalgal biomass for bioethanol production. Process Biochem. 2011;46:304–9.
15. Al-Zuhair S, Ramachandran KB, Farid M, Aroua MK, Vadlani P, Ramakrishnan S, et al. Enzymes in biofuels production. Enzyme Res. 2011;2011:2.
16. Ho S-H, Huang S-W, Chen C-Y, Hasunuma T, Kondo A, Chang J-S. Bioethanol production using carbohydrate-rich microalgae biomass as feedstock. Bioresour Technol. 2013;135:191–8.
17. Nicolai E, Preston RD. Cell-wall Studies in the chlorophyceae. I. A General survey of submicroscopic structure in filamentous species. Proc R Soc B Biol Sci. 1952;140:244–74.
18. Koyama M, Sugiyama J, Itoh T. Systematic survey on crystalline features of algal celluloses. Cellulose. 1997;4:147–60.
19. Wada M, Sugiyama J, Okano T. The crystalline phase (Iα/Iβ) system of native celluloses in relation to plant phylogenesis. Mokuzai Gakkaishi. 1995;41:186–92.
20. Michell AJ. Second-derivative FTIR spectra of native celluloses. Carbohydr Res. 1990;197:53–60.
21. Roberts AW, Roberts EM, Delmer DP. Cellulose synthase (CesA) Genes in the green alga Mesotaenium caldariorum. Eukaryot Cell. 2002;1:847–55.
22. Atalla RH, Vanderhart DL. Native cellulose: a composite of two distinct crystalline forms. Science. 1984;223:283–5.
23. VanderHart DL, Atalla RH. Studies of microstructure in native celluloses using solid-state carbon-13 NMR. Macromolecules. 1984;17:1465–72.
24. Sugiyama J, Vuong R, Chanzy H. Electron diffraction study on the two crystalline phases occurring in native cellulose from an algal cell wall. Macromolecules. 1991;24:4168–75.
25. Hayashi N, Sugiyama J, Okano T, Ishihara M. The enzymatic susceptibility of cellulose microfibrils of the algal-bacterial type and the cotton-ramie type. Carbohydr Res. 1997;305:261–9.
26. Hayashi N, Sugiyama J, Okano T, Ishihara M. Selective degradation of the cellulose Iα component in Cladophora cellulose with Trichoderma viride cellulase. Carbohydr Res. 1997;305:109–16.
27. Mihranyan A. Cellulose from cladophorales green algae: from environmental problem to high-tech composite materials. J Appl Polym Sci. 2011;119:2449–60.
28. Burczyk J, Terminska-Pabis K, Smietana B. Cell wall neutral sugar composition of chlorococcalean algae forming and not forming acetolysis resistant biopolymer. Phytochemistry. 1995;38:837–41.
29. Burczyk J, Loos E. Cell wall-bound enzymatic activities in Chlorella and Scenedesmus. J Plant Physiol. 1995;146:748–50.
30. Loos E, Meindl D. Cell wall-lytic activity in Chlorella fusca. Planta. 1984;160:357–62.
31. Loos E, Meindl D. Cell-wall-bound lytic activity in Chlorella fusca: function and characterization of an endo-mannanase. Planta. 1985;166:557–62.
32. Ogawa K, Yamaura M, Ikeda Y, Kondo S. New aldobiuronic acid, 3-O-alpha-D-glucopyranuronosyl-L-rhamnopyranose, from an acidic polysaccharide of Chlorella vulgaris. Biosci Biotechnol Biochem. 1998;62:2030–1.
33. Ogawa K, Ikeda Y, Kondo S. A new trisaccharide, α-d-glucopyranuronosyl-(1 → 3)-α-l-rhamnopyranosyl-(1 → 2)-α-l-rhamnopyranose from Chlorella vulgaris. Carbohydr Res. 1999;321:128–31.
34. Burczyk J, Smietana B, Terminska-Pabis K, Zych M, Kowalowski P. Comparison of nitrogen content amino acid composition and glucosamine content of cell walls of various chlorococcalean algae. Phytochemistry. 1999;51:491–7.
35. Rodríguez MC, Noseda MD, Cerezo AS. The fibrillar polysaccharides and their linkage to algaenan in the trilaminar layer of the cell all of Coelastrum sphaericum (Chlorophyceae). J Phycol. 1999;35:1025–31.
36. Dragone G, Fernandes BD, Abreu AP, Vicente AA, Teixeira JA. Nutrient limitation as a strategy for increasing starch accumulation in microalgae. Appl Energy. 2011;88:3331–5.
37. Choi SP, Nguyen MT, Sim SJ. Enzymatic pretreatment of Chlamydomonas reinhardtii biomass for ethanol production. Bioresour Technol. 2010;101:5330–6.
38. Nguyen MT, Choi SP, Lee J, Lee JH, Sim SJ. Hydrothermal acid pretreatment of Chlamydomonas reinhardtii biomass for ethanol production. J Microbiol Biotechnol. 2009;19:161–6.
39. Ogbonna JC, Tanaka H. Night biomass loss and changes in biochemical composition of cells during light/dark cyclic culture of Chlorella pyrenoidosa. J Ferment Bioeng. 1996;82:558–64.
40. Harun R, Danquah MK. Enzymatic hydrolysis of microalgal biomass for bioethanol production. Chem Eng J. 2011;168:1079–84.
41. Northcote DH, Goulding KJ, Horne RW. The chemical composition and structure of the cell wall of Chlorella pyrenoidosa. Biochem J. 1958;70:391–7.
42. Takeda H. Sugar composition of the cell wall and the taxonomy of Chlorella (Chlorophyceae). J Phycol. 1991;27:224–32.
43. Takeda H. Chemical composition of cell walls as a taxonomical marker. J Plant Res. 1993;106:195–200.
44. Watanabe K, Imase M, Sasaki K, Ohmura N, Saiki H, Tanaka H. Composition of the sheath produced by the green alga Chlorella sorokiniana. Lett Appl Microbiol. 2006;42:538–43.
45. Miranda JR, Passarinho PC, Gouveia L. Pre-treatment optimization of Scenedesmus obliquus microalga for bioethanol production. Bioresour Technol. 2012;104:342–8.
46. Bon E, Webb C. Glucoamylase production and nitrogen nutrition in Aspergillus awamori. Appl Biochem Biotechnol. 1993;39–40:349–69.
47. Takeda H. Classification of Chlorella strains by cell wall sugar composition. Phytochemistry. 1988;27:3823–6.
48. Pei J, Pang Q, Zhao L, Fan S, Shi H. Thermoanaerobacterium thermosaccharolyticum β-glucosidase: a glucose-tolerant enzyme with high specific activity for cellobiose. Biotechnol Biofuels. 2012;5:31.
49. Corgié SC, Smith HM, Walker LP. Enzymatic transformations of cellulose assessed by quantitative high-throughput fourier transform infrared spectroscopy (QHT-FTIR). Biotechnol Bioeng. 2011;108:1509–20.
50. Teixeira RSS, da Silva AS, Kim H, Ishikawa K, Endo T, Lee S, et al. Use of cellobiohydrolase-free cellulase blends for the hydrolysis of microcrystalline

Untreated Chlorella homosphaera biomass allows for high rates of cell wall glucan enzymatic hydrolysis...

171

cellulose and sugarcane bagasse pretreated by either ball milling or ionic liquid [Emim] [Ac]. Bioresour Technol. 2013;149:551–5.

51. Guillard RR, Lorenzen CJ. Yellow-green algae with chlorophyllide. J Phycol. 1972;8:10–4.

52. Rodrigues MA, da Silva Bon EP. Evaluation of Chlorella (Chlorophyta) as source of fermentable sugars via cell wall enzymatic hydrolysis. Enzyme Res. 2011;2011:5.

53. Gottschalk LMF, Paredes RS, Teixeira RSS, Silva AS, Bon EPS. Efficient production of lignocellulolytic enzymes xylanase, β-xylosidase, ferulic acid esterase and β-glucosidase by the mutant strain Aspergillus awamori 2B.361 U2/1. Braz J Microbiol. 2013;44:569–76.

54. Bradford MM. A rapid and sensitive method for the quantitation of microgram quantities of protein utilizing the principle of protein-dye binding. Anal Biochem. 1976;72:248–54.

55. Kwon K-S, Lee J, Kang HG, Hah YC. Detection of {beta}-glucosidase activity in polyacrylamide gels with esculin as substrate. Appl Envir Microbiol. 1994;60:4584–6.

56. Ghose TK. Measurements of cellulase activities. Pure Appl Chem. 1987;59:257–68.

57. Chow PS, Landhäusser SM. A method for routine measurements of total sugar and starch content in woody plant tissues. Tree Physiol. 2004;24:1136.

58. Ranwala AP, Miller WB. Analysis of nonstructural carbohydrates in storage organs of 30 ornamental geophytes by high-performance anion-exchange chromatography with pulsed amperometric detection. New Phytol. 2008;180:421–33.

59. Federation WE. Standard Methods for the Examination of Water and Wastewater. APHA American Public Health Association, Washington. 1999.

60. DuBois M, Gilles KA, Hamilton JK, Rebers PA, Smith F. Colorimetric method for determination of sugars and related substances. Anal Chem. 1956;28:350–6.

61. Sumner JB. Dinitrosalicylic acid: a reagent for the estimation of sugar in normal and diabetic urine. J Biol Chem. 1924;62:285–9.

62. Segal L, Creely JJ, Martin AE, Conrado CM. An empirical method for estimating the degree of crystallinity of native cellulose using the X-ray diffractometer. Text Res J. 1959;29:764–86.

12

Contribution of a family 1 carbohydrate-binding module in thermostable glycoside hydrolase 10 xylanase from *Talaromyces cellulolyticus* toward synergistic enzymatic hydrolysis of lignocellulose

Hiroyuki Inoue[*], Seiichiro Kishishita, Akio Kumagai, Misumi Kataoka, Tatsuya Fujii and Kazuhiko Ishikawa

Abstract

Background: Enzymatic removal of hemicellulose components such as xylan is an important factor for maintaining high glucose conversion from lignocelluloses subjected to low-severity pretreatment. Supplementation of xylanase in the cellulase mixture enhances glucose release from pretreated lignocellulose. Filamentous fungi produce multiple xylanases in their cellulase system, and some of them have modular structures consisting of a catalytic domain and a family 1 carbohydrate-binding module (CBM1). However, the role of CBM1 in xylanase in the synergistic hydrolysis of lignocellulose has not been investigated in depth.

Results: Thermostable endo-β-1,4-xylanase (Xyl10A) from *Talaromyces cellulolyticus*, which is recognized as one of the core enzymes in the fungal cellulase system, has a modular structure consisting of a glycoside hydrolase family 10 catalytic domain and CBM1 at the C-terminus separated by a linker region. Three recombinant Xyl10A variants, that is, intact Xyl10A (Xyl10Awt), CBM1-deleted Xyl10A (Xyl10AdC), and CBM1 and linker region-deleted Xyl10A (Xyl10AdLC), were constructed and overexpressed in *T. cellulolyticus*. Cellulose-binding ability of Xyl10A CBM1 was demonstrated using quartz crystal microbalance with dissipation monitoring. Xyl10AdC and Xyl10AdLC showed relatively high catalytic activities for soluble and insoluble xylan substrates, whereas Xyl10Awt was more effective in xylan hydrolysis of wet disc-mill treated rice straw (WDM-RS). The enzyme mixture of cellulase monocomponents and intact or mutant Xyl10A enhanced the hydrolysis of WDM-RS glucan, with the most efficient synergism found in the interactions with Xyl10Awt. The increased glucan hydrolysis yield exhibited a linear relationship with the xylan hydrolysis yield by each enzyme. This relationship revealed significant hydrolysis of WDM-RS glucan with lower supplementation of Xyl10Awt.

Conclusions: Our results suggest that Xyl10A CBM1 has the following two roles in synergistic hydrolysis of lignocellulose by Xyl10A and cellulases: enhancement of lignocellulosic xylan hydrolysis by binding to cellulose, and the efficient removal of xylan obstacles that interrupt the cellulase activity (because of similar binding target of CBM1). The combination of CBM-containing cellulases and xylanases in a fugal cellulase system could contribute to reduction of the enzyme loading in the hydrolysis of pretreated lignocelluloses.

Keywords: Enzymatic hydrolysis, GH10 xylanase, Cellulase, Family 1 carbohydrate-binding module, Lignocellulosic biomass, Rice straw, *Talaromyces cellulolyticus*

* Correspondence: inoue-h@aist.go.jp
Research Institute for Sustainable Chemistry, National Institute of Advanced
Industrial Science and Technology (AIST), 3-11-32 Kagamiyama,
Higashi-Hiroshima, Hiroshima 739-0046, Japan

Background

Lignocellulose is one of the most abundant organic compounds in the biosphere and contains large amounts of polysaccharides, such as cellulose and hemicellulose, which serve as the source of fermentable sugars used in the production of biofuels or chemicals. An efficient enzymatic hydrolysis of the cellulose and hemicellulose components to fermentable sugars is a key step in the bioconversion of lignocellulose [1,2]. The digestibility of these components is increased by pretreatment processes such as dilute acid, alkali, hot compressed water, and milling treatments that disrupt the rigid structural network consisting of cellulose-hemicellulose-lignin [3]. In the current industrial biofuel production from lignocellulose, the use of low-severity pretreatment without chemical addition is expected to have advantages that result in lowering the capital cost and minimizing waste generation [4]. However, low-severity factor results in less sugar yield. The residual hemicellulose has been known to limit the hydrolysis of cellulose, which has a β-1,4-glucan structure [5-8]. Therefore, an efficient enzymatic removal of hemicellulose is an important factor to maintain high glucose conversion from lignocellulose subjected to the low-severity pretreatment.

Xylan is a major polysaccharide of hemicellulose from agricultural residues and hardwoods, consisting of a linear backbone of β-D-xylopyranosyl residues linked by β-1,4-glycosidic bonds. Most of the xylan found is heteroxylan and its backbone could be branched due to substitution of different side groups such as L-arabinose, D-galactose, acetyl, feruloyl, p-coumaroyl, and glucuronic acid residues. Xylan forms an overlying layer through hydrogen bonding with the cellulose, while it is covalently linked with lignin, which forms an outer sheath to protect the plant [9]. Therefore, complete enzymatic hydrolysis of xylan requires a large number of hemicellulases and accessory enzymes with different specificities [9]. Xylanases (endo-β-1,4-xylanase; EC 3.2.1.8) are the main enzymes that cleave the β-1,4-glycosidic bonds in the xylan backbone. Supplementation of xylanase to cellulase mixture has been reported to enhance glucose release from the pretreated lignocellulose [6-8,10,11], which in turn also dramatically reduces cellulase loading needed to achieve reasonable glucan hydrolysis yields from the cellulose [8]. These synergistic interactions observed between the cellulases and xylanases are believed to result from an increase in the accessibility of the cellulases to cellulose due to removal of the hemicellulose [6,8,11]. The synergistic action of xylanases along with certain accessory enzymes that act on the removal of xylan side chains also contribute to further hydrolysis of cellulose by cellulase-xylanase interactions [2]. The optimization of these enzymes for the pretreated lignocellulose has great

potential not only for enhancing the glucose and xylose yields but also for reducing enzyme loading and cost [1,12,13].

Based on the amino-acid sequence similarities, most xylanases are classified into glycoside hydrolase family 10 (GH10) that have a $(\beta/\alpha)_8$ TIM-barrel structure and family 11 (GH11) that consists of a β-jelly roll structure [14-16]. GH11 xylanases have been recognized as important enzymes capable of enhancing biomass hydrolysis due to their ability to effectively hydrolyze various xylanolytic substrates [10]. On the other hand, GH10 xylanases exhibit lower substrate specificity in comparison to the GH11 xylanases and possess a higher affinity to the highly branched xylan backbone [14]. A combination of GH10 and 11 xylanases in the cellulase cocktail help in increasing both the glucose and xylose yields in ammonia fiber expansion-pretreated corn stover [12,13]. Furthermore, it has been reported that GH10 xylanases are more effective than GH11 xylanases for the synergistic glucan hydrolysis of the assessed pretreated lignocellulosic substrates [10,17]. These studies suggest that GH10 xylanases could be better candidates for enhancing the hydrolysis of more realistic substrates.

Filamentous fungi are the major source of industrial enzymes and produce a variety of cellulases and hemicellulases with high productivity and catalytic efficiency, both of which are required for low-cost enzyme supply [1]. Multiple xylanases are produced in the fungal cellulase system and a large number of GH10 xylanases found in basidiomycota and ascomycota have a modular structure that include a family 1 carbohydrate-binding module (CBM1) at N- and C-termini [18,19]. CBM1s are exclusively found in fungal cellulases and hemicellulases and generally target the enzymes to the cellulose surface. The major function of CBM is to increase the effective enzyme concentration on the polysaccharide surface and thus, enhance the enzymatic activity [20,21]. It is known that removal of the CBM1 significantly reduces the activity of cellulases toward insoluble substrates [20,22]. On the other hand, the presence of CBM1 in two GH10 xylanases from *Myceliophthora thermophila* was not always beneficial in the degradation of soluble and insoluble xylan substrates [23]. However, the role of CBM1 in xylanase in the synergistic hydrolysis of lignocellulose has not been explored in detail.

In a previous study, we identified a thermostable GH10 xylanase (Xyl10A) in the *Talaromyces cellulolyticus* cellulase system as one of the core enzymes for the hydrolysis of lignocellulose [24]. Xyl10A is a modular enzyme consisting of an N-terminal catalytic domain and CBM1 separated by a Ser/Thr-rich linker region [19]. Furthermore, an increase in the synergistic glucan hydrolysis of pretreated corn stover by the mixture of cellobiohydrolase I (Cel7A), cellobiohydrolase II (Cel6A), and Xyl10A from *T. cellulolyticus* has been observed [24]. The purpose of this study is to evaluate the role of Xyl10A CBM1 in the synergistic hydrolysis of the lignocellulose substrate. The wet disc-mill treated rice straw (WDM-RS) was used as the substrate that was subjected to synergistic hydrolysis using a mixture of cellulase monocomponents and Xyl10A, with or without the CBM1. The efficacies of the cellulose-binding ability of Xyl10A CBM1 for efficient synergistic hydrolysis of lignocellulose are discussed in this study.

Results

Expression and purification of recombinant enzymes: Xyl10Awt, Xyl10AdC, and Xyl10AdLC

To evaluate the role of CBM1 in Xyl10A from *T. cellulolyticus*, three recombinant forms of Xyl10A were designed, that is, Xyl10Awt, an intact enzyme; Xyl10AdC, a CBM1-deleted form that has Gly372 as the C-terminal residue; and Xyl10AdLC, a CBM1 and linker region-deleted form that has Leu334 as the C-terminal residue (Figure 1). These three recombinant enzymes were secreted extracellularly using the *T. cellulolyticus* homologous expression system under the control of the starch-inducible glucoamylase promoter [19]. The analysis of culture supernatant showed these recombinant proteins as major bands in the SDS-PAGE analysis (data not shown). Our earlier work has shown that the recombinant Xyl10Awt had similar physicochemical and enzymatic properties as the native Xyl10A [19]. The apparent molecular weights for purified recombinant Xyl10As were estimated to be 51 k (Xyl10Awt), 50 k (Xyl10AdC), and 37 k (Xyl10AdLC) on SDS-PAGE gel (Figure 2a). The molecular weights

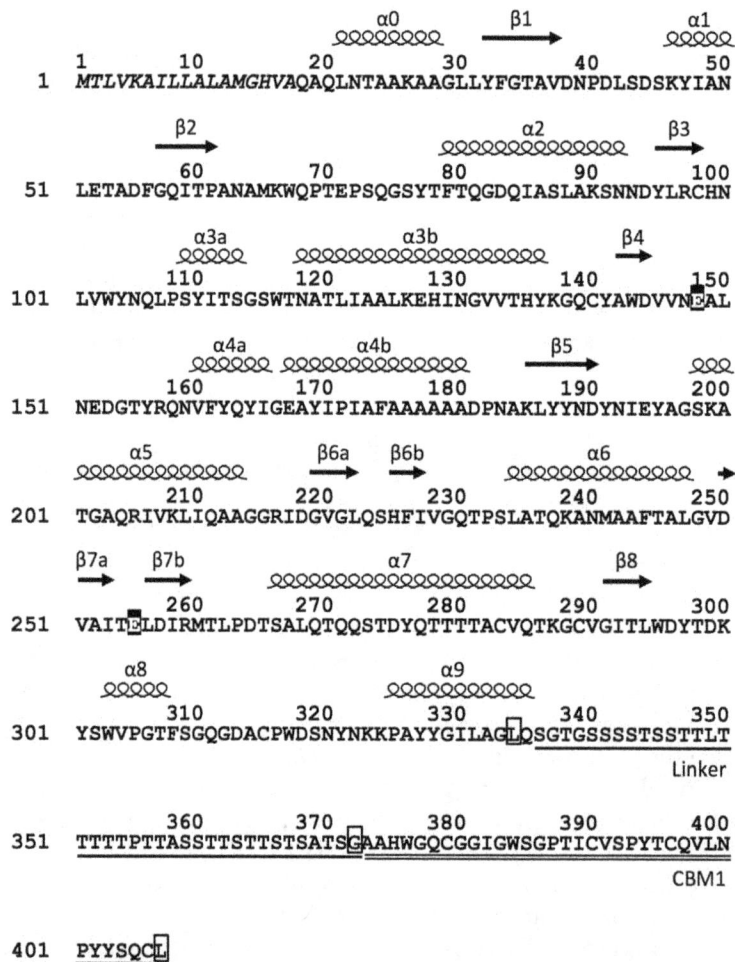

Figure 1 Amino acid sequence of Xyl10A with predicted secondary structural elements. The putative signal sequence is shown in *italics*. The putative domains in mature Xyl10A are as follows: catalytic domain (amino-acid residues 20 to 335); Ser/Thr-rich linker region (336 to 372), *underlined*; CBM1 (373 to 407), *double underlined*. C-terminal residues of Xyl10AdLC (Leu334), Xyl10AdC (Gly372), and Xyl10Awt (Leu407) are enclosed in a *box*. The predicted catalytic residues are shown in filled *black boxes*. The predicted secondary structural elements (α-helices and β-strands) based on the structure of a GH10 xylanase from *Fusarium oxysporum* (FoXyn10a, sequence identity, 50%, PDB ID, 3U7B) as template [50] are indicated above the sequence.

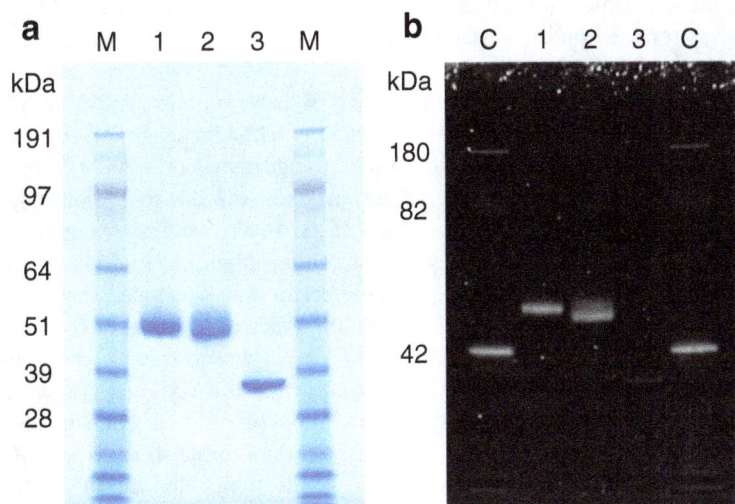

Figure 2 SDS-PAGE (4% to 12%) of purified intact and mutant Xyl10As. Approximately 5 μg **(a)** and 0.7 μg **(b)** of proteins were loaded on each gel and stained with Colloidal Coomassie G-250 and Pro-Q Emerald glycoprotein stain, respectively. Glycoproteins were visualized using a 300-nm UV transilluminator. Lanes: 1, Xyl10Awt; 2, Xyl10AdC; 3, Xyl10AdLC; M, SeeBlue Plus2 protein standard (Invitrogen); C, CandyCane glycoprotein marker (Invitrogen).

of Xyl10Awt and Xyl10AdC were higher in comparison to the calculated molecular masses, that is, 41,490 and 37,747 Da, respectively, whereas the molecular weight of Xyl10AdLC was nearly identical to the calculated value (34,245 Da). The result of glycosylation staining of these enzymes suggested that the Ser/Thr-rich linker regions of Xyl10Awt and Xyl10AdC are highly modified by O-glycosylation (Figure 2b), as reported for fungal cellulolytic enzymes possessing linker and CBM1 [25]. The similar values for molecular weight between Xyl10Awt and Xyl10AdC suggest that Xyl10AdC is slightly more glycosylated in comparison to Xyl10Awt.

Binding of Xyl10A to cellulose

CBM1 is the most common cellulose-binding module observed in fungal cellulases and various hemicellulases. CBMs in several hemicellulases, such as GH5 β-mannanase, GH74 xyloglucanase, GH54, and GH62 α-L-arabinofuranosidase, have been also experimentally demonstrated to target binding to cellulose [26-29]. The cellulose-binding ability of a GH10 xylanase CBM1 has recently been reported for an enzyme from *Malbranchea pulchella* [30]. Xyl10A CBM1 also contains the following highly conserved residues within its amino-acid sequence (Figure 1): three aromatic residues (Trp376, Tyr402, and Tyr403) that are hypothesized to form the binding face for cellulose, two polar residues (Gln378 and Asn400) that potentially form hydrogen bonds with cellulose, and four cysteine residues involved in disulfide bonds (Cys379, Cys390, Cys396, and Cys406) [29,31].

We used a quartz crystal microbalance with the dissipation monitoring (QCM-D) technique to compare the binding affinities of intact and mutant Xyl10As toward crystalline cellulose. It has been established that QCM-D is a reliable method to evaluate the binding affinity of an enzyme toward its substrate [32,33]. Figure 3 shows the results of binding assays between a cellulose nanocrystal film and the three purified Xyl10As. As expected, an increase in mass was observed for Xyl10Awt due to adsorption of the enzymes to the cellulose surface, indicating that Xyl10A CBM1 has binding ability to crystalline cellulose. Adsorption of Xyl10Awt on cellulose continued to increase under the presence of 250 μM

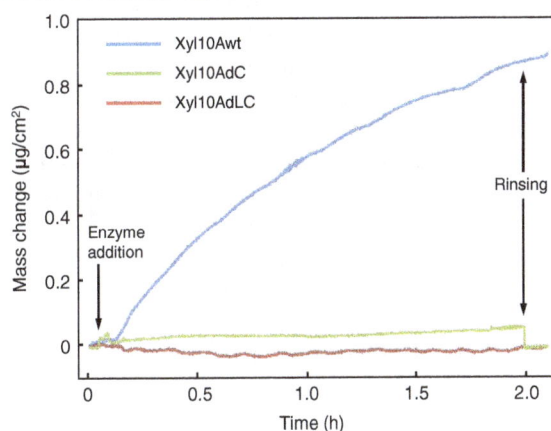

Figure 3 Adsorption of intact and mutant Xyl10As onto the cellulose nanocrystal films. Enzyme solution (250 μM) in 50 mM sodium acetate buffer (pH 5.0) was added into the QCM-D flow cell at a flow rate of 50 μL/min for 10 min. The frequency change due to adsorption on cellulose nanocrystal film was recorded and converted into adsorbed mass density (μg/cm²).

enzyme concentration during the experimental time of 2 hours. In addition, Xyl10Awt was retained on the surface of crystalline cellulose even after rinsing with buffer solution. However, the small mass change observed for Xyl10AdC decreased after rinsing and thereafter was consistent with the expected mass value of Xyl10AdLC. This observation suggests that the Ser/Thr-rich linker region in Xyl10AdC shows weak interactions with the cellulose surface.

Enzymatic properties of intact and mutant Xyl10As

The substrate specificities of Xyl10As with and without CBM1 were examined on various substrates, including both soluble and insoluble xylan and chromogenic compounds (Table 1). All Xyl10As showed the highest activities for arabinoxylan and detectable activities for the chromogenic compounds. This observation suggests that the Xyl10A catalytic domain has catalytic versatility and hydrolyzes short xylooligomers (as is seen for other GH10 xylanases) [16,18]. Further, no activity was detected when Avicel, carboxymethyl cellulose, and xyloglucan were used as substrates. Deletion of CBM1 resulted in an increase in the hydrolytic activities of Xyl10As for all the soluble substrates and insoluble oat-spelt xylan. The specific activities of Xyl10AdC and Xyl10AdLC for these substrates were found to be 1.2- to 1.6-fold higher than those of Xyl10Awt at 45°C (Table 1). These results indicate that the presence of CBM1 had no direct effects on xylan degradation.

The activity of mutant enzymes was significantly enhanced at higher temperature. Although the Xyl10Awt and Xyl10AdC have the same optimum temperature (80°C) for xylanase activity, the specific activities of Xyl10AdC were increased by 1.9-fold for birchwood xylan when the temperature increased from 70°C to 80°C (Figure 4a). The optimum activity for Xyl10AdLC was found at 75°C and dropped dramatically with increase in temperature up to 80°C. The mutant enzymes showed relatively higher activities in comparison to Xyl10Awt under acidic conditions (data not shown). The optimum pH for the hydrolysis of birchwood xylan by mutant

Xyl10As was around pH 4.5 in McIlvaine buffer, whereas that for Xyl10Awt was around pH 5.0. The relative activities of Xyl10Awt, Xyl10AdC, and Xyl10AdLC at pH 3.4 were 33%, 78%, and 67%, respectively, for the optimum pH conditions. The kinetic parameters for Xyl10As were determined using birchwood xylan as substrate. The k_{cat} for mutant enzymes was 1.3- to 1.5-fold higher in comparison to that of Xyl10Awt, whereas the apparent K_m values were almost of the same magnitude in all the Xyl10As (Table 2). Therefore, it was concluded that Xyl10As without CBM1 are more efficient (k_{cat}/K_m) in comparison to Xyl10Awt for easily degradable substrates such as purified model xylan.

In contrast, the Xyl10Awt was more effective in the degradation of residual xylan in insoluble lignocellulose substrate than mutant enzymes. When WDM-RS was used as the lignocellulose substrate, Xyl10Awt showed relatively higher xylanase activity, that is, 1.7- and 1.4-fold in comparison to Xyl10AdC and Xyl10AdLC, respectively (Table 1). Thus, this result suggests that the binding of the enzyme to cellulose through CBM1 indirectly enhances the xylan degradation in WDM-RS. Xyl10Awt had higher activity compared to mutant enzymes at all examined temperatures (Figure 4b). The activity of Xyl10AdC was found to be relatively lower compared to Xyl10AdLC (Figure 4b), suggesting that presence of the linker region without CBM1 may have a negative impact on xylan degradation in WDM-RS.

Xyl10A from *T. cellulolyticus* is a thermostable GH10 xylanase having a thermal midpoint (T_m) above 80°C [19,24]. T_m values of intact and mutant Xyl10As measured by fluorescence-based thermal shift assay were estimated to be 81.0°C (Xyl10Awt), 81.0°C (Xyl10AdC), and 75.5°C (Xyl10AdLC), and these values were in good agreement with the results of optimum temperature for xylanase activities of these enzymes (Figure 4). The reason behind the relatively low thermostability of Xyl10AdLC may be due to an incomplete α9 helix at the C-terminal region (Figure 1).

Synergistic hydrolysis of pretreated rice straw

WDM-RS is a fine fiber material generated by mechanical fibrillation of rice straw soaked in water and has

Table 1 Substrate specificities of purified intact and mutant Xyl10As toward different substrates

Substrates	Xyl10Awt (A) (U/nmol)	Xyl10AdC (B) (U/nmol)	Xyl10AdLC (C) (U/nmol)	(B)/(A)	(C)/(A)
Birchwood xylan	10.6	15.1	13.9	1.42	1.31
Wheat arabinoxylan	21.9	35.5	28.2	1.62	1.29
Insoluble oat-spelt xylan	2.53	3.85	3.66	1.52	1.45
PNP-lactose	0.048	0.064	0.061	1.34	1.28
PNP-cellobiose	0.251	0.393	0.342	1.56	1.36
PNP-xylose	0.022	0.028	0.026	1.25	1.18
WDM-RS[a]	0.094	0.054	0.066	0.57	0.70

Activities were determined at 45°C (pH 5.0). [a]Activities for WDM-RS were determined at 50°C (pH5.0). PNP-lactose, p-nitrophenyl-β-D-lactoside; PNP-cellobiose, p-nitrophenyl-β-D-cellobioside; PNP-xylose, p-nitrophenyl-β-D-xyloside; WDM-RS, wet disc-mill treated rice straw.

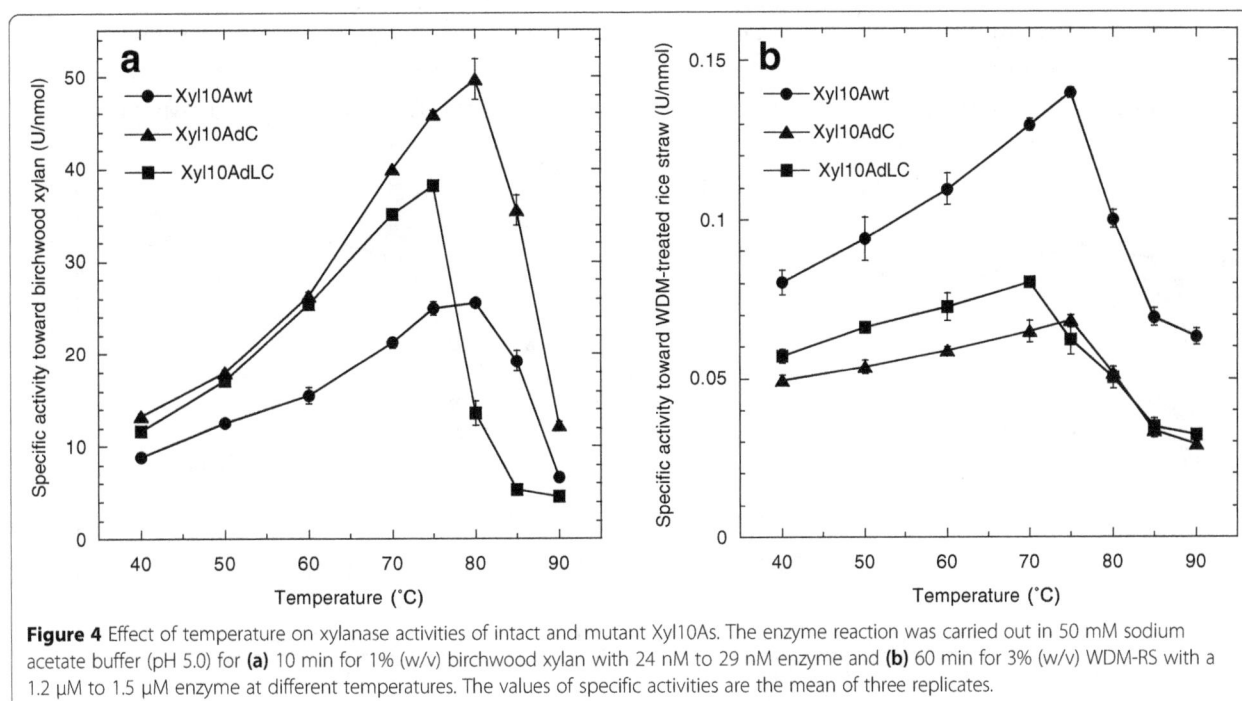

Figure 4 Effect of temperature on xylanase activities of intact and mutant Xyl10As. The enzyme reaction was carried out in 50 mM sodium acetate buffer (pH 5.0) for **(a)** 10 min for 1% (w/v) birchwood xylan with 24 nM to 29 nM enzyme and **(b)** 60 min for 3% (w/v) WDM-RS with a 1.2 μM to 1.5 μM enzyme at different temperatures. The values of specific activities are the mean of three replicates.

been reported to have improved hydrolysis yield by the cellulase mixture [34]. However, the reduction of crystallinity and removal of major hemicellulose fraction were not observed in the pretreated rice straw sample [34]. The potential of intact and mutant Xyl10As, along with cellulase monocomponents consisting of purified Cel7A and Cel6A from *T. cellulolyticus*, was evaluated for the hydrolysis of WDM-RS glucan and xylan. Optimal WDM-RS glucan hydrolysis by the mixture of Cel7A and Cel6A was observed at a mole ratio of approximately 3.9:4.5 (data not shown), and hence, the mixture of Cel7A (38.9 nmol/g glucan) and Cel6A (44.5 nmol/g glucan) were replaced with various ratios of Xyl10Awt (0 to 98.4 nmol/g glucan), Xyl10AdC (0 to 108 nmol/g glucan), and Xyl10AdLC (0 to 119 nmol/g glucan). In addition, a purified GH3 β-glucosidase (Bgl3A) from *T. cellulolyticus* was also added to the reaction mixture to prevent inhibition by cellobiose at 6.17 nmol/g glucan of protein loading.

When intact or mutant Xyl10A was used to replace small amounts of cellulases, a significant increase in the degree of synergism was observed in the hydrolysis of

Table 2 Kinetic parameters for hydrolysis of birchwood xylan by intact and mutant Xyl10As

Enzyme	K_m (mg mL^{-1})	k_{cat} (s^{-1})	k_{cat}/K_m (s^{-1} mg^{-1} mL^{-1})
Xyl10Awt	2.51 ± 0.42	232 ± 12.4	92.2 ± 16.2
Xyl10AdC	2.80 ± 0.18	359 ± 7.8	128 ± 8.9
Xyl10AdLC	2.59 ± 0.14	310 ± 5.4	120 ± 6.7

Reactions were performed at 45°C (pH 5.0).

WDM-RS glucan after 24-h reaction (Figure 5a). Further, the mixture containing Xyl10wt showed greater increase in the glucan hydrolysis by cellulases in comparison to the mutant Xyl10As. The highest glucan hydrolysis yields were observed at ratios of 0.25 (Xyl10Awt), 0.1 (Xyl10AdC), and 0.1 (Xyl10AdLC), which corresponded to a 3.1-, 1.7-, and 1.9-fold increase in the glucan hydrolysis yield, respectively, with respect to the corresponding cellulase loading without Xyl10A (Figure 5a). The synergistic enhancement of glucan hydrolysis was observed with an increase in the ratio of Xyl10A. The greatest enhancement of glucan hydrolysis was observed at a ratio of intact or mutant Xyl10A of 0.75, and the fold increases in the glucan hydrolysis yields were 4.9-fold (Xyl10Awt), 2.2-fold (Xyl10AdC), and 2.5-fold (Xyl10AdLC) with respect to the corresponding cellulase loading without Xyl10A (Figure 5a). These enzyme mixtures also showed the highest xylan hydrolysis yields for WDM-RS (Figure 5b). These results indicate that the synergistic interaction between cellulases and Xyl10A is closely correlated with the xylan removal from WDM-RS. Therefore, Xyl10A CBM1 that increases the xylan degradation in WDM-RS seems to be playing an important role in the synergistic hydrolysis of WDM-RS. The enhancement of glucan hydrolysis showed a similar trend after 96 h for WDM-RS, and the highest glucan hydrolysis was obtained at a ratio of intact or mutant Xyl10A of 0.25 (data not shown). In this case, the glucan hydrolysis yields showed an increase to 28% (Xyl10Awt), 18% (Xyl10AdC), and 21% (Xyl10AdLC).

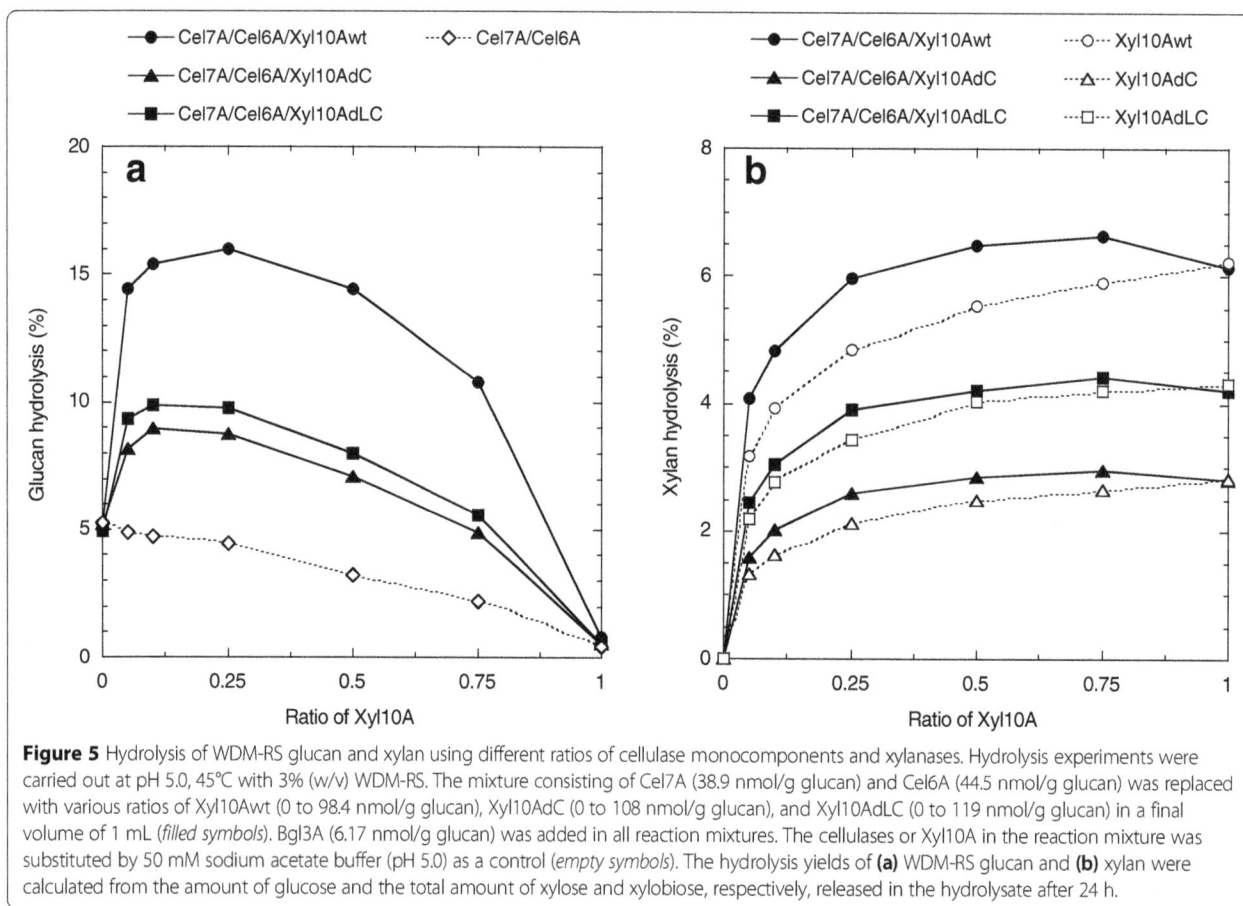

Figure 5 Hydrolysis of WDM-RS glucan and xylan using different ratios of cellulase monocomponents and xylanases. Hydrolysis experiments were carried out at pH 5.0, 45°C with 3% (w/v) WDM-RS. The mixture consisting of Cel7A (38.9 nmol/g glucan) and Cel6A (44.5 nmol/g glucan) was replaced with various ratios of Xyl10Awt (0 to 98.4 nmol/g glucan), Xyl10AdC (0 to 108 nmol/g glucan), and Xyl10AdLC (0 to 119 nmol/g glucan) in a final volume of 1 mL (*filled symbols*). Bgl3A (6.17 nmol/g glucan) was added in all reaction mixtures. The cellulases or Xyl10A in the reaction mixture was substituted by 50 mM sodium acetate buffer (pH 5.0) as a control (*empty symbols*). The hydrolysis yields of **(a)** WDM-RS glucan and **(b)** xylan were calculated from the amount of glucose and the total amount of xylose and xylobiose, respectively, released in the hydrolysate after 24 h.

The synergistic interaction between cellulases and a Xyl10A variant slightly enhanced the xylan hydrolysis of WDM-RS. A 1.1- to 1.2-fold increase in xylan hydrolysis yield was observed at a ratio of intact or mutant Xyl10A of 0.25 in comparison to the corresponding Xyl10A loading without cellulases (Figure 5b). The total sugar hydrolysis yield of both glucan and xylan was most effective at a ratio of intact or mutant Xyl10A of 0.25. The total yields after 96 h reaction were 21% (Xyl10Awt), 13% (Xyl10AdC), and 16% (Xyl10AdLC) due to relatively low xylan hydrolysis yields (4.4% to 10.4%). Glucose, xylose, and xylobiose were observed in the hydrolysates. The amounts of xylobiose released in the hydrolysates were 1.5- to 1.65-fold higher than those of xylose.

Implications of xylan removal for synergistic glucan hydrolysis in WDM-RS

To evaluate the outcomes of xylan removal from cellulose on glucan hydrolysis of WDM-RS, the experiments were carried out over a range of supplementations of Xyl10Awt (0 to 49.2 nmol/g of glucan), Xyl10AdC (0 to 54.1 nmol/g of glucan), and Xyl10AdLC (0 to 59.6 nmol/g of glucan) in the presence of fixed amounts of

the mixture of Cel7A (38.9 nmol/g of glucan), Cel6A (44.5 nmol/g of glucan), and Bgl3A (6.17 nmol mg/g of glucan) at 45°C for 24 h. Glucan and xylan hydrolysis yields increased with increasing xylanase loadings. These yields were significantly increased at a slight xylanase loading with less than 5 nmol/g of glucan, suggesting that the synergistic interactions between the cellulases and Xyl10As are very sensitive (Figure 6). The loading amounts of Xyl10Awt, Xyl10AdC, and Xyl10AdLC required to increase the glucan hydrolysis yield with Cel7A/Cel6A mixture by twofold were 0.67, 21.5, and 5.25-nmol/g of glucan, respectively. This suggests that Xyl10Awt was more effective in reducing the total enzyme loading for the degradation of WDM-RS glucan. The differences in the synergistic glucan hydrolysis among Xyl10Awt, Xyl10AdC, and Xyl10AdLC loadings seem to be related to the differences in the xylan hydrolysis yields (Figure 6). In fact, the glucan hydrolysis yield increased by the supplementation with the intact or mutant Xyl10As showed a good linear relationship with the xylan-hydrolysis yield (Figure 7).

The leverage ratio, as described by Kumar and Wyman, for Xyl10A supplementations in relation with

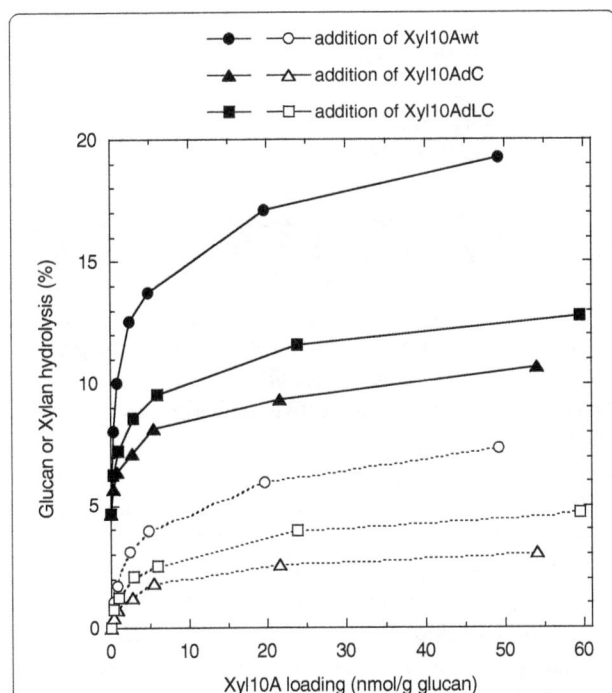

Figure 6 Effect of intact or mutant Xyl10A supplements on WDM-RS hydrolysis by cellulase monocomponents. Intact or mutant Xyl10As were added to the fixed amounts of the mixture of Cel7A (38.9 nmol/g of glucan), Cel6A (44.5 nmol/g of glucan), and Bgl3A (6.17 nmol/g of glucan). Hydrolysis experiments were carried out at pH 5.0, 45°C with 3% (w/v) WDM-RS in a final volume of 1 mL. The hydrolysis yields of WDM-RS glucan (*filled symbols*) and xylan (*empty symbols*) were calculated from the amount of glucose and the total amount of xylose and xylobiose, respectively, released in the hydrolysate after 24 h.

the enhanced glucan hydrolysis of WDM-RS can be defined as the ratio of the percent increase in the glucose release to the percent increase in xylose release [6]. Interestingly, a two-stage leverage ratio was observed in the supplementation using Xyl10Awt for the range of 0 to 49.2 nmol/g of glucan after 24 h of hydrolysis, that is, 3.17 (R^2 = 0.999) at less than 2% of xylan hydrolysis, and 1.62 (R^2 = 0.999) at more than 2% of xylan hydrolysis (Figure 7a). The increase in the leverage ratio at lower supplementations of Xyl10Awt may correlate with the binding of Xyl10Awt to cellulose; consequently, Xyl10Awt would preferentially hydrolyze a minor xylan component that interrupts glucan hydrolysis. On the other hand, the leverage ratios for the supplementations of Xyl10AdC and Xyl10AdLC were calculated to be 1.86 (R^2 = 0.993) and 1.66 (R^2 = 0.991), respectively, and showed a linear relationship with the graph passing through the origin (Figure 7a). This result suggests that Xyl10AdC and Xyl10AdLC randomly target the xylan present in WDM-RS. The leverage ratios observed after 96 h of hydrolysis for the supplements of Xyl10Awt, Xyl10AdC, and Xyl10AdLC changed to 1.29 (R^2 = 0.999, >4% of xylan hydrolysis), 2.01(R^2 = 0.993), and 1.54 (R^2 = 0.998, >2% of xylan hydrolysis), respectively (Figure 7b). Xyl10AdLC exhibited a slightly increased leverage ratio at less than 2% of xylan hydrolysis. The decreased leverage ratio observed for Xyl10Awt (>4% of xylan hydrolysis) may imply that the target for Xyl10Awt could have shifted to a larger xylan obstacle during the prolonged hydrolysis reaction.

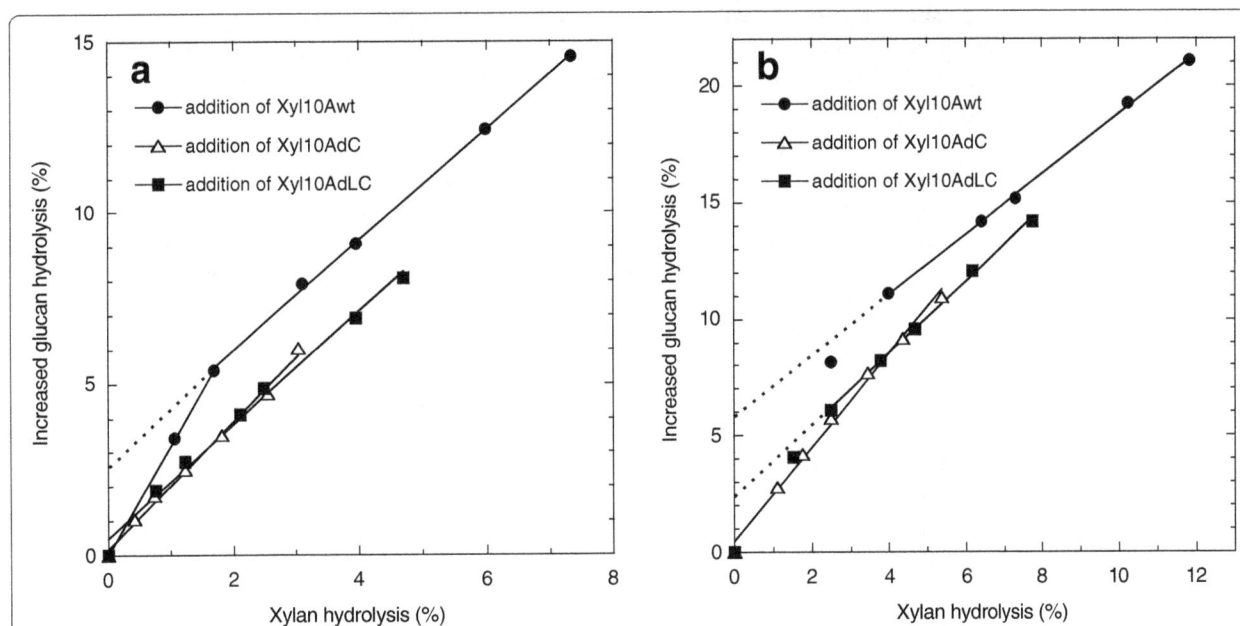

Figure 7 Glucose release and xylan removal in WDM-RS hydrolysis by cellulase monocomponents with differing xylanase supplementation. The relationship between increased glucan hydrolysis and xylan hydrolysis was analyzed using the hydrolysate after 24 h **(a)** and 96 h **(b)** based on hydrolysis experiments described in Figure 6.

Discussion

It is generally recognized that xylanases have a role in the removal of xylan obstacles that prevent glucan hydrolysis by cellulases in a synergistic hydrolysis of pretreated lignocellulose. Earlier studies on modular xylanases have been focused on bacterial enzymes possessing xylan-binding CBM, such as family 2 [35,36], family 6 [37], and family 9 [38], and the importance of CBM2 for the synergistic hydrolysis of lignocellulosic substrates by bacterial GH11 xylanases has previously been demonstrated [35,36]. However, little attention has been paid to the effect of the cellulose-binding xylanase CBM1 for the hydrolysis of lignocellulose by a fungal cellulase system. This may be because all GH10 and GH11 xylanases in the cellulase system from *Trichoderma reesei*, which is a well-known industrial cellulase producer, are non-modular enzymes and show synergistic interactions with cellulases in the case of lignocellulosic substrates [13,39]. In the present study, using the characterizations of *T. cellulolyticus* Xyl10As with and without CBM1, we evaluated the role of xylanase CBM1 in the hydrolysis of glucan in the lignocelluloses. Further, our results revealed that the binding of Xyl10A to lignocellulose through CBM1 is important for the enhancement of hydrolysis of xylan and efficient synergistic hydrolysis of glucan in the lignocellulose.

In comparison to the Xyl10Awt, the hydrolytic activities of mutant Xyl10As without CBM1 were increased by 1.2- to 1.6-fold for soluble and insoluble xylans and chromogenic substrates, whereas the activities were decreased by 0.57- to 0.70-fold for WDM-RS (Table 1). On the other hand, xylan-binding CBM-deleted bacterial xylanases have been reported to show decreased hydrolytic activity not only for xylan in lignocellulose but also for the insoluble oat xylan [35,38,40]. These differences suggest that the cellulose binding by Xyl10A CBM1 is closely related to the efficient hydrolysis of lignocellulosic xylan. In addition, we confirmed that the Xyl10A CBM1 directly binds to cellulose by QCM-D (Figure 3). These observations support the hypothesis that the function of CBM is to increase the effective enzyme concentrations on the insoluble polysaccharide surfaces, to target the catalytic domain to the substrate [20,21]. On the other hand, a xylanase without CBM1 may be more suitable for the hydrolysis of the soluble hemicellulose fraction removed from the pretreated lignocellulose. Thus, it seems reasonable that various fungal cellulase systems naturally contain multiple GH10 and GH11 xylanases with and without CBM1 [18,19,41].

The hydrolytic activity of Xyl10AdC for WDM-RS xylan was relatively lower in comparison to the Xyl10AdLC without the Ser/Thr-rich linker region. The glycosylated linker of Xyl10AdC seems to have a negative effect on the hydrolysis of WDM-RS xylan. It has been suggested that Ser/Thr-rich linker regions in modular cellulolytic enzymes act as flexible connectors between subdomains, and the presence of linker glycosylation provides protection from proteases [42]. Recently, Payne *et al.* had reported that the glycosylated linkers enhance the binding affinity of the CBM1 alone by an order of magnitude [42]. The result of QCM-D analysis for Xyl10AdC exhibited that the Xyl10A linker without CBM1 shows weak interactions with cellulose (Figure 3). These observations predict that the Xyl10A linker weakly interacts with WDM-RS. However, the lack of CBM1 may lead to non-specific interactions with WDM-RS, resulting in a negative effect on the hydrolysis of WDM-RS xylan. Hence, the linker region without CBM1 is considered to be ineffective for hydrolysis of xylan substrates including lignocellulose.

There are a number of examples of synergistic interactions observed between cellulases and xylanase required for the hydrolysis of various pretreated lignocellulose substrates using crude or purified enzymes. It has been found that glucose yield increased almost linearly with the removal of residual xylose by the enzymes, despite the substantial differences in the relative yields for individual lignocellulosic substrates [6,12,17]. In this study, we revealed that the presence of Xyl10A CBM1 is not necessarily essential both for the synergistic interactions between cellulases and xylanases and for the linear relationship between increased hydrolysis of glucan and xylan for WDM-RS (Figure 5, Figure 7). However, it should be noted that an efficient synergistic hydrolysis was found in the interactions of the substrate with Xyl10Awt that has the highest xylan hydrolysis activity. Furthermore, the high leverage ratio at lower supplementations of Xyl10Awt seems to be closely associated with the specific adsorption of CBM1 of enzyme on the cellulose. Fungal CBM1s have a highly conserved structure and bind to similar targets present along the ridges of the crystalline cellulose [43]. Xyl10Awt is expected to bind in the proximity of Cel7A (or Cel6A) on the cellulose surface in WDM-RS, which enhances the susceptibility of the substrate to the action of cellulases due to preferential removal of xylan obstacles that interrupt the initial cellulase activity. In contrast, the CBM1-lacking Xyl10A mutants hydrolyze WDM-RS xylan non-specifically, resulting in lower efficiency of glucan hydrolysis and xylan removal. These results indicate that the binding of Xyl10A CBM1 to a cellulose surface has important implications not only for the enhanced hydrolysis of lignocellulosic xylan but also for the efficient synergistic activity with the cellulases.

On the other hand, it has been reported that non-productive adsorption of modular cellulolytic enzymes on lignocellulose substrates results in reduced hydrolysis efficiency [2]. Palonen *et al.* established a clear correlation between the presence of CBM1 in cellulases and the non-productive adsorption on the surface of lignin [44]. Qi *et al.* found that higher adsorption was observed for dilute acid treated wheat straw compared to dilute alkali treated sample [45]. In this study, we confirmed that most of the Xyl10Awt enzyme was adsorbed on a WDM-RS surface using a binding assay that measures the residual activity (data not shown). However, the high synergistic hydrolysis of WDM-RS at lower supplementations of Xyl10Awt suggests that Xyl10A CBM1 is specifically adsorbed on cellulose rather than on a lignin surface in the substrate (Figure 6). WDM-RS treated with mechanical fibrillation without any chemicals may reduce the non-productive adsorption of enzyme on a lignin surface.

Várnai *et al.* reported that CBMs in cellulases were not required for an efficient hydrolysis of lignocellulose with high consistency, although CBMs were more important in the catalytic performance of cellulases at low substrate concentration (1% w/v) [22]. The hydrolytic performance of cellulases without CBMs caught up with that of cellulases with CBMs at 20% (w/v) substrate concentration [22]. It should be noted that the role of Xyl10A CBM1 was evaluated at the relatively low substrate concentrations (3% w/w) in this study. Our results seem to support the importance of CBM1 at low substrate concentration. In addition, the hydrolysis of WDM-RS using the minimal cellulases and Xyl10A mixture was compared in the relatively low conversion of the glucan and xylan. The addition of accessory enzymes that hydrolyze the side groups in heteroxylan could improve the xylan hydrolysis yield in WDM-RS. Further work would be necessary to evaluate the effect of Xyl10A CBM1 for the higher hydrolysis yield of lignocellulose at high substrate concentration.

Conclusions

We demonstrated that cellulose binding by Xyl10A CBM1 serves two roles in the synergistic hydrolysis of lignocellulose using Xyl10A and cellulases. First, it enhances the xylan hydrolysis activity by binding to cellulose, and second, it helps in the efficient removal of xylan obstacles that interrupt the cellulase activity due to a similar binding target of CBM1. Hence, the use of Xyl10A with CBM1 for the synergistic hydrolysis leads to the significant reduction in the enzyme loading used for the hydrolysis process. These findings may be applied to the synergistic hydrolysis of lignocellulose using other fungal hemicellulases and accessory enzymes possessing CBM1. The use of a combination of CBM1-containing modular cellulases, hemicellulases, and accessory enzymes present in the fungal cellulase system could contribute to further reduction of the enzyme loading for the hydrolysis of pretreated lignocellulosic substrates.

Methods

Strains, plasmids, and media

The *T. cellulolyticus* YP-4 uracil autotroph was selected for the production of recombinant proteins and was maintained on potato dextrose agar (Difco, Detroit, MI, USA) plates containing uracil and uridine at a final concentration of 1 g/L each [46]. The plasmid pANC202 [46], which contains the *pyrF* gene and the glucoamylase (*glaA*) gene promoter and terminator regions, was used in the construction of the plasmids pANC208 [19], pANC228, and pANC229 that were used for the expression of the Xyl10Awt, Xyl10AdC, and Xyl10AdLC proteins, respectively. *Escherichia coli* DH5α (Takara Bio, Kyoto, Japan) was used for the DNA procedures. The prototrophic transformants of *T. cellulolyticus* YP-4, that is, strains Y208 [19], Y228, and Y229, which express recombinant proteins Xyl10Awt, Xyl10AdC, and Xyl10AdLC, respectively, were maintained on MM agar (1% (w/v) glucose, 10 mM NH_4Cl, 10 mM potassium phosphate (pH 6.5), 7 mM KCl, and 2 mM $MgSO_4$, 1.5% (w/v) agar) plates [47]. The recombinant Xyl10As from these transformants were produced using a soluble starch medium containing 2% (w/v) soluble starch (Wako Pure Chemical Industries, Osaka, Japan) and 0.2% (w/v) urea as described in an earlier study [19].

Plasmid construction and fungal transformation

The Xyl10Awt expression plasmid pANC208, which carries the 1.45 kb *xyl10Awt* genomic region from *T. cellulolyticus* CF-2612 chromosomal DNA (DDBJ accession number: AB796434) [19] was used as the template DNA to amplify the genes *xyl10AdC* and *xyl10AdLC*. A 1.35 kb DNA fragment containing the coding region of *xyl10AdC* was amplified by PCR using the forward primer 5′-ATT*GTTAAC*AAGATGACTCTAGTAAAGGCTATTC (*Hpa*I site in italics) and the reverse primer 5′-TAA*CCTG CAGG*CTAACCTGAGGTAGCGCTTGTGCTAGTC (*Sbf*I site in italics). A 1.23 kb DNA fragment containing the coding region of *xyl10AdLC* was amplified using the same forward and reverse primers, that is, 5′-AAT*CCTG CAGG*TTATAAGCCAGCAAGGATACCATAGTATG (*Sbf*I site in italics). The *xyl10AdC* expression plasmid pANC228 and the *xyl10AdLC* expression plasmid pANC229 were constructed by introducing each PCR fragment digested with *Hpa*I/*Sbf*I into the *Eco*RV/*Sbf*I site between the *glaA* promoter and terminator regions of pANC202. All ligated gene fragments and their ligation sites were verified by sequencing.

The plasmids pANC228 and pANC229 were transformed into protoplasts of *T. cellulolyticus* YP-4 by non-homologous integration into the host chromosomal DNA [47]. Gene integration into the prototrophic transformants was verified using genomic PCR. Chromosomal DNA of the transformants was purified using the Gentra Puregene Yeast/Bact. Kit (Qiagen, Valencia, CA, USA). The strains showing high xylanase activity in the soluble starch medium were selected for the production of Xyl10AdC and Xyl10AdLC proteins and were designated as Y228 and Y229, respectively.

Purification of recombinant Xyl10As

T. cellulolyticus strains Y208, Y228, and Y229 were grown in the soluble starch medium at 30°C, 200 rpm, for 96 h in an Erlenmeyer flask. The whole broth was centrifuged and the resulting supernatant was filtered through a 0.22-μm polyether sulfone membrane (Thermo Scientific, Rockford, Illinois, USA) under sterile conditions. The culture filtrate containing recombinant Xyl10As was stored at 4°C.

Enzyme purification was carried out using an ÄKTA purifier chromatography system (GE Healthcare, Buckinghamshire, UK) at room temperature. Purification of Xyl10Awt was performed by a Source 15Q (GE Healthcare) anion-exchange chromatography and a Source 15ISO (GE Healthcare) hydrophobic interaction column chromatography as described in an earlier study [19]. Xyl10AdC and Xyl10AdLC were purified by the same procedure as Xyl10Awt, except in this case where a Source 15Phe (GE Healthcare) hydrophobic interaction column was used instead of the Source 15ISO column. The protein applied to the Source 15Phe column was eluted with a linear gradient of 1.0 M to 0.0 M $(NH_4)_2SO_4$ in 20 mM sodium acetate buffer (pH 5.5). The purity and size of the protein was analyzed by SDS-PAGE using NuPage 4-12% Bis-Tris gel (Invitrogen, Carlsbad, CA, USA). Glycosylation of proteins on an SDS-PAGE gel was detected by using the Pro-Q Emerald 300 glycoprotein stain (Invitrogen) following the manufacture's instruction. All purified enzymes were preserved in a 20-mM sodium acetate buffer (pH 5.5) containing 0.01% NaN_3 at 4°C. The protein concentration was determined with the Pierce BCA Protein Assay Kit (Thermo Scientific) using the bovine serum albumin as the standard (Thermo Scientific).

Enzyme activity assays

Xylanase activity was measured by assaying the reducing sugars released after the enzyme reaction with 1% (w/v) xylan. The concentration of reducing sugars was determined using 3,5-dinitrosalicylic acid. The enzyme reaction was carried out in 50 mM sodium acetate buffer (pH 5.0) at 45°C for 10 min. Birchwood xylan (Sigma-Aldrich, St. Louis, Missouri, USA) and wheat arabinoxylan (Megazyme, Bray, Ireland) were used as the soluble xylan substrates. The insoluble xylan was prepared by boiling oat-spelt xylan (Tokyo Chemical Industry, Tokyo, Japan) for 30 min in distilled water and recovering the residue by centrifugation as described by Moraïs et al. [35]. The enzyme activity against WDM-RS was determined by assaying the reducing sugars released after 60 min of enzyme reaction with 3% (w/v) substrate at pH 5.0 (50 mM sodium acetate buffer) and 50°C. One unit of enzyme activity was defined as the amount of enzyme

that catalyzed the formation of 1 μmol of reducing sugar per minute.

The enzyme activities against *p*-nitrophenol-based chromogenic glycosides (*p*-nitrophenyl-β-D-lactoside (PNP-lactose), *p*-nitrophenyl-β-D-cellobioside (PNP-cellobiose), and *p*-nitrophenyl-β-D-xyloside (PNP-xylose)) were determined at pH 5.0 in 50 mM sodium acetate buffer and at 45°C for 15 min using reaction mixtures including 1 mM PNP substrate. One unit of enzyme activity was defined as the amount of enzyme that catalyzed the formation of 1 μmol of *p*-nitrophenol per minute.

Kinetic parameters were determined using six concentrations of birchwood xylan (0.5 mg/ml to 16 mg/mL) at 45°C. Enzyme concentrations in the reaction mixtures (pH 5.0) were 24.1 nM (Xyl10Awt), 26.5 nM (Xyl10AdC), and 29.2 nM (Xyl10AdLC). The reaction was stopped at appropriate time intervals (2, 4, 6, 8, and 10 min), and the initial rates of xylan hydrolysis were determined accordingly. All experiments were repeated three times. The estimated kinetic parameters were obtained by fitting initial rates as a function of substrate concentration to the Michaelis-Menten equation using KaleidaGraph 4.1 (Synergy Software, Reading, PA, USA).

Protein thermal shift assay

Thermal midpoint (T_m) value of protein was determined by the fluorescence-based thermal shift assay using a real-time PCR detection system with the CFX Manager Program (CFX Connect, Bio-Rad, Hercules, CA, USA) as described in a previous study [46]. The protein sample containing the SYPRO orange dye (Invitrogen) was heated at 0.5°C per five seconds from 25°C to 95°C, and the fluorescence intensity (excitation/emission, 450 nm to 490 nm/560 nm to 580 nm) was measured every 0.5°C.

Binding assay of Xyl10A for cellulose nanocrystal films

Cellulose nanocrystal film was prepared following the methodology described by Kumagai et al. [32]. CF11 cellulose powder (GE Healthcare) was fibrillated using a grinder (Crendipitor MKCA6-2, Masuko Sangyo Co., Ltd., Saitama, Japan) and subsequently using a high-pressure homogenizer (Masscomizer MMX-L100, Masuko Sangyo). The refined cellulose powder solution was ultrasonicated to improve its dispersibility and was spin-coated as cellulose nanocrystal film on the QCM-D gold sensors (QSX 301, Biolin Scientific, Stockholm, Sweden). The sensors were subsequently heated at 80°C for 10 min in an oven.

A QCM-D (Q-Sense E1, Biolin Scientific) was used for enzyme adsorption on the cellulose nanocrystal films. The sensor coated with the cellulose nanocrystal film was placed in 50 mM sodium acetate buffer solution (pH 5.0) overnight to allow it to swell completely. The swollen sensor was mounted in the QCM-D flow cell filled with the same buffer and was allowed to swell again until no appreciable frequency shifts were observed. The 250 µM enzyme solution in 50 mM sodium acetate buffer (pH 5.0) was introduced into the QCM-D flow cell with a peristaltic pump at a flow rate of 50 µL/min. The pump was stopped after 10 min when the buffer solution was completely replaced by the enzyme solution. The adsorption of enzymes onto the cellulose nanocrystal films was monitored for 2 h under static conditions. After that, only the buffer solution was introduced to rinse the system and follow the enzyme desorption. The temperature of the QCM-D flow cell was maintained at 40 ± 0.02°C during the measurements. The frequency changes at the fundamental resonance frequency (5 MHz) and its overtone frequencies (15, 25, 35, 55, and 75 MHz) were monitored simultaneously, and the third overtone (15 MHz) was used in the data evaluation. The frequency changes were converted into adsorbed mass density (µg/cm^2) according to the Sauerbrey equation using the QTools software (Biolin Scientific) [48].

Hydrolysis of pretreated rice straw

The WDM-treated rice straw was prepared using a Super-masscolloider MKZA10 (Masuko Sangyo), based on a previous report [34]. The rice straw slurry was introduced into the equipment, and the operation was repeated 20 times. The WDM-treated sample was freeze-dried and kept in a desiccator cabinet at room temperature until use. The composition (% dry weight) of structural carbohydrates in the pretreated rice straw was analyzed based on the standard NREL laboratory analytical procedure (LAP) [49] and was found to consist of 32.7% glucan, 19.4% xylan, 2.5% arabinan, and 1.5% galactan.

The hydrolysis of pretreated rice straw was carried out at 3% (w/v) solids loading in 50 mM sodium acetate buffer (pH 5.0). In the standard assay, the mixture consisting of Cel7A (38.9 nmol/g glucan) and Cel6A (44.5 nmol/g glucan) were replaced with various ratios of Xyl10Awt (0 to 98.4 nmol/g glucan), Xyl10AdC (0 to 108 nmol/g glucan), and Xyl10AdLC (0 to 119 nmol/g glucan) in a final volume of 1 mL. Bgl3A (6.17 nmol/g glucan) was added in all reaction mixtures. The total protein loading was 4.59 mg/g glucan. The reaction mixture was incubated at 45°C on a rotator. Cel7A, Cel6A, and Bgl3A were purified from *T. cellulolyticus* CF-2612 as described previously [24]. The hydrolysate of the pretreated rice straw was analyzed using a HPLC system equipped with a refractive index detector (RI-2031Plus, JASCO, Tokyo, Japan), an Aminex HPX-87P column (7.8 mm I.D. x 30 cm, Bio-Rad), and a Carbo-P micro-guard cartridge (Bio-Rad). The samples were eluted with water at a rate of 1 mL/min at 80°C. The glucan hydrolysis yield was defined as the ratio of the total equivalents of glucan hydrolyzed (glucose) to the total potential glucan available in the pretreated solids. The xylan hydrolysis yield was defined as the ratio of total equivalents of xylan hydrolyzed (xylose and xylobiose) to the total potential xylan available in the pretreated solids. All WDM-RS hydrolysis experiments were run in duplicates.

Abbreviations

Bgl3A: glycosyl hydrolase family 3 β-glucosidase from *Talaromyces cellulolyticus*; Cel6A: cellobiohydrolase II from *Talaromyces cellulolyticus*; Cel7A: cellobiohydrolase I from *Talaromyces cellulolyticus*; CBM: carbohydrate-binding module; GH: glycosyl hydrolase; PNP-cellobiose: *p*-nitrophenyl-β-d-cellobioside; PNP-lactose: *p*-nitrophenyl-β-d-lactoside; PNP-xylose: *p*-nitrophenyl-β-d-xyloside; QCM-D: quartz crystal microbalance with dissipation monitoring; T_m: thermal midpoint; WDM-RS: wet disc-mill treated rice straw; Xyl10A: glycosyl hydrolase family 10 xylanase from *Talaromyces cellulolyticus*; Xyl10Awt: recombinant intact xylanase; Xyl10AdC: CBM1-deleted xylanase; Xyl10AdLC: CBM1 and linker region-deleted xylanase.

Competing interests

The authors declare that they have no competing interests.

Authors' contributions

HI and SK designed the study, analyzed the results, and drafted the manuscript. AK and MK participated in the enzymatic characterization and QCM-D analysis. TF participated in the experiments related to molecular biology and reviewed and commented on the manuscript. KI reviewed and commented on the manuscript. All authors have read and approved the final form of the manuscript.

Acknowledgements

The authors thank Miho Yoshimi and Benchaporn Inoue for their technical assistance. This work was supported by the Japan-US Cooperation Project for Research and Standardization of Clean Energy Technologies, Ministry of Economy, Trade and Industry, Japan.

References

1. Merino ST, Cherry J. Progress and challenges in enzyme development for biomass utilization. Adv Biochem Eng Biotechnol. 2007;108:95–120.
2. Van Dyk JS, Pletschke BI. A review of lignocellulose bioconversion using enzymatic hydrolysis and synergistic cooperation between enzymes-Factors affecting enzymes, conversion and synergy. Biotechnol Adv. 2012;30:1458–80.
3. Alvira P, Tomás-Pejó E, Ballesteros M, Negro MJ. Pretreatment technologies for an efficient bioethanol production process based on enzymatic hydrolysis: a review. Bioresour Technol. 2010;101:4851–61.
4. Harris PV, Xu F, Kreel NE, Kang C, Fukuyama S. New enzyme insights drive advances in commercial ethanol production. Curr Opinion Chem Biol. 2014;19:162–70.
5. Selig MJ, Knoshaug EP, Adney WS, Himmel ME, Decker SR. Synergistic enhancement of cellobiohydrolase performance on pretreated corn stover by addition of xylanase and esterase activities. Bioresour Technol. 2008;99:4997–5005.
6. Kumar R, Wyman CE. Effect of xylanase supplementation of cellulase on digestion of corn stover solids prepared by leading pretreatment technologies. Bioresour Technol. 2009;100:4203–13.
7. Bura R, Chandra R, Saddler J. Influence of xylan on the enzymatic hydrolysis of steam-pretreated corn stover and hybrid poplar. Biotechnol Prog. 2009;25:315–22.
8. Hu J, Arantes V, Saddler JN. The enhancement of enzymatic hydrolysis of lignocellulosic substrates by the addition of accessory enzymes such as xylanase: is it an additive or synergistic effect? Biotechnol Biofuels. 2011;4:36.
9. Beg QK, Kapoor M, Mahajan L, Hoondal GS. Microbial xylanases and their industrial applications: a review. Appl Microbiol Biotechnol. 2001;56:326–38.
10. Hu J, Arantes V, Pribowo A, Saddler JN. The synergistic action of accessory enzymes enhances the hydrolytic potential of a "cellulase mixture" but is highly substrate specific. Biotechnol Biofuels. 2013;6:112.
11. Öhgren K, Bura R, Saddler J, Zacchi G. Effect of hemicellulose and lignin removal on enzymatic hydrolysis of steam pretreated corn stover. Bioresour Technol. 2007;98:2503–10.

12. Gao D, Uppugundla N, Chundawat SP, Yu X, Hermanson S, Gowda K, et al. Hemicellulases and auxiliary enzymes for improved conversion of lignocellulosic biomass to monosaccharides. Biotechnol Biofuels. 2011;4:5.

13. Banerjee G, Car S, Scott-Craig JS, Borrusch MS, Bongers M, Walton JD. Synthetic multi-component enzyme mixtures for deconstruction of lignocellulosic biomass. Bioresour Technol. 2010;101:9097–105.

14. Collins T, Gerday C, Feller G. Xylanases, xylanase families and extremophilic xylanases. FEMS Microbiol Rev. 2005;29:3–23.

15. Kataoka M, Akita F, Maeno Y, Inoue B, Inoue H, Ishikawa K. Crystal structure of *Talaromyces cellulolyticus* (formerly known as *Acremonium cellulolyticus*) GH family 11 xylanase. Appl Biochem Biotechnol. 2014;174:1599–612.

16. Cantarel BL, Coutinho PM, Rancurel C, Bernard T, Lombard V, Henrissat B. The Carbohydrate-Active EnZymes database (CAZy): an expert resource for glycogenomics. Nucleic Acids Res. 2009;37:D233–8.

17. Zhang J, Tuomainen P, Siika-Aho M, Viikari L. Comparison of the synergistic action of two thermostable xylanases from GH families 10 and 11 with thermostable cellulases in lignocellulose hydrolysis. Bioresour Technol. 2011;102:9090–5.

18. Ustinov BB, Gusakov AV, Antonov AI, Sinitsyn AP. Comparison of properties and mode of action of six secreted xylanases from *Chrysosporium lucknowense*. Enzyme Microb Technol. 2008;43:56–65.

19. Kishishita S, Yoshimi M, Fujii T, Taylor II LE, Decker SR, Ishikawa K, et al. Cellulose-inducible xylanase Xyl10A from *Acremonium cellulolyticus*: Purification, cloning and homologous expression. Protein Expr Purif. 2014;94:40–5.

20. Várnai A, Mäkelä MR, Djajadi DT, Rahikainen J, Hatakka A, Viikari L. Carbohydrate-binding modules of fungal cellulases: occurrence in nature, function, and relevance in industrial biomass conversion. Adv Appl Microbiol. 2014;88:103–65.

21. Boraston AB, Bolam DN, Gilbert HJ, Davies GJ. Carbohydrate-binding modules: fine-tuning polysaccharide recognition. Biochem J. 2004;382:769–81.

22. Várnai A, Siika-Aho M, Viikari L. Carbohydrate-binding modules (CBMs) revisited: reduced amount of water counterbalances the need for CBMs. Biotechnol Biofuels. 2013;6:30.

23. Van Gool MP, Van Muiswinkel GC, Hinz SW, Schols HA, Sinitsyn AP, Gruppen H. Two GH10 endo-xylanases from *Myceliophthora thermophila* C1 with and without cellulose binding module act differently towards soluble and insoluble xylans. Bioresour Technol. 2012;119:123–32.

24. Inoue H, Decker SR, Taylor II LE, Yano S, Sawayama S. Identification and characterization of core cellulolytic enzymes from *Talaromyces cellulolyticus* (formerly *Acremonium cellulolyticus*) critical for hydrolysis of lignocellulosic biomass. Biotechnol Biofuels. 2014;7:151.

25. Harrison MJ, Nouwens AS, Jardine DR, Zachara NE, Gooley AA, Nevalainen H, et al. Modified glycosylation of cellobiohydrolase I from a high cellulase-producing mutant strain of *Trichoderma reesei*. Eur J Biochem. 1998;256:119–27.

26. Hägglund P, Eriksson T, Collén A, Nerinckx W, Claeyssens M, Stålbrand H. A cellulose-binding module of the *Trichoderma reesei* β-mannanase Man5A increases the mannan-hydrolysis of complex substrates. J Biotechnol. 2003;101:37–48.

27. Ishida T, Yaoi K, Hiyoshi A, Igarashi K, Samejima M. Substrate recognition by glycoside hydrolase family 74 xyloglucanase from the basidiomycete *Phanerochaete chrysosporium*. FEBS J. 2007;274:5727–36.

28. Guais O, Tourrasse O, Dourdoigne M, Parrou JL, Francois JM. Characterization of the family GH54 α-L-arabinofuranosidases in *Penicillium funiculosum*, including a novel protein bearing a cellulose-binding domain. Applied Microbiol Biotechnol. 2010;87:1007–21.

29. De La Mare M, Guais O, Bonnin E, Weber J, Francois JM. Molecular and biochemical characterization of three GH62 α-L-arabinofuranosidases from the soil deuteromycete *Penicillium funiculosum*. Enzyme Microb Technol. 2013;53:351–8.

30. Ribeiro LFC, De Lucas RC, Vitcosque GL, Ribeiro LF, Ward RJ, Rubio MV, et al. A novel thermostable xylanase GH10 from *Malbranchea pulchella* expressed in *Aspergillus nidulans* with potential applications in biotechnology. Biotechnol Biofuels. 2014;7:115.

31. Beckham GT, Matthews JF, Bomble YJ, Bu L, Adney WS, Himmel ME, et al. Identification of amino acids responsible for processivity in a Family 1 carbohydrate-binding module from a fungal cellulase. J Phys Chem B. 2010;114:1447–53.

32. Kumagai A, Lee SH, Endo T. Thin film of lignocellulosic nanofibrils with different chemical composition for QCM-D study. Biomacromolecules. 2013;14:2420–6.

33. Josefsson P, Henriksson G, Wågberg L. The physical action of cellulases revealed by a quartz crystal microbalance study using ultrathin cellulose films and pure cellulases. Biomacromolecules. 2008;9:249–54.

34. Hideno A, Inoue H, Tsukahara K, Fujimoto S, Minowa T, Inoue S, et al. Wet disk milling pretreatment without sulfuric acid for enzymatic hydrolysis of rice straw. Bioresour Technol. 2009;100:2706–11.

35. Moraïs S, Barak Y, Caspi J, Hadar Y, Lamed R, Shoham Y, et al. Contribution of a xylan-binding module to the degradation of a complex cellulosic substrate by designer cellulosomes. Appl Environm Microbiol. 2010;76:3787–96.

36. Pavón-Orozco P, Santiago-Hernández A, Rosengren A, Hidalgo-Lara ME, Stålbrand H. The family II carbohydrate-binding module of xylanase CflXyn11A from *Cellulomonas flavigena* increases the synergy with cellulase TrCel7B from *Trichoderma reesei* during the hydrolysis of sugar cane bagasse. Bioresour Technol. 2012;104:622–30.

37. Mangala SL, Kittur FS, Nishimoto M, Sakka K, Ohmiya K, Kitaoka M, et al. Fusion of family VI cellulose binding domains to *Bacillus halodurans* xylanase increases its catalytic activity and substrate-binding capacity to insoluble xylan. J Mol Catal B Enzym. 2003;21:221–30.

38. Ali MK, Hayashi H, Karita S, Goto M, Kimura T, Sakka K, et al. Importance of the carbohydrate-binding module of *Clostridium stercorarium* Xyn10B to xylan hydrolysis. Biosci Biotechnol Biochem. 2001;65:41–7.

39. Gao D, Chundawat SP, Krishnan C, Balan V, Dale BE. Mixture optimization of six core glycosyl hydrolases for maximizing saccharification of ammonia fiber expansion (AFEX) pretreated corn stover. Bioresour Technol. 2010;101:2770–81.

40. Sun JL, Sakka K, Karita S, Kimura T, Ohmiya K. Adsorption of *Clostridium stercorarium* xylanase A to insoluble xylan and the importance of the CBDs to xylan hydrolysis. J Ferment Bioeng. 1998;85:63–8.

41. Watanabe M, Inoue H, Inoue B, Yoshimi M, Fujii T, Ishikawa K. Xylanase (GH11) from *Acremonium cellulolyticus*: homologous expression and characterization. AMB Express. 2014;4:27.

42. Payne CM, Resch MG, Chen L, Crowley MF, Himmel ME, Taylor II LE, et al. Glycosylated linkers in multimodular lignocellulose-degrading enzymes dynamically bind to cellulose. Proc Natl Acad Sci U S A. 2013;110:14646–51.

43. Lehtiö J, Sugiyama J, Gustavsson M, Fransson L, Linder M, Teeri TT. The binding specificity and affinity determinants of family 1 and family 3 cellulose binding modules. Proc Natl Acad Sci U S A. 2003;100:484–9.

44. Palonen H, Tjerneld F, Zacchi G, Tenkanen M. Adsorption of *Trichoderma reesei* CBH I and EG II and their catalytic domains on steam pretreated softwood and isolated lignin. J Biotechnol. 2004;107:65–72.

45. Qi B, Chen X, Su Y, Wan Y. Enzyme adsorption and recycling during hydrolysis of wheat straw lignocellulose. Bioresour Technol. 2011;102:2881–9.

46. Inoue H, Fujii T, Yoshimi M, Taylor II LE, Decker SR, Kishishita S, et al. Construction of a starch-inducible homologous expression system to produce cellulolytic enzymes from *Acremonium cellulolyticus*. J Ind Microbiol Biotechnol. 2013;40:823–30.

47. Fujii T, Iwata K, Murakami K, Yano S, Sawayama S. Isolation of uracil auxotrophs of the fungus *Acremonium cellulolyticus* and the development of a transformation system with the *pyrF* gene. Biosci Biotechnol Biochem. 2012;76:245–9.

48. Rodahl M, Höök F, Krozer A, Brzezinski P, Kasemo B. Quartz-crystal microbalance setup for frequency and Q-factor measurements in gaseous and liquid environments. Rev Sci Instrum. 1995;66:3924–30.

49. Sluiter A, Hames B, Ruiz R, Scarlata C, Sluiter J, Templeton D, et al. Determination of structural carbohydrates and lignin in biomass. Laboratory Analytical Procedure NREL Technical Report TP-510-42618. National Renewable Energy Laboratory (NREL); 2008. http://www.nrel.gov/biomass/pdfs/42618.pdf.

50. Dimarogona M, Topakas E, Christakopoulos P, Chrysina ED. The structure of a GH10 xylanase from *Fusarium oxysporum* reveals the presence of an extended loop on top of the catalytic cleft. Acta Crystallogr D Biol Crystallogr. 2012;68:735–42.

13

Redesigning the regulatory pathway to enhance cellulase production in *Penicillium oxalicum*

Guangshan Yao[1], Zhonghai Li[1,3*], Liwei Gao[1], Ruimei Wu[1], Qinbiao Kan[1], Guodong Liu[1] and Yinbo Qu[1,2*]

Abstract

Background: In cellulolytic fungi, induction and repression mechanisms synchronously regulate the synthesis of cellulolytic enzymes for accurate responses to carbon sources in the environment. Many proteins, particularly transcription regulatory factors involved in these processes, were identified and genetically engineered in *Penicillium oxalicum* and other cellulolytic fungi. Despite such great efforts, its effect of modifying a single target to improve the production of cellulase is highly limited.

Results: In this study, we developed a systematic strategy for the genetic engineering of *P. oxalicum* to enhance cellulase yields, by enhancing induction (by blocking intracellular inducer hydrolysis and increasing the activator level) and relieving the repression. We obtained a trigenic recombinant strain named 'RE-10' by deleting *bgl2* and *creA*, along with over-expressing the gene *clrB*. The cellulolytic ability of RE-10 was significantly improved; the filter paper activity and extracellular protein concentration increased by up to over 20- and 10-fold, respectively, higher than those of the wild-type (WT) strain 114-2 both on pure cellulose and complex wheat bran media. Most strikingly, the cellulolytic ability of RE-10 was comparable with that of the industrial *P. oxalicum* strain JU-A10-T obtained by random mutagenesis. Comparative proteomics analysis provided further insights into the differential secretomes between RE-10 and WT strains. In particular, the enzymes and accessory proteins involved in lignocellulose degradation were elevated specifically and dramatically in the recombinant, thereby confirming the importance of them in biomass deconstruction and implying a possible co-regulatory mechanism.

Conclusions: We established a novel route to substantially improve cellulolytic enzyme production up to the industrial level in *P. oxalicum* by combinational manipulation of three key genes to amplify the induction along with derepression, representing a milestone in strain engineering of filamentous fungi. Given the conservation in the mode of cellulose expression regulation among filamentous fungi, this strategy could be compatible with other cellulase-producing fungi.

Keywords: *Penicillium oxalicum*, Cellulase, Transcription factor, Beta-glucosidase, Genetic engineering, Secretome

Background

Plant biomass-based fuels and chemicals offer an appealing and long-term solution as a replacement to fossil fuels [1]. Enzymatic hydrolysis of biomass to fermentable sugar is a key step for biofuel refinery. However, high-cost cellulases are the major bottlenecks to economically competitive cellulose-to-biofuel conversion [2]. *Trichoderma reesei* has always been used as the workhorse to produce cellulase cocktails, and multiple strategies have been applied to improve its enzyme yields to lower production costs [3,4].

Fungi from *Penicillium* genus have recently attracted a great deal of research attention, and they are considered as potential alternatives to *T. reesei* for second-generation biofuel production [5].

Penicillium oxalicum (previously named as *Penicillium decumbens*) was selected for its strong cellulolytic ability in saprophytic condition. It has been under investigation for more than 30 years in China [6]. Whole genome sequencing revealed that the fungus obtains a unique lignocellulose-degrading enzyme arsenal during evolution. Data from comparative genomics analysis demonstrates that this fungus has higher number and more types of genes encoding hemicellulases, as well as genes encoding cellulose binding domain (CBM) containing

* Correspondence: lzhlzh@vip.126.com; quyinbo@sdu.edu.cn
[1]State Key Laboratory of Microbial Technology, Shandong University, Jinan City, Shandong Province 250100, China
Full list of author information is available at the end of the article

proteins, than other cellulolytic fungi, such as *T. reesei* and *Aspergillus niger* [6]. *P. oxalicum* also exhibits higher beta-glucosidase activity in its enzyme system than that of *T. reesei*, which is shared by many other *Penicillium* species [7]. Random mutagenesis and process engineering had been successfully applied to *P. oxalicum* for improving cellulase production [8-10]. One of these mutants, JU-A10-T, with a maximum volume productivity of 160 U L^{-1} h^{-1}, was developed and utilized in industrial scale cellulase production since 1996 in China [10]. Similar to *T. reesei*, breakthroughs are required for maximizing the yields and minimizing the cost to make the biomass-based fuel a powerful competitor to fossil fuel [4].

In recent years, studies based on omics and system biology have provided us with crucial information to understand the biology of lignocellulose-degrading enzyme production in cellulolytic fungi [10,11]. It is widely accepted that the expressions of almost all genes encoding lignocellulose-degrading enzymes are triggered by inducer released from complex plant polysaccharides and regulated by transcription factors in a coordinated manner [12,13]. Systematic investigation of the regulatory network for cellulase gene expression has led to the identification of many transcription factors. Some of these transcription factors are found in all cellulolytic fungi, but others are genus- or species-specific. Clr2/ClrB protein, containing Zn2Cys6-type DNA binding domain, functions as a positive regulator for cellulase gene expression in ascomycete fungi, and its deletion incurs defects in growth and cellulase activity to nearly undetectable level when cultured on Avicel medium in *Neurospora crassa* and *Aspergillus nidulans* [14]. Similarly, a homolog of *clr-2/clrB* in *P. oxalicum* was found to be necessary for efficient cellulase production, and its over-expression resulted in a significant increase in cellulases at both transcriptional and protein levels (unpublished data).

In addition to the transcriptional activation mechanism, carbon catabolite repression (CCR) triggered by glucose and other easily metabolized carbon sources exists widely in *Saccharomyces cerevisiae* and filamentous fungi [15]. In cellulolytic fungi, the CCR mechanism is mediated mainly by the transcription factor CreA/Cre1, which suppresses the expression of a majority of cellulase and hemicellulase genes in *A. niger*, *T. reesei*, *N. crassa*, and *P. oxalicum* [9,12,16,17].

Recently, investigations on the induction mechanism uncover the potential targets in the upstream of transcriptional regulators. Cellobiose, a hydrolysate from plant cell wall materials, is assumed to be the natural inducer for cellulase gene expression in *N. crassa* and *T. reesei* when growing on cellulose medium [18,19]. The level of intracellular cellobiose is balanced by the import by cellobiose transporter and intracellular beta-glucosidase hydrolysis. In *P. oxalicum*, a previous study demonstrated that the deletion of gene *bgl2* encoding the major intracellular beta-glucosidase results in significantly improved cellulase production [20].

Genetically modifying its regulatory factor rather than the target gene is an efficient and promising strategy in the improvement of complex cellulase mixture in filamentous fungi. It is reported that the deletion or replacement of gene *creA* with truncated mutant variant enhances the cellulolytic enzyme production capacity in *T. reesei* [21]. Point mutation of the activator *xyr1* in *T. reesei* or misexpression of the *clr-2* in *N. crassa*, respectively, exhibits inducer-independent production of cellulolytic enzymes, but not in *A. nidulans* [22,23]. Taken together, although some of these aforementioned regulatory factors have been genetically engineered in cellulolytic fungi, their effects, however, in improving the productivity of cellulolytic enzymes are highly limited.

In the present study, we developed a systematic strategy to redesign the regulatory network (RE) to enhance cellulase production by combinatorial manipulation of three important regulators in *P. oxalicum*: over-expressed *clrB* for enhancing induction, deleted *bgl2* for inducer accumulation, and deleted *creA* for derepression. As a result, the cellulolytic ability of the triple-gene recombinant RE-10 was significantly improved. Comparative analysis of the secretomes between RE-10 and wild type (WT) provided more insights into the cellulase system of *P. oxalicum* and alterations caused by the strain rational engineering.

Results

Designing a systematic strategy to genetically modify the cellulolytic fungus *P. oxalicum* for improving cellulolytic enzyme expression

To improve the production of the lignocellulolytic enzymes, a systematic approach was developed for genetically modifying the cellulolytic fungus *P. oxalicum* (Figure 1). Clr-2, its ortholog in *P. oxalicum* is ClrB, is essential for inducing cellulase expression and conserved in ascomycete fungi [14]. Therefore, as the first target, the level of ClrB was increased by constitutively over-expressing the gene with the promoter *gpdA* from *A. nidulans* [24]. Multiple transformants with *clrB* expression cassette insertion and resistant to pyrithiamine were screened, and one of them was verified by PCR and Southern blot (Additional file 1: Figure S1A). As expected, the transcript level of gene *clrB* increased by up to 100 and 12 times at cellulose induction 4 and 22 h, respectively, compared with those of WT (Additional file 2: Figure S2A).

Previous work in our laboratory demonstrated that the major intracellular beta-glucosidase BGL2 plays a negative role in the induction of cellulases and xylanase, thereby leading to a hypothesis that the deletion of *bgl2*

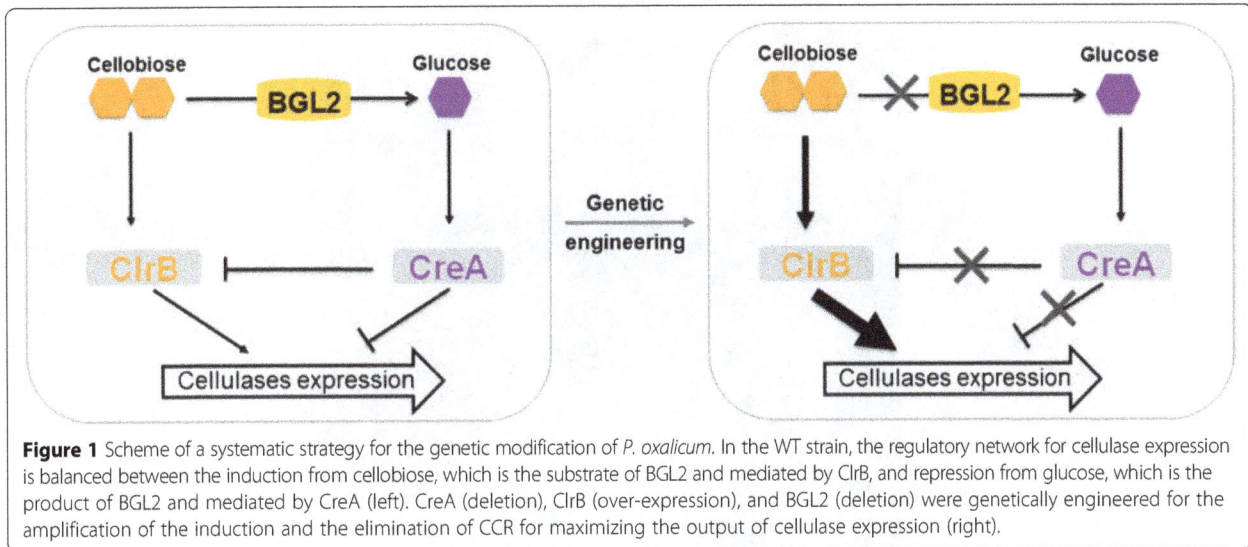

Figure 1 Scheme of a systematic strategy for the genetic modification of *P. oxalicum*. In the WT strain, the regulatory network for cellulase expression is balanced between the induction from cellobiose, which is the substrate of BGL2 and mediated by ClrB, and repression from glucose, which is the product of BGL2 and mediated by CreA (left). CreA (deletion), ClrB (over-expression), and BGL2 (deletion) were genetically engineered for the amplification of the induction and the elimination of CCR for maximizing the output of cellulase expression (right).

facilitates the accumulation of intracellular cellobiose [20], which is the natural stimuli for inducing the expression of cellulolytic enzyme genes. As a consequence, a knockout cassette was constructed to replace the gene *bgl2* with the marker gene *hph* in the above mutant with *clrB* over-expression. More than ten transformants resistant to hygromycin B were obtained and verified by PCR (data not shown). One of those transformants, confirmed by Southern blot (Additional file 1: Figure S1B), was selected for further genetic engineering. No transcription of *bgl2* was further confirmed by real-time quantitative polymerase chain reaction (qRT-PCR) (Additional file 2: Figure S2B).

In addition to the induction mechanism, the expressions of cellulase and hemicellulase genes are repressed when the preferred carbon sources are available, which is mediated by transcription factor CreA. To overcome repression, we deleted the gene *creA* by using the gene *bar* as the marker. Three transformants with resistance to herbicide bialaphos on a selectable plate were selected [25] and then verified by PCR. One of three transformants was confirmed by Southern blot (Additional file 1: Figure S1C) and named RE-10 in the following research. The lack of transcription of the gene *creA* was further confirmed by qRT-PCR (Additional file 2: Figure S2C).

In summary, the regulatory pathway was modified to reconstruct the regulatory network of expression of cellulolytic enzyme genes, by simultaneously strengthening induction and relieving repression. Three key factors involved in this network were genetically modified (Figure 1).

Regulatory pathway redesigning substantially enhanced cellulolytic enzyme production in the mutant RE-10

Comparative genomics analysis between *P. oxalicum* WT strain 114-2 and cellulolytic enzyme hyper-producer

JU-A10-T demonstrates that functional mutation occurred in the transcription factor CreA, and the promoter of *cbh1* contributes greatly to the high-producing phenotype [9]. Three key factors which control the expression of genes encoding lignocellulose-degrading enzymes were deleted or over-expressed in this study to investigate whether these manipulations could improve the cellulase production as classical mutagenesis. To determine the effect of the genetic modifications on *P. oxalicum*, phenotypic and cellulolytic ability analyses were conducted. Equivalent fresh spores of the WT, RE-10, and JU-A10-T strains were inoculated on plates with 2% glucose or 2% cellulose as the sole carbon source for 4 or 8 days, respectively. As shown in Figures 2A and B, slightly smaller colonies and less conidia were observed in RE-10 than those in WT strain when grown on glucose medium. However, the morphological defects became less evident when cultured on cellulose medium. These morphological changes coincided with our previous observations of the repressor gene *creA* deletion in *P. oxalicum* WT strain 114-2 [9]. Similar morphological alterations caused by *cre1/creA* mutant were widely reported in other filamentous fungi. For example, deletion of the gene *cre-1* in *N. crassa* led to grow slower and denser than the parental strain under glucose, sucrose, or xylose conditions [16]. Moreover, the *creA* mutant displayed reduced growth, conidia under repressing conditions (including glucose, ethanol, and galactose) in *A. nidulans* [26], as well as in *Acremonium cellulolyticus* [27].

As expected, a considerably great and clear cellulolytic halo was observed around the colonies of RE-10 but not in WT after 4 days of incubation, and this halo was much more pronounced after 8-day growth (Figures 2C and D). Beyond we expected, the diameter of the cellulolytic halo of RE-10 was much greater than that of the

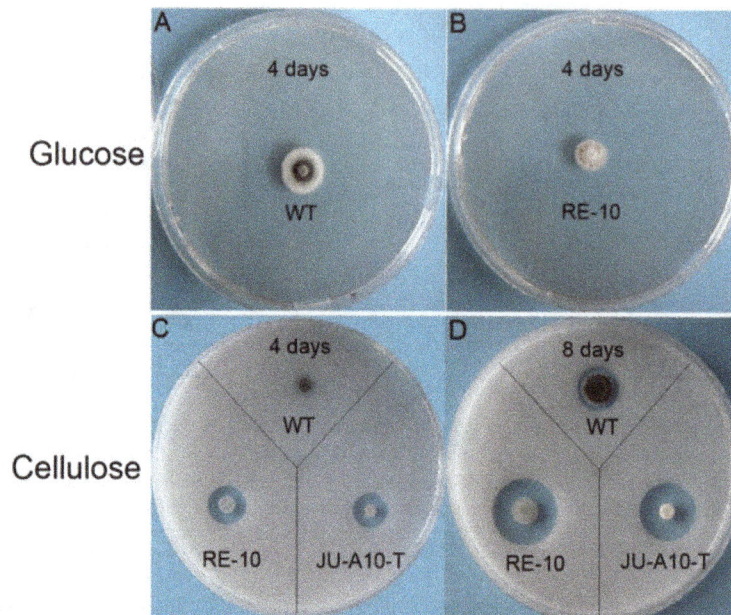

Figure 2 Cellulolytic phenotypic analysis of the recombinant RE-10. Equally harvested conidia of WT and RE-10 or JU-A10-T were inoculated on both glucose medium (**A** (WT), **B** (RE-10)) or cellulose medium (**C and D**) for 4 and 8 days, respectively, and then photographed.

industrial strain JU-A10-T (1.1 ± 0.1 and 1.9 ± 0.1 cm at days 4 and 8 for RE-10 compared with 0.9 ± 0.1 and 1.6 ± 0.1 cm at days 4 and 8 for JU-A10-T) (Figures 2C and D).

On the other hand, the cellulolytic enzyme activities were evaluated comprehensively for RE-10 and WT when cultured on cellulose medium, along with JU-A10-T when cultured on wheat bran medium. The later medium is a cheap and ideal mixture optimized for cellulase production by *P. oxalicum*. The levels of filter paper activity (FPA, representing overall cellulase activity), CMCases (endoglucanase), *p*NPCases (cellobiohydrolase), *p*NPGases (beta-glucosidase), xylanase activities, and total extracellular protein concentration in the culture supernatant of all strains, including WT, RE-10, and JU-A10-T were determined. The results (as shown in Additional file 3: Figure S3) showed that almost all the enzymes assayed were elevated substantially in RE-10 compared with those in WT on cellulose media, which were consistent with the phenotypic analyses above, except beta-glucosidases. As for the comparison between WT and RE-10, the levels of FPA, *p*NPCase, CMCase, xylanase activities, and extracellular total protein were approximately 20-, 50-, 50-, 16-, and tenfold higher, respectively, in RE-10 than those in the wild-type strain (Additional file 3: Figures S3A-E). Notably, a remarkable decrease in FPA was observed in RE-10 at 108 h than that at 96 h, which may be due to product feedback inhibition caused by reduction of *p*NPGase activity in the late stage of fermentation (Additional file 3: Figures S3A and S3D). This founding

was in accordance with a common consensus that beta-glucosidases play an important role in eliminating product inhibition of cellobiohydrolases and endoglucanases [28,29]. However, the similar results were not detected in wheat bran medium (Figure 3A).

Complex carbon sources from plant materials are more efficient than pure cellulose in induction of expression of lignocellulose-degrading enzymes. Expectedly, when cultured on wheat bran medium, much higher cellulase activities were detected for both WT and RE-10. In particular, FPA reached 6.94 ± 0.57 U/mL in the RE-10 strain, which was 27-fold higher than that of WT (*P* value <0.01). Moreover, *p*NPCase, CMCase, xylanase activities, and extracellular protein level increased by 10-, 16-, 5-, and 10-fold in RE-10 compared with those of WT, respectively (Figure 3).

Inspiringly, our results demonstrated that the cellulolytic activities, including FPA, xylanase, CMCase, *p*NPCase, and *p*NPGase, of the recombinant strain RE-10 were comparable to those of the industrial strain JU-A10-T (Figures 3A to E), which were coherent with the above cellulolytic halo assay. It is generally viewed that the production of hydrolytic enzymes is closely associated with the fungal biomass. Thereby, we questioned that whether the improvement in cellulase synthesis in RE-10 was due to its higher biomass level. Then, we examined both the growth kinetics and glucose utilization of WT, RE-10, and JU-A10-T strains, and the results were showed in Figure 4. We observed that JU-A10-T accumulated the slightly higher level of biomass (approximately 8 g/L), relative to the WT (7.5 g/L) in the

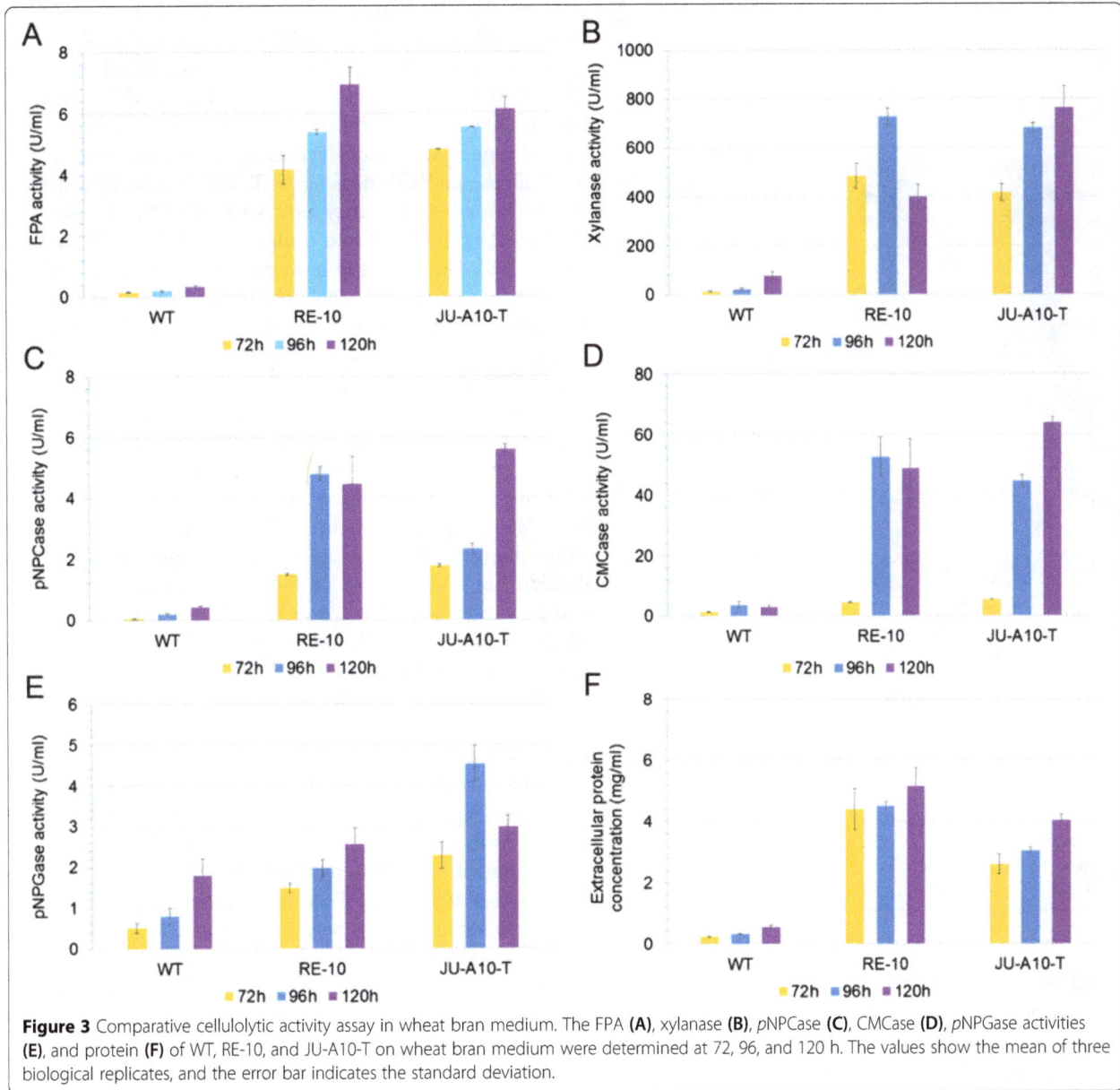

Figure 3 Comparative cellulolytic activity assay in wheat bran medium. The FPA **(A)**, xylanase **(B)**, pNPCase **(C)**, CMCase **(D)**, pNPGase activities **(E)**, and protein **(F)** of WT, RE-10, and JU-A10-T on wheat bran medium were determined at 72, 96, and 120 h. The values show the mean of three biological replicates, and the error bar indicates the standard deviation.

end (Figure 4A), although at slightly low growth and glucose utilization rates (Figure 4B). Clearly, RE-10 (about 6 g/L) had a lower biomass formation ($P <0.05$) than that of WT or JU-A10-T (Figure 4A), although consumed the glucose at the same rate as the WT (Figure 4B). Collectively, our results confirmed that higher cellulolytic activities in RE-10 did not correlate with the biomass level.

Subsequently, the supernatants from the WT, RE-10, and JU-A10-T strains cultured in wheat bran medium were profiled by SDS-PAGE. When equal protein loading, the protein profile secreted by RE-10 resembled that of WT, differed from that of JU-A10-T (Figure 5A). Significantly, much more protein bands were detected in RE-10 compared with WT, especially at the range of 45 to 116 kDa, which showed good correlation with

the extracellular protein concentration measurement (Figure 3F, Additional file 4: Figure S4). More importantly, dynamic zymography analysis by using 4-methylumbeliferyl-β-d-cellobioside [30] as the substrate revealed that RE-10 exhibited a cellobiohydrolase activity pattern similar to WT, and the cellobiohydrolase activity (especially band 2) significantly enhanced in 120 h relative to WT (Figure 5B). However, a significant difference could be seen in the cellobiohydrolase zymography between WT/RE-10 (two bright bands) and JU-A10-T (only one bright band).

Substantial up-regulation in transcript levels of major cellulolytic enzyme genes analyzed by qRT-PCR

Compared with the WT strain 114-2, a significant improvement in cellulase yields was observed in RE-10. We

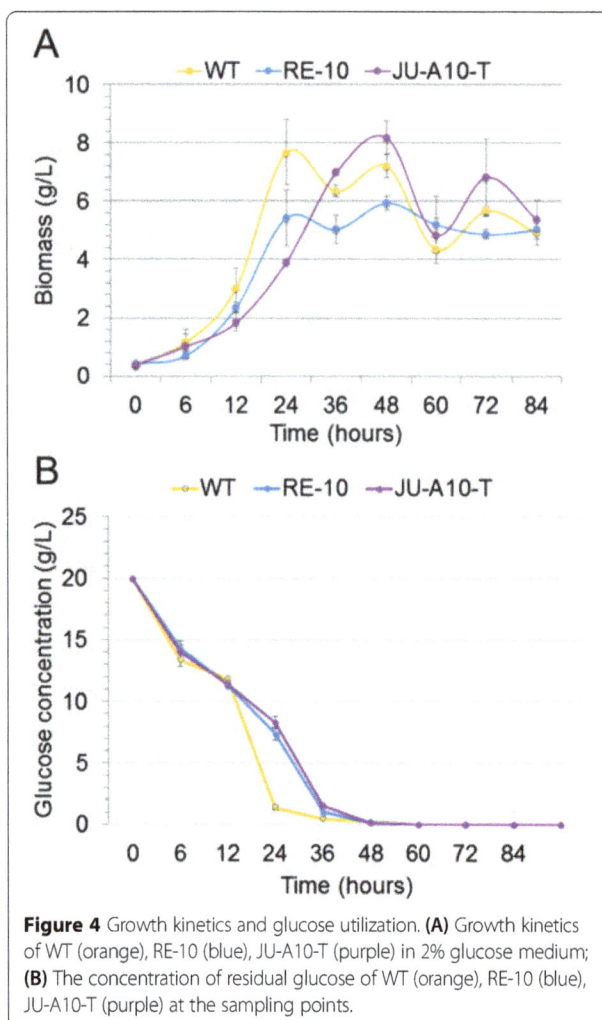

Figure 4 Growth kinetics and glucose utilization. **(A)** Growth kinetics of WT (orange), RE-10 (blue), JU-A10-T (purple) in 2% glucose medium; **(B)** The concentration of residual glucose of WT (orange), RE-10 (blue), JU-A10-T (purple) at the sampling points.

wondered whether this improvement was the result from the up-regulation in transcription level. Therefore, the transcripts of major cellulases Cel7A-2, Cel5B, BGL1, Cel61A, and accessory protein swollenin are selected for qRT-PCR analyses. *P. oxalicum* 114-2, RE-10, and JU-A10-T mutant strains were pre-cultured on glucose medium for 20 h, starved for 2 h under no carbon source conditions, and transferred to cellulose medium for further induction for 4 and 22 h, respectively. The results indicated that the transcript abundance of genes encoding major cellulolytic enzymes and synergetic protein except *bgl1* (PDE_02736) in RE-10 was up-regulated significantly than that of WT under both starvation and induction conditions (Figure 6A to E), which was consistent with enzyme activity analysis above. The transcript level of *cel7A-2* (PDE_07945) increased by 27- and tenfold (P value <0.01) under starvation condition and 4 h of induction, respectively (Figure 6A). Meanwhile, the transcription expression level of major endoglucanase gene *cel5B* (PDE_09226) increased by over 80-fold

(P value <0.001) both on no carbon source and 4 h of induction (Figure 6B). Especially, the transcript abundance of *cel61A* (PDE_05633) was up-regulated up to 2929-fold in RE-10 compared with that of WT strain under starvation conditions. Moreover, the transcript level of gene *PDE_02102*, which encodes swollenin destroying cellulose and enhances lignocellulose hydrolysis when supplemented to the enzyme system [31,32], was also increased by over 100-fold under starvation conditions (Figure 6E). However, the transcription expression level of the major extracellular beta-glucosidase encoding gene *bgl1* decreased significantly in RE-10 under both starvation and induction conditions (Figure 6C), which was consistent with the aforementioned enzyme activity assay, and previous reports that extracellular beta-glucosidases were regulated by a different regulatory pathway, at least in part, from other induced cellulases. Moreover, JU-A10-T substantially up-regulated the expression levels of most cellulolytic genes in both starvation condition (fold change >100, P value <0.01) and post-22 h induction, but not in post-4 h induction. Although it exhibited an expression pattern, which is distinct from that of RE-10 (Figure 6A to E), both RE-10 and JU-A10-T enhanced transcription expression level of most of cellulolytic genes at the same order of magnitude relative to WT.

Comparative secretome analysis
The cellulolytic ability of the triple-gene recombinant RE-10 was significantly enhanced at both the transcript and enzyme activity levels. We assumed that a great change may occur in the secretome for RE-10 compared with that of WT strain. To explore the different profile of total secreted proteins between *P. oxalicum* 114-2 and RE-10 on cellulose medium, we performed high-resolution proteomics to dissect their secretomes. The supernatants of RE-10 and WT strain cultures were collected at hour 96 and analyzed by liquid chromatography-tandem mass spectrometry (LC-MS/MS). We identified a total of 157 proteins in the secretome of WT with confidence while 144 proteins were detected in that of RE-10 (Additional file 5: Table S1). Among the proteins tested, 79 of 157 had the predicted secretion signals for WT strain, while 67 of 144 were predicted to be secreted for RE-10. However, we could not rule out the possibility that the proteins without predicted secretion signals were extracellularly located; like that, 18% of secretome without classic secretion signals are detected in the secretome of *A. niger* [33]. Among those proteins with predicted secretion signals, 39 proteins were shared by both RE-10 and WT, 28 proteins were unique to RE-10, and 40 proteins were specific to WT (Figure 7A). The majority of the secretomes shared by both RE-10 and WT were cellulolytic enzymes and accessory proteins (Table 1). Core cellulases were included in the secretome induced

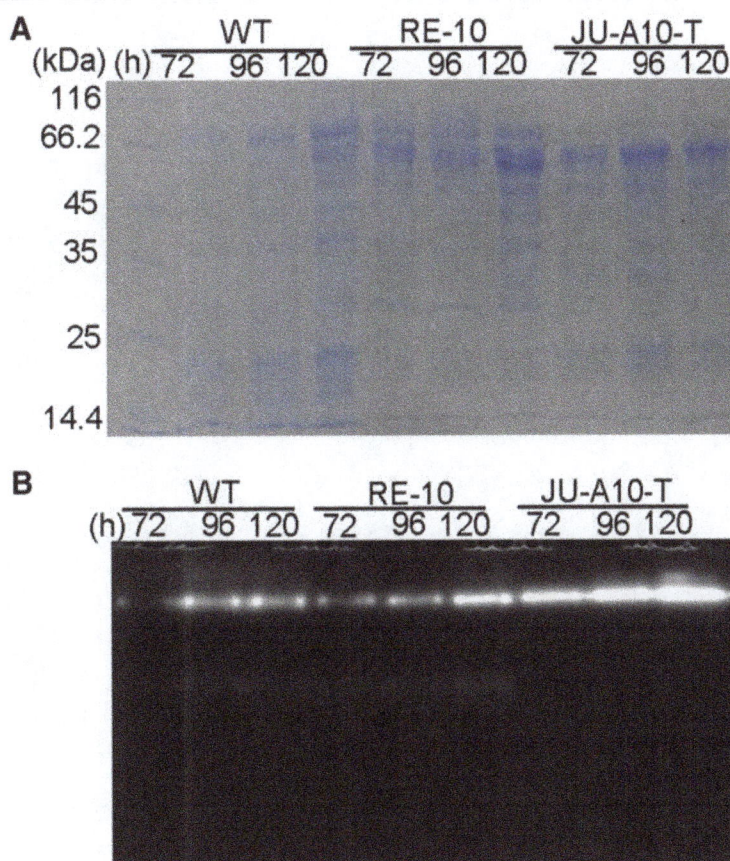

Figure 5 SDS-PAGE and zymography analysis of the secreted protein. **(A)** SDS-PAGE analysis of the culture supernatants of the WT, RE-10, and JU-A10-T. **(B)** Activity staining was used to measure the CBH activity of the WT (left), RE-10 (middle), and JU-A10-T (right).

by cellulose, which is almost consistent with the result released recently in *P. oxalicum* GZ-2 under the cellulose medium [34]. It included all the three predicted cellobiohydrolases, Cel7A-1 (PDE_5445), CBH7A-2 (PDE_07945), Cel6A (PDE_07124), and four of 11 predicted endoglucanases, Cel7B (PDE_07929), Cel5B (PDE_09226), Cel5C (PDE_09969) and Cel12A (PDE_06439). Unsurprisingly, seven hemicellulases, Xyn10A (PDE_08094), Xyn11A (PDE_02101), Xyn11B (PDE_02682), Axe1A (PDE_09278), Axe5A (PDE_04182), Man5A (PDE_06023), and Aga27A (PDE_02514), were also detected in the overlapped secretomes. It also contained one pectinase, Pga48A (PDE_04162), an amylase (PDE_01354), a swollenin (PDE_02102), a putative rhamnogalacturonan alpha-L-rhamnopyranohydrolase (PDE_09285), a putative polygalacturonase (PDE_07938), lysophospholipase (PDE_05537), a cell wall integrity and stress response component (PDE_01796), and 17 other proteins with uncharacteristic functions (Table 1). Subsequently, functional annotation of the WT- or RE-10-specific secretomes showed that a large amount of them were proteins with unidentified functions. It is worth noting that the protein PDE_00507 (an ortholog of EGII from *T. reesei*, which

belongs to the GH5 family) was exclusive to RE-10, which showed relatively higher activity on cellulose than other endoglucanases from *P. oxalicum* [35]. Furthermore, a protein encoding superoxide dismutase (PDE_09399) was detected in the secretome of RE-10. Likewise, this protein was also detected in cellulase preparations from *T. reesei* (unpublished data), but its function in cellulose degradation has not yet been reported. Especially, two GH61 proteins (PDE_05633 and PDE_01261), which recently re-classified into auxiliary activities family 9 (AA9), were detected uniquely in RE-10.

Most of core cellulases were common to both WT and RE-10; thus, we speculated that the difference in secretomes between WT and RE-10 was mainly reflected in proportion and quantity. In the above, one-dimension PAGE and protein concentration analyses revealed the extracellular protein from RE-10 was up-regulated substantially when compared with WT strain (Figures 3F, and 5). Therefore, we then compared the proportion of major cellulolytic enzymes in the secretome between different strains. As expected, the proportion of almost all major cellulases, expected BGL1 was significantly elevated in RE-10 than that of WT, which was consistent with the results

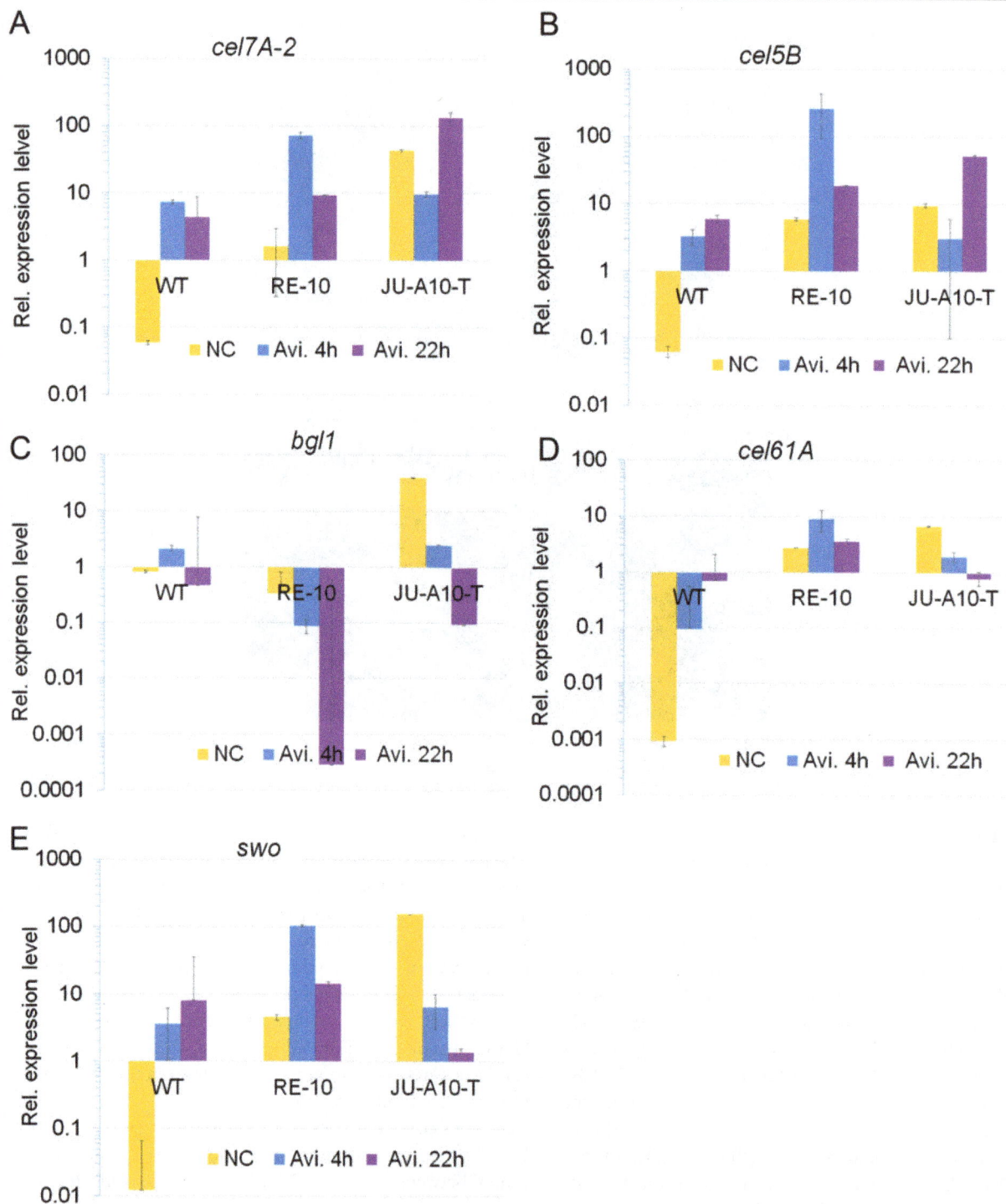

Figure 6 qRT-PCR analysis of the transcripts. Transcript levels of genes encoding cellulases and accessory protein including *cel7A-2* **(A)**, *cel5B* **(B)**, *bgl1* **(C)**, *cel61A* **(D)**, *swo* **(E)** in the WT (orange), RE-10 (blue), and JU-A10-T (purple) were analyzed. The values show the mean of three replicates, and the error bar indicates the standard deviation.

from cellulolytic enzyme activities and transcriptional level assays. Specifically, Cel7A-1, Cel7A-2, Cel6A, Cel5B, Cel5C, and Cel7B were up-regulated by 1.39-, 1.26-, 1.68-, 1.3-, 1.99-, and 1.49-fold, respectively, in the secretome of RE-10 compared with those in that of WT (Figure 7B). Furthermore, the ratio of swollenin in RE-10 increased by up to nine times compared with that in WT strain, which was consistent with the qRT-PCR results.

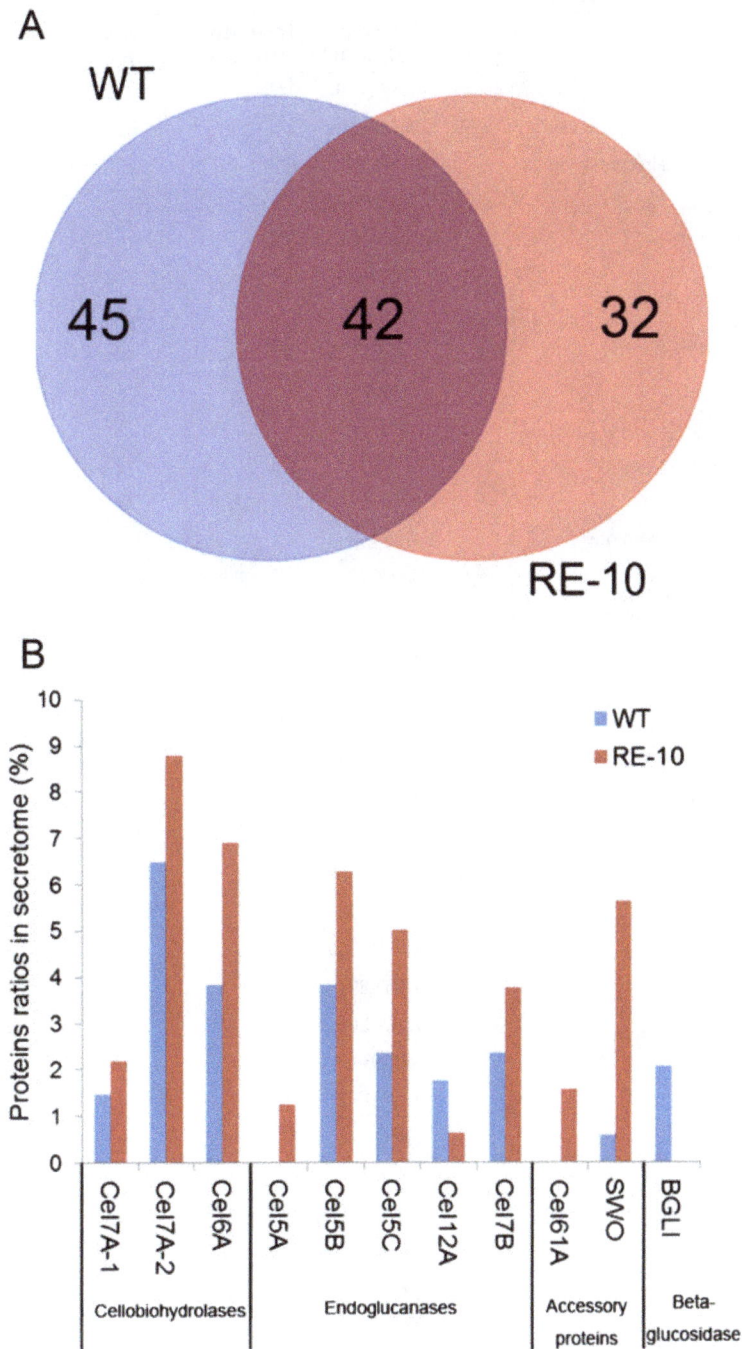

Figure 7 Comparative proteomics analysis between WT and RE-10. Venn diagram shows the shared secretome (overlap) and specific secretome for WT (blue) and RE-10 (red) **(A)**. The result of comparative analysis of the ratios of major cellulases and accessory proteins between WT and RE-10 is shown **(B)**.

On the contrary, the major extracellular beta-glucosidase BGL1 was undetectable in the secretome of RE-10, and its content was about 2% in the secretome of WT strain. A similar phenomenon was observed in the comparative secretome analysis between industrial strain JU-A10-T and 114-2 [9]. The decline of BGL1 content in hour 96 was consistent with the result from pNPGase activity assay on cellulose medium (Additional file 3: Figure S3D), also in agreement with the transcriptional level assay (Figure 6E). However, replacing the cellulose medium with wheat bran medium could alleviate the decrease, suggesting the expression of beta-glucosidases was affected by carbon sources.

Table 1 Secreted proteins shared by WT and RE-10

Gene ID	Function description	Signal peptides	Coverage (WT/RE-10)	Total peptides (WT/RE-10)	Unique peptides (WT/RE-10)	MW [kDa]	Calc. pI
PDE_01261	Putative cellulose monooxygenase	Y	4.79/24.79	1/3	1/1	36.2	5.9
PDE_01335	Hypothetical protein	Y	26.16/22.09	3/2	3/2	18.4	4.97
PDE_01354	Starch and chitin binding domain-containing	Y	6.7/23.82	6/8	6/7	42.8	6.38
PDE_01796	Hypothetical protein	Y	36.27/36.27	4/6	4/6	19.9	4.92
PDE_02101	Putative endo-beta-1,4-xylanase	Y	10.65/10.65	3/3	3/3	31.3	6.96
PDE_02102	Putative swollenin	Y	8.62/45.29	2/18	1/16	51.9	5.05
PDE_02392	Hypothetical protein	N	10.26/6.62	3/2	1/1	64.4	7.56
PDE_02514	Putative alpha-galactosidase	Y	11.64/14.69	3/3	1/2	55	6.24
PDE_02536	Hypothetical protein	Y	24.87/24.87	7/5	7/4	18.6	5.01
PDE_02682	Putative endo-beta-1,4-xylanase	Y	3.46/3.46	1/1	1/1	29.5	6.28
PDE_03112	Putative exo-beta-1,3-glucanase	Y	18.62/7.29	7/3	7/3	50.3	4.79
PDE_03255	Hypothetical protein	Y	82.31/40.82	7/3	7/3	15	5.26
PDE_03292	Chitin binding domain-containing protein	Y	4.43/4.43	1/1	1/1	27.6	4.41
PDE_03437	Hypothetical protein	Y	9.33/6.22	3/2	2/2	23.4	5.45
PDE_03452	Putative beta-1,3-glucanosyltransglycosylase	Y	2.43/4.85	1/2	1/1	57	5.22
PDE_03466	Putative rhamnogalacturonase	Y	17.01/27.13	5/7	3/5	47	6.6
PDE_03659	Hypothetical protein	N	12.17/4.52	4/2	1/1	86.7	6.61
PDE_03916	Hypothetical protein	Y	15.17/20.3	3/4	1/1	51.2	5.64
PDE_03934	Hypothetical protein	Y	19.73/18.37	3/1	2/1	15.7	6.95
PDE_04162	Putative endopolygalacturonase	Y	12.89/14.74	3/3	2/3	38.4	5.85
PDE_04182	Putative acetyl xylan esterase	Y	59.83/51.28	11/6	11/5	24	7.15
PDE_04519	Hypothetical protein	Y	29.63/8.47	3/1	3/1	19.6	4.81
PDE_05305	Hypothetical protein	Y	8.14/10.98	1/2	1/1	59.7	5.17
PDE_05445	Putative cellobiohydrolase	Y	14.57/20.09	5/7	4/1	48.1	4.91
PDE_05537	Hypothetical protein	N	2.34/3.89	1/2	1/1	68.5	5.01
PDE_06023	Putative beta-1,4-mannanase	Y	7.01/41.86	1/7	1/6	47.4	5.33
PDE_06089	Hypothetical protein	Y	30.12/39.76	4/4	4/4	18.1	7.37
PDE_06128	Hypothetical protein	Y	19.21/19.21	4/3	4/3	16.6	5.22
PDE_06138	Ecm33 domain-containing protein	Y	37.66/7.73	17/3	16/3	41.9	5.12
PDE_06252	Hypothetical protein	N	16.42/16.22	8/12	1/1	115.7	7.09
PDE_06439	Putative endo-beta-1,4-glucanase	Y	8.86/37.97	2/6	2/5	25.5	6.52
PDE_06649	Putative feruloyl esterase	Y	11.44/3.23	1/1	1/1	35.7	7.74
PDE_06677	Hypothetical protein	Y	9.26/27.31	2/1	2/1	23.1	5.07
PDE_06697	Hypothetical protein	N	8.65/11.78	4/2	1/1	128.2	7.28
PDE_07033	Hypothetical protein	Y	14.99/4.5	3/1	1/1	50.4	5.34
PDE_07073	Hypothetical protein	N	6.27/2.53	3/2	1/1	118.1	5.86
PDE_07106	Hypothetical protein	Y	44.14/71.03	5/6	4/6	15.1	8.47
PDE_07124	Cellobiohydrolase Cel6A	Y	55.39/56.25	13/22	10/22	48.5	5.67
PDE_07928	Endo-beta-1,4-glucanase Cel45A	Y	9.77/8.65	3/2	3/2	26.9	4.84
PDE_07929	Endo-beta-1,4-glucanase Cel7B	Y	26.37/36.29	8/12	7/10	49.3	5.26
PDE_07938	Putative polygalacturonase	Y	23.72/25.07	6/6	6/5	37.6	5.14
PDE_07945	Cellobiohydrolase CBHI/Cel7A-2	Y	45.05/45.6	22/28	19/25	56.9	5.12
PDE_08075	Hypothetical protein	Y	21.18/5.9	4/1	3/1	27.7	4.02

Table 1 Secreted proteins shared by WT and RE-10 *(Continued)*

PDE_08094	Putative endo-beta-1,4-xylanase	Y	23.41/61.95	5/20	5/19	43.5	6.42
PDE_08650	Hypothetical protein	N	6.43/8.55	11/13	1/1	359.1	7.17
PDE_09226	Putative endo-beta-1,4-glucanase	Y	43.1/49.15	13/20	13/20	44	5.21
PDE_09278	Putative acetyl xylan esterase	Y	15.91/31.82	3/7	2/7	41.3	7.84
PDE_09285	Putative alpha-L-rhamnopyranohydrolase	Y	41.08/31.4	12/10	10/9	51.4	4.92
PDE_09289	Hypothetical protein	Y	9.64/9.64	1/2	1/1	18	4.83
PDE_09417	Glucoamylase Amy15A	Y	41.1/51.02	30/25	30/25	67.2	6.01
PDE_09969	Endo-beta-1,4-glucanase Cel5C	Y	21.47/41.5	8/16	7/16	65	5.82

Discussion

In this study, a novel route was developed to rationally redesign the expression regulatory network of cellulolytic fungus *P. oxalicum* for improving cellulase production. Traditionally, random mutagenesis as the main approach is used to construct industrial fungal cellulase producers, which is time-consuming and laborious. Importantly, luck is not always sufficient to obtain the desired mutant by random mutagenesis. Especially, undesired phenotypes accompanied during mutagenesis process were another disadvantage. In *P. oxalicum*, 30 years were devoted to mutagenesis and screening until the high-producer JU-A10-T was obtained. Similarly, the *T. reesei* strain QM6a was discovered during Second World War; however, the cellulase hyper-producer RUT-C30 derived from which was screened until 1979 [36]. By contrast, only about 6 months were spent to construct the triple mutant RE-10, whose cellulolytic ability was parallel to that of industrial strain JU-A10-T (Figures 2, 3, 5, and 6).

We have performed the production of cellulase by RE-10 in 7.5-L bioreactor, and the results exhibited that the strain RE-10 produced higher FPA activity (12 to14 U/mL) than that in flasks, which is expected to be further improved by medium composition and fermentation parameter optimization. Undoubtedly, this study is the most efficient example of the application of genetic modification for achieving cellulase hyper-producer in fungi of the genus of *Penicillium*. To the best of our knowledge, this was also the first report on engineering native filamentous fungus up to the level of an industrial producer in the field of enzyme mixture production.

A growing body of evidence points out that filamentous fungi employ both conserved and unique mechanisms to regulate the production of cellulolytic enzymes. Genomic analysis unveiled that the orthologs of multiple components, including Cre1, ACEI, and LaeA in the cellular mechanism, which control the cellulase production in *T. reesei*, were also present in *P. oxalicum* [6]. On the other hand, species or genus-specific regulators, such as ACEII, Xpp1, ENVOY, and GRD1, were lost in *P.*

oxalicum. Lactose, which was always considered as the efficient inducer for cellulase expression in *T. reesei*, also induced the expression of lignocellulolytic genes in *P. oxalicum* at lower concentration [37]. However, sophorose, which is the transglycosylation product of beta-glucosidase, was unable to induce the expression of cellulolytic genes in *P. oxalicum* [20] but intensely stimulated the expression of cellulases in *T. reesei* [38]. Recently, gene perturbation studies in *P. oxalicum* revealed the regulatory role of several proteins in cellulase expression. CreA and its orthologs, as the major repressors for negatively regulating cellulolytic gene expression in *P. oxalicum* [9], and other filamentous fungi, including *T. reesei* and *N. crassa* [16,39], and their deletions are essential to rationally construct cellulase hyper-producer of almost all the cellulolytic fungi. In *P. oxalicum*, our unpublished data found that deletion of the activator ClrB almost block the expression and secretion of cellulases, which is highly similar to the role of its ortholog Clr-2 in *N. crassa* [14], and over-expression of ClrB could lead to a drastic increase in cellulase expression and activities. Resemble the results observed in *T. reesei* [40], the transglycosylation activity of BGL2 was only observed at high substrate concentration in *P. oxalicum* (data not show), and the regulatory role of BGL2 in cellulase production in *P. oxalicum* was basically similar to *N. crassa*, however, clearly different from that of *T. reesei* [18-20]. Our recent studies demonstrated that cellodextrin transporters exhibited functional redundancy in *P. oxalicum*, and deletion of any one of the three (CdtC, CdtD, and CdtG) proteins did not affect cellulase expression [41]; however, lack of either cellodextrin transporter CDT-1 or CDT-2 in *N. crassa* led to notable defects in cellulose utilization [42]. Overall, three regulators in this studies were basically conserved in the three species; the regulatory mechanisms clarified in one fungi can be helpful for genetic engineering of other species for cellulase production improvement.

Just as observed in *A. nidulans* [22], single over-expression of *clrB* in *P. oxalicum* was not sufficient to get rid of the dependence on inducer (data not show). Likewise, a

strain carrying *creA/cre1* deletion was only partially dere-
pressed in both *P. oxalicum* and *T. reesei* [43], albeit it improved FPA activity by 1.5-fold under induction condition
in *P. oxalicum* [9]. Furthermore, deletion *creA* coupling
with over-expression activator *clrB* was necessary to get
rid of the dependence of inducer, and other double gene
mutations including over-expressing *clrB* coupling with
deleting *bgl2* and the *bgl2* and *creA* double deletions were
able to induce the expression and secretion of cellulase to
a higher level (data in our another submitting manuscript); however, their effects were still far less than triple-
gene mutants in RE-10, underlining the cumulative effects
of the three mutants. Similarly, further lack of Cre-1 in
N. crassa Δ3βG strain showed higher concentration of
secreted active cellulases in Δ3βGΔcre versus Δ3βG in
response to induction with cellobiose but not in Avicel
[18]. Therefore, we hypothesized that there was a significant synergistic reinforcement effect among the three-step
genetic engineering in the induction of cellulolytic gene
expression in *P. oxalicum.*

As expected, the mutant RE-10 demonstrated greatly
enhanced cellulase activity than WT. However, the extracellular beta-glucosidase activity of RE-10 was less than
that of WT, and it was almost undetected at both transcript and protein levels (Figure 6C and Additional file 3:
Figure S3D). This phenomenon was only observed under
cellulose conditions but not under wheat bran conditions
(Figure 3D). The expression of beta-glucosidases is generally believed to be regulated by a unique system that differs from that of other cellulases. BglR, a specific regulator
of beta-glucosidases, was recently identified as an activator
for efficient beta-glucosidase expression in *T. reesei* [44],
but no homolog of *bglR* exists in the genome of *P. oxalicum*. Thus, other novel regulators remain further identification and engineering applications in the near future.
Alternatively, the expression of beta-glucosidases may also
be improved by placing genes under the promoter from
cellulase genes in the future, resulting in co-regulated expression with other cellulases.

Comparative secretome analysis between RE-10 and
WT suggested that an up-regulation specifically in
lignocellulose-degrading enzymes in the mutant was a
consequence of rational and directed engineering. In
the secretome of *N. crassa* induced by Avicel, the core
cellulase mixture involving cellobiohydrolases, endoglucanase, and β-glucosidase dominated the secretome and
comprised 63% to 65% by weight [45], and those homologs in *T. reesei* RUT-C30 account for a larger proportion
when induced by lactose [46]. The ratio of the core cellulase mixture was only approximately 24.2% in WT 114-2,
and rose to 34.8% in RE-10 when induced by cellulose,
which was still lower than that of *N. crassa* or *T. reesei.*
Therefore, with the help of recently developed marker gene
recycling system (unpublished data), the recombinant

RE-10 still holds great potential in the further improvement regarding the content of cellulases in the secretome of *P. oxalicum.*

Strikingly, the protein Cel61A, contained a CBM domain,
oxidatively deconstructing cellulose by the aid of redox-
active cofactor [47,48], could only be detected in high-
producer RE-10 but not in WT. Our results implied that
the gene *cel61A* was co-regulated with other core cellulases
by the same regulatory network, which accorded with a recent report that a *T. reesei* strain with the gene *cre1* deletion
lead to twofold higher over-expression of *cel61A* under
cellulose inducing condition [49]. Four GH61 family
protein encoding genes were identified in the genome of
P. oxalicum; two of them were specially detected in
RE-10, which were attractive targets in the investigation of
alternative lignocellulose-degrading mechanism in filamentous fungi or to supplement the industrial cellulase
preparations to improve the efficiency of hydrolysis. In
addition, we further confirmed that the protein swollenin
was co-regulated with cellulolytic enzymes, as a similar
result was previously reported in *T. reesei* [49]. Further,
many proteins which are involved in hemicellulose degradation were specially present in the secretome of RE-10, including a putative alpha-L-arabinofuranosidase (PDE_00076),
putative beta-1,3-glucanosyltransglycosylase (PDE_03134),
putative endo-beta-1,4-xylanase (PDE_05900), and a
beta-xylosidase (PDE_02716) (Additional file 5: Table S1).
These data conformed to a common sense that expressions of the majority of cellulase and many other plant cell
wall-degrading enzymes such as hemicellulase are coordi-
nately regulated in cellulolytic fungi. In addition, a chitin
glucanosyltransferase (PDE_05831), a pectin methylester-
ase (PDE_00923), and a beta-galactosidase were unique
to the secretome of RE-10. The latter was previously reported to participate in cellulase induction in *T. reesei*
[50]. A comparative analysis of the secreted proteins between WT and cellulase hyper-producers in different fungi
will be highly instructive for the identification of conserved
and important proteins in lignocellulose degradation.
Taken together, all the above knowledge laid a foundation for understanding the unique cellulase system of *P.
oxalicum* and provided novel targets for further engineering of *P. oxalicum* and other industrial producers.

Conclusions

In this study, we genetically modified three key regulators
to redesign the regulatory network of the expression of
cellulolytic enzymes. The recombinant strain exhibited
remarkably strong cellulolytic ability and high cellulase
expression and secretion compared with WT. The performance of the recombinant was even comparable with
that of industrial strain. Data from comparative proteomics analysis revealed that the lignocellulose-degrading
enzymes were elevated specifically in RE-10. Given the

conservation of the regulation of cellulolytic enzymes, this novel strategy could be compatible with other cellulolytic fungi.

Methods

Strains and culture conditions

P. oxalicum 114-2 (CGMCC 5302) and all mutants from it maintained on malt extract agar. Cellulose medium comprised 1× Vogel's salts [51] and 2% microcrystalline cellulose. The wheat bran medium was composed of corn cob residue (2.0000%), Avicel (0.6000%), wheat bran solid (4.6571%), soybean cake powder plate (1.0000%), $(NH_4)_2SO_4$ (0.2000%), $NaNO_3$ (0.2789%), urea (0.1000%), KH_2PO_4 (0.3000%), and $MgSO_4$ (0.0500%). Cellulase production induced by wheat bran medium and cellulose was performed in a 500 mL flask with 100 mL fluid medium at initial pH 5.5, 30°C, and 200 rpm, using 20 h mycelia pre-grown on 1× Vogel's medium with glucose (2%, w/v) as a sole carbon source. Cultures were collected by a method of vacuum drum filtration, and 0.5 g vegetative mycelia were added into the above inducing medium. Samples were collected at the time points indicated in the text. Microcrystalline cellulose (CB0279) was purchased from Sangon (Shanghai, China).

Construction of RE-10 (*gpdA*(p)::*clrB*::*ptra*-Δ*bgl2*::*hph*-Δ*creA*::*bar*) mutant

The cassettes were constructed for gene deletion or overexpression using double-joint PCR [52]. Fungal transformation was conducted by using PEG-mediated protoplast method developed by Li *et al.* [30]. We replaced the *clrB* promoter with *gpdA* (glyceraldehyde-3-phosphate dehydrogenase) promoter from *A. nidulans* [24], using the gene *ptrA* from *A. oryzae* as the selective marker for transformant screening [53]. Using primers PgpdA-F1 and PgpdA-R1, 1314 base pair (bp) of *gpdA* promoter was amplified from plasmid pAN7-1. The 2008 bp *ptra* selectable marker cassette was PCR-amplified with primer pair PtraF1and PtraR1. 3148 bp of *clrB* coding region and terminator was amplified with primer pair clrB-Fa and clrB-Ra, and this fragment overlapped with *gpdA* promoter and *ptra* fragment by 25 bp at its ends, respectively. These 6375 bp PCR products were fused in the order *gpdA* (p)-*clrB* (coding region)-*ptra* by double-joint PCR with nest primers PgpdA-F2 and PtraR1. One of these transformants, which was verified by PCR with primers PgpdA-F2 and PtraR1, was used for further genetic engineering.

Δ*bgl2*::*hph* knockout cassette with the *hph* cassette was amplified from Δ*bgl2*::*hph* mutant genome DNA with primer pair Bgl2-F2 + Bgl2-R2 [20]. The Δ*bgl2*::*hph* knockout cassette was used to transform protoplasts of the *gpdA*(p)::*clrB* mutant and obtained Δ*bgl2*-*gpdA*(p)::*clrB* mutants.

Double-joint PCR was performed to construct knockout cassette with the *bar* cassette flanked by 1.5 kb upstream (CreA-F1 and Crebar-R) and 1.5 kb downstream (Crebar-F and CreA-R1) of the *creA* ORF. 1.038 kb of *bar* cassette was amplified from plasmid pUC-bar by using prime pair Bar-F and Bar-R [54]. The final deletion cassette fragment Δ*creA*::*bar* was obtained by using the primer pair (Crenest-F + Crenest-R) with the three fragments above as a template for PCR. The Δ*creA*::*bar* cassettes were transformed into the *P. oxalicum gpdA*(p)::*clrB-ptra* and Δ*bgl2* strain protoplasts and obtained the mutant RE-10 *gpdA*(p)::*clrB*::*ptra*-Δ*bgl2*::*hph*-Δ*creA*::*bar*. Transformants were verified by PCR with primer pairs (CreA-F1 + Bar-R, Bar-F + CreA-R1 and Crenest-F + Crenest-R). The primers used in this study were listed in Additional file 6: Table S2.

Southern blot

The 0.89 kb probe for determining the copy numbers of *clrB* was PCR-amplified from genomic DNA using primers clrb-pF and clrb-pR. The enzymes *Pst*I and *Xba*I were used to digest the genome DNA. Only one 4.3 kb band should be visualized in WT after hybridization, and more than one band should observe in RE-10 (other than the native one). To determine the deletion of *creA*, previous probes and the same restriction enzymes were used [55]. To verify the deletion of *bgl2*, probe was amplified with primers Pbgl2-F and Pbgl2-F, then cleaved the genomes of WT (2.5 kb) and RE-10 (2.9 kb) with enzymes *Eco*RI and *Cla*I, respectively. Labeling of probe, hybridization, and color detection were performed by using DIG High Prime labelling kit (Roche) according to the manufacturer's methods.

Phenotype analysis

Microcrystalline cellulose was milled with beads for 6 days at room temperature. For phenotypic analysis, equal volume of conidia (10^8 per mL) of JU-A10-T, RE-10, and WT were spotted on medium contained 1× Vogel's salts with 2% glucose or ball-milled cellulose above at 30°C for 4 or 8 days. Canon EOS 600D (Canon, Japan) was used for photographing. We repeated the results and all the above analysis were performed in triplicates.

Cellulase activity assays

Supernatants were collected by centrifugation for removing mycelia and residual medium. The concentration of total extracellular protein was measured by a Bradford Kit (Sangon, Shang Hai, China), according to the manufacturer's instructions. FPA and CMCase activities were assayed with filter paper Whatman No. 1 (Shanghai, China) and CMC-NA (Sigma, USA) as the substrates, respectively. The enzyme reactions were performed in 0.2 M acetate buffer (pH 4.8) at 50°C for 60 and 30 min, respectively, using DNS method to quantify the released reducing sugar. Xylanase activity were assayed according to the method described by Sun *et al.* [56]. The *p*NPCase and *p*NPGase activities were measured in the above-used acetate buffer at 30°C for 30 min with *p*NPC and *p*NPG (Sigma, USA) as substrates. One enzyme activity unit was defined as the amount of enzyme required for producing 1 μmol glucose or *p*NP per minute under the assayed conditions. Three biological triplicates were performed in all analyses.

Biomass and glucose utilization determination

The WT, RE-10, and JU-A10-T strains were pre-grown in 1× Vogel's medium with glucose (2%, *w/v*) at 30, 200 rpm for 20 h. Equal amount of collected mycelia from each strain was transferred to the same media freshly prepared, for another 0, 6, 12, 24, 36, 48, 60, 72, and 84 h, respectively. Buchner funnel was used to separate the mycelia and supernatant. All sampled mycelia trapped into filter paper were drying at 65°C to constant weight. The concentration of glucose was determined by using the biosensor. All analyses were performed in biological triplicates.

qRT-PCR analysis

Spores of RE-10 and WT strains were washed by the solution containing 0.09 NaCl and 0.1% Twain-20, then inoculated into 1× Vogel's medium with glucose (2%, w/v) as a sole carbon source at 30°C for 20 h with shaking. Mycelia were collected by filtration and transferred to 1× Vogel's medium without any carbon source for 2 h, then transferred to inducing medium containing 2% cellulose as a sole carbon source for another 4 or 22 h. Samples of 2 h after starvation, 4 or 22 h after induction were collected. The total RNA extraction and cDNA synthesis using the RNAiso™ reagent (TaKaRa, Japan) and PrimeScript RT Reagent Kit (TaKaRa, Japan) were performed according to the manufacturer's instructions. qRT-PCR was performed on the LightCycler instrument (Roche, Germany) with software Version 4.0 (Roche, Germany) as previously described [41]. The primers used are shown in Additional file 6: Table S2. At least two biological triplicates were performed, and qRT-PCR of each gene was performed in three triplicates. The expression of actin was chosen as the reference gene for data normalization. The relative expression level was defined as follows: Rel. expression level (gene X) = copy number of gene X/copy number of gene action.

Protein gel electrophoresis

Unconcentrated supernatants were added to loading buffer, boiled for 5 min for degeneration, and loaded onto a 12% Tris-HCl polyacrylamide gel. Coomassie blue stain reagent was used for staining.

Zymography analysis

The 4-methylumbeliferyl-β-d-cellobioside [30] (Sigma) was used as the substrate to detect the cellobiohydrolase activities. Culture supernatant (equal protein) was separated on the polyacrylamide gels (SDS-PAGE) on ice at 138 V for 1 h. After electrophoresis, the gel was washed in 0.2 M acetate buffer (pH 4.8) with 5% Triton X-100 at least for three times to remove the SDS to renature the protein, then washed twice in acetate buffer at room temperature. The gel was directly soaked into acetate buffer containing 0.1% MUC and shaked at 50°C for 1 h, then visualized under UV illumination. And the software Image J (BioRad, USA) was used to obtain the pictures.

Proteomics analysis

Fresh spores were inoculated into 100 mL of 1× Vogel's salts supplemented with 2% glucose in a 500 mL Erlenmeyer flask for 24 h and transferred to 2% cellulose medium for 96 hour. The supernatant was collected by filtration through a 0.22 μm PES membrane, concentrated, and desalted by a centrifugal concentrator with a molecular cut-off of 10 kDa (Pall Corporation). The samples were precipitated by acetone and trichloroacetic acid (20:1). Dry protein powders were dissolved in denaturation buffer (0.5 M Tris-HCl, 2.75 mM EDTA, 6 M guanidine-HCl) and reduced by 1 M dithiothreitol at 37°C for 1 h. Alkylation was performed using iodoacetamide for 2 h in the dark. The alkylated samples were desalted and collected by a Microcon YM-10 Centrifugal Filter (Millipore Corporation, USA) according to the manufacturer's instruction. The collected protein samples were digested by trypsin (Sigma, USA) at 37°C overnight. Digested peptides were desalted and collected by a ZipTip C18 column (Millipore Corporation, USA). The collected secretome samples were separated on a C18 reversed phase column (15 cm long, 75 μm inner diameter, packed in-house with ReproSil-Pur C18-AQ 3 μm resin, provided by Dr. Maisch) directly mounted on the electrospray ion source of a mass spectrometer. The peptides were subjected to nanoelectrospray ionization, followed by tandem mass spectrometry (MS/MS) in an LTQ Orbitrap Velos Pro (Thermo Scientific™, USA) coupled inline to HPLC. Intact peptides were detected in the Orbitrap at a resolution of 60,000. Peptides were selected for MS/MS using collision-induced dissociation operating mode with a normalized collision energy setting of 35%. Ion fragments were detected in the LTQ. A data-dependent procedure that alternated between one MS scan followed by ten MS/MS scans was applied for the ten most abundant precursor ions above a threshold ion count of 5,000 in the MS survey scan with the following dynamic exclusion settings: repeat counts, 2; repeat duration, 30 s; and exclusion duration, 120 s. The applied electrospray voltage was 2.2 kV. For MS scans, the m/z scan range was 350 Da to 1,800 Da. MS data processing was performed using Mass-Lynx software (version 4.1, Waters, USA). Proteins with high confidence ($P < 0.01$) or at least two peptides detected were collected for further analysis. Data resulting from LC-MS/MS analysis of trypsin-digested proteins were searched against the *P. oxalicum* protein database as previously described [6]. Functional matching of identified proteins was conducted using SEQUEST. SignalP 4.1 (http://www.cbs.dtu.dk/services/SignalP/#citations), SecretomeP 2.0 (http://www.cbs.dtu.dk/services/SecretomeP/), and WoLF PSORT (http://www.genscript.com/psort/wolf_psort.html) with default cut-off values were used to predict the secreted proteins. The ratio of individual protein was calculated by the number of peptides divided by the total peptides of all proteins in the secretome.

Statistical analysis

A *t*-Student one-tail test for paired samples was performed with the software Microsoft Office 2013 Excel (Microsoft, USA). The mean values, standard deviations, and *P* values were calculated in all quantitative analysis.

Accession numbers

The GenBank accession numbers for the three proteins manipulated in this study are as follows: ClrB, EPS31045; BGL2, EPS25645; CreA, EPS28222.

Additional files

Additional file 1: Figure S1. Southern blot analysis of all mutants of RE-10. (A) *clrB* over-expression, (B) *bgl2* deletion, (C) *creA* deletion.

Additional file 2: Figure S2. qRT-PCR analysis of the transcripts of manipulated regulator genes (A) *clrB*, (B) *bgl2*, (C) *creA*.

Additional file 3: Figure S3. Comparative cellulolytic activity assay in cellulose medium. The FPA (A), *p*NPCase (B), CMCase (C), *p*NPGase (D), xylanase activities (E), and protein (F) of WT and RE-10 on cellulose medium were determined.

Additional file 4: Figure S4. SDS-PAGE analysis of the secreted protein.

Additional file 5: Table S1. Proteins identified in the secretome of WT and RE-10.

Additional file 6: Table S2. Primers used in this study.

Abbreviations

Bgl2: beta-glucosidase 2; CBM: cellulose binding domain; CDT/Cdt: cellodextrin transporter; ClrB: cellulase regulator B; CMC: carboxymethylcellulose; CreA: carbon catabolite repressor A; FPA: filter paper activity; *p*NPC: 4-Nitrophenyl-β-D-cellobioside; *p*NPG: 4-Nitrophenyl-β-D-glucopyranoside; WT: wild type.

Competing interests

The authors declare that they have no competing interests.

Authors' contributions

GSY designed the work, performed the experiments, analyzed the data, and drafted the manuscript. ZHL designed the work and performed the experiments and revised the manuscript. LWG and QBK participated in the enzyme activity assay experiment and collected and analyzed the data. RMW carried out the over-expression experiment and analyzed the data. GDL conceived the

experiment and helped revise the manuscript. YBQ designed the work and revised the manuscript. All authors read and approved the final manuscript.

Acknowledgements

We wish to thank Meng Liu and Piao Yang for helping in our enzyme activity assay experiments. We thank a lot professor Luying Xun from Washington State University (American) for paper revision. This study was supported by grants from the National Basic Research Program of China (Grant no. 2011CB707403) and the National Natural Science Foundation of China (Grant nos. 31030001, 31370086, and 31200065). Zhonghai Li also thanks the Foundation of State Key Laboratory of Microbial Technology of Shandong University (No. M2014-07).

Author details

[1]State Key Laboratory of Microbial Technology, Shandong University, Jinan City, Shandong Province 250100, China. [2]National Glycoengineering Research Center, Shandong University, Jinan City, Shandong Province 250100, China. [3]Qingdao Vland Biotech Group Co. Ltd., Shandong Expressway Mansion, Miaoling Road, Qingdao, Shandong, China.

References

1. Himmel ME, Bayer EA. Lignocellulose conversion to biofuels: current challenges, global perspectives. Curr Opin Biotechnol. 2009;20(3):316–7.
2. Klein-Marcuschamer D, Oleskowicz-Popiel P, Simmons BA, Blanch HW. The challenge of enzyme cost in the production of lignocellulosic biofuels. Biotechnol Bioeng. 2012;109(4):1083–7.
3. Dashtban M, Schraft H, Qin W. Fungal bioconversion of lignocellulosic residues; opportunities & perspectives. Int J Biol Sci. 2009;5(6):578–95.
4. Seiboth B, Herold S, Kubicek CP. Metabolic engineering of inducer formation for cellulase and hemicellulase gene expression in *Trichoderma reesei*. Subcell Biochem. 2012;64:367–90.
5. Gusakov AV. Alternatives to *Trichoderma reesei* in biofuel production. Trends Biotechnol. 2011;29(9):419–25.
6. Liu G, Zhang L, Wei X, Zou G, Qin Y, Ma L, et al. Genomic and secretomic analyses reveal unique features of the lignocellulolytic enzyme system of *Penicillium decumbens*. PLoS One. 2013;8(2):e55185.
7. Gusakov AV, Sinitsyn AP. Cellulases from *Penicillium* species for producing fuels from biomass. Biofuels. 2012;3(4):463–77.
8. Qu YB, Zhao X, Gao PJ, Wang ZN. Cellulase production from spent sulfite liquor and paper-mill waste fiber. Scientific note Appl Biochem Biotechnol. 1991;28–29:363–8.
9. Liu G, Zhang L, Qin Y, Zou G, Li Z, Yan X, et al. Long-term strain improvements accumulate mutations in regulatory elements responsible for hyper-production of cellulolytic enzymes. Sci Rep. 2013;3:1569.
10. Liu G, Qin Y, Li Z, Qu Y. Improving lignocellulolytic enzyme production with Penicillium: from strain screening to systems biology. Biofuels. 2013;4(5):523–34.
11. Kubicek CP. Systems biological approaches towards understanding cellulase production by *Trichoderma reesei*. J Biotechnol. 2013;163(2):133–42.
12. Glass NL, Schmoll M, Cate JH, Coradetti S. Plant cell wall deconstruction by ascomycete fungi. Annu Rev Microbiol. 2013;67:477–98.
13. Amore A, Giacobbe S, Faraco V. Regulation of cellulase and hemicellulase gene expression in fungi. Curr Genomics. 2013;14(4):230–49.
14. Coradetti ST, Craig JP, Xiong Y, Shock T, Tian C, Glass NL. Conserved and essential transcription factors for cellulase gene expression in ascomycete fungi. Proc Natl Acad Sci U S A. 2012;109(19):7397–402.
15. Ronne H. Glucose repression in fungi. Trends Genet. 1995;11(1):12–7.
16. Sun J, Glass NL. Identification of the CRE-1 cellulolytic regulon in *Neurospora crassa*. PLoS One. 2011;6(9):e25654.
17. Stricker AR, Mach RL, de Graaff LH. Regulation of transcription of cellulases- and hemicellulases-encoding genes in Aspergillus niger and *Hypocrea jecorina* (*Trichoderma reesei*). Appl Microbiol Biotechnol. 2008;78(2):211–20.
18. Znameroski EA, Coradetti ST, Roche CM, Tsai JC, Iavarone AT, Cate JH, et al. Induction of lignocellulose-degrading enzymes in *Neurospora crassa* by cellodextrins. Proc Natl Acad Sci U S A. 2012;109(16):6012–7.
19. Zhou Q, Xu J, Kou Y, Lv X, Zhang X, Zhao G, et al. Differential involvement of beta-glucosidases from *Hypocrea jecorina* in rapid induction of cellulase genes by cellulose and cellobiose. Eukaryot Cell. 2012;11(11):1371–81.

20. Chen M, Qin Y, Cao Q, Liu G, Li J, Li Z, et al. Promotion of extracellular lignocellulolytic enzymes production by restraining the intracellular beta-glucosidase in *Penicillium decumbens*. Bioresour Technol. 2013;137:33–40.

21. Nakari-Setala T, Paloheimo M, Kallio J, Vehmaanpera J, Penttila M, Saloheimo M. Genetic modification of carbon catabolite repression in *Trichoderma reesei* for improved protein production. Appl Environ Microbiol. 2009;75(14):4853–60.

22. Coradetti ST, Xiong Y, Glass NL. Analysis of a conserved cellulase transcriptional regulator reveals inducer-independent production of cellulolytic enzymes in *Neurospora crassa*. Microbiologyopen. 2013;2(4):595–609.

23. Derntl C, Gudynaite-Savitch L, Calixte S, White T, Mach RL, Mach-Aigner AR. Mutation of the Xylanase regulator 1 causes a glucose blind hydrolase expressing phenotype in industrially used *Trichoderma* strains. Biotechnol Biofuels. 2013;6(1):62.

24. Itoh Y, Scott B. Effect of de-phosphorylation of linearized pAN7-1 and of addition of restriction enzyme on plasmid integration in *Penicillium paxilli*. Curr Genet. 1997;32(2):147–51.

25. Avalos J, Geever RF, Case ME. Bialaphos resistance as a dominant selectable marker in *Neurospora crassa*. Curr Genet. 1989;16(5–6):369–72.

26. Roy P, Lockington RA, Kelly JM. CreA-mediated repression in *Aspergillus nidulans* does not require transcriptional auto-regulation, regulated intracellular localisation or degradation of CreA. Fungal Genet Biol. 2008;45(5):657–70.

27. Fujii T, Inoue H, Ishikawa K. Enhancing cellulase and hemicellulase production by genetic modification of the carbon catabolite repressor gene, *creA*, in *Acremonium cellulolyticus*. AMB Express. 2013;3(1):73.

28. Lynd LR, Weimer PJ, van Zyl WH, Pretorius IS. Microbial cellulose utilization: fundamentals and biotechnology. Microbiol Mol Biol Rev. 2002;66(3):506–77.

29. Phitsuwan P, Laohakunjit N, Kerdchoechuen O, Kyu KL, Ratanakhanokchai K. Present and potential applications of cellulases in agriculture, biotechnology, and bioenergy. Folia Microbiol. 2013;58(2):163–76.

30. Liu D, Li J, Zhao S, Zhang R, Wang M, Miao Y, et al. Secretome diversity and quantitative analysis of cellulolytic *Aspergillus fumigatus* Z5 in the presence of different carbon sources. Biotechnol Biofuels. 2013;6(1):149.

31. Gourlay K, Hu J, Arantes V, Andberg M, Saloheimo M, Penttila M, et al. Swollenin aids in the amorphogenesis step during the enzymatic hydrolysis of pretreated biomass. Bioresour Technol. 2013;142:498–503.

32. Saloheimo M, Paloheimo M, Hakola S, Pere J, Swanson B, Nyyssonen E, et al. Swollenin, a *Trichoderma reesei* protein with sequence similarity to the plant expansins, exhibits disruption activity on cellulosic materials. Eur J Biochem. 2002;269(17):4202–11.

33. Tsang A, Butler G, Powlowski J, Panisko EA, Baker SE. Analytical and computational approaches to define the *Aspergillus niger* secretome. Fungal Genet Biol. 2009;46 Suppl 1:S153–60.

34. Liao H, Li S, Wei Z, Shen Q, Xu Y. Insights into high-efficiency lignocellulolytic enzyme production by *Penicillium oxalicum* GZ-2 induced by a complex substrate. Biotechnol Biofuels. 2014;7(1):162.

35. Wei XM, Qin YQ, Qu YB. Molecular cloning and characterization of two major endoglucanases from *Penicillium decumbens*. J Microbiol Biotechnol. 2010;20(2):265–70.

36. Peterson R, Nevalainen H. Trichoderma reesei RUT-C30 - thirty years of strain improvement. Microbiology. 2012;158(Pt 1):58–68.

37. Wei X, Zheng K, Chen M, Liu G, Li J, Lei Y, et al. Transcription analysis of lignocellulolytic enzymes of *Penicillium decumbens* 114-2 and its catabolite-repression-resistant mutant. C R Biol. 2011;334(11):806–11.

38. Nogawa M, Goto M, Okada H, Morikawa Y. L-Sorbose induces cellulase gene transcription in the cellulolytic fungus *Trichoderma reesei*. Curr Genet. 2001;38(6):329–34.

39. Portnoy T, Margeot A, Linke R, Atanasova L, Fekete E, Sandor E, et al. The CRE1 carbon catabolite repressor of the fungus *Trichoderma reesei*: a master regulator of carbon assimilation. BMC Genomics. 2011;12:269.

40. Saloheimo M, Kuja-Panula J, Ylosmaki E, Ward M, Penttila M. Enzymatic properties and intracellular localization of the novel *Trichoderma reesei* beta-glucosidase BGLII (cel1A). Appl Environ Microbiol. 2002;68(9):4546–53.

41. Li J, Liu G, Chen M, Li Z, Qin Y, Qu Y. Cellodextrin transporters play important roles in cellulase induction in the cellulolytic fungus *Penicillium oxalicum*. Appl Microbiol Biotechnol. 2013;97(24):10479–88.

42. Galazka JM, Tian C, Beeson WT, Martinez B, Glass NL, Cate JH. Cellodextrin transport in yeast for improved biofuel production. Science. 2010;330(6000):84–6.

43. Mello-de-Sousa TM, Gorsche R, Rassinger A, Pocas-Fonseca MJ, Mach RL, Mach-Aigner AR. A truncated form of the Carbon catabolite repressor 1 increases cellulase production in *Trichoderma reesei*. Biotechnol Biofuels. 2014;7(1):129.

44. Nitta M, Furukawa T, Shida Y, Mori K, Kuhara S, Morikawa Y, et al. A new Zn(II)(2)Cys(6)-type transcription factor BglR regulates beta-glucosidase expression in *Trichoderma reesei*. Fungal Genet Biol. 2012;49(5):388–97.

45. Phillips CM, Iavarone AT, Marletta MA. Quantitative proteomic approach for cellulose degradation by *Neurospora crassa*. J Proteome Res. 2011;10(9):4177–85.

46. Herpoel-Gimbert I, Margeot A, Dolla A, Jan G, Molle D, Lignon S, et al. Comparative secretome analyses of two *Trichoderma reesei* RUT-C30 and CL847 hypersecretory strains. Biotechnol Biofuels. 2008;1(1):18.

47. Beeson WT, Phillips CM, Cate JH, Marletta MA. Oxidative cleavage of cellulose by fungal copper-dependent polysaccharide monooxygenases. J Am Chem Soc. 2012;134(2):890–2.

48. Quinlan RJ, Sweeney MD, Lo Leggio L, Otten H, Poulsen JC, Johansen KS, et al. Insights into the oxidative degradation of cellulose by a copper metalloenzyme that exploits biomass components. Proc Natl Acad Sci U S A. 2011;108(37):15079–84.

49. Castro Ldos S, Antonieto AC, Pedersoli WR, Silva-Rocha R, Persinoti GF, Silva RN. Expression pattern of cellulolytic and xylanolytic genes regulated by transcriptional factors XYR1 and CRE1 are affected by carbon source in *Trichoderma reesei*. Gene Expr Patterns. 2014;14(2):88–95.

50. Seiboth B, Hartl L, Salovuori N, Lanthaler K, Robson GD, Vehmaanpera J, et al. Role of the bga1-encoded extracellular {beta}-galactosidase of *Hypocrea jecorina* in cellulase induction by lactose. Appl Environ Microbiol. 2005;71(2):851–7.

51. Hu Y, Liu G, Li Z, Qin Y, Qu Y, Song X. G protein-cAMP signaling pathway mediated by PGA3 plays different roles in regulating the expressions of amylases and cellulases in *Penicillium decumbens*. Fungal Genet Biol. 2013;58–59:62–70.

52. Yu JH, Hamari Z, Han KH, Seo JA, Reyes-Dominguez Y, Scazzocchio C. Double-joint PCR: a PCR-based molecular tool for gene manipulations in filamentous fungi. Fungal Genet Biol. 2004;41(11):973–81.

53. Kubodera T, Yamashita N, Nishimura A. Transformation of *Aspergillus sp.* and *Trichoderma reesei* using the pyrithiamine resistance gene (ptrA) of *Aspergillus oryzae*. Biosci Biotechnol Biochem. 2002;66(2):404–6.

54. Qin Y, Ortiz-Urquiza A, Keyhani NO. A putative methyltransferase, mtrA, contributes to development, spore viability, protein secretion and virulence in the entomopathogenic fungus *Beauveria bassiana*. Microbiology. 2014;160(Pt 11):2526–37.

55. Li ZH, Du CM, Zhong YH, Wang TH. Development of a highly efficient gene targeting system allowing rapid genetic manipulations in *Penicillium decumbens*. Appl Microbiol Biotechnol. 2010;87(3):1065–76.

56. Sun X, Liu Z, Qu Y, Li X. The effects of wheat bran composition on the production of biomass-hydrolyzing enzymes by *Penicillium decumbens*. Appl Biochem Biotechnol. 2008;146(1–3):119–28.

Permissions

The contributors of this book come from diverse backgrounds, making this book a truly international effort. This book will bring forth new frontiers with its revolutionizing research information and detailed analysis of the nascent developments around the world.

We would like to thank all the contributing authors for lending their expertise to make the book truly unique. They have played a crucial role in the development of this book. Without their invaluable contributions this book wouldn't have been possible. They have made vital efforts to compile up to date information on the varied aspects of this subject to make this book a valuable addition to the collection of many professionals and students.

This book was conceptualized with the vision of imparting up-to-date information and advanced data in this field. To ensure the same, a matchless editorial board was set up. Every individual on the board went through rigorous rounds of assessment to prove their worth. After which they invested a large part of their time researching and compiling the most relevant data for our readers.

The editorial board has been involved in producing this book since its inception. They have spent rigorous hours researching and exploring the diverse topics which have resulted in the successful publishing of this book. They have passed on their knowledge of decades through this book. To expedite this challenging task, the publisher supported the team at every step. A small team of assistant editors was also appointed to further simplify the editing procedure and attain best results for the readers.

Apart from the editorial board, the designing team has also invested a significant amount of their time in understanding the subject and creating the most relevant covers. They scrutinized every image to scout for the most suitable representation of the subject and create an appropriate cover for the book.

The publishing team has been an ardent support to the editorial, designing and production team. Their endless efforts to recruit the best for this project, has resulted in the accomplishment of this book. They are a veteran in the field of academics and their pool of knowledge is as vast as their experience in printing. Their expertise and guidance has proved useful at every step. Their uncompromising quality standards have made this book an exceptional effort. Their encouragement from time to time has been an inspiration for everyone.

The publisher and the editorial board hope that this book will prove to be a valuable piece of knowledge for researchers, students, practitioners and scholars across the globe.

List of Contributors

Wuttichai Mhuantong
Enzyme Technology Laboratory, Bioresources Technology Unit, National Center for Genetic Engineering and Biotechnology (BIOTEC), Thailand Science Park, Pathumthani 12120, Thailand

Varodom Charoensawan
Department of Biochemistry, Faculty of Science, Mahidol University, Bangkok 10400, Thailand
Integrative Computational BioScience (ICBS) Center, Mahidol University, Nakhon Pathom 73170, Thailand

Pattanop Kanokratana
Enzyme Technology Laboratory, Bioresources Technology Unit, National Center for Genetic Engineering and Biotechnology (BIOTEC), Thailand Science Park, Pathumthani 12120, Thailand

Sithichoke Tangphatsornruang
Genome Institute, National Center for Genetic Engineering and Biotechnology (BIOTEC), Thailand Science Park, Pathumthani 12120, Thailand

Verawat Champreda
Enzyme Technology Laboratory, Bioresources Technology Unit, National Center for Genetic Engineering and Biotechnology (BIOTEC), Thailand Science Park, Pathumthani 12120, Thailand

Magdalena Calusinska
Centre for Protein Engineering, Bacterial Physiology and Genetics, University of Liège, Allée de la Chimie 3, B-4000 Liège, Belgium
Environmental Research and Innovation Department, Luxembourg Institute of Science and Technology, Rue du Brill 41, L-4422 Belvaux, Luxembourg

Christopher Hamilton
Walloon Centre of Industrial Biology, University of Liège, Boulevard du Rectorat 29, B-4000 Liège, Belgium

Pieter Monsieurs
Microbiology Unit, Expertise Group for Molecular and Cellular Biology, Institute for Environment, Health and Safety, Belgian Nuclear Research Centre (SCK-CEN), Boeretang 200, B-2400 Mol, Belgium

Gregory Mathy
Bioenergetics Laboratory, University of Liège, Boulevard du Rectorat 27, B-4000 Liège, Belgium

Natalie Leys
Microbiology Unit, Expertise Group for Molecular and Cellular Biology, Institute for Environment, Health and Safety, Belgian Nuclear Research Centre (SCK-CEN), Boeretang 200, B-2400 Mol, Belgium

Fabrice Franck
Bioenergetics Laboratory, University of Liège, Boulevard du Rectorat 27, B-4000 Liège, Belgium

Bernard Joris
Centre for Protein Engineering, Bacterial Physiology and Genetics, University of Liège, Allée de la Chimie 3, B-4000 Liège, Belgium

Philippe Thonart
Walloon Centre of Industrial Biology, University of Liège, Boulevard du Rectorat 29, B-4000 Liège, Belgium

Serge Hiligsmann
Walloon Centre of Industrial Biology, University of Liège, Boulevard du Rectorat 29, B-4000 Liège, Belgium

Annick Wilmotte
Centre for Protein Engineering, Bacterial Physiology and Genetics, University of Liège, Allée de la Chimie 3, B-4000 Liège, Belgium

Ajaya K Biswal
Department of Biochemistry and Molecular Biology, University of Georgia, B122 Life Sciences Bldg., Athens, GA 30602, USA
Complex Carbohydrate Research Center, University of Georgia, 315 Riverbend Road, Athens, GA 30602, USA
DOE-BioEnergy Science Center (BESC), Oak Ridge, USA

Zhangying Hao
Department of Plant Biology, University of Georgia, 2502 Miller Plant Sciences, Athens, GA 30602, USA
Complex Carbohydrate Research Center, University of Georgia, 315 Riverbend Road, Athens, GA 30602, USA
DOE-BioEnergy Science Center (BESC), Oak Ridge, USA

Sivakumar Pattathil
Complex Carbohydrate Research Center, University of Georgia, 315 Riverbend Road, Athens, GA 30602, USA
DOE-BioEnergy Science Center (BESC), Oak Ridge, USA

Xiaohan Yang
DOE-BioEnergy Science Center (BESC), Oak Ridge, USA
Bioscience Division, Oak Ridge National Laboratory, Oak Ridge, TN 37831, USA

Kim Winkeler
DOE-BioEnergy Science Center (BESC), Oak Ridge, USA
ArborGen Inc., 2011 Broadbank Ct, Ridgeville, SC 29472, USA

Cassandra Collins
DOE-BioEnergy Science Center (BESC), Oak Ridge, USA
ArborGen Inc., 2011 Broadbank Ct, Ridgeville, SC 29472, USA

Sushree S Mohanty
Complex Carbohydrate Research Center, University of Georgia, 315 Riverbend Road, Athens, GA 30602, USA
DOE-BioEnergy Science Center (BESC), Oak Ridge, USA

Elizabeth A Richardson
Department of Plant Biology, University of Georgia, 2502 Miller Plant Sciences, Athens, GA 30602, USA

Ivana Gelineo-Albersheim
Complex Carbohydrate Research Center, University of Georgia, 315 Riverbend Road, Athens, GA 30602, USA
DOE-BioEnergy Science Center (BESC), Oak Ridge, USA

Kimberly Hunt
Complex Carbohydrate Research Center, University of Georgia, 315 Riverbend Road, Athens, GA 30602, USA
DOE-BioEnergy Science Center (BESC), Oak Ridge, USA

David Ryno
Complex Carbohydrate Research Center, University of Georgia, 315 Riverbend Road, Athens, GA 30602, USA
DOE-BioEnergy Science Center (BESC), Oak Ridge, USA

Robert W Sykes
DOE-BioEnergy Science Center (BESC), Oak Ridge, USA
National Renewable Energy Laboratory, 15013 Denver West Parkway, Golden, CO 80401-3305, USA

Geoffrey B Turner
DOE-BioEnergy Science Center (BESC), Oak Ridge, USA
National Renewable Energy Laboratory, 15013 Denver West Parkway, Golden, CO 80401-3305, USA

Angela Ziebell
DOE-BioEnergy Science Center (BESC), Oak Ridge, USA
National Renewable Energy Laboratory, 15013 Denver West Parkway, Golden, CO 80401-3305, USA

Erica Gjersing
DOE-BioEnergy Science Center (BESC), Oak Ridge, USA
National Renewable Energy Laboratory, 15013 Denver West Parkway, Golden, CO 80401-3305, USA

Wolfgang Lukowitz
Department of Plant Biology, University of Georgia, 2502 Miller Plant Sciences, Athens, GA 30602, USA

Mark F Davis
DOE-BioEnergy Science Center (BESC), Oak Ridge, USA
National Renewable Energy Laboratory, 15013 Denver West Parkway, Golden, CO 80401-3305, USA

Stephen R Decker
DOE-BioEnergy Science Center (BESC), Oak Ridge, USA
National Renewable Energy Laboratory, 15013 Denver West Parkway, Golden, CO 80401-3305, USA

Michael G Hahn
Department of Plant Biology, University of Georgia, 2502 Miller Plant Sciences, Athens, GA 30602, USA
Complex Carbohydrate Research Center, University of Georgia, 315 Riverbend Road, Athens, GA 30602, USA
DOE-BioEnergy Science Center (BESC), Oak Ridge, USA

Debra Mohnen
Department of Biochemistry and Molecular Biology, University of Georgia, B122 Life Sciences Bldg., Athens, GA 30602, USA
Complex Carbohydrate Research Center, University of Georgia, 315 Riverbend Road, Athens, GA 30602, USA
DOE-BioEnergy Science Center (BESC), Oak Ridge, USA

Yu-Chien Chuang
Taiwan International Graduate Program in Molecular and Cellular Biology, Academia Sinica, Taipei 115, Taiwan
Institute of Life Sciences, National Defense Medical Center, Taipei 115, Taiwan
Institute of Molecular Biology, Academia Sinica, Taipei 115, Taiwan

Wan-Chen Li
Institute of Molecular Biology, Academia Sinica, Taipei 115, Taiwan
Institute of Genome Sciences, National Yang-Ming University, Taipei 112, Taiwan

Chia-Ling Chen
Institute of Molecular Biology, Academia Sinica, Taipei 115, Taiwan

Paul Wei-Che Hsu
Institute of Molecular Biology, Academia Sinica, Taipei 115, Taiwan

Shu-Yun Tung
Institute of Molecular Biology, Academia Sinica, Taipei 115, Taiwan

Hsiao-Che Kuo
Institute of Molecular Biology, Academia Sinica, Taipei 115, Taiwan
Department of Forest Sciences, University of Helsinki, Helsinki, Finland

Monika Schmoll
Austrian Institute of Technology, Health and Environment Department, Bioresources, University and Research Center, UFT Campus Tulln, Tulln/Donau 3430, Austria

Ting-Fang Wang
Taiwan International Graduate Program in Molecular and Cellular Biology, Academia Sinica, Taipei 115, Taiwan Institute of Molecular Biology, Academia Sinica, Taipei 115, Taiwan

Daria Feldman
Department of Plant Pathology and Microbiology, The R.H. Smith Faculty of Agriculture, Food and Environment, The Hebrew University of Jerusalem, POB 12, Rehovot 76100, Israel

David J Kowbel
Department of Plant and Microbial Biology, University of California at Berkeley, 111 Koshland Hall, Berkeley, California 94720, USA

N Louise Glass
Department of Plant and Microbial Biology, University of California at Berkeley, 111 Koshland Hall, Berkeley, California 94720, USA

Oded Yarden
Department of Plant Pathology and Microbiology, The R.H. Smith Faculty of Agriculture, Food and Environment, The Hebrew University of Jerusalem, POB 12, Rehovot 76100, Israel

Yitzhak Hadar
Department of Plant Pathology and Microbiology, The R.H. Smith Faculty of Agriculture, Food and Environment, The Hebrew University of Jerusalem, POB 12, Rehovot 76100, Israel

Ausra Peciulyte
Department of Biology and Biological Engineering, Division of Industrial Biotechnology, Chalmers University of Technology, Kemivägen 10, Gothenburg SE-412 96, Sweden

Katarina Karlström
Innventia AB, Drottning Kristinas väg 61, Stockholm SE-114 86, Sweden

Per Tomas Larsson
Innventia AB, Drottning Kristinas väg 61, Stockholm SE-114 86, Sweden
Wallenberg Wood Science Center, KTH Royal Institute of Technology, Teknikringen 56-58, Stockholm SE-100 44, Sweden

Lisbeth Olsson
Department of Biology and Biological Engineering, Division of Industrial Biotechnology, Chalmers University of Technology, Kemivägen 10, Gothenburg SE-412 96, Sweden
Wallenberg Wood Science Center, Chalmers University of Technology, Kemigården 4, Gothenburg SE-412 96, Sweden

Jeffrey G Linger
National Bioenergy Center, National Renewable Energy Laboratory, 16253 Denver West Parkway, Golden, CO 80401, USA

Larry E Taylor II
Biosciences Center, National Renewable Energy Laboratory, 16253 Denver West Parkway, Golden, CO 80401, USA

John O Baker
Biosciences Center, National Renewable Energy Laboratory, 16253 Denver West Parkway, Golden, CO 80401, USA

Todd Vander Wall
Biosciences Center, National Renewable Energy Laboratory, 16253 Denver West Parkway, Golden, CO 80401, USA

Sarah E Hobdey
Biosciences Center, National Renewable Energy Laboratory, 16253 Denver West Parkway, Golden, CO 80401, USA

Kara Podkaminer
Biosciences Center, National Renewable Energy Laboratory, 16253 Denver West Parkway, Golden, CO 80401, USA

Michael E Himmel
Biosciences Center, National Renewable Energy Laboratory, 16253 Denver West Parkway, Golden, CO 80401, USA

Stephen R Decker
Biosciences Center, National Renewable Energy Laboratory, 16253 Denver West Parkway, Golden, CO 80401, USA

Noppon Lertwattanasakul
Applied Molecular Bioscience, Graduate School of Medicine, Yamaguchi University, Ube 755-8505, Japan
Department of Microbiology, Faculty of Science, Kasetsart University, Bangkok 10900, Thailand

Tomoyuki Kosaka
Department of Biological Chemistry, Faculty of Agriculture, Yamaguchi University, Yamaguchi 753-8515, Japan

Akira Hosoyama
National Institute of Technology and Evaluation, Shibuya-ku, Tokyo 151-0066, Japan

Yutaka Suzuki
Department of Medical Genome Sciences, The University of Tokyo, Chiba 277-8562, Japan

Nadchanok Rodrussamee
Applied Molecular Bioscience, Graduate School of Medicine, Yamaguchi University, Ube 755-8505, Japan

Minenosuke Matsutani
Department of Biological Chemistry, Faculty of Agriculture, Yamaguchi University, Yamaguchi 753-8515, Japan

Masayuki Murata
Applied Molecular Bioscience, Graduate School of Medicine, Yamaguchi University, Ube 755-8505, Japan

Naoko Fujimoto
Applied Molecular Bioscience, Graduate School of Medicine, Yamaguchi University, Ube 755-8505, Japan

Suprayogi
Applied Molecular Bioscience, Graduate School of Medicine, Yamaguchi University, Ube 755-8505, Japan

Keiko Tsuchikane
National Institute of Technology and Evaluation, Shibuya-ku, Tokyo 151-0066, Japan

Savitree Limtong
Department of Microbiology, Faculty of Science, Kasetsart University, Bangkok 10900, Thailand

Nobuyuki Fujita
National Institute of Technology and Evaluation, Shibuya-ku, Tokyo 151-0066, Japan

Mamoru Yamada
Applied Molecular Bioscience, Graduate School of Medicine, Yamaguchi University, Ube 755-8505, Japan
Department of Biological Chemistry, Faculty of Agriculture, Yamaguchi University, Yamaguchi 753-8515, Japan

Stjepan Krešimir Kračun
Department of Plant and Environmental Sciences, Thorvaldsensvej 40, Frederiksberg C 1871, Denmark

Julia Schückel
Department of Plant and Environmental Sciences, Thorvaldsensvej 40, Frederiksberg C 1871, Denmark

Bjørge Westereng
Department of Plant and Environmental Sciences, Thorvaldsensvej 40, Frederiksberg C 1871, Denmark
Department of Chemistry, Biotechnology and Food Science, Norwegian University of Life Sciences, Chr. M. Falsens vei 1., Aas 1432, Norway
University of Copenhagen, Faculty of Science, Rolighedsvej 23, Frederiksberg C 1958, Denmark

Lisbeth Garbrecht Thygesen
University of Copenhagen, Faculty of Science, Rolighedsvej 23, Frederiksberg C 1958, Denmark

Rune Nygaard Monrad
Novozymes A/S, Krogshoejvej 36, Bagsværd 2880, Denmark

Vincent G H Eijsink
Department of Chemistry, Biotechnology and Food Science, Norwegian University of Life Sciences, Chr. M. Falsens vei 1., Aas 1432, Norway

William George Tycho Willats
Department of Plant and Environmental Sciences, Thorvaldsensvej 40, Frederiksberg C 1871, Denmark

Dominic Pinel
Department of Biology, Centre for Structural and Functional Genomics, Concordia University, 7141 Sherbrooke Street West, Montréal, Québec H4B 1R6, Canada
Current address: Energy Biosciences Institute, University of California, Berkeley, Berkeley, CA 94704, USA

David Colatriano
Department of Biology, Centre for Structural and Functional Genomics, Concordia University, 7141 Sherbrooke Street West, Montréal, Québec H4B 1R6, Canada

Heng Jiang
Department of Biology, Centre for Structural and Functional Genomics, Concordia University, 7141 Sherbrooke Street West, Montréal, Québec H4B 1R6, Canada
Current address: Crabtree Nutrition Laboratories, McGill University Health Center, Montreal, Quebec H3A 1A1, Canada

Hung Lee
School of Environmental Sciences, University of Guelph, Guelph, Ontario N1G 2 W1, Canada

Vincent JJ Martin
Department of Biology, Centre for Structural and Functional Genomics, Concordia University, 7141 Sherbrooke Street West, Montréal, Québec H4B 1R6, Canada

Marcoaurélio Almenara Rodrigues
Federal University of Rio de Janeiro, Institute of Chemistry, Department of Biochemistry, Applied Photosynthesis Laboratory, Athos Avenida da Silveria Ramos, 149-Technology Centre, Block A, Room 532, University City, Rio de Janeiro, RJ 21941-909, Brazil

Ricardo Sposina Sobral Teixeira
Federal University of Rio de Janeiro, Institute of Chemistry, Department of Biochemistry, Enzyme Technology Laboratory, 21941-909 Rio de Janeiro, RJ, Brazil

Viridiana Santana Ferreira-Leitão
National Institute of Technology - Ministry of Science, Technology and Innovation, Biocatalysis Laboratory, 20081-312 Rio de Janeiro, RJ, Brazil
Federal University of Rio de Janeiro, Institute of Chemistry, Department of Biochemistry, Enzyme Technology Laboratory, 21941-909 Rio de Janeiro, RJ, Brazil

Elba Pinto da Silva Bon
Federal University of Rio de Janeiro, Institute of Chemistry, Department of Biochemistry, Enzyme Technology Laboratory, 21941-909 Rio de Janeiro, RJ, Brazil

Hiroyuki Inoue
Research Institute for Sustainable Chemistry, National Institute of Advanced Industrial Science and Technology (AIST), 3-11-32 Kagamiyama, Higashi-Hiroshima, Hiroshima 739-0046, Japan

Seiichiro Kishishita
Research Institute for Sustainable Chemistry, National Institute of Advanced Industrial Science and Technology (AIST), 3-11-32 Kagamiyama, Higashi-Hiroshima, Hiroshima 739-0046, Japan

Akio Kumagai
Research Institute for Sustainable Chemistry, National Institute of Advanced Industrial Science and Technology (AIST), 3-11-32 Kagamiyama, Higashi-Hiroshima, Hiroshima 739-0046, Japan

Misumi Kataoka
Research Institute for Sustainable Chemistry, National Institute of Advanced Industrial Science and Technology (AIST), 3-11-32 Kagamiyama, Higashi-Hiroshima, Hiroshima 739-0046, Japan

Tatsuya Fujii
Research Institute for Sustainable Chemistry, National Institute of Advanced Industrial Science and Technology (AIST), 3-11-32 Kagamiyama, Higashi-Hiroshima, Hiroshima 739-0046, Japan

Kazuhiko Ishikawa
Research Institute for Sustainable Chemistry, National Institute of Advanced Industrial Science and Technology (AIST), 3-11-32 Kagamiyama, Higashi-Hiroshima, Hiroshima 739-0046, Japan

Guangshan Yao
State Key Laboratory of Microbial Technology, Shandong University, Jinan City, Shandong Province 250100, China

Zhonghai Li
State Key Laboratory of Microbial Technology, Shandong University, Jinan City, Shandong Province 250100, China
Qingdao Vland Biotech Group Co. Ltd., Shandong Expressway Mansion, Miaoling Road, Qingdao, Shandong, China

Liwei Gao
State Key Laboratory of Microbial Technology, Shandong University, Jinan City, Shandong Province 250100, China

Ruimei Wu
State Key Laboratory of Microbial Technology, Shandong University, Jinan City, Shandong Province 250100, China

Qinbiao Kan
State Key Laboratory of Microbial Technology, Shandong University, Jinan City, Shandong Province 250100, China

Guodong Liu
State Key Laboratory of Microbial Technology, Shandong University, Jinan City, Shandong Province 250100, China

Yinbo Qu
State Key Laboratory of Microbial Technology, Shandong University, Jinan City, Shandong Province 250100, China
National Glycoengineering Research Center, Shandong University, Jinan City, Shandong Province 250100, China